The Story of Semiconductors

The Story of Semiconductors

John Orton

Emeritus Professor, University of Nottingham, UK

OXFORD
UNIVERSITY PRESS

Great Clarendon Street, Oxford OX2 6DP

Oxford University Press is a department of the University of Oxford.
It furthers the University's objective of excellence in research, scholarship,
and education by publishing worldwide in

Oxford New York

Auckland Bangkok Buenos Aires Cape Town Chennai
Dar es Salaam Delhi Hong Kong Istanbul Karachi Kolkata
Kuala Lumpur Madrid Melbourne Mexico City Mumbai Nairobi
São Paulo Shanghai Taipei Tokyo Toronto

Oxford is a registered trade mark of Oxford University Press
in the UK and in certain other countries

Published in the United States
by Oxford University Press Inc., New York

First published 2004

A catalogue record for this title is available from the British Library

Library of Congress Cataloging in Publication Data

(Data available)

ISBN 0 19 853083 8 (hbk)

10 9 8 7 6 5 4 3 2 1

Typeset by Newgen Imaging Systems (P) Ltd., Chennai, India
Printed in Great Britain
on acid-free paper by
Antony Rowe Ltd., Chippenham

Preface

My wife and I bought our first television set in 1966, a major family decision, which just happened to coincide with England's soccer World Cup success at Wembley Stadium. It cost us about £100 out of my then salary of £2000 a year. Thirty years later, when I retired on a salary some twenty times greater, the purchase of an infinitely superior colour set priced at little more than £500 could be contemplated with considerably less heart-searching. Indeed, the financial outlay involved in watching England's rugby World Cup success in 2003 gave us scarcely a qualm, one measure, perhaps, of the quite remarkable trend in consumer friendliness inherent in the modern electronics industry. In this we see one of the great successes of capitalist philosophy—a highly competitive business environment yielding previously unimaginable value for the consumer, while providing relatively comfortable employment for a very large workforce and (in spite of recent setbacks, exemplified by the misfortunes of the Marconi company) a satisfactory return on invested capital for its shareholders. But, more significantly from the viewpoint of this book, we also see a business based fairly and squarely on investment in scientific research. With the possible exception of the pharmaceutical industry, there has never been such a commitment to *organized* R&D and never before has the marriage between science and industry been so prolific in its progeny.

More specifically, this remarkable commercial success owes its existence largely to discoveries in semiconductor physics, which blossomed during the first half of the twentieth century and to developments in semiconductor technology and device concept, which followed the exciting events of Christmas 1947 when Bell scientists realized the World's first successful solid-state amplifier. Here was vindication for Bell's commitment to basic solid-state research in an *industrial* laboratory, which set the pattern for a rapidly expanding commercial activity, an activity which has continued to grow at a remarkably consistent rate into the present, truly worldwide industry we know today. It began with germanium, which was immediately replaced by silicon, then gradually drew in an amazing cohort of compound semiconductor materials required to meet the rapidly diversifying range of device demands, based on an equally diverse range of applications. Today, we take for granted the involvement of visible light and both infrared and ultraviolet radiation as well as that of electrons. This expansive industry is concerned with lighting, display, thermal imaging, solar electricity generation, optical communications, compact disc audio systems, DVD video systems, and a quite remarkable array of other uses for semiconductor lasers, as well as the more conventional electronic applications typified by the personal computer. All in all, this omnipresent electronics industry represents an annual turnover of about 5×10^{11} US$, a figure that compares not unfavourably with the $\$2 \times 10^{12}$ of the trend-setting automobile industry.

It is little more than 50 years since the inception of transistor electronics, a period which has seen quite dramatic developments in semiconductor devices, an activity with which I was personally involved for over 30 years. Having, during this time, written a number of books and specialist review articles, I felt it worthwhile, on reaching retirement, to attempt some kind of summary of the field in which I had worked. It had seemed to me for some time that, in spite of the numerous excellent texts which describe the physics and technology of semiconductor devices, there was a distinct lack of any coherent account of just how these devices came into being. What were the driving forces, what the difficulties to be overcome, what determined why a particular development occurred when it did, where was the work undertaken and by whom—in other words, how did the history of the subject develop. As I became more and more interested in such questions, it occurred to me that other workers in the field might appreciate a reasonably concise account of its history, as background to their current endeavours, also that there might be a wider audience of scientists who would find a non-specialist account of this epoch-making activity of general interest and, finally, that undergraduate students should be encouraged to understand not only semiconductor physics and device technology but also the background story of their advent. It is a human story and, as such, may surely illuminate the technical aspects of the subject to advantage. It is also a rapidly moving story and much of what I have written will very soon be superseded, so I have made no attempt to include the very latest developments. The account stops roughly (and perhaps appropriately) at the millennium, my intention all along, having been to write a history, not an up-to-date text book.

In most academic studies we expect to know something of the people involved—who painted such and such a picture, who developed such and such a philosophical idea, who was responsible for certain political innovations—and it seems no less appropriate in science. The difficulty here lies in the nature of modern scientific research, which has become much more of a team activity, rather than that associated with any one specific individual; so, in many cases, I have found it appropriate to refer to laboratories, rather than individuals. In attempting a broad overview of the subject, it is scarcely possible to give an accurate account of exactly how each individual scientist contributed to any particular discovery and I have not even tried to do so. This must be the task of the professional historian—and I make no pretence of being one. Perhaps this can be taken as encouragement to serious historians to become involved in the intricacies of scientific and technological history. It is a vital part of modern culture and, as such, demands considerably more attention than it currently receives. I only hope that the present broad-brush account may serve as a stimulus to further, more detailed studies.

Having said this, I should acknowledge that one or two detailed studies do exist. I think, particularly, of the excellent 'Crystal Fire' by Michael Riordan and Lillian Hoddeson, which describes the early work on transistors and integrated circuits, the 'Electronic Genie' by Frederick Seitz and Norman Einspruch, covering somewhat similar ground and the admirable survey of fibre optics provided by Jeff Hecht in his 'City of Light'. Charles Townes has also given us valuable insights into the origins of the laser in 'How the Laser Happened', though with rather little reference to semiconductor lasers. All these I have found helpful, as I have acknowledged in the relevant parts of my own account. There is, though, considerable scope for other studies, as anyone reading this book will appreciate. At present, we are far better informed as to the details of Michael Faraday's researches in the early years of the nineteenth century than we are to the development of group III-V semiconductors in the twentieth.

I have already outlined the audience to whom I have addressed this book, and it covers, I accept, a rather broad spectrum. This has influenced the format of the book in one important respect, the inclusion of 'Boxes' which contain the more specialized and mathematical detail supporting the basic account given in the main text. The book may be read without reference to these boxes, the text being complete in itself. Only readers interested in gaining deeper understanding need apply themselves to the boxes and this they may do either while reading the text or, if preferred, treat them as appendices to be read separately. I imagine that most readers interested primarily in the historical aspect of the subject will be happy with the basic text, while students, in particular, should find the additional insight provided by these boxes of value. I should nevertheless emphasize that the book is not, in any sense, to be seen as a substitute for the various standard texts on semiconductor physics and devices but rather as complementary to them, serving to provide a human slant to much that is otherwise purely technical. I hope and believe that many students will find this background information extremely helpful in satisfying their natural curiosity about how and why things came to pass and help them to appreciate the nature of the process of device development. Being a human activity, it should preferably be understood in that context, complete with all its human foibles.

The approach I have adopted throughout is essentially an interdisciplinary one. I have tried always to set device development in the context of relevant applications, providing, for instance, a fairly thorough account of the development of optical fibres by way of introduction to long wavelength semiconductor lasers and photodetectors. I have, similarly, outlined several applications of semiconductor power devices before describing the relevant devices. In all cases, the technical material is presented in terms of the relevant timescale and I have devoted considerable attention to the importance of semiconductor materials, their development in response to device demands and the vital cross-links with semiconductor physics. All three strands are well represented and can only be properly understood as a trinity. The book should therefore be of interest to physicists, electrical engineers, and to materials specialists, alike. Indeed, if I have been able to impart the essential message that real human activities, such as this, inevitably cross pedagogic boundaries, I shall be well-satisfied. It is clearly apparent that, without these interdisciplinary interactions, the electronics industry would not be where it is today and it would be well that its future workforce (i.e., today's students) should start their careers with an adequate understanding of this essential truth. While it is common to present scientific learning, at both School and University levels, in tidy and coherent packages, the real world shows little respect for such neat subdivisions—the successful inventor or entrepreneur must frequently demonstrate powers of imagination that transcend conventional boundaries.

Anyone familiar with the subject of semiconductor physics or device development will appreciate that an account of their history, contained within a book of modest size, must inevitably be highly selective, and I make no apology for the fact that my own account lays itself wide open to such criticism. As Norman Davies remarked, in the preface to his (relatively thick!) work *Europe—A History*: 'This volume—is only one from an almost infinite number of histories of Europe that could be written. It is the view of one pair of eyes, filtered by one brain and translated by one pen.'

Apart form the fact that I *typed* my thoughts directly onto my computer, I could make an identical statement here. This book represents one person's view of the semiconductor story. Its emphases are my own, based on my own involvement and inevitably coloured by my own experiences—and prejudices. But I certainly believe it to represent *one* history and one which I hope can be read with much enjoyment. Should others wish to write *their* histories, I shall be delighted to read them with equal enjoyment, secure in the knowledge that I may possibly have stimulated them to improve on my prototype.

Finally, I am happy to acknowledge the considerable debt which I owe to many colleagues. My wife, Joyce, suffered patiently the long hours of separation (even while we existed under the same roof!) and still found it possible to offer words of encouragement. Specific help was provided (in no particular order) by Professor Nick Holonyack of the University of Illinois, Urbana; Dr Frank James of the Royal Institution, London; Dr Sunao Ishihara of NTT, Kanagawa; Dr Tony Hartland of the National Physical Laboratory, Teddington; Dr Hirofumi Matsuhata of the Electrotechnical Laboratory, Tsukuba; Professor Sir Roger Elliott of Oxford University; Professor Tom Foxon and Dr Richard Campion, University of Nottingham; Mr Brian Fernley of Siemens, Professor Rodney Loudon, University of Essex; and Professor Martin Green, University of New South Wales, Sidney. More generally I must thank those many colleagues with whom I worked during my years at the Mullard (later Philips) Research Laboratories, Redhill and their counterparts in the Philips Nat Lab in Eindhoven. They are far too numerous for individual mention but I owe them a huge debt of gratitude for innumerable stimulating interactions as a result of which my imperfect understanding of semiconductor physics became gradually less blatant. It is with great affection that I dedicate this book to them—without their help, I could scarcely even have contemplated writing it. However, while their contribution made it all possible, the errors and obscurities that almost certainly remain are, of course, my personal responsibility.

Orchard Cottage
Cotgrave

December 2003

CONTENTS

CHAPTER 1

Perspectives

1.1 The 'Information Age'

Sitting contemplatively, in front of my computer screen, seeking inspiration on how best to make a convincing start to the writing of this book, my mind wandered over the technical changes which had occurred in my life since I had last been engaged on a similar enterprise. That was in the 1980s when I collaborated with Peter Blood on, what turned out (to us) to be, an awesomely lengthy summary of experimental techniques for semiconductor characterization. I reflected that, between us, we had written all 1026 pages of these two volumes by hand, delivering countless longhand pages to long-suffering typists who performed the near impossible task of rendering them comprehensible to the typesetters of Academic Press. How things have changed! Today, I compose everything, I write on my own word processor, and submit it in electronic form (now the errors really are all my own!), the major problem of filing and collating huge quantities of information being taken care of more or less automatically. I need do no more than take care to back up my long-suffering hard disk with an array of carefully labelled floppies—or, better still, a single CD.

Of course, similar possibilities existed 20 years ago, when Peter and I began our collaboration but the fact that we could choose to ignore them then serves to emphasize the change in working practices which those 20 years have ushered in. It reflects, though, only one incidence of the influence the ubiquitous silicon chip has had on our lives over the past few decades—remember, the transistor, itself, was invented as recently as 1947, little more than 50 years ago—and, then, it was made from germanium—silicon had not even been heard of (!) (at least, not outside the laboratory and certain esoteric military applications). It is hardly an original thought, but the pace of change in our times is certainly remarkable, if not (to many) actually frightening.

Once launched, then, on my train of thought, I have little difficulty in following this observation with others in a similar vein. Not only do I have a facsimile machine as an integral function of my office telephone, but my computer serves me also as basis for e-mail communication with colleagues and friends all round the world. Similarly, I can obtain technical information (or the times of trains to London) in huge variety from the latest wonder of the modern world, the Internet. On a more mundane level, the central heating system which is even now keeping me in bodily comfort against the external chill is controlled by a microprocessor, the family washing machine in an adjacent room makes available to us a wide range of washing programmes under the supervision of a similar tiny piece of suitably processed silicon. While my wife and I would never claim to be in the vanguard of the electronic revolution, we routinely use either of two audio systems based on the wonders of optoelectronics, not to mention the inevitable television set, with its accompanying video recorder which graces a corner of our living room, where it can, of course, be remotely controlled. The 1990s vacuum cleaner we use without much serious thought has a motor with electronic control, so too does the food mixer which relieves us of much of the physical hard work in our kitchen, not to mention several power tools which languish unused in my workshop while I am otherwise engaged in writing. We even have a mobile phone, though it stays firmly closeted in the car glove box for emergency use only. Our mass-produced car, in common with most of its competitors, boasts an engine monitoring system and electronic ignition, the stop lights are augmented by bright-red light emitting diodes (which everyone, nowadays, refers to as LEDs, such is their comonplace nature!), the instrument panel display is also based largely on LEDs and we are able to control the central locking with an infrared device buried conveniently within the ignition key. There is nothing remarkable in any of this, of course. Out of the house, one of my favourite pastimes is hill walking and my daughters recently bought me a wonderful satellite navigation sytem which tells me exactly where I am at any moment and where I need to aim in order to reach my next way point. I find it truly amazing but, no doubt, by the time this book reaches the market, everyone will have one and it will be seen as just another of those modern aids to civilized existence which we all tend to take for granted.

I could extend this line of thought considerably but I have probably said more than enough already—we are all aware of the information technology (IT) revolution—the press is full of the latest possibilities for home working, life on the Internet Waves, electronic home buying, etc. and we are rapidly becoming accustomed to the virtues (?) of smart telephones and smart cards which seem able to do most things without human intervention, these days. However, most people are, perhaps, less familiar with the origins of their new-found information skills—how is it all possible?, what depth of technical expertise has been employed in order to produce the necessary hardware and software?, what are the limitations to further progress?, etc. One reason for our relative ignorance is the enormous size of the subject and its daunting technical complexity, not made easier by the difficulty many scientists have in writing for the non-specialist. What follows here, therefore, is an attempt to describe just one aspect of this exciting story in narrative form so that the efforts of countless scientists and technologists may be more widely appreciated, while furthering better understanding on the part of non-specialists and specialists alike. For, what is more, it makes a wonderful story, every bit as fascinating as the sagas of earlier technological breakthroughs (of which we actually know far less) such as copper, bronze, and iron.

1.2 Early materials technology

All hardware aspects of mankind's many technologies are based on, and are limited by materials, so obvious a truism that we are prone to overlook it. Early man made use of stone for millennia, before discovering the wider possibilities of copper. It took a mere thousand years to acquire the greater capability of bronze and perhaps a further thousand for iron to make its entry into the evolutionary stream. However, things move at a greater pace nowadays—semiconductors, the materials of the information age, took just a hundred years to develop from the status of ill-understood and totally uncontrolled materials with certain mysterious properties, to their present position as some of the most thoroughly explored and well understood of all mankind's conquests. It represents a success story of which we should be proud, ranking alongside impressionism, Concord, mediaeval cathedrals, Burgundy wines, the Beethoven symphonies, and modern medicine, to name but a few (largely European!) highlights. No fewer than fourteen semiconductor scientists have been honoured by the Nobel Committee since the inception of the Prize in 1901 (when the discovery of X-rays by Wilhelm Conrad Rontgen was formally acknowledged). In 1909 Carl Braun shared the prize with Marconi for the development of wireless telegraphy (Braun's contribution included the discovery of semiconductor rectification), there was then a considerable gap until 1956, when John Bardeen, Walter Brattain, and William Shockley were recognized for their world-shaking invention of the transistor, followed much more briskly by awards to Leo Esaki (1973) for his discovery of tunnelling in semiconductors, Sir Nevill Mott and Philip Anderson (1977) for discoveries in amorphous semiconductors, Klaus von Klitzing (1985) for discovering the quantum Hall effect in a metal oxide silicon structure, Robert Laughlin, Horst Stormer, and Daniel Tsui (1998) for their work on the fractional quantum Hall effect in a gallium arsenide 'low dimensional structure' and, finally in 2000, Zhores Alferov, Herb Kroemer, and Jack Kilby for various contributions to the fields of electronics and optoelectronics. Clearly, my eulogistic statement can be supported with some sound references!

The history of mankind's discovery and taming of materials is a long one, stretching back to the stone age, some 5000 years before the birth of Christ. While this is no place for a detailed analysis of our, generally slow, progress (even if the present author were competent to undertake it), there are similarities with the recent development of semiconductors that make it worthwhile to look briefly at one or two aspects of the story. In all such material development activity one recognizes certain common features—first, the discovery of the material in its crude form, its isolation, perhaps from suitable ores, its initial application to 'practical' problems, the realization of limitations, the discovery of means for modifying the raw material properties, and a gradual struggle to gain control over and perfect each material in turn. Thus, we see that (very roughly) about 4000 BC copper was first employed in making small items of jewellery, possibly as a by-product of attempts to obtain a suitable pigment for making green eye-shadow! (human vanity plays its part in a multitude of ways). Some small amounts of metallic copper were probably found in proximity to the ores used as pigment and this was followed by the discovery that copper could be hammered into desirable shapes. However, work-hardening must have been a serious problem and it is only with the application of heat that our ancestors could begin to gain satisfactory control over the material. Again, this probably happened as a side-effect of early attempts (*c.*3000 BC) to make an artificial form of lapis lazuli for the cheaper end of the jewellery trade, a process involving copper-blue colour in decorative glass, widely known as Egyptian faience. This probably represents the first serious attempt to develop a materials technology based on the application of heat, in this case to glass formation, and represents a particularly important step, controlled heating being an essential feature of the majority of technologies which followed, not least the development of semiconductors.

The next major development came with the discovery that copper could be melted in a crucible by raising the temperature sufficiently (as we now know to 1083°C for the pure metal), a requirement which implies the use of some form of forced air flow, probably by fanning or use of a blow-pipe. This led to the use of moulds to form a more sophisticated range of shapes, including tools and (not surprisingly, perhaps!) weapons. The advantages of technological superiority in warfare were realized long before our high-tech age and there could be no denying their political influence, even in 3000 BC (though increasing considerably in value as populations grew and mobility became greater). Important though this was, there were still problems with obtaining adequately free flow of molten copper for accurate moulding and, as we know well, copper is a trifle soft in relation to the need for maintaining sharp cutting edges. The answer to this difficulty emerged, eventually(!)—round about 2000 BC it was found that the addition of small amounts of tin to the copper melt resulted in three major improvements. First, the 'alloys' melted at significantly lower temperatures, making the firing process easier to manage, second, the resulting melt was considerably less viscous, allowing more accurate and finer moulding, and third, the final material was harder and less prone to (uncontrolled) work-hardening. (The success of the Assyrian armies in the period prior to 1000 BC can be attributed in no small degree to these particular properties.) Thus began the Bronze Age and, over the ensuing centuries, it was established just how the properties of this superior alloy could be adjusted, by incorporating various controlled proportions of tin (typically between 5 and 15%), to optimize its performance against specific requirements. Finally, in the period round about 1000 BC, it was discovered that iron could also be applied to many of the more demanding tasks, such as the manufacture of weaponry and 'industrial' tools. This was not so much the result of iron's superiority, as of its greater availability but it brought with it the need for even higher furnace temperatures (iron melts at 1535°C) and the application of hot forging techniques, the invention of the bellows at about this time being an essential co-requisite.

We shall now skip conveniently past the intermediate centuries in which mankind gradually gained increasing control over the technology of iron-based materials and simply note the importance of recent developments in steels based on the addition of small amounts of suitable 'impurities' into the molten iron (shades of the early Bronze Age?). The production of tool steel is just one example of this, demonstrating the importance of obtaining a highly purified basic material which may then be modified in a number of desirable ways (by the addition of small amounts of chromium, vanadium, or nickel) to meet wide ranging requirements. We also note the importance of controlling the atmosphere surrounding the molten charge—it is no longer adequate merely to heat in air—carefully controlled oxidizing or reducing conditions are frequently essential. We shall see many parallels in semiconductor processing in the following pages.

1.3 What makes a semiconductor?

Semiconductors have been, and are, used in various forms, as mechanically cut slices from a single crystal 'boule', as single crystal thin films deposited on a suitable substrate by a more or less complex chemical or physical process, as glass-like elements, and as polycrystalline or glassy thin films deposited on (typically) a glass substrate. In the great majority of applications, an essential part of the process is concerned with growing a high quality bulk single crystal either to serve directly as the active material or to act as a substrate for epitaxial film growth (see Box 1.1). This, therefore, has generally required crystal growth from a crucible of the molten semiconductor, a technique demanding similar care and attention to appropriate atmosphere as those encountered in

Box 1.1. Epitaxy

The word 'epitaxy' is derived from two Greek words 'epi' = 'on' and 'taxis' = 'arranged'. It implies that appropriate atoms or molecules may be placed in some convenient way on a supporting surface or 'substrate' so as to produce a thin film of the desired material. However, in crystal growth lore, it has further been taken to imply that the atomic arrangement of the deposited material conforms precisely to that of the substrate. The most straightforward case to consider is that of 'homoepitaxy' where the substrate and growing film consist of the selfsame material, for example, a single crystal GaAs film growing on a single crystal GaAs substrate.

This process is very widely utilized in semiconductor technology, in the numerous situations where bulk single crystals are available but where their electrical quality is inadequate for direct application to device fabrication. A frequent method of avoiding the difficulty thus created is, then, to grow a high quality epitaxial film on a carefully prepared bulk crystal slice which serves merely as a mechanical support. This, of course, adds to the complexity (and cost!) of the overall process but is almost certainly preferable to a direct attack on the almost impossible task of growing sufficiently high grade bulk crystals. In many cases, the concept has been extended into 'heteroepitaxy' where the grown film differs, chemically, from its substrate but where there is close similarity between their crystal structures. An excellent example of this is the growth of AlAs films on GaAs substrates. Not only do both materials crystallize in the same form, but their natural lattice parameters (essentially, the separation between neighbouring atoms) are closely similar. The further extension to cases where the two materials have significantly different lattice parameters often introduces serious difficulties and the extreme case of growing a film of one structure on a substrate which crystallizes in a different structure can only be justified when no other substrate is available. It has occasionally been done with remarkable success but is, without doubt, the last resort of desperate men—for example, those crystal growers whose managers have decreed that compound XYZ_3 is the only answer to the managing director's urgent request for a solution to his latest marketing problem, bulk crystals of XYZ_3 being impossible to grow, except at temperatures of 3500°C under hydrogen vapour pressures in excess of 20 kbar!

steel production. However, the purity levels required for semiconductor preparation turn out to be enormously more stringent—impurities in steel typically demand control at the percentage level (1 in 10^2), whereas a typical semiconductor will be susceptible to impurity levels measured in parts per billion (1 in 10^9). Crystal perfection is another critical parameter which raises demands on the crystal grower to levels unheard of in most metallurgical applications. So, in summary, we see that, though there are qualitative similarities between the material technologies of metals and semiconductors (which will surely act as helpful guides in many cases), the quantitative differences raise possibly formidable problems.

This having been said, by way of introduction, it is now time to come to terms with the nature of these intriguing materials on which so much of our lives depend. What, exactly, is a semiconductor? The usual dictionary definition that it is a material which conducts electricity with a facility somewhere between those of metals and insulators (as the name clearly suggests) certainly provides a convenient starting point but leaves an awful lot unsaid. However, it is useful first to quantify the above definition. Most metals are found to be good electrical conductors,

Box 1.2. Electrical resistivity

Entertain conjecture (as Shakespeare once put it) of a regular, uniform cube of silicon with sides each 1 m in length and having metallic electrical contacts covering one pair of opposite faces. ('Conjecture' is appropriate here—in spite of the truly amazing feats off bulk crystal growth demonstrated of late, no such volume of crystalline material has yet been seriously contemplated.) If a small current I (ampere) is passed through this massive block, from one contact to the other and the voltage drop V (volts) across the sample measured, the resistance R (ohms) obtained, $R = V/I$ is, by definition, the *resistivity* ρ of the silicon material. Its units are ohm-metres (Ω m). Note that, because the geometry of the measurement is specified, ρ is a material parameter, that is, a property of the silicon alone. It depends on the density of free carriers within the silicon and on their 'mobility', that is, their ability to move through the crystal, but not on any external features. For example, if we change the geometry to a slightly more general one of a rectangular brick of length L and cross-sectional area A, the *resistance* of this sample is given by $R = \rho L/A$. Thus, resistance depends on both geometry and material.

having resistivities (see Box 1.2) in the order of 10^{-7}–10^{-8} Ω m, whereas, at the other end of the scale, we encounter insulating materials such as certain oxide films, mica, glass, plastics, etc. where the corresponding quantity ranges between 10^{10} Ω m and 10^{14} Ω m. This huge variation of resistivity between metals and insulators is remarkable in itself but we are more interested at present in where typical semiconductors lie in the scheme of things—they cover quite a range themselves—10^{-6}–10^{2} Ω m being typical resistivities for silicon, for example, whereas inclusion of the so-called 'semi-insulating' gallium arsenide with resistivity near 10^{7} Ω m extends the range upwards by a further five orders of magnitude. Clearly, the values appropriate to semiconductors do lie between those of metals and insulators but, perhaps the more striking observation is that semiconductor resistivities, themselves vary so much (roughly 13 orders of magnitude!) that it is hard to see this as a suitable parameter with which to 'pin them down'. We need an explanation for the origin of the resistivity if we wish to know what these numbers really mean, and this can only be obtained by referring to the band theory of solids developed during the late 1920s and early 1930s. It was this which laid the foundation for our present understanding of semiconductors and how they relate to metals and insulators.

The band theory represented an important application of the recently developed (and highly exciting) quantum theory of atomic structure. The first major success of quantum theory was its explanation of atomic spectra, particularly that of the simplest atom, hydrogen. An important concept introduced by quantum mechanics was the notion that electrons in atoms could occupy only certain well-defined energy states (in contrast to classical mechanics which allows all possible energy values) and, in single (i.e. isolated) atoms these energy states were extremely sharp. The resulting spectral emission lines which corresponded to electrons jumping from one 'allowed' energy state to another (of lower energy) showed correspondingly narrow line widths. The rather simple, but very important, equation which defines this emission process was found to be:

$$h\nu = E_2 - E_1, \tag{1.1}$$

where ν is the frequency of the light emitted, E_2 is the upper and E_1 the lower of the two energy states and h is the (now) famous Planck's constant, one of the most important fundamental constants of modern physics. In semiconductor work, it is customary to refer to energies in units of 'electron volts' (the energy an electron gains when it is accelerated through a voltage drop of 1 V) so we can define the Planck's constant in terms of the unit 'electron volts per Hertz', in which case, it takes the value 4.136×10^{-15} eV s—a value of $\Delta E = (E_2 - E_1) = 1$ eV corresponds to a frequency of 2.418×10^{14} Hz which lies in the near-infrared region of the spectrum. Another very useful relation which can be obtained from this connects the emission *wavelength* λ with the energy difference, as follows:

$$\lambda = 1.240/\Delta E \tag{1.2}$$

in which the wavelength is measured in microns ($1\ \mu\text{m} = 10^{-6}$ m) and the energy difference in electron volts. In this book, we shall make much use of these equations (and the corresponding physical concepts) which is why they have been spelled out in detail here.

In a crystalline solid, the atoms of copper, aluminium, silicon, germanium, gallium, and arsenic (in GaAs), for example, cannot be treated as isolated—in fact, they are in close proximity and nearest neighbours are chemically bonded to one another. This means that an electron on one atom 'sees' the electric field due to electrons on other atoms and the nature of the chemical bond implies that electrons on close-neighbour atoms are able to exchange with one another. Two important results follow: the sharp *atomic* energy states are broadened into energy 'bands' in the solid and these bands are associated, no longer with single atoms, but with the crystal as a whole. In other words, electrons may appear with equal probability on atoms anywhere in the crystal. This implies that these negatively charged electrons are able to move through the crystal lattice and we have, at least in principle, the possibility of electrical conduction (which is simply the flow of electric charge).

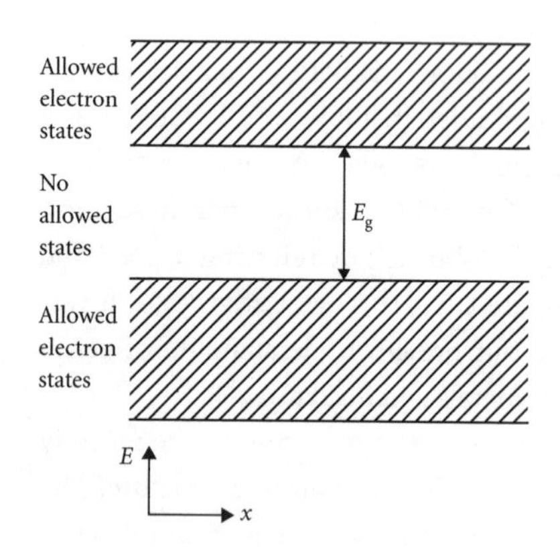

Figure 1.1. Schematic diagram representing the valence and conduction bands in a semiconductor. The vertical axis represents energy, while the horizontal axis represents a spatial coordinate. Semiconductors are characterized by the forbidden energy gap which separates the two allowed bands of states, the gap being typically 0.5–2.0 eV wide. No allowed states for electrons exist within the gap of a pure semiconductor.

Figure 1.1 provides a schematic illustration of the energy bands in a semiconducting crystal. It is an essential feature of any semiconductor that these two bands are separated by an 'energy gap', that is, there is a range of energies which is not available to electrons, and this gap is known variously as 'the fundamental energy gap', the 'band gap', the 'energy gap', or the 'forbidden gap'. By whichever name, it is the most important property of any semiconducting material, as we shall see. In a perfect, pure semiconductor at the absolute zero of temperature, the lower band, known as the 'valence' band is completely full of electrons—that is, every available energy state is occupied—while the upper band, the 'conduction' band is entirely empty. At first sight, the presence of the filled valence band seems to imply that the material might act as an electrical conductor but deeper insight supports the opposite conclusion. In order that *net* charge can flow, there must be empty states available for electrons to move into, which is not the case in a completely filled band—the *exchange* of electrons between any pair of states does not, of course, change the overall electron distribution, so such a process (which *is* possible) does not represent a flow of charge. Needless to say, the empty conduction band is equally ineffective, but for a more clearly obvious reason. Under these, rather special circumstances, therefore, our 'semiconductor' behaves as a perfect insulator! Its true semiconductor properties only become apparent if we lift the restriction on its temperature.

At temperatures above absolute zero, thermal energy is freely available in the crystal in the form of 'lattice vibrations'—that is, the atoms of the crystal oscillate about their mean lattice positions, the amplitude of the oscillation increasing in sympathy with the rise in temperature—and some of this energy may be transferred to the valence band electrons, so as to 'excite' a small proportion of them into the empty conduction band. Immediately, it is apparent that these 'free' electrons (they have been *liberated* from the confines of the valence band) can act as charge carriers—if we apply an electric field across the sample (by connecting the terminals of a battery across it) these conduction electrons will be able to move through the crystal—there being large numbers of empty states available for them in the conduction band. Hey presto! a current flows. We call it an 'electron current' because it is carried by free electrons. What may be less immediately obvious, the empty states created in the valence band can also carry current—their existence is all that is required to permit a net flow of charge in the valence band, too. In practice, semiconductor scientists have chosen to call this current a 'hole current'—though we must be clear that it is the electrons which physically move in the valence band, the net effect can equally well be represented as a flow of 'positive holes' in the opposite direction to the electron flow. Because the number of these holes is identically equal to that of the free conduction band electrons, it is convenient to think of the holes as the charge carriers. Figure 1.2 will make clear that, though the hole and electron flows are in opposite directions, the net *charge* flows are in the *same* sense—the two currents add together. A flow of *negative* electrons from right to left represents a *positive* charge flow from left to right, while the *positive* holes, flowing from left to right, also constitute a *positive* current in the same direction.

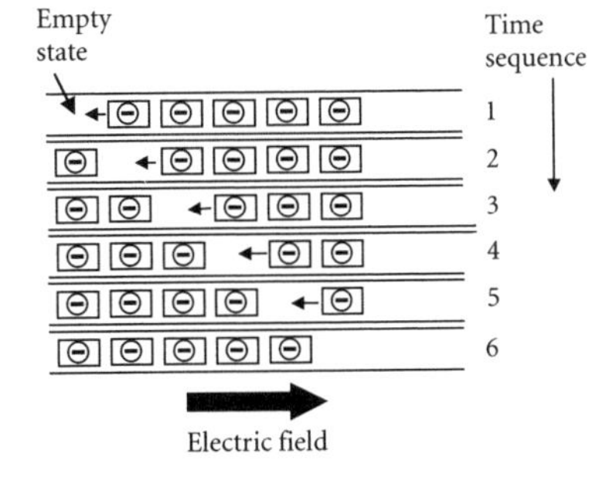

Figure 1.2. Pictorial representation of a positive hole current in the valence band of a semiconductor. Electrons move, under the influence of an applied electric field, into available empty states in the valence band, leaving behind new empty states in the sites they have abandoned. When negatively charged electrons move from right to left, the resulting holes move from left to right. Positive charge flow is in the direction of hole flow.

Following that subtle piece of sleight of hand, we may now begin to understand the large range of resistivities observed between different semiconductors. It originates with the fact that different semiconductors are characterized by different band gaps—for example, indium arsenide has a band gap of 0.354 eV, germanium 0.664 eV, silicon 1.12 eV, gallium arsenide 1.43 eV, (cubic) zinc sellenide 2.70 eV, GaN 3.43 eV, and all the way to diamond with a gap of 5.5 eV. What is more, indeed very much more, the numbers of free carriers in the conduction and valence bands due to thermal excitation are related to the energy gap by the following exponential expression:

$$n = p = \surd\{N_C N_V\} \exp\{-E_g/2kT\}. \tag{1.3}$$

In this important equation, n and p are the *densities* of free electrons ('n'egative) and holes ('p'ositive), E_g is the energy gap, k is Boltzmann's constant and T the temperature of the semiconductor sample in degrees absolute. (Readers familiar with the kinetic theory of gases will recall the significance of kT in relation to the kinetic energy of gas molecules—in our case, because of the strong interaction between atoms in a solid, this energy cannot be associated with single atoms, but, rather, with their collective motion.) N_C and N_V are parameters, referred to as 'effective densities of states' for the conduction and valence bands, respectively. At room temperature, the prefactor $\surd\{N_C N_V\}$ has a typical value of about 10^{25} m^{-3}. Again, at room temperature, the 'thermal energy' $kT = 0.026$ eV so equation (1.3) may then be written as:

$$n = p = 10^{25} \exp\{-E_g(\text{eV})/0.052\} \tag{1.4}$$

and the reader who can lay hands on a pocket calculator will have no difficulty in confirming the following (approximate) values for n and p at room temperature:

Germanium 2.85×10^{19} m^{-3}, silicon 4.43×10^{15} m^{-3},
GaAs 1.14×10^{13} m^{-3}, ZnSe 2.82×10^{2} m^{-3}.

The range of values for free carrier density is clearly enormous, and, if we bear in mind that the resistivity ρ is inversely proportional to n (or p), we can readily appreciate the likelihood of ρ varying strongly from one pure semiconductor to another. Suffice it, for the moment, to say that the values of resistivity corresponding to these free carrier densities vary (very roughly) between 1 Ω m (germanium) and 10^{17} Ω m (zinc sellenide). In other words, pure zinc sellenide behaves as a very good insulator, indeed! (Compare the typical values of measured resistivity quoted above.) We shall see in a moment that this account overlooks another vital aspect of the semiconductor repertoire so we should be careful not to treat our newly acquired numbers as quite Gospel truth, but they do, nevertheless, represent a major step forward in our understanding of semiconductor behaviour. It is clear now, that wide band gap materials *tend* to behave as insulators—true semiconductors appear to be those materials which have band gaps in the region of 0.3–2.0 eV—this is still a very imprecise definition but somewhat more manageable than our earlier attempt to use resistivity, which varied over some thirteen orders of magnitude!

Before leaving equation (1.3), it would be well to point out another important feature, namely the form of the temperature-dependence of free carrier density. Intuitively, it should be obvious that, as more thermal energy becomes available at higher temperatures, the larger will be the density of free carriers excited. This, indeed, is consistent with equation (1.3) (see Box 1.3). In turn, it implies that semiconductor resistivities *decrease* with increasing temperature, that is, the temperature coefficient of resistivity is negative, a property which can be taken to

Box 1.3. Temperature coefficient of resistivity

The temperature coefficient of resistivity α of a metal or semiconductor is defined as follows:

$$\rho(T) = \rho_0\{1 + \alpha(T - T_0)\}, \tag{B1.1}$$

where $\rho(T)$ is the resistivity at temperature T, ρ_0 is the resistivity at the reference temperature T_0 (say room temperature). It assumes a linear variation of resistivity with temperature, and, as such, is usually only valid over a very limited range of temperature. If we now differentiate equation (B1.1) with respect to temperature, we obtain:

$$\alpha = (1/\rho_0)\,\mathrm{d}\rho/\mathrm{d}T. \tag{B1.2}$$

Referring now to equation (1.3), and bearing in mind that $\rho \propto 1/n$, we can write:

$$\rho = C\exp\{E_g/2kT\}, \tag{B1.3}$$

where C is a constant. Differentiating equation (B1.3), $\mathrm{d}\rho/\mathrm{d}T = -(CE_g/2kT)(1/T)\exp\{E_g/2kT\}$ and writing $\rho_0 = C\exp\{E_g/2kT\}$, it then follows that:

$$\alpha = -(E_g/2kT)(1/T), \tag{B1.4}$$

which demonstrates that α is negative but also that it decreases with increasing temperature. The relationship between resistivity and temperature is not a linear one so we expect α to depend on T, as it does. Evaluating α at room temperature for a semiconductor with a band gap of 1 eV, we obtain: $\alpha = -(1/0.052)(1/300) = -6.4 \times 10^{-2}\ \mathrm{K}^{-1}$. This is a large negative coefficient, compared with a value typical for a metal of $+4 \times 10^{-3}\ \mathrm{K}^{-1}$.

distinguish semiconductors from metals, which are characterized by positive temperature coefficients of resistivity with magnitudes typically about ten times smaller. (Conduction in metals takes place in a permanently unfilled band, from which it follows that the density of free carriers remains constant with increasing temperature—the increase in resistivity arises as a result of increased thermal motion of the metal atoms which makes it harder for the electrons to move through the lattice—this also happens in semiconductors but it is a very much smaller effect than the variation of free carrier density described above.)

1.4 Semiconductor doping

At this point we are in a position (and, indeed, are duty-bound) to unfold one of the most intriguing and important properties of semiconductors, namely that their conductivity can be strongly influenced and, more importantly, *controlled* by the introduction of relatively small amounts of certain impurity atoms, called 'dopants'. It is this aspect of their behaviour which gives semiconductors their power—which, indeed, makes solid state electronics (and optoelectronics) possible. The '*p-n* junction', which plays a central role in so many electronic devices, and has dominated the field of solid state electronics, is made possible by this phenomenon of 'doping'. Conveniently, it also enables us to complete our answer to the question posed at the beginning of this discussion—what *exactly* is a semiconductor?

To understand doping, we must first look in greater detail at the way in which semiconductor atoms are bound together in a crystal and this, in turn, obliges us to know something of the electronic structure of atoms (see the periodic table of the elements in Table 1.1). Because it constitutes the simplest case, we shall base our discussion on silicon, leaving other materials and complications until later chapters. Holding our technical noses, we plunge in, feet first. Silicon is an element

Table 1.1. The periodic table of the elements

H^{1}									He^{2}
Li^{3}	Be^{4}		B^{5}	C^{6}	N^{7}	O^{8}	F^{9}	Ne^{10}
Na^{11}	Mg^{12}		Al^{13}	Si^{14}	P^{15}	S^{16}	Cl^{17}	Ar^{18}
K^{19}	Ca^{20}	[Sc^{21} Ti^{22} V^{23} Cr^{24} Mn^{25} Fe^{26} Co^{27} Ni^{28} Cu^{29} Zn^{30}]	Ga^{31}	Ge^{32}	As^{33}	Se^{34}	Br^{35}	Kr^{36}
Rb^{37}	Sr^{38}	[Y^{39} Zr^{40} Nb^{41} Mo^{42}Tc^{43} Ru^{44} Rh^{45}Pd^{46}Ag^{47} Cd^{48}]	In^{49}	Sn^{50}	Sb^{51}	Te^{52}	I^{53}	Xe^{54}
Cs^{55}	Ba^{56}	[La^{57}Hf^{72}Ta^{73} W^{74} Re^{75} Os^{76} Ir^{77} Pt^{78} Au^{79} Hg^{80}]	Tl^{81}	Pb^{82}	Bi^{83}	Po^{84}	At^{85}	Rn^{86}
		{Ce^{58} Pr^{59} Nd^{60} Pm^{61} Sm^{62} Eu^{63} Gd^{64} Tb^{65} Dy^{66} Ho^{67} Er^{68} Tm^{69} Yb^{70} Lu^{71}}						
Fr^{87}	Ra^{88}	[Ac^{89}						
		{Th^{90} Pa^{91} U^{92} Np^{93} Pu^{94} Am^{95} Cm^{96} Bk^{97} Cf^{98} Es^{99} Fm^{100} Md^{101} Lu^{103}}						

which occupies the second row of Group IV in the periodic system, its electron configuration, being $1s^2 2s^2 2p^6 3s^2 3p^2$. As quantum chemists discovered during the exciting 1920s, the $1s^2 2s^2 2p^6$ configurations represent 'closed shells' of electrons which are chemically inert—it is the 'outer' electrons $3s^2 3p^2$ which take part in chemical bonding. The bond in silicon is a purely 'covalent' one, electrons being *shared* between neighbouring atoms in such a way as to provide each Si atom with a complete shell $3s^2 3p^6$ (the next closed shell configuration). This is achieved by an arrangement whereby each Si atom in the crystal is bonded to four other atoms, each of which shares one of its outer electrons with the central Si atom, resulting in the desired (i.e. low energy) configuration for this particular atom, while the central atom shares its four outer electrons with the four neighbouring atoms, one electron to each. (More sleight of hand!). Figure 1.3 makes this clear but is misleading in one important sense—it is a two-dimensional model, whereas reality is three dimensional, the necessary symmetry being provided by a tetrahedral arrangement of the atoms, each Si atom lying at the centre of a regular tetrahedron of neighbouring atoms (to which it is bonded), as shown in Figure 1.4. The beauty of this arrangement is that every Si atom in the crystal sees precisely the same environment (with the exception of those atoms at the outer surfaces—but more of them anon), consistent with the overall crystal being made up of a regular array of atoms. It is the structure which minimizes the total energy of the system and this explains why silicon crystallizes in this particular form.

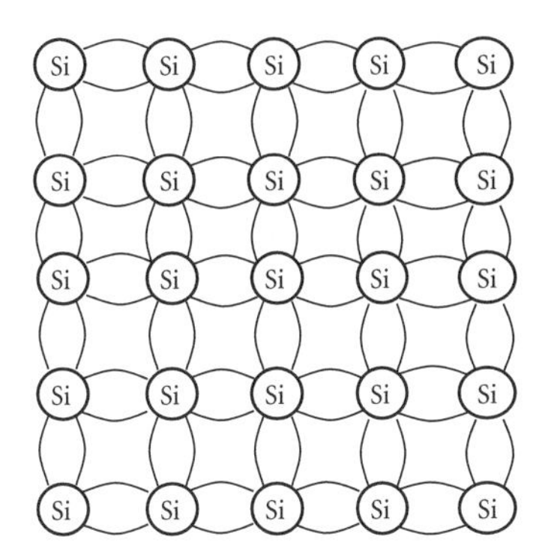

Figure 1.3. Schematic diagram, illustrating the covalent bonding of Si atoms in a single crystal of silicon. The four outer (i.e. bonding) electrons from each Si atom are shared with each, of four, neighbouring atoms. In turn, each neighbouring atom shares one electron with the 'central' Si atom. In this way, each Si atom acquires a closed shell of eight electrons ($3s^2 3p^6$) which represents the minimum energy configuration of the system.

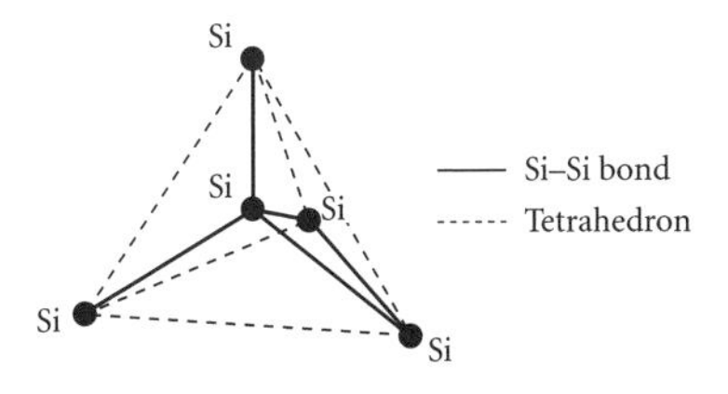

Figure 1.4. The cubic, tetrahedral arrangement of Si atoms which represents the actual three-dimensional structure in a silicon crystal. Each Si atom is bonded to four nearest neighbours, occupying the corners of a regular tetrahedron, with the first Si atom at its centre.

Once this crystal structure and bonding scheme are grasped, doping is easy! Let us suppose that one Si atom is replaced in the crystal lattice by a phosphorus atom. Phosphorus is next to Si in the periodic table, being in Group V, which means that it has one more outer (3p) electron than Si, the key to its behaviour as a dopant atom. Four of the five outer electrons of P (P = phosphorus—not to be confused with p = hole density!) will be taken up with bonding to the four neighbouring Si atoms but this leaves, as it were, a single electron unemployed and this electron is easily detached from its host atom (P), becoming free to wander through the crystal and, if encouraged by an applied electric field, to take part in electrical conduction. The P atom is said to have 'donated' a free electron to the crystal, and the P atom is therefore known as a 'donor'. The silicon crystal has been 'doped' by the P atom. A small amount of thermal energy (roughly 50 meV) is required to free the electron (see Box 1.4) but this is very much smaller than the 1.12 eV required to excite an electron across the forbidden energy gap. Of course, one P atom, donating one free electron is unlikely to generate a significant (measurable) effect in the crystal but if, say, 10^{20} Si atoms per cubic metre are replaced (roughly 1 in 10^9 of the total), this will generate a free electron density of $n = 10^{20}\ \mathrm{m}^{-3}$, already vastly greater than the thermally generated carrier density in pure silicon which we calculated above. In other words, this (chemically speaking)

Box 1.4. The hydrogen model of shallow donor states

The concept of *n*-type doping with impurity atoms such as phosphorus (P) in silicon depends on the fact that these donor atoms possess one outer electron more than is needed for bonding in the silicon tetrahedral crystal lattice. This extra electron may readily by removed from the donor and then becomes free to take part in electrical conduction. We can estimate the amount of energy required to free it (known as the ionization energy) by using a simple analogy. The electron is loosely bound to the P atom by a Coulomb potential in the same manner as the electron in a hydrogen atom is bound to its core proton. The P atom, itself, is electrically neutral but, if we remove one electron, it becomes positively charged and this positively charged core acts as an 'effective proton' in binding the electron. The quantum mechanical calculation of ionization energy (or binding energy) proceeds along identical lines to that for the hydrogen atom but with two modifications: first, we must take account of the fact (see Section 3.3) that the electron in a semiconductor behaves as though its mass differs from that of a free electron (i.e. an electron in free space) and, second, we note that the medium in which our pseudo-hydrogen atom is located is characterized by a dielectric constant which reduces the Coulomb force between the two 'particles'. The expression for the binding energy of the electron in the real hydrogen atom is:

$$E_H = me^2/8\varepsilon_0^2h^2, \qquad \text{(B1.5)}$$

where m is the free electron mass, ε_0 is the permitivity of free space, and h is Planck's constant. Inserting values for the parameters results in $E_H = 13.6$ eV. The corresponding expression for our pseudo-hydrogen atom is obtained by replacing m with the effective electron mass m_e and multiplying ε_0 by the *relative* dielectric constant of silicon ε. Thus, we arrive at our final result:

$$E_D^H = \{(m_e/m)/\varepsilon^2\}E_H. \qquad \text{(B1.6)}$$

Taking an approximate electron mass of $m_e = 0.3m$ and a relative dielectric constant for Si of $\varepsilon = 11.7$, we obtain $E_D^H = 30$ meV. E_D is known as the 'donor binding energy' and represents the amount of energy required to remove an electron from the donor atom and place it in the conduction band. Notice that, by the nature of the calculation, E_D^H is the same for all donors—we have taken no account of differences in electronic structure of the different donors, simply representing the core by a positive point charge $q = (+)e$. This is a good first approximation because the electron orbit (the so-called Bohr orbit) is relatively large (radius $a^H = 2.1$ nm, that is, about eight times the Si–Si bond length). Unfortunately, because of its complicated conduction band structure, silicon is not a particularly good subject for such simple modelling and experimental values do show significant departures from the value of E_D^H we obtained above: P (45 meV), As (54 meV), and Sb (43 meV). The hydrogen model works much better for other materials such as GaAs and GaN, as we shall see later.

extremely low doping level completely dominates the electrical conductivity of the silicon crystal and, what is more, this conductivity can readily be varied over some seven orders of magnitude by the simple (?) expedient of controlling the amount of phosphorus added to the crystal during its growth. By contrast, there is no possibility of modulating *metallic* conduction to this degree.

Also of considerable importance, is the fact that P-doping results in the generation of only free electrons—no holes are produced in this process—and, to emphasize this, conduction is said to be '*n*-type' (i.e. electron-type).

Alternatively, the silicon is said to have been doped *n*-type. What, then, about holes? Can we also generate '*p*-type' conduction? The answer, of course, is yes—all we need to change is the chemical nature of the doping atoms employed. If, instead of substituting Si by P atoms, we use an element such as B or Al from Group III of the periodic table (i.e. having only three outer electrons), it is readily apparent that this will result in missing bonds in the crystal structure, in other words, holes in the valence band. The Al or B atoms are known as 'acceptors' because they are 'keen' to accept a free electron in order to complete a closed shell bonding configuration. So, not only can the electrical conductivity of silicon be varied almost at will, but in two fundamentally distinct ways, *n*- and *p*-type, using electrons and holes, respectively. This is what distinguishes semiconductors from all other electrical conductors and, at last, allows us to say just what semiconductors really are. However, this is so important that we need to put it in a separate paragraph!

We have learnt that semiconductors are materials characterized by a forbidden energy gap, separating valence and conduction bands, and which can be doped *n*-type or *p*-type to give a wide range of controlled electrical conductivities. We have also learnt to distinguish 'intrinsic' conduction due to thermal excitation of carriers across the forbidden gap (which is important in narrow gap materials) from 'extrinsic' conduction which results from doping semiconductors with suitable donor or acceptor atoms. These facts encapsulate the essential properties of semiconductors and can reasonably be taken as defining them.

It is clear that our ability to dope them makes semiconductors more flexible than metals (for instance) but, of even greater significance, it opens the way to a range of other phenomena associated with the combination of both *n*- and *p*-type doping in the same structure. Rectifiers, radio detectors, transistors, thyristors, LEDs, laser diodes, photodetectors all rely on the properties of *p*-*n* junctions, already referred to above, which result from the introduction of contiguous *p*- and *n*-regions in a single semiconductor sample. While this does not, as we shall see, represent the full extent of semiconductor wizardry, it certainly added a totally new dimension to the application of materials in the field of electronics. All this will become clear as the individual chapters of our story unfold but we shall postpone further details, for the moment, in order to examine the incidence of semiconductors within the periodic table (Table 1.1).

1.5 How many semiconductors are there?

We have already made use of the table in discussing semiconductor doping, but it is also instructive to see how the materials themselves are distributed and, in the course of doing so, to discover just how wide a range of materials, in fact, show semiconducting properties.

Our discussion of doping explains why, for example, we can regard ZnSe as a semiconductor, even though, in its pure form it clearly behaves as an insulator. The thermal energy involved in removing the free electron from a donor atom is a small fraction of that required to excite an electron from the valence band. Similarly, it explains why many other potential insulators behave as semiconductors—even diamond, with a band gap of 5.5 eV, can be included in the list because it is possible to dope it successfully—and this considerably broadens the range of relevant materials, well beyond most people's expectations. However, we should sound one important note of qualification—the usual trend is for donor energies to increase with increasing band gap which means that many wide gap materials, even if doped, show insulating behaviour at room temperature. If this were not so, we might find difficulty in explaining the occurrence of good insulators—every non-metal would behave as a semiconductor, which is certainly not the case.

The silicon chip is probably almost as often featured as the potato chip, though probably much less widely recognized, but few people can name more than one or two other semiconductors—germanium, perhaps, GaAs very occasionally. It may, therefore, come as something of a surprise to realize that upwards of 600 semiconducting materials are known to exist, ranging from germanium and silicon in Group IV to such exotic compounds as PbSnTe which is used as a narrow gap material for infrared detectors and light sources. For reference, Table 1.2

Table 1.2. Band gaps (in eV) of some common semiconductors

C	SiC	Si	Ge
5.5	2.86 (6H)	1.12	0.664
AlN	GaN	InN	
6.2 (WZ)	3.43 (WZ)	1.95 (WZ)	
AlP	GaP	InP	
2.51	2.27	1.34	
AlAs	GaAs	InAs	
2.15	1.43	0.354	
AlSb	GaSb	InSb	
1.62	0.75	0.17	
ZnO	CdO		
3.40	0.55 (NaCl)		
ZnS	CdS		
3.78 (3.68 ZB)	2.49		
ZnSe	CdSe		
2.83 (2.70 ZB)	1.75		
ZnTe	CdTe		
2.28 (ZB)	1.49 (ZB)		

shows the energy gaps of many of the common semiconductors. Without attempting any detailed explanation, it is significant to observe that, within Group IV, diamond (the tetrahedrally coordinated form of carbon, having the same structure as Si) is a wide gap semiconductor, whereas, as we move up the column, Si and Ge have progressively decreasing gaps while Sn and Pb are both metals (in a rough sense, having zero gap). It is interesting that, whereas Si and Ge form an alloy system with band gaps intermediate between those of the elements, Si and C form a well-defined compound, SiC which has a band gap of about 2.86 eV, roughly midway between Si and diamond (SiC actually exists in a number of different crystal forms—2.86 eV is the band gap of the most common one, having the so-called 6H structure).

The next interesting Group of materials is the III-V group of binary compounds, being made from a Group III metal atom combined with a Group V non-metal atom. The best known example is, undoubtedly GaAs, both Ga and As lying in the same row as Ge. Below them, in the first row (containing C) we find BN which is a wide gap compound, then in row 2 (Si), AlP and, in row 4 (Sn), InSb. Again, the band gaps show a clear trend, decreasing as we proceed up the table. However, many other III-V combinations are possible, as each Group V atom may be combined with any of the Group IIIs. In the sequence GaN, GaP, GaAs, GaSb we again see a strong reduction of band gap as the cation moves up the table. Similar trends occur for the Al and In compounds. Keeping the cation fixed while changing the anion, shows a similar effect. Working outwards again, we reach the II-VI compounds, Zn, Cd, and Hg, combining with O, S, Se, and Te, though there is a complication here—due to the presence of the 3d transition group, there are effectively two columns of divalent metals. Be, Mg, Ca, Sr, and Ba are the true Group II elements and some of their compounds (e.g. MgS, MgSe, CaS,) are certainly semiconducting, but the better known II-VI compounds are those containing Zn, Cd, and (to a lesser extent) Hg. The I-VII compounds, such as NaCl, are all strongly ionic and act as electronic insulators. This represents the limiting case—II-VI compounds tend to be more ionic and show generally larger gaps than the III-V materials, similarly, the III-Vs are more ionic than the Group IV materials and also show larger gaps. Thus, in the third row, the gaps of Ge, GaAs, and ZnSe are, as we have seen, 0.66, 1.43, and 2.70 eV, respectively which illustrates this general trend. Finally, we should note one other trend, the fact that crystal structure also changes as we move from Group IV materials to the I-VI compounds. These latter generally crystallize in the octahedral cubic form rather than the tetrahedral cubic structure appropriate to Si and Ge—the III-V materials are mainly tetrahedral (the structure being known as zinc blende—ZB), apart from the nitrides which crystallize in a hexagonal modification known as wurtzite (WZ), while the II-VI semiconductors vary between tetrahedral and hexagonal structure, some compounds being known in both modifications. A moment's thought suggests that we can reasonably expect the nature of the chemical bond to be closely related to both the preferred crystal type and to the corresponding band gap so that these trends should not altogether surprise us. However, their detailed understanding is not essential to our present purpose.

These are the common semiconductors but there are many more. In particular, it is possible to form continuous ternary and quaternary alloys in many of these material systems, for example, InGaAs, InGaP, GaAsP, AlGaN, AlGaInN, InGaAsP, etc. in the III-V group, CdHgTe, MgZnSe, CdZnSe, and MgZnSSe in the II-VI group and this increases the number of possibilities considerably. Interestingly, it also allows one to select a desired band gap by selecting an appropriate alloy—the gaps usually vary smoothly between the values appropriate to the end members of the system. Other semiconductors are formed from less obvious combinations of elements, such as CuInSe, AgInS, PbSnSe, GaSe but we shall not attempt a comprehensive listing. Enough has been said to illustrate the extremely wide range of materials which show semiconducting properties. We shall deal in more detail with *some* of the more common ones in subsequent chapters—it would obviously be a Herculean task for anyone to compile a complete account (and not, I suspect, a lot of fun to read, either!). Any reader wishing to familiarize himself or herself with the list of known semiconductors may consult the recent compilation 'Semiconductors—Basic Data' edited by Professor Madelung.

Bibliography

Hodges, H. (1971) *Technology in the Ancient World*, Penguin Books Ltd, Harmondsworth, Middlesex, UK.

Levinshtein, M., Rumyantsev, S., and Shur, M. (Vol 1 1996, Vol 2 1999) *Handbook Series on Semiconductor Parameters*, World Scientific Publishing Co., Singapore.

Madelung, O. (ed.) (1996) *Semiconductors—Basic Data*, Springer, Berlin.

Weber, R. L. (1980) *Pioneers of Science*, The Institute of Physics, Bristol and London.

CHAPTER 2

The cat's whiskers

2.1 Early days

Even within the ranks of modern-day semiconductor research groups, it is probably not widely recognized that semiconductor research began as long ago as 1833 when Michael Faraday published his observations on the electrical conductivity of silver sulfide. In fact, he made one particularly important discovery, observing, for the first time, the negative temperature coefficient of resistivity which we have already met in our first survey of semiconductor properties. Thus, it was established very early on that there existed materials whose electrical properties were essentially different from those of the better known metals, though, needless to say, there was no possibility, at that time, of a proper understanding of the phenomenon. This had to wait for very nearly a hundred years, until the development of the quantum theory of solids in the late 1920s and early 1930s. Nor should it be supposed that even this basic observation was free of controversy. Faraday was, at the time, interested in studying the changes in conductivity associated with change of state from solid to liquid phase and several of the materials he measured were probably ionic conductors (which also show negative temperature coefficients of resistivity), rather than electronic. In fact, it was not until a hundred years after the original experiments that the controversy was finally laid to rest (in favour of Ag_2S being an electronic semiconductor), a clear illustration of the difficulties inherent in understanding complex material behaviour in the absence of any well-established theory of electronic conduction in solids. At the same time, uncertainties concerning material quality could only add further confusion. Nevertheless, Faraday's work set the scene for a gradual accumulation of experimental data which was to cover a surprising range of both materials and phenomena.

The rate of progress was, by today's standards, extremely sedate but it must be borne in mind that there was none of the urgent commercialism which we now take for granted as part and parcel of semiconductor development, nor was there very much in the way of *organized* research. Many discoveries were made by self-motivated and often self-financed individuals who enjoyed the challenge of unravelling nature's secrets as a matter of intellectual satisfaction—no more. Nevertheless, as the nineteenth century yielded to the twentieth, the situation was to change considerably—with the advent of radio, the pressure to apply these discoveries grew stronger and the beginnings of the twentieth-century electronics industry could be seen struggling to emerge. But this is to look too far ahead—the next important discovery was reported in 1839 by Becquerel who observed the generation of a 'photovoltage' when he shone light on one of the electrodes in an electrolytic cell. Though this probably represents the first stirring of the science of opto-electronics, it was some 37 years before the effect was reported for a dry semiconductor, when Adams and Day (1876) observed a photovoltage on illuminating the electrical contact on a sample of selenium. Only slightly earlier than this, two other important discoveries had been reported. First, in 1873, Warren Smith described an alternative effect of illuminating a selenium sample, the reduction in its bulk resistivity, the phenomenon of 'photoconductivity', and, in 1874, 'rectification' was reported by Carl Ferdinand Braun. Braun studied the electrical behaviour of various metallic contacts to naturally occurring crystals of certain sulfides, such as lead sulfide (galena) and iron sulfide (pyrites), and discovered that their current–voltage characteristics were distinctly non-linear (in contrast to those of metal–metal contacts). In particular, he noted that the amount of current flowing in one direction through the circuit was considerably greater than that with the applied voltage reversed. Somewhat about the same time, Schuster reported similar results from his studies of the contact between wires of clean and oxidized copper—copper oxide acting, in this instance, as a semiconductor.

It is interesting to note, therefore, that, well before the end of the nineteenth century, four significant phenomena which can all be taken as characteristic semiconductor properties—negative temperature coefficient of resistivity, photoconductivity (both bulk effects), photovoltage, and rectification (contact effects), had been clearly identified. These were, of course, purely empirical observations and there was much controversy concerning their interpretation. In particular, the origin of rectification was disputed for quite a long time—was it a bulk effect or did it originate at the metal–semiconductor contact? There was no satisfactory theory available to provide guidance, and experimental data showed considerable scatter—indeed, not all samples even showed the effect, an observation which inevitably caused doubts in the minds of some workers!

Braun, himself, was convinced that his observations were the result of contact properties. His experiments were conducted with crystals to which, on one side, he attached a large area metal contact, while the other contact took the form of a fine metal wire which he pressed into the free face of his sample, and it was at this 'point contact' that he believed rectification took place. Such structures were later to become the basis of the famous 'cat's whisker' radio detector which revolutionized the practice of receiver design (and helped, no doubt, in persuading the Nobel Prize committee to reward Braun with a half share in the 1909 award). As anyone who has personal experience of using a cat's whisker detector will know, such devices can be little short of exasperating, in respect of their inconsistency and irreproducibility—it was often necessary to play for several minutes before a satisfactory rectifying contact could be achieved and even a moderate degree of vibration might destroy the desired performance! Little wonder, then, that many attempts to reproduce Braun's results were fraught with frustration. In terms of the interpretation of the effect, there was also controversy as to whether its origin was electronic (and related to the properties of the specific material used) or thermal—that is, associated with heat generated at the fine contact when electric current flowed through. Nevertheless, the phenomenon was undoubtedly real and led others to seek alternative and more reliable structures with which to pursue their investigations. Practical exigencies also made it desirable to increase the current handling capacity of the rectifiers and this necessitated larger area contacts and, therefore, an alternative technology.

2.2 First applications

Our historical outline has so far taken us roughly to the end of the nineteenth century, the story being concerned largely with the discovery of a number of basic semiconductor properties, though with very little in the way of real understanding (a considerable element of hindsight allows us now to adopt this superior viewpoint—it should not be seen as reflecting badly on those concerned at the time!). The early years of the twentieth century saw efforts being concentrated more towards the application of some of these new phenomena and it is interesting to recognize the interaction with other developments which took place during this same period, in particular that of the thermionic valve (or vacuum tube). The main driving force was, in both cases, the discovery by Hertz (1888) of electromagnetic 'radio' waves which, as demonstrated by Marconi in 1901, could be transmitted considerable distances through empty space—in this case, the space separating Cornwall from Newfoundland. Their use as a practical means of long-range communication demanded convenient methods of generating and detecting these waves and it was here that, first, the cat's whisker and, later, the vacuum tube were to make a major impact. In Hertz's original work, he used a spark gap for generation which resulted in a broad band of frequencies, rather than the desired single frequency which could be selected by a tuned receiver, and, for detection, a 'coherer' which consisted of a glass tube filled with metal filings, an inefficient and unreliable solution.

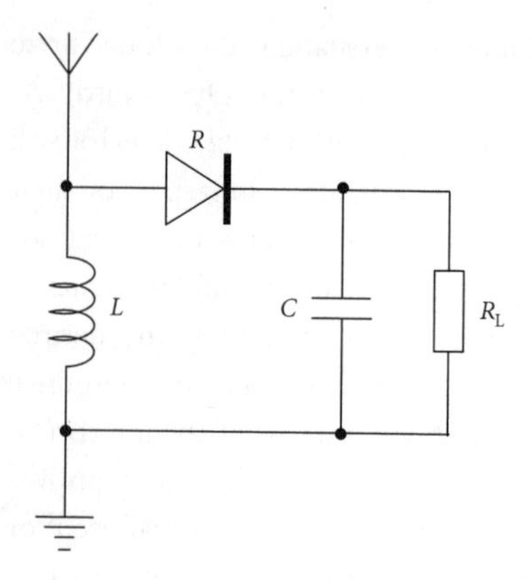

Figure 2.1. Simplified circuit of a cat's whisker crystal radio receiver. The radio signal is collected by the aerial which applies a radio frequency (RF) voltage across the inductor *L*. Rectified (i.e. unidirectional) current flows through the parallel combination of the load resistor R_L and the capacitor *C*. *C* provides an effective short circuit for the high frequency component of the current, while the audio signal appears across R_L.

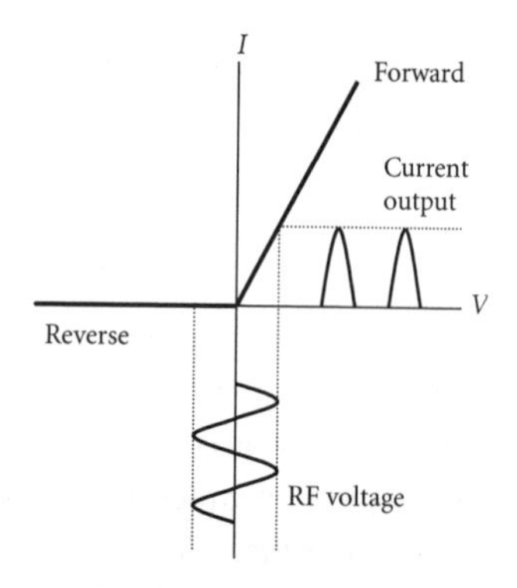

Figure 2.2. Highly idealized current–voltage characteristic of a rectifier, showing how an applied alternating voltage results in a unidirectional current flow in the output circuit. The output takes the form of a sequence of current pulses with a repetition frequency which corresponds to the frequency of the applied RF voltage.

Braun's major contributions to the newly discovered technology were two—he developed a 'tuned circuit', consisting of a capacitor and an inductor which facilitated the important feature of wavelength selection and, of greater significance for our present purpose, in 1904 he applied his cat's whisker to the detection problem. A variety of crystals was tried, including PbS, SiC, Te, and Si, PbS generally giving the best results. The improvement was dramatic and radio communications were firmly established from that moment on. We should pause briefly, therefore, to understand the basic principle behind his use of a rectifying device in this application. The radio waves were collected by a long aerial wire whose function was to generate a small, rapidly oscillating voltage (typically in the range 10^5–10^6 Hz) and feed it to the detecting circuit (see Figure 2.1). At that time, there were no detectors capable of responding directly to the high frequency signals so the function of the point contact rectifier was to convert the high frequency alternating voltage to a direct current (DC) which could readily be recorded as a DC voltage across the resistor *R*. The mechanism for this is illustrated in Figure 2.2. The current–voltage characteristic of an idealized rectifier is shown here, from which it is apparent that the positive half-cycles of the applied voltage give rise to current pulses in the rectifier circuit, while the negative half-cycles are suppressed. Thus, the *average* output current from the circuit is positive and represents the desired output signal, whereas, without the rectifier the average current would be zero (equal positive and negative contributions cancelling one another). Notice that the signal actually takes the form of a DC level with a superimposed high frequency oscillation but the DC component may be separated by using a capacitor (*C* in Figure 2.1) to shunt the output, thus presenting a very low impedance for the high frequency component, whereas the DC flows through the resistive load R_L and develops the required DC signal voltage. This procedure is referred to as 'smoothing' of the DC level and we shall meet it again in connection with rectifier power supplies.

An interesting (and important) subtlety came into play in these early radio experiments. The high frequency waves generated by the spark gap consisted of repeated pulses of energy with a repetition frequency which lay in the audio range, so the output from the crystal rectifier was, actually, not a continuous, steady DC level but rather a 'chopped' DC level which effectively represented an audio signal which could be heard in a pair of earphones. This led to the use of the Morse code system for transmitting information, the keyed dots and dashes being heard as short and long audio 'bleeps'. Thus the use of a spark transmitter (inadvertently!) produced the first example of an audio-modulated radio frequency (RF) signal, while the radio transmission with which we are now familiar employs deliberate modulation of a complex audio signal (such as music) onto a steady (i.e. continuous) RF transmitter wave. Such are the vagaries of technological development!

The crystal rectifier made commercial radio communication possible though, as pointed out earlier, it left something to be desired in terms of convenience and reliability. It should evince little surprise, therefore, that it was fairly soon replaced. The thermionic valve can trace its origins to 1883 when Thomas Edison took out a patent for a vacuum diode but it was not until 1904 that Fleming began experiments to employ a similar device as a radio detector. Its superior stability and reproducibility soon made it first choice for this application and the crystal rectifier rapidly faded from the picture—though, as we shall see, it enjoyed a resurgence as a radar detector in the early years of the Second World War. The detector diode was soon followed (in 1906) by the triode valve, the 'audion' invented by Lee de Forest which, because of its ability to amplify electronic signals, completely revolutionized radio technology. A few years later, in 1912, the device was taken up by Bell Laboratories and in the following year Bell demonstrated its use in repeater stations which formed a vital part of their first long distance telephone transmission. At this point, semiconductors probably seemed totally redundant, though this was, as we know, only a temporary setback. Valve technology had certainly won an important battle but semiconductors, with the invention of the transistor in 1947, were surely set to win the war. In fact, even before this, semiconductor rectifiers still had an important part to play, as we shall see in the next section.

2.3 Commercial semiconductor rectifiers

The ability of certain metal–semiconductor contacts to rectify—that is, change alternating currents into direct currents (AC–DC conversion) had obvious implications for a quite different application in the new field of radio. Thermionic valves required a steady positive potential to be applied to the anode in order to draw current from the cathode (which took the form of a heated filament) and therefore demanded a DC power supply of, typically, about 100–200 V. Public electricity supplies were based on AC generators (for reasons of efficient power distribution) and this placed a premium on the development of practical rectifiers, either as built-in sources of DC power or as battery chargers (many radio sets being battery-powered).

In the context of this 'commercial' requirement, it is interesting to note that a suitable large area selenium rectifier had been demonstrated by Fritts as early as 1886, though it apparently lay dormant until the end of the 1920s when consumer demand stimulated a much greater level of effort into the development of practical current rectifiers. First into serious contention was the copper–copper oxide rectifier demonstrated by Grondahl and Geiger in 1927, followed a few years later by the selenium rectifier which gradually became accepted as the standard for a majority of applications. Another significant outcome of 'commercial pull' was the much increased effort devoted to gaining a deeper understanding of basic semiconductor properties which we discuss in a later section. We see here the beginnings of *directed*, as opposed to purely 'blue skies' research—if research into the fundamentals of selenium or copper oxide could lead to the development of an improved performance from the corresponding device or a reduction in manufacturing costs, there was clearly good reason to pursue it. In this regard, it is also significant that much of this *applied* research was being done in the United States where, at that time, the culture of the practical, rather than the theoretical probably held far stronger sway. As is clearly revealed by Box 2.1, up to the Second World War, most of the 'new physics' was being done in Europe while American efforts were more focused on developing commercial enterprises. However, the huge surge in US research investment after the war can probably be seen as justification for this earlier focus.

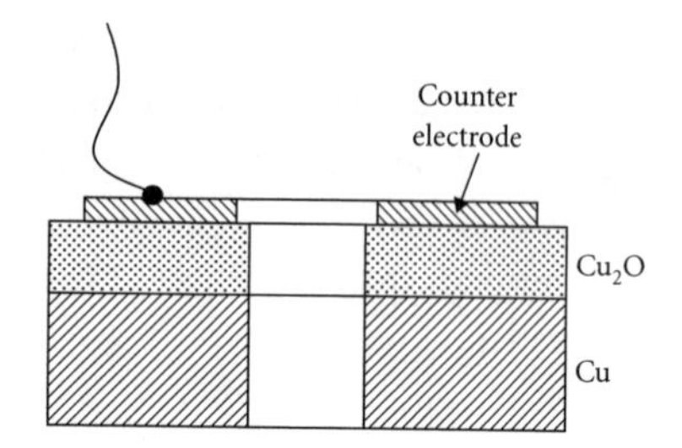

Figure 2.3. Structure of a typical copper oxide plate rectifier. The copper washer is oxidized to produce a thin layer of Cu_2O, then a counter electrode is applied to contact the upper surface of the oxide. The rectifying junction is at the interface between the copper and the copper oxide. The hole through the centre of the disc allows several rectifiers to be stacked together on an insulating rod.

We shall now briefly examine the structures of the two competing rectifier devices and comment on their performance. The copper oxide rectifier (see Figure 2.3) consisted, typically, of a copper disc about 1 mm thick which was oxidized on its upper surface to form a Cu_2O film some 1000 μm thick, the oxide surface being provided with a counter electrode in the form of a pressed metallic contact, such as lead or, alternatively a deposited metal film which may be sputtered, evaporated, or electroplated onto the oxide. The precise form was mainly a matter of convenience, the critical rectifying interface being that between the copper and its oxide layer, though the counter electrode had to be optimized for satisfactory adhesion and to achieve minimum series resistance. Oxidation of the copper was achieved by heating in air at temperatures of 1000–1030°C, followed by annealing at a temperature in the vicinity of 500°C. Before applying the counter electrode, it was necessary to remove a thin layer of cupric oxide (an insulator) which formed on the surface of the Cu_2O by suitable etching treatments. We may note, in passing, the early use of thermal and chemical treatments which have parallels in today's silicon processing. Though details inevitably differ, it is interesting to observe the similarity between the oxidation process required to produce the semiconducting Cu_2O and the silicon oxidation process involved in making metal oxide silicon (MOS) transistors (though the oxide, in this case acts as an insulator). Chemical etching similarly plays an essential role in the manufacture of silicon integrated circuits.

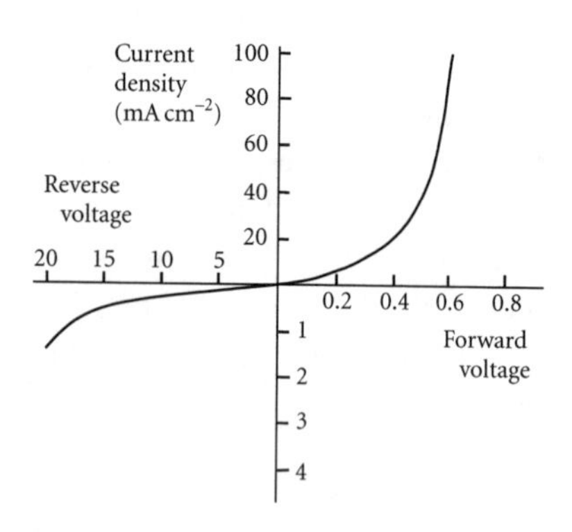

Figure 2.4. Typical current–voltage characteristic of a copper oxide plate rectifier (following Henisch 1957: 115). Note that the current and voltage scales differ considerably between forward and reverse directions. The reverse current begins to increase significantly for applied voltages greater than about 20 V. By permission of Oxford University Press.

Figure 2.4 shows a typical current–voltage characteristic of a copper oxide rectifier. Notice the rapid increase of current in the 'forward' direction, contrasting with the relatively flat 'reverse' characteristic—though notice, too, the fact that, as the reverse voltage increases beyond about 20 V, the reverse current also begins to increase significantly. This was

Box 2.1. Modern physics—up to Second World War

Given the historical nature of much of this chapter, it may be helpful to readers to have available an outline of other developments in physics which were contemporary with the events described here. Lack of space makes it impossible to include even a modicum of detail on each of these topics but the following table may serve to set semiconductor research in some kind of context

Author	Location	Subject	Year
Maxwell	UK	Electromagnetic wave theory	1864
Hertz	Germany	Discovery of the photoelectric effect	1887
Roentgen	Germany	Discovery of X-rays	1895
Bequerel and Curie	France	Discovery of radioactivity	1896
Thompson	UK	Discovery of the electron	1897
Lenard	Germany	Photoelectric effect associated with electrons	1900
Planck	Germany	The quantum hypothesis—black body radiation	1901
Einstein	Germany	Quantum theory of photoemission	1905
Einstein	Germany	Special theory of relativity	1905
Millikan	USA	Measurement of the electron charge	1909
Rutherford	UK	The nuclear atom	1911
Hess and Kolhorster	Germany	Discovery of cosmic rays	1913
Bohr	Denmark	'Old' quantum theory of the atom	1913
Thompson	UK	Discovery of isotopes	1913
Einstein	Germany	General theory of relativity	1916
Einstein	Germany	The radiation laws—light emission from atoms	1917
Pauli	Germany	The exclusion principle	1925
de Broglie	France	The wave nature of matter—postulate	1925
Heisenberg	Germany	Matrix mechanics—quantum mechanics	1925
Schrodinger	Germany	Wave mechanics—alternative formulation	1926
Uhlenbeck and Goudsmit	Germany	Proposal of electron spin	1926
Davisson and Germer	USA	Discovery of electron diffraction	1927
Heisenberg	Germany	The Uncertainty principle	1927
Dirac	UK	Theory of electron spin	1928
Sommerfeld	Germany	Quantum theory of metals	1928
Wilson	UK	Quantum theory of semiconductors	1931
Chadwick	UK	Discovery of the neutron	1932
Cockroft and Walton	UK	Artificial nuclear transmutation	1932
Anderson	USA	Discovery of the positron	1932
Hahn and Strassmann	Germany	Nuclear fission	1939
Fermi and Szilard	USA	First nuclear reactor	1942

important in so far as the application to rectifying AC power to generate a DC voltage of, say, 150 V required the rectifier to hold off a peak reverse voltage of about 250 V without passing significant current. This could only be achieved by stacking upwards of a dozen individual rectifiers in series and the structure illustrated in Figure 2.3 therefore included a central hole to facilitate stack mounting on an insulating rod. It is well to be aware, too, that reverse currents generally increase with increasing temperature so allowance had to be made for this in designing for any specific application. Yet, other problems arose in respect of so-called reverse current 'creep' which depends in complex fashion on the time for which reverse voltages are applied and on the length of the rest period during which the reverse voltage is removed.

The selenium rectifier which took over many applications from the copper oxide device shows many similarities in overall structure, though its manufacture involved important differences. In fact, the methods used to make selenium rectifiers have evolved considerably over the years and many variations have been demonstrated. Basically, a layer of selenium was deposited on a metal-backing plate, then a metallic counter electrode was deposited on top of the selenium. It is the junction between the selenium and the counter electrode which is responsible for rectification though the details are complex and difficult to understand. The backing plate might take the form of a steel disc, cleaned, sandblasted, and plated with nickel or, alternatively, it could be aluminium. The selenium layer might be applied as a powder which was heated under pressure to form a uniform film or it might be painted on or thermally evaporated under vacuum conditions and subsequently heat-treated. Various additional processes were involved, such as coating the aluminium with a very thin layer of bismuth and the final rectifier performance was found to depend on parameters such as heating and cooling rates, maximum temperature, time of heating, etc. The counter electrode often consisted of a low melting point alloy of lead, tin, bismuth, and cadmium (in various proportions) which was applied in a spray process under a carbon dioxide atmosphere. Other trace metals might be included to enhance device performance but details are obscure, sometimes a very thin insulating layer was used between the selenium and the counter electrode to improve the characteristics under high reverse voltages, usually it was necessary to apply a 'forming' voltage to enhance (or even produce) the rectification behaviour, all of which illustrates the highly empirical nature of the manufacturing process. Final rectifier characteristics differ rather little from those of the copper oxide device and similar remarks concerning reverse current leakage and 'creep' effects also apply. There is not, perhaps, a great deal here which smacks of the later *scientific* approach to semiconductor developments but we may note one significant feature—rather than simply using naturally occurring crystals (as, for example, in the cat's whisker), attempts were being made to synthesize the semiconductor material in a form convenient to the specific application—the problem was that material preparation methods were relatively crude and characterization techniques were very much in their infancy. More importantly, the basic understanding of what was required was significantly lacking.

Before leaving the topic of copper oxide and selenium, we should not overlook two other important applications. We have already referred to the observation of photoconductivity and the photovoltage effect—both effects were discovered in the 1870s, and in 1886, Siemens made the important observation that the photovoltaic effect represented a direct conversion of light energy into electrical energy. However, serious attempts to apply them to practical problems were not made until the 1930s, following the demonstration by Grondahl and Geiger (1927) of a large area copper oxide rectifier. Lange first reported large photocurrents (of order milliamperes) in 1931 and applications soon followed, though the understanding of the photovoltage effect did not emerge until 1939, when Mott proposed an explanation in terms of his barrier model of rectification. The principal uses of these devices, up to the Second World War, were in the development of photocells for photographic exposure meters and, in the film industry, for converting the sound markings on cine-film into electric signals which could, in turn, be converted into sound. Both devices act as light detectors—that is, they convert light signals into electrical signals which can then be used to activate, in the first instance, a sensitive electric meter or, in the second, a loud speaker. The photovoltage (or photovoltaic) effect, in particular, is associated with the non-linear characteristic of a metal–semiconductor contact, whereas photoconductivity is a bulk effect but we shall leave discussion of the details until Section 2.4 which is concerned with the development of our basic understanding of semiconductor properties.

Though copper oxide and selenium rectifiers have proved of considerable practical usefulness, they clearly suffer from a number of undesirable defects which can probably be traced to imperfections in the semiconductor properties, imperfections which are extremely difficult to control, given the nature of the manufacturing process. As we shall see, the need for considerably greater control of material preparation came into prominence during the development of the transistor and these developments in germanium and silicon materials gradually led to the introduction of much more reproducible and reliable rectifiers based on silicon and the phasing out of the earlier devices. Crystal perfection and purity at the parts per billion level could never be features of the crude manufacturing processes previously employed, and, as we saw in Chapter 1, these are now recognized as essential to well-controlled semiconductor device work. Nevertheless, serious attempts were made to understand the properties of the earlier materials such as selenium and cuprous oxide and we should now explore the nature of this more basic semiconductor work which was taking place in parallel with many of the empirical device developments described above.

2.4 Early semiconductor physics

The scientific understanding of semiconductor behaviour which has been a central feature of the later burgeoning of semiconductor work makes two essential, and inextricable demands—the establishment of adequate measurement, or 'characterization' techniques and their application in parallel with a well-founded theory. Not only does theory serve to provide a proper understanding of experimental data but, more importantly, it offers guidance in the selection of *appropriate* measurements. As we shall see from numerous examples, the choice of appropriate measurements is determined by the nature of a particular semiconductor device—only when we understand the basis of device operation, can we make an intelligent selection. If device function depends crucially on free electron density, then it must be our prime concern to measure it—if hole mobility is vital, we aim to measure that parameter, instead. The difficulty faced in the early work was, of course, that this kind of understanding simply did not exist and progress was inevitably constrained by it. The realization that semiconductors were characterized by a band gap came only with the advent of a quantum theory of solids in 1931 and the understanding of the nature of metal–semiconductor contacts had to wait until 1939 for the theory of barrier potentials. The extent of the progress actually made was therefore all the more remarkable, and depended largely on the application of the 'Hall effect'.

Hall discovered his 'effect' in metals as early as 1879 but it was only with the discovery of the electron 18 years later that its significance could be fully appreciated. It was applied to a wide range of materials during the first half of the twentieth century, in particular, to semiconductors in a burst of activity between 1930 and 1933. The mathematical description of the Hall effect can be found in Box 2.2 but the basic measurement is easily grasped—a sample of material in the form of a rectangular bar (Figure 2.5) is provided with electrical contacts at either end, a current I_x is passed along it and a magnetic field B_z is applied normal to the plane of the bar. This field deflects the moving charge carriers to the sides of the bar, setting up a transverse *electric* field (known as the 'Hall field') across the bar. A measurement of this field, together with that of the current flow along the bar can be used to determine the sign of the charge carriers in the material and to provide a value for their density. In semiconductor terms, this means we can distinguish holes from electrons and measure the parameters n or p (see Section 1.3 of Chapter 1). What is more, if we also determine the resistivity of the material by measuring the voltage drop along the bar and comparing this with the current I, we can obtain a value for the free carrier mobility μ. Even without the knowledge of band gaps or any understanding of semiconductor doping, this information proved invaluable to the semiconductor physicist and led to a rapid improvement in the classification of a wide range of materials.

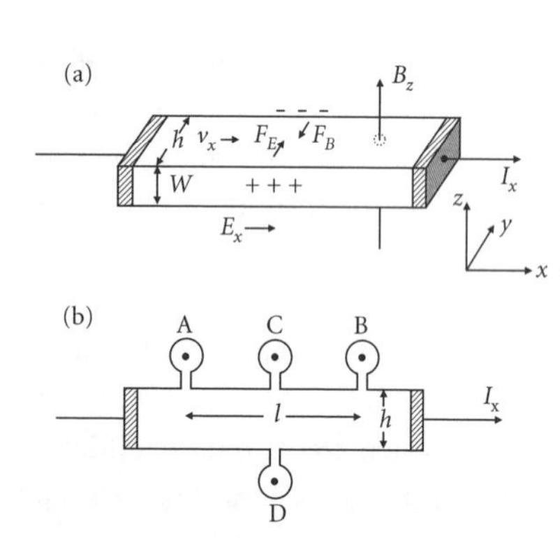

Figure 2.5. Schematic drawing of a Hall bar used to measure free carrier density and carrier mobility of a semiconductor. A current I_x is passed along the bar, while a magnetic field B_z can be applied in a direction normal to the plane of the bar. The side arms A and B are used to measure the voltage drop along the length l of the bar, using a high impedance voltmeter. C and D are used to measure the Hall voltage which results from the Lorentz force acting to deflect carriers along the y-direction. The thickness of the sample is W and the width is h. (From Blood, P. and Orton, J. W. (1992) *The Electrical Characterization of Semiconductors: Majority Carriers and Electron States*, Academic Press, London p. 96). Reprinted with permission form Elsevier.

Box 2.2. The Hall effect in semiconductors

The Hall effect is one of the most valuable and commonly used methods of characterizing semiconductor samples so it is worthwhile examining the background theory in moderate detail. We refer to the bar sample shown in Figure 2.5 and assume throughout that the material of the bar is uniform. For convenience, we shall assume that the material is *p*-type so that current is carried by positive holes. We choose a system of coordinates such that the hole current density $J_x = pev_x$ (where v_x is the hole drift velocity) flows from left to right along the positive *x*-direction, under the influence of an electric field E_x. A uniform magnetic induction B_z is applied along the positive *z*-direction (normal to the plane of the bar) and gives rise to a Lorentz force $F_B = -eB_zv_x$, acting on the holes, which tends to push them in the negative *y*-direction (i.e. across the bar). The resulting deflection of the positive charge sets up an *electric* field E_y across the bar which opposes the Lorentz force and a steady state is reached when there is no transverse current, that is, $J_y = 0$. Under these conditions, the current I_x is undisturbed and the two transverse forces are in balance so we can write:

$$eB_zv_x = eE_y \tag{B2.1}$$

The pair of contacts on either side of the bar (C and D in Figure 2.5(b)) differ in potential by the Hall voltage $V_H = E_yh$ which can be measured with a high resistance voltmeter.

From equation (B2.1), we see that the Hall field E_y is proportional to B_z and to J_x so we can write:

$$E_y = R_H J_x B_z, \tag{B2.2}$$

where R_H is a constant of proportionality known as the 'Hall coefficient'. We can now combine equations (B2.1) and (B2.2), together with the relation $J_x = epv_x$, to give:

$$R_H = E_y/B_zJ_x = v_x/J_x = 1/ep. \tag{B2.3}$$

This is an important relation, showing that a measurement of R_H yields a value for the hole density p. Similarly, if we consider an *n*-type sample, a parallel argument shows that:

$$R_H = -1/en. \tag{B2.4}$$

The sign of the Hall coefficient therefore tells us whether conduction occurs by hole or electron flow and thus tells us the conductivity type of the sample in question. The value of Hall effect measurements is now clear—all we have to do is explain how the Hall coefficient is measured in practice. This involves the sample thickness as well as the electrical parameters, as shown by the following manipulations:

$$R_H = -E_y/B_zJ_x = (V_H/h)/(B_zI_x/Wh) = V_HW/B_zI_x = (V_D - V_C)W/B_zI_x. \tag{B2.5}$$

Finally, we note that a concomitant measurement of the sample resistivity ρ allows us to derive a value for the free carrier mobility μ. It is usual to work with the conductivity $\sigma = \rho^{-1}$, as follows:

$$\sigma = J_x/E_x = (I_x/Wh)/[(V_A - V_B)/l] = [I_x/(V_A - V_B)] \times [l/Wh]. \tag{B2.6}$$

Box 2.2. Continued

It follows that σ can be measured using the contacts A and B, instead of C and D (note that this measurement does not require a magnetic field). All that remains is to obtain an expression for μ, which we can readily do, bearing in mind that the conductivity of a p-type sample $\sigma_p = pe\mu_p$. Thus:

$$\mu_p = \sigma_p/ep = \sigma_p R_H. \quad \text{(B2.7)}$$

So, once R_H and σ have been determined, μ is readily found. These two measurements, of conductivity and Hall coefficient, give us all the information we need about the transport properties of an *extrinsic* semiconductor sample.

The above treatment is not, as it stands, appropriate to an *intrinsic* semiconductor because we have assumed that current is carried entirely by one kind of carrier, whereas intrinsic samples contain equal densities of holes and electrons. However, a slightly more sophisticated treatment shows that, when both holes and electrons are present (a condition known as 'mixed conduction'), the following relations hold:

$$R_H = (1/e)[p\mu_p^2 - n\mu_n^2]/[(p\mu_p + n\mu_n)^2], \quad \text{(B2.8)}$$

$$\sigma = \sigma_n + \sigma_p = ne\mu_n + pe\mu_p, \quad \text{(B2.9)}$$

where R_H and σ are the parameters measured in the same way as in equations (B2.5) and (B2.6) above. In intrinsic material, where $n = p$, these simplify to:

$$R_H = (1/en_i)[(\mu_p - \mu_n)/(\mu_p + \mu_n)] \quad \text{(B2.10)}$$

and

$$\sigma = en_i(\mu_n + \mu_p), \quad \text{(B2.11)}$$

where n_i $(=p_i)$ is known as the intrinsic carrier density. Notice that it is not possible to derive values for all the unknown parameters in this case because we have three unknowns (four in the case of mixed conduction) and only two measured quantities, R_H and σ. The best way to proceed then is to make measurements over a range of temperatures and attempt to infer values from the changes in relative importance of the various terms.

The first application of the Hall effect to semiconductors appears to have been by Koenigsberger and co-workers during the period 1907–20. They discovered that the free carrier densities were generally much smaller than in metals (where the figure is of order 10^{28} m^{-3}, representing roughly one electron per atom) and that mobilities tended to be rather larger. Lacking the understanding provided by band theory, it came as a surprise when they found not only negative Hall fields, characteristic of electron conduction, but also, in several cases, positive values. Even more disconcerting was the observation of sign changes when measurements were performed over a range of temperatures. Nevertheless, they were able to classify many materials as semiconductors and gain an intuitive understanding of semiconductor properties. The work of Baedeker (1912) on CuI was important here because he demonstrated that controlling the stoichiometry, by addition of varying amounts of iodine, strongly influenced the free carrier density. This represented an effective doping of the material and was later observed in other similar compounds such as Cu_2O and ZnO. In the case of cuprous oxide the Hall effect showed a positive sign and led to the designation of Cu_2O as being a 'defect' semiconductor (i.e. defective of electrons), the vacant metal sites acting as acceptors. On the contrary, ZnO, which showed a negative Hall effect, was recognized as an 'excess' semiconductor, the density of metal atoms being slightly greater than that of oxygen, with the excess acting as donors. Our discussion of doping in Chapter 1 should make it clear that only very small departures from stoichiometry were necessary to effect significant changes in free carrier density.

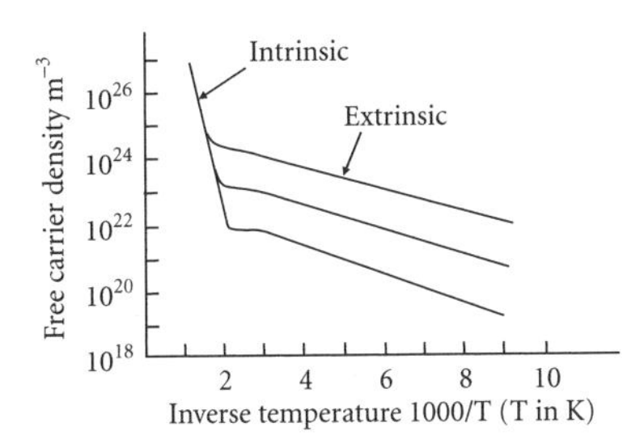

Figure 2.6. Example of measured free carrier densities in a range of samples of the same semiconductor material, as a function of inverse temperature. Two distinct regions are found, representing intrinsic conductivity, with a steep temperature-dependence (i.e. large activation energy) and extrinsic conductivity which varies from sample to sample and which shows a much smaller activation energy (corresponding to the energy needed to ionize the donor or acceptor involved).

Further progress was made when it became clear that many materials showed distinctly different behaviour in different temperature ranges (see Figure 2.6). In particular, it was observed in many cases that an individual material showed strongly temperature-dependent carrier density at high temperatures which was characteristic of all samples measured, while, at lower temperatures, the variation was not only much smaller, but also depended strongly on the specific sample measured. We now recognize these two regimes as corresponding to 'intrinsic' and 'extrinsic' conduction, the strong dependence at high temperatures resulting from thermal generation across the band gap, while the low temperature behaviour is dominated by (random—that is, uncontrolled) doping effects with much smaller activation energies. It was observed that, in the high temperature regime, the Hall effect normally had a negative sign but only later could it be appreciated that this was a reflection of the fact that mobilities of electrons are usually larger than those of holes. It is apparent from equation (B2.7) that, when both electrons and holes are present, the sign of the Hall effect depends on the sign of the term $(p\mu_p^2 - n\mu_n^2)$ and when $n = p$ (as is always the case for intrinsic conduction), this implies a negative sign when $\mu_n > \mu_p$. It is also possible to explain sign changes—a p-type semiconductor will show a negative Hall effect at high temperature because $\mu_n > \mu_p$ but a positive effect at low temperature where $p \gg n$, the sign changing at the point when $p{\mu_p}^2 = n{\mu_n}^2$.

The fact that the Hall effect allowed the experiment to distinguish between the behaviour of free carrier density and free carrier mobility led to an improved understanding of the positive temperature coefficient of resistivity in metals. It was found that the density of free electrons in metals remained approximately independent of temperature—it was the mobility which varied due to the fact that the metal atoms vibrated with increasing amplitude as the temperature increased. This made it harder for electrons to pass through the crystal lattice (they are 'scattered' more strongly by the lattice vibrations), thus decreasing the mobility and, with it, the conductivity (i.e. increasing the resistivity). This same effect occurs also in semiconductors but is generally swamped by the much greater changes in carrier density. This distinguishing feature of semiconductor behaviour thus became established well before it could be properly understood in terms of the band theory.

Yet another empirical observation which became established at an early stage was the correlation between the results of Hall effect measurements and those of thermoelectric power. We shall not discuss thermoelectric effects here but merely note that the correlation added substance to the growing volume of 'facts' which constituted a valuable catalogue of semiconductor properties, all of which was to fall neatly into place with the advent of the band theory. In two papers published in the Proceedings of the Royal Society in 1931, A. H. Wilson presented the first adequate theory of semiconductor properties. In the first he developed the theory of intrinsic conduction based on the realization that semiconductors are characterized by a forbidden energy gap and in the second he extended his ideas to explain extrinsic conduction, resulting from impurity doping. We are already familiar with these concepts and will find little difficulty in accepting their consequences but it was not so when Wilson's ideas were first published. It took quite a long time before they were widely understood—as we shall see, even as the work leading to the realization of the first transistor was under way at Bell Labs after the Second World War, these ideas were still being assimilated. John Bardeen was to apply them with consummate skill to the understanding of transistor action but not without considerable effort to come to terms with their unfamiliar basis. It is also salutary for us to note, in passing, that Wilson makes a firm statement in his second paper that silicon should be regarded as a metal. The lack of high quality, pure semiconductor materials made such attributions extremely difficult to make with any degree of accuracy and it was not for another 10 years that the need for much greater effort towards improving material quality became appreciated.

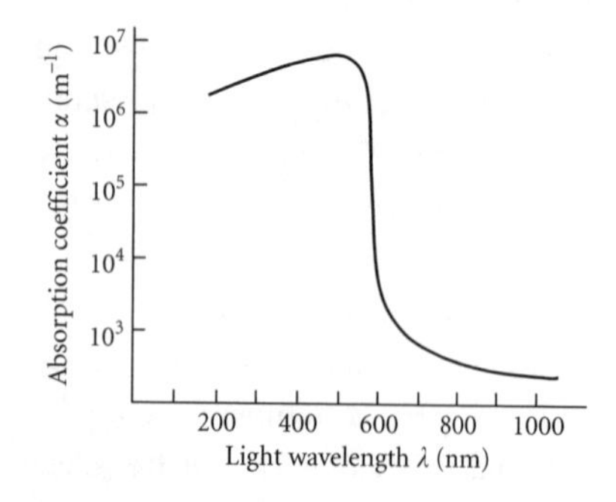

Figure 2.7. A typical example of an optical absorption curve for a semiconductor, illustrating the absorption edge which occurs at a wavelength of approximately 0.6 μm (energy gap E_g of approximately 2.1 eV).

Not only were 'transport' problems (i.e. transport of electrons and holes through a crystal lattice) elucidated by the new theory but the optical properties of semiconductors, too, depend fundamentally on the concept of a band gap. We have already met the concept of thermal excitation of electrons from valence band to conduction band, giving rise to intrinsic conduction, but this is not the only excitation process available. Shining light on a semiconductor also results in enhanced conduction, the phenomenon of 'photoconductivity' (see Box 2.3) in which the bulk resistivity of a sample is reduced in proportion to the intensity of the exciting light. However, there is one important feature of this process which we have yet to make clear—it is characterized by an 'absorption edge' (see Figure 2.7). If the wavelength of the light

Box 2.3. Photoconductivity

Provided the photon energy of light incident on a semiconductor exceeds the band gap energy, light will be absorbed in the semiconductor, each photon absorbed generating an electron–hole pair. Generally speaking, light is strongly absorbed, its intensity being reduced by an order of magnitude within a few microns of the semiconductor surface. In mathematical terms, the intensity I at a depth x below the surface is related to the incident intensity I_0 by the equation:

$$I(x) = I_0 \exp\{-\alpha x\}, \qquad \text{(B2.12)}$$

where α is the 'absorption coefficient' which, well above the band edge, has a typical value of $3 \times 10^6\ \mathrm{m}^{-1}$ (or, more conveniently, $3\ \mu\mathrm{m}^{-1}$) showing that the intensity is reduced by a factor of $e^3 = 20$ over a distance of $1\ \mu\mathrm{m}$ (Notice that the product αx must be dimensionless which explains why α has dimensions of $(\mathrm{length})^{-1}$). To work out how much light is absorbed in a depth x we can rewrite equation (B2.12) in the form:

$$\Delta I(x) = I_0 - I(x) = I_0[1 - \exp\{-\alpha x\}]. \qquad \text{(B2.13)}$$

In the special case of light with photon energy just below the band edge, where α is small enough to make $\alpha x << 1$, we can expand the exponential as $e^{\alpha x} = (1 - \alpha x)$ and obtain the simple result:

$$\Delta I(x) = I_0 \alpha x, \qquad \text{(B2.14)}$$

showing that an equal amount of light is absorbed in any increment of depth x, which is another way of saying that the absorption is uniform throughout the thickness of the material. But we must emphasize that this only holds when the light is absorbed close to (and on the low energy side) of the semiconductor band edge.

It is often convenient to refer to light intensity in terms of the photon flux N which is the number of photons crossing unit area of surface in unit time (units of $\mathrm{m}^{-2}\mathrm{s}^{-1}$) and we can relate this to intensity by remembering that each photon carries an amount of energy equal to $h\nu$. Thus the energy crossing unit area per second is given by $I = Nh\nu$ so, for any particular frequency, we see that I is proportional to N and we could therefore write all the above equations in terms of N, rather than I. Note that for a typical visible light frequency $\nu = 10^{15}$ Hz, $h\nu = 6 \times 10^{-19}$ J, so one watt of light per square metre ($1\ \mathrm{J\,m^{-2}s^{-1}}$) implies $N = 1/6 \times 10^{-19} = 1.7 \times 10^{18}\ \mathrm{m^{-2}s^{-1}}$. This represents a rather low intensity of illumination—contrast it with the milliwatt of laser light focused on an area of $10^{-12}\ \mathrm{m}^2$ in the compact disc player, for example, that is, $10^9\ \mathrm{J\,m^{-2}s^{-1}}$, or about 2×10^{27} photons per square metre per second.

Clearly, the generation of free carriers must have the effect of increasing the conductivity of the semiconductor and thoughtful readers may begin to worry that this increase will continue to grow indefinitely, as long as the light continues to shine. That this is not the case results from another phenomenon which we have so far overlooked—electrons and holes can 'recombine'. Imagine an electron wandering through the crystal, finding itself in the vicinity of a valence band hole—there is a Coulomb attraction between the pair which may draw the electron into the hole, thus satisfying the chemical bonding requirement at that point and, in the process, causing the mutual elimination of *both* electron and hole. This process is just the opposite of the generation process which we have already encountered and produces a balance between generation and recombination which represents a 'steady state' condition. The balance will occur at some specific value of electron–hole pair density. To calculate it we note that it must depend on the strength of the generation process (i.e. on the light intensity and absorption coefficient) and on the probability for recombination (which depends on the densities of electrons and holes). We shall illustrate this with a specific example.

Box 2.3. Continued

Suppose that we do a 'thought experiment'. We have a *p*-type sample of semiconductor and suppose that the hole density p_0 is always much larger than the density of light-generated carriers (this condition is referred to as 'low injection'). We shine on light for a few moments until the free electron density reaches its steady state value $n(0)$, then we switch off the light so that only the recombination process remains. The excess free carriers will recombine until the system returns to its *dark* condition in a time determined by the recombination rate. The probability that an electron and a hole will recombine is proportional to the product of their respective densities n and p (where $p \approx p_0$) so we can write:

$$-\mathrm{d}n/\mathrm{d}t = Anp = Anp_0, \quad \text{(B2.15)}$$

where A is a constant. Now, because p_0 is also a constant, we can easily solve this differential equation to obtain an expression for the excess carrier density at any time t after the light was switched off:

$$n(t) = n(0)\exp(-t/\tau), \quad \text{(B2.16)}$$

where τ is known as the 'recombination lifetime' and is given by $\tau = (Ap_0)^{-1}$. It characterizes the time taken for the excess carriers to disappear and is a function of the particular semiconductor sample involved. In germanium τ may be as long as a few milliseconds, whereas in GaAs it might be measured in nanoseconds (10^{-9} s).

Returning, now, to our original quest to estimate the density of excess carriers generated by the light source, we have to compare the recombination rate with the generation rate, which, from our earlier discussion, can be written as $G \approx N\alpha$. Equating the two rates gives:

$$N\alpha = n/\tau, \quad \text{(B2.17)}$$

so the steady state electron density generated by the light is simply $n = N\alpha\tau$ and the photoinduced conductivity is given by:

$$\Delta\sigma = eN\alpha\tau\mu, \quad \text{(B2.18)}$$

where μ is the electron mobility. (Note that we have neglected the contribution to σ from holes on the grounds that the hole mobility is likely to be much smaller than the electron mobility.) Finally, then, the photoconductivity is proportional to the light flux (as we should expect) and to the product $\mu\tau$ which is a property of the semiconductor involved, and which may vary over many orders of magnitude, according to the material used. For high sensitivity, we obviously need a large value of $\mu\tau$ which implies a long lifetime. However, a long lifetime has the additional, and less desirable effect of making the photoconductor very slow to respond to changes in light intensity. As in most aspects of life, we must seek the compromise best suited to the application in hand.

is very long, a semiconductor is found to be transparent—no light is absorbed and no photoconduction occurs. However, as the wavelength of the light is decreased, a wavelength is reached at which absorption begins, a sharp increase takes place in both absorption and photoconduction, then, for further decrease in wavelength, little further change occurs. This absorption edge can be readily understood in terms of the discussion of atomic energy levels in Section 1.3—reducing the wavelength (i.e. increasing the frequency ν) has the effect of increasing the photon energy $h\nu$ of the light and it is only when the photon energy becomes equal to the semiconductor band gap that light can be absorbed. What is more, the absorption process involves the excitation of an electron from the valence band to the conduction band, for every photon absorbed. Just as for the thermal excitation process, light absorption generates both free electrons and free holes which are responsible for the photoconduction effect. This simple picture of photoconductivity enables us not only to understand the basic effect but also to choose a suitable semiconductor material for a specific application. By selecting materials with appropriate band gap, we can obtain a long wavelength cut-off at whatever wavelength we require. If our interest happens to be in the detection of ultraviolet (UV) radiation, we should select a semiconductor with a wide gap, such as GaN or ZnS—if in detecting the infrared (IR) radiation used in optical-fibre communications, a narrow-gap material such as InAs or InGaAs. Similarly, this is the basis for the application of the narrow-gap alloy HgCdTe for thermal imaging systems which detect radiation of wavelengths in the 3–5 μm or 7–10 μm range.

Similar considerations apply to the wavelengths to which photovoltaic detectors are sensitive though to understand their operation requires some knowledge of how a metal–semiconductor rectifier works and this only became available in the late 1930s when, more or less coincidentally, Mott, Schottky, and Davydov published barrier models for the metal–semiconductor contact. As we have remarked earlier, there was controversy concerning the origin of rectification in, for example, the copper–copper oxide rectifier and final confirmation that the effect was associated with the contact, rather than with bulk Cu_2O, came only from these barrier theories. The overall effect is complex (we provide a simplified account in Box 2.4) but we can understand the principles easily enough. The essential idea depended on the difference in work function between the metal and semiconductor concerned, the work function being the amount of energy required to remove completely an electron from the metal (or semiconductor). The result is band-bending within the semiconductor as shown in Figure 2.8 which implies the existence of an electric field just below the semiconductor surface, extending into the bulk over a very small distance of typically about 1 μm or less. Application of a 'forward bias' (i.e. a voltage difference between metal and semiconductor which makes the metal positive and the semiconductor negative) reduces the size of the barrier and makes it easier for electrons from the semiconductor to flow into the metal, whereas application of a 'reverse bias' has no effect on the barrier which prevents electrons flowing from metal to semiconductor. Thus, the current in the forward direction is much larger than in the reverse and the contact acts as a rectifier.

Box 2.4. The barrier rectifier

The existence of a barrier to the flow of electrons between a semiconductor and a metal in intimate contact results from the difference in their respective work functions. The work function is the amount of energy required to remove an electron from the material and take it to infinity. If we, therefore, consider an *n*-type semiconductor and a metal close together, but not yet in contact, we have the situation shown in Figure 2.8(a). The so-called 'vacuum level' which is the electron energy at infinite distance is the same for both materials so it will be clear that, if the work functions are different, the conduction band in the semiconductor and the top of the occupied states in the metal will be misaligned. For example, the work functions of silicon and gold are 4.1 and 5.1 eV, respectively so the misalignment is just 1 eV in this case.

Suppose that the two materials are brought into contact (Figure 2.8(b)). Electrons will always tend to move from a region of high potential energy to one of lower energy so, in this case, they will flow into the metal. This increases the negative charge in the gold and decreases it in the silicon, thus setting up an electric dipole near the interface which opposes the electron flow. As more and more electrons cross the interface, the dipole builds up until it is strong enough to balance the original potential step due to the work function difference. At this point, the flow ceases and we have an equilibrium (or steady) state situation (Figure 2.8(c)) in which the abrupt potential step at the interface is balanced by a gradual change in potential within the silicon, that is, we have 'band-bending' in the silicon and a barrier region which is empty (depleted) of free electrons. The width of this barrier region is determined by the silicon doping density, being greater, the lower the doping level. Typical widths are in the range 0.01–1.0 μm. Strictly, there is similar band-bending within the gold but, because of the very high charge density in a metal, this takes place over such a short distance as to be negligible.

Two consequences of this interface barrier can be deduced. First, because the barrier region is depleted of free electrons (it is often referred to as the 'depletion region'), the barrier behaves like a parallel plate capacitor with capacitance C, given by:

$$C = \varepsilon\varepsilon_0 A/d, \qquad \text{(B2.19)}$$

where ε is the relative dielectric constant of the silicon ($\varepsilon = 11.7$ for Si), ε_0 is the permitivity of free space ($\varepsilon_0 = 8.85 \times 10^{-12}$ F m^{-1}), A is the area of the contact, and d is the depletion width of the barrier. Thus, if $A = 10^{-6}$ m^2 and $d = 0.1$ μm, $C = 1.04 \times 10^{-9}$ F, or about 1000 pico-Farads. Second (and rather less simply!), the barrier is responsible for rectification. If a voltage is applied across the junction, it appears almost entirely on the semiconductor side (because the

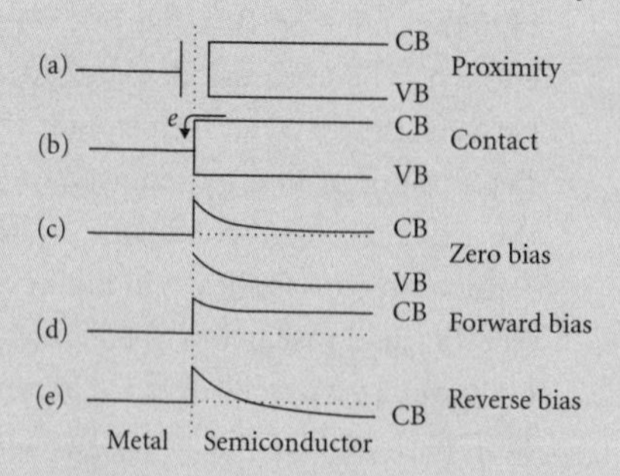

Figure 2.8. Band diagrams of a metal–semiconductor contact under various conditions. In (a) the metal and semiconductor are close together but not actually in contact, in (b) they are in contact and electrons are flowing from semiconductor to metal under the influence of the potential difference due to the difference in work functions of the two materials, while (c) illustrates the semiconductor band-bending which occurs when equilibrium has been established. In (d) a forward bias has been applied, making the semiconductor more negative and reducing the barrier height on the semiconductor side of the junction. In (e) a reverse bias has been applied which increases the semiconductor barrier height. The potential step at the metal is unaffected by the applied voltage. Note: CB=Conduction Band, VB=Valence Band.

Box 2.4. Continued

semiconductor resistivity is much greater than that of the metal) and, if applied in such a way as to make the silicon more positive, this pushes the silicon conduction band up in energy, reducing the height of the barrier on the silicon side. It thus becomes easier for electrons to cross the interface and enter the gold (see Figure 2.8(d)), though having no effect on the barrier at the gold side of the interface. This allows current to flow in the forward direction, increasing rapidly as the applied voltage increases. However, if the applied voltage has the opposite sense, it increases the barrier on the silicon side (Figure 2.8(e)), but still has no effect on the metal side, so the reverse current (i.e. electron current from metal to semiconductor) remains constant.

To understand this effect in detail we need to look more carefully at the nature of the current flow. In the case of zero applied voltage, there is no net current across the interface but this is not to say there is no current at all. In fact there are two currents which just cancel one another. In the simplest model of current flow, electrons cross the interface as a result of thermionic emission over the top of the barrier (electrons gain enough energy from lattice vibrations, in the same way as they do in intrinsic conductivity to reach the conduction band from the valence band) but, at zero bias, the barrier is the same in both directions. If we apply a forward bias so as to reduce the barrier on the semiconductor side, the electron flow from semiconductor to metal is enhanced so the net current flow is in this direction. However, if we apply a reverse bias which increases the barrier on the semiconductor side, the flow into the metal is reduced, leaving that from metal to semiconductor unchanged. As the reverse bias increases, the semiconductor–metal flow rapidly becomes negligible so the net current is simply the thermionic emission over the barrier from metal to semiconductor, which is constant. This is the 'saturation current' I_0 in equation (2.1) in the text. This overall argument (hopefully!) makes equation (2.1) for the current–voltage characteristic plausible.

Detailed calculation shows that a good approximation for the current flowing is given by equation (2.1):

$$I = I_0\{\exp(eV/kT) - 1\}, \tag{2.1}$$

where V is the applied voltage (positive in the forward direction) and kT represents the thermal energy which we encountered in Chapter 1. At room temperature, $kT/e = 0.026$ V, so it becomes clear that for forward applied voltages greater than about 0.1 V, $I \propto \exp\{38\ \mathrm{V}\}$ and therefore increases very rapidly. However, for reverse bias (i.e. *negative* applied voltages), the exponential term rapidly becomes small and $I \approx I_0$, that is, the reverse current 'saturates' at the constant (small) value I_0. I_0 depends, in practice, on the barrier height and on the doping level in the semiconductor and is therefore affected by the choice of both semiconductor and of metal. It may be minimized by minimizing the doping level which suggests that making a good rectifier demands a relatively pure semiconductor crystal and accounts, at least in part, for the variability of practical rectifiers where doping is outside the control of the manufacturer. (However, this is not the whole story, as we shall see when we discuss the point contact rectifier in Section 2.5.) The ideal 'diode' characteristic, represented by equation (2.1), is shown in Figure 2.9. It clearly represents only a rough approximation to the measured characteristics

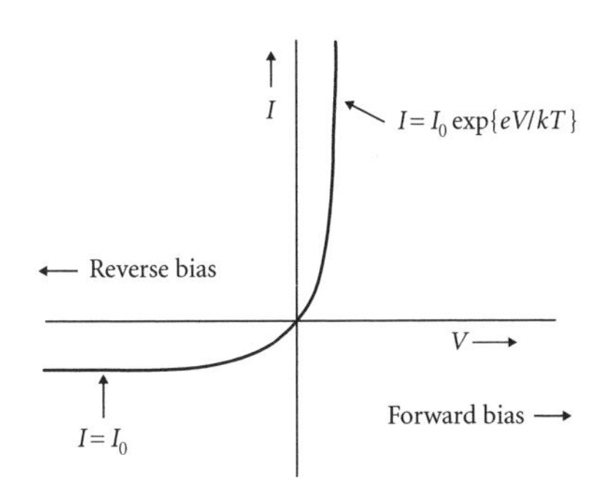

Figure 2.9. Ideal current–voltage characteristic of a metal–semiconductor contact, as represented by equation (2.1). This illustrates the rapid (exponential) increase in forward current when the applied voltage exceeds a few times kT/e and the saturation of the reverse current at a value of $I = I_0$.

shown for the copper oxide rectifier in Figure 2.4 but it provides an excellent starting point from which to develop more realistic models of rectifier contacts.

Finally, we are now in a position to explain the operation of a photovoltaic light detector. Light absorbed in the vicinity of a semiconductor contact generates electron–hole pairs which experience the effect of the electric field in the surface region. This accelerates electrons into the body of the semiconductor and holes towards the metal and this charge separation sets up an electric dipole which opposes the original field. The result is a voltage (the 'photovoltage') which appears between the metal and the semiconductor back contact—the device therefore generates a voltage in an external circuit which is proportional to the light intensity, a direct consequence of the surface barrier responsible for rectification. Note that this argument depends on the assumption that the device is effectively open circuit (or, at least, shunted by a very high impedance). If the shunt impedance were low, we should see a light-induced current, rather than a voltage. These are simply two methods of using the device—either method results in conversion of light into an electric signal which is the desired effect.

We can try now to summarize the state of semiconductor research at the beginning of the Second World War. The most important development was, of course, the application of quantum mechanics to provide the essential understanding of semiconductor behaviour—on this would be based the next 50 years of success. A great deal had also been achieved in classifying semiconductor properties and relating them to effective doping levels. The significance of the forbidden band gap was recognized and the distinction between intrinsic and extrinsic conductivity was clear. The distinction between metals and semiconductors was also clear and the mechanism of hole conduction was properly understood. The basic understanding of optical absorption was appreciated, as were the phenomena of photoconductivity and photovoltage. A basic theory of rectification was in place which enabled future workers to design the much improved semiconductor diodes based on better characterized materials. What was still generally lacking, however, was the availability of high purity, well-characterized single crystals which would allow carefully controlled doping, most samples up to this time having been polycrystals with levels of random impurities far too high to permit this. The significance of this fundamental requirement was, nevertheless, beginning to be appreciated, as we shall see in our final section.

2.5 The cat's whisker reborn

To a large extent, science was on hold during the war years, apart from those aspects which had direct relevance to the war effort. Scientists found themselves being drafted into unfamiliar territory and expected to perform in a manner few had ever experienced. Not only were they working in new fields, under external (rather than self-induced) pressures but the familiar luxury of open publication of results was no longer available to them. *Organized* science suddenly became the norm, an activity of particular urgency being the development of microwave radar for use on the ground, at sea, and in the air. On this, in particular, hung the result of the aerial battles which raged over Europe during the early 1940s and scientists were heavily involved in their outcome. The concept of radar was already well established but, in practice, using relatively long wavelengths which severely limited the spatial resolution of the resulting images—what was desperately needed was the application of microwave techniques (i.e. centimetre wavelengths) to improve resolution and yield much more precise target location. Two major improvements were essential—the development of a high power microwave source and of a sensitive and reliable detecting device. The cavity magnetron was to satisfy the former and our (now discredited) old friend, the cat's whisker rectifier, the latter. This section is dedicated, therefore, to the rejuvenation of the cat's whisker.

We have already met the application of the rectifier diode in detecting radio waves—the requirement for radar was no different, except for the very much higher frequencies involved. This, however, was critical because it demanded a rectifier which could respond very much faster than anything previously contemplated (a wavelength of 10 cm—the so-called S-band—corresponds to a frequency of 3×10^9 Hz, or 3 GHz). Why was this a serious problem? There are, in fact, two answers to this question. First, we need to remember that the semiconductor rectifier had been replaced by the thermionic diode as a detector of radio waves and, though this worked well at modest radio frequencies, it was found to be completely useless at very high frequencies. This depends on the fact that electrons take a finite time t_{transit} to traverse the distance (about a millimetre) between cathode and anode, leading to a maximum possible operating frequency given by the condition:

$$\omega_{\text{max}} \sim 1/t_{\text{transit}} \tag{2.2}$$

and, for a typical thermionic valve, this turns out to be roughly 10^8 Hz, some 30 times too small to be useful at microwave frequencies. However, in the case of the semiconductor rectifier, the distance over which electrons must travel (the width of the surface barrier region) is very much smaller, typically less than a micron and therefore some four orders of magnitude smaller than for the vacuum diode, quite small enough to allow operation at 3 GHz. Unfortunately, this is not the whole story!

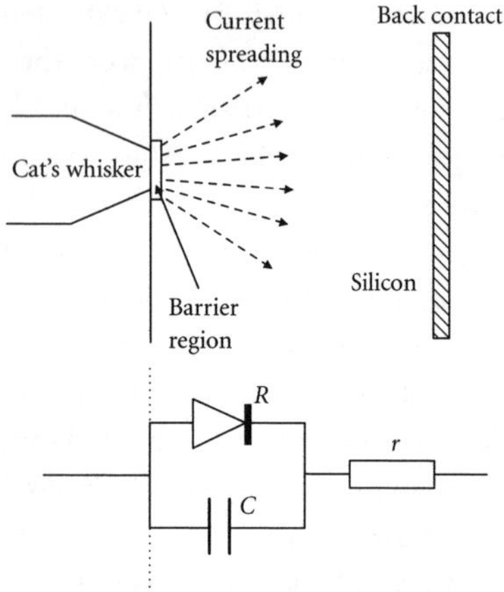

Figure 2.10. Diagram to illustrate the derivation of a simple equivalent circuit for a point contact rectifier. *R* represents the rectifier resistance, *r* the spreading resistance within the bulk semiconductor, and *C* is the barrier (depletion) capacitance at the contact.

The second answer to the question posed above is: 'because of the presence of capacitance'. Associated with the semiconductor barrier is a capacitance *C* which appears effectively in parallel with the rectifier resistance (see the 'equivalent circuit' in Figure 2.10). This capacitance offers an impedance to the flow of a high frequency electric current which is inversely proportional to the operating frequency, that is, $Z_C = 1/j\omega C$ (where ω is the angular frequency $2\pi\nu$ and j is the square root of -1, indicating a 90° phase lag). Thus, at high frequencies Z_C becomes small and acts as an effective short circuit—in other words, at sufficiently high frequencies, all the RF current would flow through the capacitor and the rectifier would no longer function. Because of this, even a semiconductor rectifier might not work at microwave frequencies—it all depends on the actual values of the various impedances in Figure 2.10. Clearly, we require that the rectifier resistance *R* be smaller than (i.e. $\omega CR < 1$) but, at the same time, there is another consideration, the series resistance *r* must be smaller than the rectifier resistance *R*, otherwise the current will be limited by *r* (rather than by *R* which is the condition for rectification to occur). The situation is very complex because all these parameters depend on the doping level in the semiconductor—*r* gets smaller as the doping level increases (which is good) but *C* gets larger (which is bad). We should also expect *R* to get smaller (which is both good and bad!), though experimental evidence appears to be less clear in this case. In practice, it was found to be a close run thing, though the reader will appreciate that I should hardly be telling the story if the outcome had been negative! However, we need first to look at some relevant details.

The old cat's whisker, as we have seen, could be based on a number of semiconducting crystals but more often than not on PbS which gave the best sensitivity. Its main drawback was the difficulty in obtaining reproducible performance, in particular the tendency for vibration to move the precise point of contact and destroy the rectifier characteristic, not at all desirable in a device which might be carried in a fast manoeuvring aircraft or a warship under fire! The first requirement was for a rectifier which could be sealed into a convenient capsule and relied upon to continue performing under mechanical stress and it was found that a silicon crystal combined with a tungsten whisker was better able to meet this aspect of the specification. Typical capsules are shown in Figure 2.11. Stabilization of the contact was, at first, achieved by tapping the device to optimize the reverse/forward resistance ratio, then either sealing the capsule with wax, or polymerizing a small dot of cement round the point of contact. It was later found that using a ground (rather than polished) silicon surface gave adequate stability without the need for sealing. But much more significant for the long-term development of semiconductor technology was the discovery that naturally occurring silicon was far too variable in performance, leading to serious effort being devoted to its purification. This

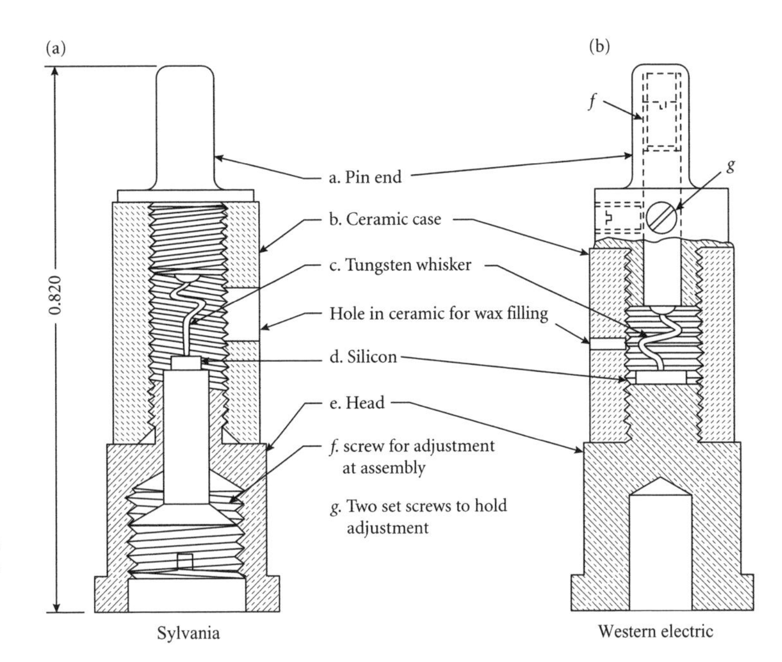

Figure 2.11. Examples of encapsulated cat's whisker silicon diodes, as used for microwave radar detection. The construction is such as to allow adjustment of the contacts for optimum rectifier performance before the cartridges are sealed and the contacts stabilized. (From Torrey and Whitmer 1948: p 16). Courtesy of McGraw-Hill.

involved vacuum melting (at 1410°C) of silicon powder in berylia crucibles and achieved silicon resistivities of order $5 \times 10^{-2}\ \Omega\,\mathrm{m}$ (free carrier densities of order 10^{21}–$10^{22}\ \mathrm{m}^{-3}$). However, this relatively pure material proved unsuitable for rectifiers and further led to the discovery that doping with some 0.002% of boron was necessary, reducing the resistivity by a factor of a hundred (increasing the free carrier density—presumably holes—to about $10^{24}\ \mathrm{m}^{-3}$). All this appears to have been done largely empirically but we can recognize the first clear application of controlled doping to the development of a semiconductor device, a highly significant advance.

Armed with this knowledge, we can estimate the values of the series resistance r and the barrier capacitance C based on the geometry of the point contact. Typically, the tip radius was about 100 μm and the corresponding contact radius about 5 μm (area $A \sim 10^{-10}\ \mathrm{m}^2$) which leads to $C \sim 3 \times 10^{-13}$ F and $Z_C \sim 150\ \Omega$. This is just about equal to measured values of R so our first criterion is just met. To calculate the series resistance r we first note that, due to the use of a point contact, r is the so-called 'spreading resistance' $r = \rho/4a \sim 15\ \Omega$, so our second criterion is also well satisfied. (Note that, if the contact radius had been greater than the silicon crystal thickness L, we should need to calculate r from $r = \rho L/A$ which gives a significantly larger value, thus illustrating the importance of the point contact device for high frequencies.)

In summary, not only did the rejuvenated and refined cat's whisker satisfy an urgent military need, but we, with our sophisticated semiconductor knowledge, can understand (more or less!) how it did so. What is more, in achieving their targets, the mobilized ranks of 'boffins' laid the foundations of future semiconductor technology which was applied so splendidly, postwar, to what was probably the greatest single breakthrough in semiconductor science—the development of the transistor. But that, of course, is another story.

2.6 Postscript—how things happen

We have made numerous references in this chapter to advances made by named individuals (such as the discovery of the cat's whisker rectifier by Braun) and we have also discussed something of the early history of radio to which Braun's discovery was applied. It is, perhaps, convenient at this point to pause briefly and consider the manner of these contributions. Human nature appears bent on seeing such developments in a competitive light and willing for there always to be a *winner* in the competition—some individual who can be credited with being *first*. We delight in setting someone apart as the *inventor* or *discoverer* even though things rarely happen quite like that. In practice, many people may contribute to the invention or discovery and it is almost always true that the individual given the credit was greatly indebted to others for important leads and helpful ideas on which to build. In one respect, this is simply to emphasize the importance of publication of results as a spur to scientific endeavour—as true in today's commercial world as it was in the nineteenth century when modern science was beginning to flex its muscles. It is also highly instructive to examine the relationship between *pure* and *applied* science in the development of practical ideas—we shall see many examples of such interaction in the following chapters and it may be well that we recognize an early example which rather beautifully illustrates the point, namely, the early history of radio, itself.

The story has been very clearly presented by Garratt (1994) in his excellent monograph, 'The Early History of Radio—from Faraday to Marconi' and we shall begin with a quotation from his introduction:

> even quite simple inventions are generally the outcome of a chain of prior developments which have been spread, not infrequently, over a considerable period of time and to which a number of individuals have separately contributed.
>
> Never was this more true than in the field which forms the subject of this book, for although there is a widespread belief that the invention of wireless telegraphy was the work of a young Italian, Guglielmo Marconi, and although it is perfectly true that the first patent ever to be granted for a system of wireless telegraphy stands in his name, the fact is that his achievement was only (sic!) the practical application of scientific developments and discoveries which had been spread over a period of almost eighty years.

All our sources credit Heinrich Hertz with being the discoverer of radio waves in the years 1886–90 but it is certainly true that at least one other physicist, Oliver Lodge, was close to making the same discovery and that valuable contributions were also made by others—Garratt cites the work of Faraday, Maxwell, FitzGerald, and Lodge, not to mention the remarkable observations of a Professor of Music, one David Hughes, who probably made the original discovery of radio waves in 1879 but was persuaded against such an interpretation by a trio of 'eminent scientists'! What is of greater significance from our present viewpoint, is the background to Hertz's discovery, for which we must refer back to Maxwell's theory of electromagnetic waves. Maxwell had been deeply concerned with trying to understand the nature of light and it was in this quest that he developed his electromagnetic wave equations in 1864. However, the theory was left in a kind of limbo for many years on account of the apparent impossibility of confirming (or otherwise) the existence of long wavelength electromagnetic waves which represented an essential prediction of Maxwell's work. The difficulty lay in the lack of any suitable detector of such waves (the eye having been excellently adapted to the detection of light) and it was not until Hertz's work with spark gaps that the problem could be overcome. Both Hertz and Lodge were aware of the vital need for experimental support for Maxwell's theory and made it a central goal of their work to find one—Hertz, in particular, showed quite remarkable experimental skill in using a finely adjusted spark gap between two small copper spheres, forming part of an aerial system, as a detector of the radiation emitted from a second spark gap when excited by the discharge from a 'Ruhmkorff induction coil'. With this delicately adjusted secondary spark, he was able to demonstrate unequivocally that long wavelength 'radio' waves passed between source and detector even when separated by a few metres of air. Lodge was later to employ a 'coherer' in the form of a glass tube filled with metal filings as a more reliable and much more sensitive detector with which to confirm Hertz's initial success. (The coherer had been invented in 1890 by the French scientist Edouard Branly.)

At this juncture, we should note yet another significant aspect of Hertz's discovery—it was made in a context where well-informed people were available who could immediately and fully appreciate the significance of what he had done (i.e. Lodge and FitzGerald). They welcomed Hertz's achievement and ensured its rapid acceptance by the scientific community as a whole, an important, perhaps essential requirement. The rapid promulgation of any new idea requires not only a competent discoverer but also an audience sympathetic to its presentation, which is why science and technology flourish in an environment where several different groups of researchers work in closely related (even identical) fields. It is the fascinating interplay between the *competitive* and *collaborative* elements in human endeavour which most effectively stimulates progress.

It is fascinating that neither Hertz nor Lodge harboured any thought that these electromagnetic waves might have practical importance, for them it was more than sufficient that they had, at last, been able to bridge the yawning gap between Maxwell's theory and its experimental verification. It was left to Marconi to provide the 'commercial' vision which led to the multimillion dollar (Euro?) industry we know today. However, even this was not immediately straightforward—Marconi's first attempt to raise the British Government's interest in (and funding for) the practical application of radio waves was in a system for remote guidance of torpedoes! (It probably required less in the way of imagination to appreciate this rather smaller step into the unknown.) Such is frequently the manner of our stumbling progress towards technological success, many aspects of this remarkable story having been repeated many times over in the development of semiconductor devices and the systems based on them. It is also beyond doubt that recent work has depended more and more on teams of scientists, rather than the inspired individual, which makes it increasingly difficult to pick 'winners'—though, of course, we still do it!

It would, of course, be totally misleading to imply that individual personalities are not important. Science is first and foremost a *human* activity and both gains and suffers from it. It is, for example, a matter of human perspective how much significance an individual gives to the entrepreneurial aspects of a new discovery, whether he regards the purity of pure science as possessing that innate superiority over the mere application of ideas to new technology which forbids his further involvement in commercial development, whether he feels comfortable working alone in semi-secrecy or prefers the stimulus of open intercourse with like-minded colleagues, whether he is happy to be a member of a team and share the kudos attached to the team's success or whether his need to prove himself dominates his performance. We are all different, and it would be an inadequate (and ultimately unsuccessful!) research manager who overlooked this basic and incontrovertible fact. The impact of personality is, once again, well illustrated by the early history of radio.

When Guglielmo Marconi first arrived in England in 1896, Oliver Lodge was in his Professorial prime at the University of Liverpool and probably possessed a better understanding of electromagnetic waves than any man alive at that time (Hertz having tragically died in 1894 at the early age of 36). It seems, therefore, that a combination of Lodge's academic understanding with Marconi's undoubted entrepreneurial enthusiasm would have been an obvious world beater. That it failed to happen may have been due to Lodge's tendency to be disdainful of the younger man's lack of academic distinction, allied with his disapproval of the then Engineer-in-Chief at the Post Office, Sir William Preece with whom Marconi set up an early collaboration. Whatever the reason, it must surely have delayed quite considerably the commercial development of wireless telegraphy and certainly prevented Lodge from being seen as its joint inventor. While it is true that Lodge did collaborate with Alexander Muirhead in developing an alternative, and very well-engineered system, it turned out to be too little, too late—the Marconi company had, by then, established a monopoly position which allowed it to dominate the commercial scene. In fairness, Lodge was undeniably very busy with his academic responsibilities and academics were not, at that time, encouraged (even cajoled!) into taking up entrepreneurial activities—the point I wish to make is nonetheless valid, human foibles can never be dismissed from any serious historical account, whether it be the field of science or any other.

The point is further emphasized in a recent account (Lomas 1999) of the life of yet another brilliant inventor, Nikola Tesla who was working in the United States during the same period. Tesla (whose name is perpetuated in the SI unit of magnetic induction) is widely recognized as the inventor of the 'Tesla coil' in 1891, though his contribution to the development of AC electric power systems was probably of considerably greater practical importance. However, in terms of the history of radio, it is even more startling to discover that Tesla described, in a lecture to the National Electric Light Association in 1893, all the features necessary for a wireless communication system, including a tuning circuit to provide selectivity. He had already demonstrated in his private laboratory the ability to transmit electric power from a high frequency generator to a light bulb without the need for connecting wires and this was 3 years before Marconi began his own experiments in Italy. Why, then, we might ask, was Tesla not credited with the invention of radio?, not even mentioned as an early contributor? The reasons are complex, partly involving Tesla's personality, partly the nature of technical advance. Tesla was not only commercially naive but was extremely reluctant to publish his research findings until he had dotted all the 'i's and crossed every 't' so his contribution was not widely available until the year 1900. In addition, his driving interest happened to be in the direction of power distribution, rather than telegraphy so, even then, it was the 'wrong' people who learned of his results, in particular George Westinghouse, who had a vested interest in AC power transmission along *wires* and was all too keen to keep Tesla's work from reaching practical application! The gods who control technological innovation also move in mysterious ways!

Having said all this, we are now excellently prepared for our next major topic, the discovery of transistor action and its subsequent development into the all-pervasive silicon chip.

Bibliography

Bleaney, B., Ryde, J. W., and Kinman, T. H. (1946) 'Crystal valves', *J. Inst. Elect. Engrs. IIIA*, 93, 847–854.

Garratt, G. R. M. (1994) *The Early History of Radio*, Institute of Electrical Engineers, London.

Henisch, H. K. (1957) *Rectifying Semi-Conductor Contacts*, Oxford University Press, Oxford.

Lark-Horovitz, K. (1954) 'The new electronics', in *The Present State of Physics* (ed. F. S. Brackett) American Association for the Advancement of Science, pp. 57–127.

Levinshtein, M. E. and Simin, G. S. (1992) *Getting to Know Semiconductors*, World Scientific Publishing Co Pte Ltd, Singapore.

Lomas, R. (1999) *The Man Who Invented the Twentieth Century*, Headline Book Publishing, London, ch. 9.

Pearson, G. L. and Brattain, W. H. (1955) 'History of semiconductor research', *Proc. IRE*, 43, 1794–1806.

Richtmyer, F. K. and Kennard, E. H. (1950) *Introduction to Modern Physics* McGraw-Hill, New York.

Seitz, F. and Einspruch, N. G. (1998) *Electronic Genie—The Tangled History of Silicon*, University of Illinois Press, Urbana.

Smith, R. A. (1959) *Semiconductors*, Cambridge University Press.

Torrey, H. C. and Whitmer, C. A. (1948) 'Crystal rectifiers', in *MIT Radiation Laboratory Series* (eds. S. A. Goudsmit, J. L. Lawson, L. B. Linford, and A. M. Stone) McGraw-Hill, New York.

CHAPTER 3

Minority rule

3.1 The transistor

If one had to choose a single event which truly put semiconductors on the international map, it would surely be the invention of the transistor at Bell Telephone Laboratories in late 1947. Without this dramatic step, the leap into 'information technology' which has so radically changed all our lives may never have occurred. What is more, the stimulus which led to so many wide-ranging advances in semiconductor physics and technology would have been lacking and this book may not have been worth writing! The successful development of the transistor can be seen, not only as an enabling technology in its own right but as once-and-for-all justification for the huge financial investment in semiconductor research which was a noteworthy feature of the second half of the twentieth century.

It is interesting, therefore, to examine in some detail the events which led up to this highly significant event, and we do this before attempting to trace the further technical developments leading to the information technology explosion which has affected the social organization of our lives more traumatically, even, than the two world wars which preceded it. Several contributing strands can be distinguished. The philosophy of the *research* laboratory (as distinct from that of a teaching facility) was, by then, well established but it was the Second World War which not only made influential people aware of the huge contribution science and technology could make to the most deadly serious of human endeavours, it also established, the concept of scientists working together as research teams, rather than as inspired individuals. Additionally, there had arisen in the United States an appreciation of the importance of pure science as a basis for technological advance, in preference to the older 'cut and try' methods of entrepreneurial progress which dominated the prewar scene. (On the other hand, the lesson to be learned in Europe had been the importance of *applying* science in the interests of such progress!) More specifically, AT&T were determined to expand their telecommunications links, worldwide and were keen to maintain technological advantage in any way possible. The thermionic valve, important though it was, could be seen to suffer certain disadvantages and (another lesson from the war) semiconductor scientists were now well aware of the essential need to control their material technology to a far higher degree than could previously have been imagined. All these factors played their part but we should certainly not overlook the dogged determination and inspiration of the Bell scientists which brought ultimate success. Perhaps the principal lesson to be learned from the exercise was the need for well-*directed* and suitably *motivated* human beings.

The widespread occurrence of scientific research laboratories is now so commonplace that it is easy to overlook the fact that this is a phenomenon largely to be associated with the second half of the twentieth century, though its origins were certainly much earlier. There is, in fact, a rough parallel with the rise of industry—the first British example was that of the Royal Institution, established in 1799, at the time of the (British) Industrial Revolution—though there was no obvious rush to replicate it elsewhere. In fact, science laboratories were seen essentially as teaching establishments where students could observe and copy demonstration experiments performed by their tutors. An early example was that of the Glasgow University chemistry laboratory which rose to fame in the late eighteenth century but it was only *c*.1820 that research began to be taken seriously. The Liebig chemistry laboratory, established in Giessen (near Frankfurt) in 1824 represented a similar example of a laboratory dedicated to research, while, as far as the Americas were concerned, it was not until 1844 that their first research laboratory came into being, in the form of the Franklin Institute. In 1808 and 1810 we have documentary evidence for (Sir) Humphrey Davy's efforts to raise funding for his research programmes (interestingly with a patriotic emphasis based on the need for Britain to maintain technological superiority over its French enemies) and no one can be in any doubt that he and his protégé Michael Faraday undertook serious research in the Royal Institution during the period 1810–60. What is, perhaps, less widely known is the extent to which Faraday, in particular, accepted commissions from industry and from the British Government to apply his expertise in the interests of applied science. Thus, the realization that science could, indeed, be *useful* was certainly not new in 1947 (as has been admirably detailed by J.D. Bernal (1953) in his monograph *Science and Industry in the Nineteenth Century*)—the essential difference was one of *organization* and *dedication* such as became the hallmark of the Industrial Research Laboratory in the twentieth century. University and Government (e.g. standards) laboratories were established both in Europe and in America before the end of the nineteenth century—for example, the Jefferson Physical Laboratory at Harvard was built in 1884 with one wing specifically devoted to research and the now famous Cavendish and Clarendon Laboratories in Cambridge and Oxford, respectively, date from the 1870s while the National Physical Laboratory and its American counterpart the

National Bureau of Standards were also established in the late nineteenth century.

The world's first *industrial* research laboratory was probably that set up by Thomas Edison in Menlo Park round about 1870 but it was not until *c*.1920 that the concept of an Industrial Research Laboratory became widely accepted. The GEC laboratory in Wembley was opened in 1919, the Philips Research Laboratory in Eindhoven dates from 1914, and Bell Telephone Laboratories was established as a separate entity in 1925 so, by the outbreak of war in 1939, industrial research organizations were well able to take up the challenge of urgent research and development projects demanded by the appropriate military authorities. It was then only natural that, on the cessation of hostilities, such laboratories were eager to apply the lessons learned during the war years to the equally demanding requirements of expanding industrial endeavour. Such a one was the newly purpose-built Bell Telephone Laboratory at Murray Hill in New Jersey.

More important than buildings, of course, are the people who work in them and the philosophy which guides them. The essential factor in the successful development of an all-solid-state amplifier was, without doubt, the conviction shared by management and staff that the pursuit of pure science could lead to a useful practical outcome. This was a philosophy in line with current US policy at the highest level, as illustrated by an important memorandum from Vannevar Bush, Chairman of the National Defense Research Committee, to President Truman in 1945, urging him strongly to support basic research in the postwar world—this was to be the beginning of large scale government support for basic science on the understanding that it would eventually lead to successful (i.e. commercially successful) practical results. However, Bell Labs was already ahead of the field—in 1943 Mervin Kelly, Director of Research had sent an internal memorandum to Bell management emphasizing the future importance of semiconductor research for Bell, a message which clearly bore fruit in the formation of a strong solid state group in the Murray Hill laboratory. Headed by William Shockley (a physicist) and Stanley Morgan (a chemist), it contained an important semiconductor group containing not only physicists but also circuit engineers and chemists, both experimentalists and theoreticians, a truly interdisciplinary team such as had been successful in wartime research projects. It included the other two future Nobel Prize winners, John Bardeen and Walter Brattain, and all were motivated by the desire to find applications for their work, even while they were exploring the very frontiers of semiconductor understanding. Here was a marked contrast with so much 'pure' research undertaken in the past, particularly in Europe, very much for its own sake—the story is told that when J. J. Thompson discovered the electron in 1897, the 'victory celebrations' were concluded with the hope: 'and may it never be useful to anyone!'.

The Bell researchers were in no doubt that their work *should* be useful and so, indeed, it was—though (importantly) Bell were by no means the only beneficiaries. By licensing the invention to other 'interested parties' they ensured the rapid growth of solid state electronics, to the benefit of all concerned. As we shall see, the integrated circuit was invented not by Bell but by a small new company known as Texas Instruments (TI).

Another important legacy from wartime research on microwave semiconductor diodes was the realization that, if semiconductor devices were to show reproducible characteristics, it was necessary to control the semiconductor material itself to a previously unheard of degree. With the gradual understanding of semiconductor properties in both theoretical and experimental aspects, it became clear that impurity levels in the parts per billion regime could have serious consequences and materials scientists struggled (successfully!) to find ways of purifying silicon and germanium to this order of accuracy. It is significant, too, that these efforts were concentrated on elemental semiconductors, not only because they possessed suitable electronic properties but also because, being the simplest possible materials, they were most easily controlled—binary compounds, for example, like copper oxide and gallium arsenide, have another degree of freedom (freedom to go wrong!) in respect of possible departure from the correct stoichiometric proportions. The fact that purification was achieved by melting, resulted in Ge (MP 937°C) being the first semiconductor to be successfully purified. Si (MP 1412°C) is significantly less easy to process and also much more chemically reactive, which accounted for the choice of Ge for much of the early transistor work. This was really a follow-up to wartime development at Purdue University of 'high back voltage' microwave detectors which demanded especially high purity Ge crystals.

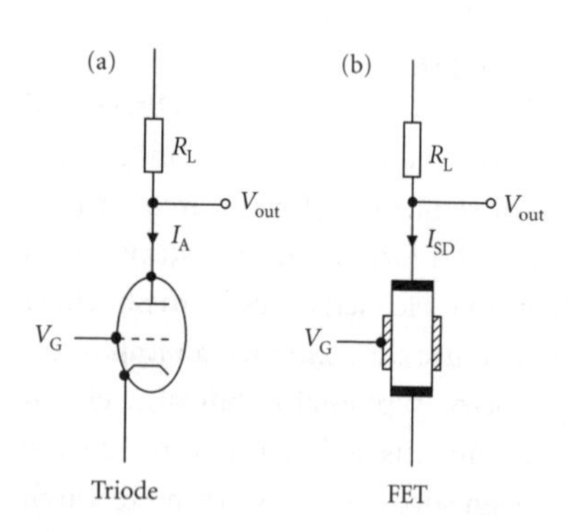

Figure 3.1. Comparison between (a) a triode valve and (b) a field effect transistor, illustrating the parallel between the triode grid electrode and the gate of the FET. A voltage applied to either one controls the amplitude of the current flowing through the device, the anode current I_A or the drain current I_{SD}, respectively. In both cases this current flows through a load resistor (R_L), the voltage drop representing the output of the system.

Given this well-planned and motivated activity, it is perhaps not surprising that some exciting results were obtained at Murray Hill during the immediate postwar years (1945–7) but the manner of their discovery makes an intriguing story. It has been splendidly told in the book by Riordan and Hoddeson (1997), *Crystal Fire* and the interested reader will find much fascinating detail there (see also the somewhat more discursive account in *Electronic Genie—The Tangled History of Silicon* by Seitz and Einspruch 1998)—here we can do no more than pick out a few highlights. The first point to be appreciated is that the idea of a solid state amplifier to rival the triode valve had been exercising a number of technical minds for several years and, in particular, William Shockley's was one of them. It is probably not too strong a statement to say that Shockley had an obsession with the idea for what later became known as a 'field effect transistor' (FET), a device in which the conductivity of a sliver of semiconducting material is strongly modified by the application of a control voltage to a 'field electrode' or 'gate' placed alongside it (see Figure 3.1(b)). The gate could be likened to the grid electrode in the triode valve (see Figure 3.1(a)), in that it controlled the flow of current through the device without any significant amount of power being dissipated in the gate circuit. By including a load resistor in series, a small AC voltage on the gate would induce a much larger AC voltage across the load, resulting in considerable voltage gain and, because very little current flowed in the gate circuit, this also corresponded to power gain. Shockley performed calculations (Box 3.1) which appeared to prove that quite modest gate voltages would be sufficient to modulate the semiconductor conductivity adequately and had, himself, tried to demonstrate the effect experimentally but without success—not even the smallest modulation could be detected. It was this prior frustration (and Shockley's powerful personality!) which biased the Bell team's efforts towards studies of the surface properties of germanium and silicon. Clearly, there was something important going on at the semiconductor surface which prevented the operation of a promising device and the search for a viable field effect was never far from the team's collective conscience. It was ironic, therefore, that their first great success employed a quite different physical phenomenon, known as minority carrier injection!

The first significant development was Bardeen's theory of 'surface states', which he put forward as an explanation for the failure of Shockley's field effect experiments. No matter how pure and perfect the bulk semiconductor crystal might be, there could still exist immobile electron states on the surface as a result of either surface impurity atoms (e.g. oxygen, nitrogen, or sulfur) or by virtue of the fact that the surface atoms are not bonded in the same way as those in the bulk. So-called 'dangling bonds' at the surface may capture free electrons and render them immobile, thus reducing the overall semiconductor conductivity. In this way, Bardeen explained how the electrons induced by the gate electrode could make no contribution to electrical conduction—it required only about $10^{16}\ \mathrm{m}^{-2}$ of such states (i.e. roughly one for every thousand surface atoms) to mask completely the predicted field effect. Brattain then demonstrated that these surface states did, indeed, exist (in more than sufficient densities) on both Si and Ge by detecting a surface photovoltage, using a probe electrode close to the semiconductor surface. Light absorbed near the surface generated electron–hole pairs which were then separated by the surface electric field due to the trapped charge. In *n*-type material, the holes were swept to the surface while the electrons entered the bulk, the net effect being to neutralize the effect of trapped surface charge and modulate the surface potential. It was this which Brattain's probe detected. This was progress, indeed and placed the Bell group well ahead of the world in terms of their basic understanding.

The next step was of particular interest, it was decided to explore the surface photovoltage behaviour over a range of temperatures which

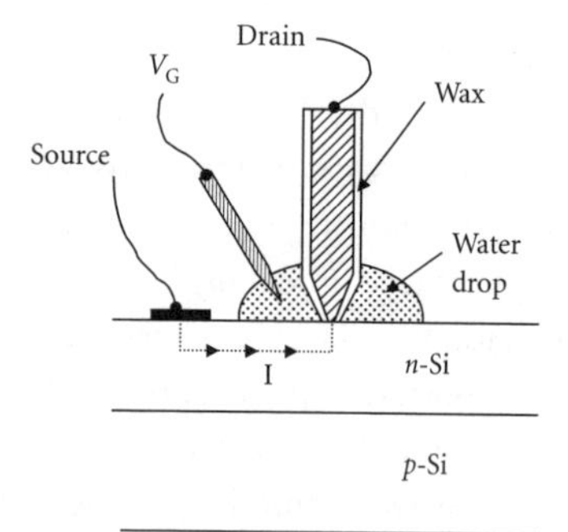

Figure 3.2. Schematic diagram of the experiment performed at Bell Labs in 1947, resulting in the first semiconductor amplifier. The voltage V_G applied to the water drop 'gate' controlled the current flowing to the point contact, which acted as the drain electrode of a FET. Compare this diagram with Figure 3.1(b). Though net gain was achieved, the operating frequency was limited to about 10 Hz by ionic conduction in the liquid.

implied performing the experiment with the sample in a low temperature cryostat. Initial results were erratic due to moisture condensing on the sample so, rather than rebuilding the whole apparatus in a vacuum which would have taken perhaps a month, Brattain decided to compromise by immersing the sample in a liquid such as alcohol, acetone, or toluene. This overcame the condensation problem but, more importantly introduced a new dimension—the photovoltage effect was considerably enhanced and, what was more, it could be strongly influenced by changing the voltage on the probe. In particular, it was observed to go to zero and then change sign, an exciting result because it meant that the introduction of the liquid had enabled Brattain to *neutralize* the effect of the surface states. The field effect was no longer 'screened' by them and the way was open for a real field effect device which the Bell team demonstrated very shortly afterwards. This was serendipity of a high order, a chance observation resulting from what was, in many ways, an ill-advised experimental short cut! But the device worked, beyond a doubt. The structure is shown in Figure 3.2, based on a metal point contact surrounded by a drop of liquid but insulated from it by a film of wax. The liquid (water in this case) acted as the gate electrode and effectively modulated the current flowing from the point contact into the *n*-type Si surface layer, a positive voltage producing an increase in the conductivity. (A positive gate voltage attracts additional electrons into the Si.) Even better results were obtained with Ge, and glycol borate which eventually produced power gains of several thousand but, even more surprising, the gate voltage required was of opposite sign(!), an observation suggesting that the charged ions in the electrolyte had induced an 'inversion layer' in the Ge surface, the field being so strong as to make the surface *p*-type. (The concept of a field-induced inversion layer was to prove invaluable in their future work.) However, there was a serious snag—the response was extremely slow, being limited, by the low mobility of the ions in the gate electrolyte, to frequencies below 10 Hz.

The development of a useful amplifier clearly required that the electrolyte be replaced by a material which did not rely on ionic conductivity and in the next experiments it was proposed to use a thin film of GeO_2, together with an evaporated Au contact on its upper surface to act as the gate—a thin film and a modest gate voltage would still generate the necessary high electric field to neutralize surface states. What happened next is even more bizarre than Brattain's first serendipitous breakthrough—in washing the sample before attempting to make measurements, he inadvertently dissolved the oxide film and ended up with not one but two metal contacts directly on the Ge surface. And, again the result was favourable—by applying a positive voltage to the Au dot and negative to the point contact he observed voltage amplification *at frequencies up to 10 kHz!* There was no power gain but Brattain realized

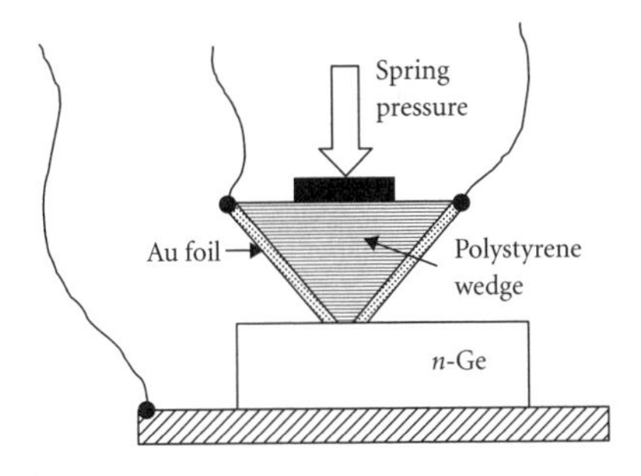

Figure 3.3. Schematic of the first germanium point contact transistor made by Walter Brattain. Two metal point contacts (emitter and collector) were arranged on the *n*-type Ge sample with separation of order 50 μm. To achieve this, Brattain covered a polystyrene wedge with gold foil, then slit the foil at the apex of the wedge with a razor blade, before pressing the wedge against the germanium surface.

that this was due to the fact that the gate electrode was relatively very large—to improve the efficiency he needed to use a second point contact close to the first. Bardeen calculated that the separation should be no more than 50 μm and, in order to obtain such a minute separation, Brattain employed the ingenious structure shown in Figure 3.3, based on a polystyrene wedge covered in Au foil which he carefully slit along the apex with a razor blade! The rest is history (!)—the device gave power gains of up to 4.5 at a frequency of 1 kHz and, if further proof of its amplifying capability were needed, application of suitable feedback produced an oscillator. On 24 December 1947, the 'transistor' was finally born—though the name was invented somewhat later, when Bell unveiled the new device to a still uncomprehending world.

What was actually happening in the device was the subject of some debate but Bardeen realized that the positively biased gate contact was again inducing an inversion layer in the *n*-type Ge, thereby 'injecting' holes which were being collected by the negatively biased 'collector' contact, thus enhancing the current flowing into it. By varying the voltage on the gate, it was possible to control the collector current with very little power dissipation in the gate circuit (the requisite condition for power gain) *but*, as hinted earlier, the mechanism was not that of field effect modulation but depended purely on the injection of 'minority carriers' (i.e. holes) into the *n*-type Ge film by the gate point contact. (This was later proved by numerous further experiments which allowed measurement of the mobility and 'diffusion length' of the injected holes.) By using pure Ge, containing a rather low density of free electrons ($n \sim 10^{21}$ m^{-3}), the injected hole density was large enough to dominate the collector current, thus making the gate control almost perfect. In fact, the recognition of this new gain mechanism led to the immediate renaming of the various contacts as 'emitter', 'base', and 'collector', as shown in Figure 3.4 (taken from the first publication in *Physical Review* 1948), the nomenclature with which we are now very familiar. This diagram also illustrates the way in which the transistor could be used as a circuit element to provide voltage gain. A small signal voltage in the emitter–base circuit modulates the collector current I_C which flows through the load resistor R_L in the collector–base circuit, setting up an output voltage $\Delta V_C = R_L \Delta I_C$ (where Δ implies a small change in the appropriate quantity). Because the collector contact is reverse biased, it presents a high resistance to current flow so the load resistor can be made relatively high also, without seriously influencing I_C, and the output signal voltage ΔV_C can therefore be made very much larger than the input signal, by suitable choice of R_L.

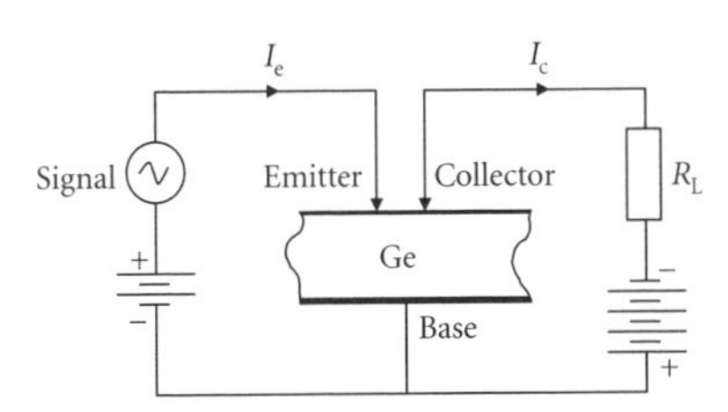

Figure 3.4. Circuit diagram to illustrate the use of the point contact transistor in an amplifier circuit. The signal voltage modulated the collector current which flowed through the load resistor R_L, thus generating an amplified version of the input voltage. (This figure is a copy of that appearing in the first transistor publication in *Physical Review*, Bardeen, J. and Brattain, W. H. (1948) *Phys. Rev.* 74, 230). Reprinted with permission from the American Physical Society.

It is important to understand that the ease with which an inversion layer can be induced in Ge is a consequence of the rather small band gap of this material (0.66 eV)—its choice as a suitable material for these pioneering experiments, based, as it was, on melting point rather than band gap, was again somewhat fortuitous. If, then, the whole process appears to have owed as much to serendipity as to foresight, I should emphasize that the very essence of good research frequently comes down to just this—the ability to ride one's lucky breaks and capitalize on them is paramount—the invention of the transistor was a wonderful example and surely no one (least of all the present author) could possibly wish to belittle it. But, this said, it is frequently a valuable exercise to analyse the manner in which such breakthroughs are made—only then can we adequately learn from them. Scientific progress can rarely be pre-planned! However, it is important to have well-specified goals, as the Bell work clearly demonstrated, and to keep them always in mind, no matter where the immediate problem may appear to be leading. What was truly remarkable about the invention of the transistor was its short timescale—untargeted research may eventually have produced a similar result but certainly not within 2 years! It is also well to emphasize that the essence of their success lay in the fact that the Bell team was encouraged to concentrate on basic research—in making an *invention*, they were obliged to generate a wealth of new physical *understanding* which could only have been achieved by application of the most sophisticated and advanced ideas. This is well illustrated by the publication in 1950 (just 3 years later) of Shockley's book *Electrons and Holes in Semiconductors*, which represents, in itself, a significant achievement, providing not only a thorough account of the research leading to the invention of the transistor but also of the quantum–mechanical background needed for a detailed understanding of its operation. Moreover, in the Foreword, Ralph Bown, the then Director of Research, demonstrates the importance of a management committed to basic research. I quote: 'If there be any lingering doubts as to the wisdom of doing deeply fundamental research in an industrial research laboratory, this book should dissipate them.' Bell proved their commitment by setting up a new team to develop a manufacturable version of the transistor while encouraging the original inventors to continue their fundamental work. We shall return to this, but first, we need to look at the developments in material science and technology which made it all possible.

3.2 Ge and Si technology

It will be apparent, from our discussion of the development of the microwave diode and the transistor, that the key to their success lay with improvements in Ge and Si material technology, by which we mean the preparation of extremely pure and highly perfect single crystal samples. In particular, Ge for the so-called 'high back voltage diodes' was carefully selected to have very low background doping levels and it was this material which was also employed by the Bell scientists who first demonstrated transistor action. For transistors, three material parameters are important: the doping level of the base material should be low, while the injected minority carriers should have high drift mobilities and long lifetimes. The first of these requires material of high purity, the second demands highly perfect single crystals, and the third depends on both these factors. In this section, therefore, we shall say something about the techniques used to prepare such material, including, as a matter of convenience, both Ge and Si in our discussion, their technologies and applications having much in common. It is also true to say that they were the first well-controlled semiconductor materials. As we have already seen, the quality of materials used for earlier device work, such as Cu_2O, Se, and PbS, was hardly, if at all, under control, while later chapters will reveal that other semiconductors, such as GaAs, InSb, and ZnSe, proved far more difficult to tame. Without doubt, Ge and Si technology rapidly achieved a standard difficult to emulate.

The importance of purity has already been emphasized in previous discussion but we should also note that, for most applications, optimum performance of any semiconductor requires that it be in the form of a good quality single crystal. Much of the early transitor work made use of polycrystalline material, consisting of an agglomeration of relatively small crystals connected through 'grain boundaries' at which there occur lattice misalignments and 'dislocations'—that is, lines of atoms not in their correct positions relative to those in the bulk crystal below. These grain boundaries may, in some cases, be electrically charged, which results in band-bending within the individual crystallites, and this, together with the associated lattice misalignment, results in 'scattering' of free electrons (or holes) in their transport through the material. In other words, the effective mobility of these free carriers is reduced with respect to the values appropriate to good single crystals, sometimes by a large factor. In addition, grain boundaries can act as very effective recombination regions for minority carriers which considerably shorten their lifetimes. Both effects reduce the distance over which minority carriers can be transported and thus degrade transistor performance. These effects were particularly serious in the evaporated films of Ge used in experiments where a very thin layer of material was required but even the bulk Ge routinely employed for much of the work was polycrystalline—the essential difference lay in the size of the individual grains. In the case of evaporated films, grain sizes were measured in microns, rather than millimetres, whereas in bulk material the grains were large enough that it was frequently possible to cut adequate single crystals from a polycrystal ingot. Even so this cut-and-try methodology was not one to be welcomed by those responsible for commercial exploitation—there was clearly a need to improve crystal quality, as well as the control of purity. Two crucial techniques were, therefore developed to solve these fundamental material problems, known, respectively, as 'zone refining' and 'crystal pulling', the former to achieve the necessary purity, the latter the desired structural quality. Zone refining was introduced by Pfann in 1952, while the pulling of single crystals of Ge and Si from the melt was first reported for Ge in 1950 (Teal and Little) and for Si in 1952 (Teal and Buehler), all this pioneering work being performed at Bell Labs.

Zone refining (Pfann 1957) can be seen as a sophisticated form of fractional crystallization (used for many years by chemists to purify a wide range of compounds), in which only a part of the ingot is molten at any particular time. It depends for its success on the same basic phenomenon of impurity segregation at a solid–liquid interface. In particular, if an ingot of molten Ge is slowly cooled so that it begins to solidify (this is known as 'normal freezing'), most of the unwanted impurities tend to segregate into the still molten fraction of the ingot, a process that continues until solidification is complete and results in a distribution of impurities throughout the newly solidified ingot that is low at one end, while correspondingly greater towards the opposite end. There is no overall loss of impurity atoms so purification of the sample implies that the high concentration end of the ingot must be rejected, in favour of the purified end (hence the term 'fractional crystallization'—only a fraction of the ingot is purified). Each impurity species is characterized by a 'segregation coefficient' $k = C_s/C_l$ (C_s and C_l being the impurity concentrations in the solid and liquid phases at the solid–liquid interface) and, in most cases of interest to us, k is significantly smaller than unity. Typical values for Ge are, Al = 0.1, Ga = 0.1, In = 0.001, P = 0.12, As = 0.04, Sb = 0.003, Cu = 1.5×10^{-5}, Ag = 10^{-4}, Au = 3×10^{-5}, and Ni = 5×10^{-6}, though it should be emphasized that, in practice, k depends on the experimental conditions. For example, if the freezing interface moves too rapidly, k tends towards unity and no segregation occurs. Mixing of the liquid enhances the effect, so, ideally, freezing should be slow and the liquid should, if possible, be well stirred (not mechanically, but by arranging to have a large temperature gradient in the plane of the interface to encourage convection currents).

As a matter of interest, the normal freezing method was applied to silicon and germanium during the 1940s and it was in this way that the *p-n* junction was first observed at Bell Labs. A Si ingot was melted in a vertical quartz tube and slowly frozen from the top downwards in an effort to produce a single crystal but the resulting rod was remarkable not for its crystallinity but for showing a very large photovoltaic effect. Brattain, having seen similar photovoltages when light was shone on the junction region of a copper oxide rectifier, suggested that a similar junction must therefore be present in this Si rod, a prediction later confirmed by further experiment. When the rod was etched in nitric acid, the junction could be seen under a microscope and there was clear evidence of p-type behaviour from the top end of the rod and n-type behaviour from the bottom. The explanation of these surprising observations was that the rod contained both B and P impurities which are characterized by significantly different segregation coefficients—the phosphorus ($k = 0.04$) was swept very effectively to the bottom of the freezing ingot, doping it n-type, while the boron, with a value of k close to unity ($k = 0.8$), tended to remain fairly uniformly distributed. At the top of the rod, therefore, the concentration of B was greater than that of P, resulting in p-type behaviour, while the converse was true at the bottom.

In principle, the process may be repeated by re-melting and re-freezing the ingot but there is inevitably the possibility of impurity redistribution during the re-melt which negates the segregation effect from the previous freezing step. To avoid this and to offer the advantage of combining multiple freezing steps in a single experiment, the zone refining process employs one or more 'molten zones' along the length of the ingot, rather than resorting to complete melting. These zones are then moved along the ingot by mechanical translation of the ingot past the local heaters which are used to keep them molten. This process effectively sweeps impurity atoms from one end of the ingot to the other and, if N heaters are employed, N sweeps are effected in a single translation of the ingot. A typical zone refining equipment for purifying Ge is illustrated in Figure 3.5. The Ge ingot, some 30 cm long by 3 cm diameter, is contained in a horizontal high purity graphite tube. Six molten zones are maintained by induction heaters, each zone being about 2.5 cm wide, with 10 cm separation between zones and the ingot is translated along its axis at a rate of, perhaps, 10 cm h^{-1}, corresponding to roughly 10 h for a complete scan, which is sufficient to reduce most impurity concentrations to below 1 part in 10^{10}.

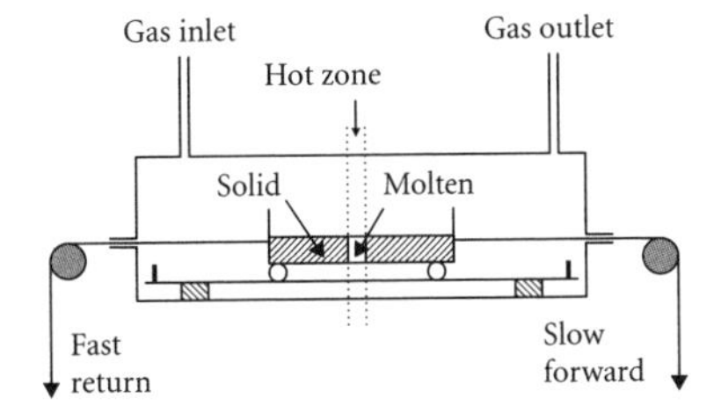

Figure 3.5. Schematic diagram of a typical zone refining equipment for purifying germanium. A series of six molten zones (only one of which is shown in the figure) is moved along the Ge ingot, contained in a high purity graphite boat, resulting in six simultaneous zone refining sweeps. The molten zones are approximately 2.5 cm and the Ge ingot 30 cm in length. Oxidation of the Ge is prevented by enclosing the apparatus within a quartz tube through which an inert gas is flowing.

Silicon, too, was first purified in similar manner but, because of its much higher chemical reactivity, the ingot was contained in a thin quartz boat—thin in order to prevent cracking due to differential thermal contraction during the freezing stage. The major disadvantage of this process was the reaction between SiO_2 and Si to release oxygen, some of which became dissolved in the Si ingot, resulting in concentrations of parts per million of oxygen (though this appeared to be surprisingly benign in its effect on electrical properties) and lesser, but significant, amounts of other impurities from the quartz. Significantly better results were later achieved by the ingenious method of employing a 'floating zone'. A vertical Si ingot was supported at either end, while a molten zone was passed along its length by vertical translation of a heater coil, the zone being stabilized by the surface tension of the molten Si, thus alleviating the need for the *molten* Si to be in contact with any container. In practice, the process usually proceeds in an atmosphere of inert gas, such as helium. The reader will appreciate

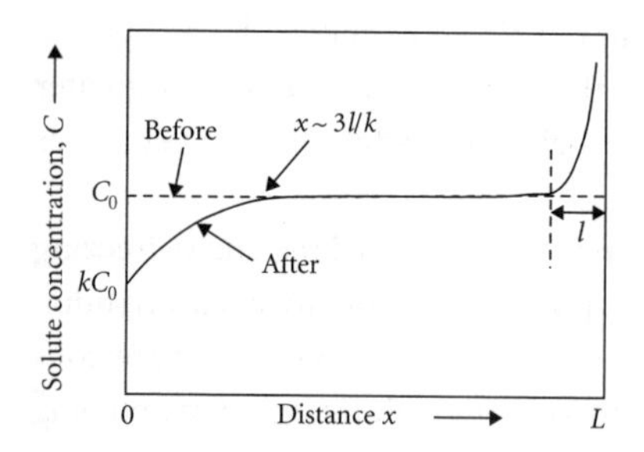

Figure 3.6. Illustrative plot of the concentration of solute in a semiconductor rod after a single zone pass (assuming the segregation coefficient k is less than unity). L and l are the lengths of the ingot and molten zone, respectively. Because no solute is lost, the areas over (under) the curve at the ends are equal. (From Pfann 1957, vol. 4: 423, fig. 7.) Reprinted with permission from Elsevier.

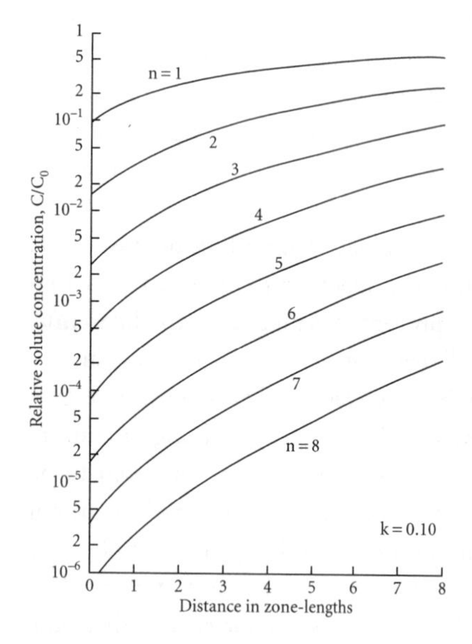

Figure 3.7. Calculated plots of residual solute concentration after n zone passes for the case of $k = 0.1$. This illustrates very clearly how multiple passes may result in large improvements in final purity. (From Pfann 1957: 446; see Figure 3.6.) Reprinted with permission from Elsevier.

that, in this configuration, only a single zone may be passed through the rod so multiple pass experiments involve multiple transit times. Unfortunately, zone melting is not effective in removing boron from silicon, on account of its unfavourable segregation coefficient, so B had first to be removed by being oxidized in a stream of water vapour, before zone refining to remove the other impurities.

Having briefly outlined the process as applied to Si and Ge, we should now look in greater detail at the nature of the resulting impurity profile. The comment made above, to the effect that normal freezing does not actually remove impurities but merely redistributes them, applies equally to zone refining. Let us consider a single zone moving through a cylindrical rod of impure semiconductor and assume a uniform concentration C_0 of an impurity with a segregation coefficient significantly less than unity. At the beginning of the scan, the zone sees a concentration C_0 of impurity at its melting interface, while leaving a lower concentration in the solid at the freezing interface, implying that, as the zone moves, it will gradually become more heavily contaminated. At a certain point, it will contain a concentration C_0/k of impurity, which means the concentration left in the freezing solid will be C_0, while the take-up from the melting interface will also be C_0—in other words, the overall process has reached an equilibrium with concentration C_0 in the solid and a higher concentration C_0/k in the molten zone. This is maintained until the zone reaches the end of the ingot, when the zone itself solidifies, retaining a high concentration of impurity. The resulting profile is shown in Figure 3.6 which emphasizes the point that only the initial section of material is actually purified—to design a process which is efficient in terms of impurity removal requires further thought concerning the relative lengths of the three sections of the curve.

The mathematical shape of the distribution curve for a single pass, up to, but excluding the end section can be expressed by the simple equation:

$$C(x)/C_0 = 1-(1-k)\exp\{-kx/l\} \tag{3.1}$$

where x represents the distance from the starting end and l is the length of the molten zone. For effective purification over most of the ingot (i.e. $C(x) \ll C_0$) this equation demonstrates that not only should k be small but, in addition, $kL/l < 1$. Thus, if $k = 0.1$, we require $L/l < 10$, whereas, if $k = 0.01$, it suffices for $L/l < 100$ which is trivially easy to arrange. (Note that the practical Ge zone refiner described above was characterized by $L/l = 12$.) It is easy to see that multi-pass schemes will improve the degree of purification achieved, though it is no longer possible to describe the distribution in simple mathematical terms. Figure 3.7, which shows distribution curves for the case $k = 0.1$, with the number of passes as a parameter, shows how the degree of purity can be improved by this method. Even impurities with *k*-values close to unity can be removed by employing large numbers of passes—typically several hundred—though there can be no doubt that this is a very time-consuming process.

In summarizing this account, two points should be emphasized: (1) the last zone length to be solidified will always contain a high density of impurities and must be discarded and (2) no matter how pure the purified section, it can be expected to show a non-uniform distribution of impurities and this may frequently be regarded as undesirable. It is therefore of interest to note that zone refining can also be used to obtain a uniform doping level throughout the length of an ingot, this modification of the process being known as 'zone levelling'. Several variants are possible—we shall refer to two, in particular. First, consider the fact that, in principle, the standard single pass process gives rise to an initial transition region, followed by a region of constant composition. In the transition region the impurity concentration in the molten zone is gradually building up until it reaches C_0/k, at which point the concentration in the solid remains constant at C_0, but suppose the concentration in the first zone length were artificially raised to the level C_0/k—there would then be no transition region and the concentration in the solid would be C_0 right from the word go, exactly what is required. The only problem is to arrange the correct impurity level in the first section but this can be solved by the ingenious stratagem of tacking on a separate rod in front of the original ingot, having the appropriate impurity level C_0/k and starting the scan with the molten zone in this new ingot, fusing the two rods together as the scan proceeds. This method works particularly well for impurities with k of order 0.01 or less and enables a relatively pure ingot to be deliberately doped at a pre-selected level (e.g. if Ge is to be doped *n*-type, Sb ($k = 0.003$) is used as donor, while the acceptor In ($k = 0.001$) is preferred for *p*-type doping). The effectiveness of this method is readily understood when one recognizes that the molten zone contains an impurity level some 10^3 times that in the solid and that it leaves only very small fraction of this behind as it traverses the ingot. It is therefore only very slightly depleted even at the end of its travel. A quite different approach is to use several zone passes in *either direction* through a straight ingot. This results in a uniform concentration at an impurity level corresponding to the mean level produced by a previous multi-pass process which will, generally, be considerably lower than that appropriate to the first method.

So much for purity—our other concern is with crystal quality. How is it possible to obtain large single crystals, with a minimum of defects? In particular, high carrier mobility demands that grain boundaries be sensibly absent and that dislocation densities be minimized. In the case of Si and Ge, one answer has been to 'pull' crystals from a melt using the method originally devised by the Polish scientist J. Czochralski in 1918, a typical apparatus for the growth of Ge single crystals being shown in

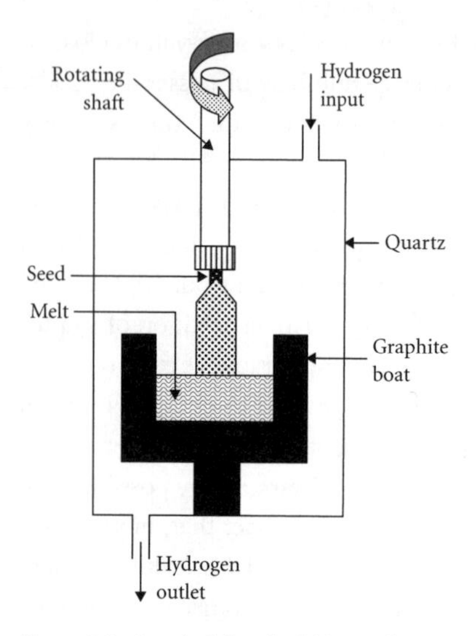

Figure 3.8. A typical Czochralski crystal puller for the growth of single crystals of germanium. Ge is melted by RF heating in a graphite crucible and a small seed crystal dipped into the melt, then slowly withdrawn as the melt crystallizes. The diameter of the pulled crystal depends on the rate of withdrawal. Rotation of the rod improves uniformity and also helps to stir the melt. Pure hydrogen is used to exclude oxygen from the growing crystal.

Figure 3.8. The molten Ge is held in an inductively heated graphite crucible contained within a quartz envelope through which flows high purity hydrogen gas to provide a clean atmosphere. A small seed crystal, held on the end of a vertical rod is dipped into the melt and slowly withdrawn at a controlled rate (typically of order 10 cm h^{-1}) so as to grow a bulk crystal of considerably larger diameter than that of the seed. The rod is simultaneously rotated about its axis which has the effect of stirring the melt and smoothing out effects due to non-uniformities in temperature and geometry. The temperature of the melt is accurately controlled (to the order of ± 0.1°C, that is, to about 1 part in 10^4) using a thermocouple embedded in the crucible. The chosen pulling rate controls the diameter of the grown crystal but may also influence the impurity distribution through the fact that the impurity segregation coefficient k tends toward unity as the growth rate increases. Careful programming of the pulling rate was actually used to produce regions of constant doping in some Czochralski-grown Ge crystals.

Similar techniques were also applied to the growth of Si single crystals, though Si is more difficult on account of its higher melting temperature and increased reactivity. The latter necessitates the use of a quartz liner in the graphite crucible, though, even then, impurities may be taken up from the quartz (even high purity quartz being well below the standards of purity necessary for semiconductor materials). The best Czochralski-grown Si had a background doping level (p-type) of 5×10^{19} m^{-3} which, though certainly useful, is over three orders of magnitude greater than the room temperature intrinsic level. This led to a reassessment of the floating zone approach (which avoids direct contact with any crucible material) when it was found possible to seed the molten zone at its starting position and thereby produce single crystals with at least an order of magnitude lower doping and having minority carrier lifetimes as long as 500 μs. Finally, we may note that this method was also used very successfully for Ge (see, for example, the article by Cressell and Powell 1957), achieving near-intrinsic background doping and lifetimes as long as several milliseconds. What is more, careful control of growth rate and axial temperature gradient resulted in dislocation densities as low as 10 cm^{-2} (cf. more usual values of order 10^3–10^4 cm^{-2}).

3.3 The physics of Ge and Si

One of the themes which emerges time and again in the development of semiconductors and semiconductor devices is the interplay between three aspects: device development, material technology, and physics. Once it became clear that devices required better material in order to function properly, material scientists, as we have just seen, applied themselves to producing purer and more nearly perfect single crystals in support of their colleagues' efforts to optimize device performance. But this immediately begs the question as to which physical parameters are important for the perfection of any particular device—free carrier mobility?, minority carrier lifetime?, the temperature dependence of free carrier density?, the density of states at the semiconductor surface?, the optical absorption coefficient?, the band gap?, etc. And, whichever of these may be critical to device performance, there is an implication that we understand the physics of its involvement—precisely how does the device actually work? The development of the first transistors provided a good illustration of the kind of uncertainties which may arise in attempts to answer such questions, minority carrier injection, for example, being (at the time) a completely new concept. Then, having reached a view as to which parameters must be optimized, we need an understanding of the relation between material properties and device parameters—what is the physics involved?, what, for example, determines the mobility of free carriers?, are we able to influence this by modifying material quality?, how does minority carrier lifetime depend on crystal structure or purity?, and, if the latter, which impurities are important?, how do impurities actually influence these parameters? Many such questions must be answered, all of which imply that we understand the physics of the materials we are working with. Progress in one aspect inevitably depends on corresponding progress in the others so it is appropriate that, at this point, we consider the development of semiconductor physics and how it influenced device development.

However, before dipping so much as a toe into the unfathomable depths of solid state physics, one further point of a general nature should be appreciated—the proper understanding of semiconductor physics, itself depends on having good material with which to work. It was the improvement in material quality generated by the interest in microwave diodes and transistors which enabled the semiconductor physicist to perform meaningful measurements in order to understand the way in which these devices actually work. If material properties are dominated by uncontrolled impurities or unknown defects of one kind or another, the physicist, in making measurements, however carefully, will learn little about the fundamentals of his subject—only when the material is properly under control will physical measurements be of real worth. Pfann (1957) expressed the point succinctly in his review *Techniques of Zone Melting and Crystal Growing*: 'Of what value is a detailed investigation of some substance which contains unspecified impurities or crystalline imperfections that later are shown to have a critical bearing on its behaviour?' But, as he then goes on to say: 'Solid state physicists are paying more attention to such matters. Not only has the single crystal more often become the sine qua non of the experiment, but also the details of its composition and perfection are increasingly taken into account.' This was in 1957. Time has confirmed and amplified the importance of this modest statement.

Hopefully, the reader will now have some appreciation of the intimate relationship between devices, materials, and physics which will be a recurring theme throughout our discourse. Good devices require good material and good understanding—good understanding requires good material and good devices—good material also requires good understanding. In particular, the present section is based on the fact that the availability of high quality Ge and Si led immediately to a rapid expansion in semiconductor physics. Fully to appreciate this, we need only to remind ourselves that in 1931, A. H. Wilson, in his seminal paper on the nature of semiconductors, firmly classified silicon as a metal! By the 1950s, science knew better. By the twenty-first century, so does half the world!

By the end of the 1950s the basic physics behind the electronic properties of Ge and Si was very largely understood. We now take up the challenge of tracing the development of this understanding, which requires us, of course, to study a certain amount of semiconductor physics, ourselves. We shall look at the so-called 'band structure' of Ge and Si, introduce the concept of 'effective mass' for electrons and holes and go on to examine the nature of electrical conduction in these materials, pausing briefly to learn about the various scattering mechanisms which limit the mobility of free carriers. Our next topic will be that of 'minority carriers' which we have already alluded to on several occasions—we shall look in slightly more detail at minority carrier injection and electron–hole recombination and introduce the concept of minority carrier diffusion. Finally, we shall say a little more about surface states and their influence on semiconductor properties such as 'surface recombination' and Schottky barrier heights. These topics encompass a huge range of semiconductor physics, and the best we can do will be to introduce the basic ideas and, hopefully, make them plausible—but, there can be no mistake, *some* understanding is essential for the proper appreciation of the rest of this book.

The band structure of a semiconductor represents an extension of the idea with which we are already familiar of a conduction band (in which electrical conduction may occur by way of free electrons) and a valence band (in which free holes fulfil a similar role). These bands, we have said, are separated by an energy gap containing no electron states, whereas the bands may contain free carriers with thermal motion which causes them to occupy states at energies of order kT above the bottom (below the top) of the conduction (valence) band. Energy is clearly an important parameter in defining the appropriate bands but when considering the motion of free electrons, for example, their momentum is also of significance. In fact, it is only when we can establish the relationship between energy and momentum of an electron that we are able to properly define its motion. Band structure is nothing more than a representation of this relationship. Its practical importance will become apparent when we discuss properties such as electrical conductivity, optical absorption, and 'hot electron' effects such as the Gunn effect (Chapter 5).

In Box 3.1 we demonstrate the rather simple parabolic relationship between energy E and momentum p for a totally free electron—that is, an electron in a vacuum. For an electron in the conduction band of a semiconductor crystal, we might look for a similar relation but should not be surprised to discover that it differs considerably. In a word, as our electron is accelerated through the crystal by an applied electric field, it is pulled and pushed in all manner of different directions by the electrostatic forces associated with the electric charge on each semiconductor atom. Inevitably, as it moves through the crystal lattice, it must draw closer to a particular atom, then separate from it again and approach the next atom in line. At each stage it experiences varying electric forces (in three dimensions, of course) which significantly influence its motion and, in so doing, modify the relation between energy and momentum. What is more, the extent of this influence depends on the direction of its motion within the crystal (in relation to the crystallographic direction, or, more simply, with respect to the crystal axes). In principle, we can still plot a curve of E vs p but it takes a much more complicated form and (perhaps not too surprisingly?) it takes a lot of skilled theoretical effort to calculate it. In fact, even the best available theories of band structure are unable to reproduce the curve exactly and we are obliged to combine such calculations with experimental data in order to obtain the best working approximation and, because the curve takes different forms for different directions, it is usual to provide values only for specific crystal axes. Moreover, even for the same direction, the shapes of the conduction and valence bands differ considerably.

As an example of such a 'band structure', we show in Figure 3.10 the shape of the conduction and valence bands in Si appropriate to the (0 0 1) crystallographic direction, a principal axis in the cubic (tetrahedral) structure of the Si lattice. We note a number of features of this 'band diagram'. First, energy E is plotted against the 'wave vector' k (see Box 3.1) which is proportional to p, rather than p, itself. Second, as also explained in Box 3.1, k takes values between 0 and $2\pi/a$ (where 'a' is the Si lattice parameter) which is sufficient to define the whole range of behaviour. Third, there are three curves for the valence band—or, as it is usually expressed, there are three valence bands. Two of these have the same energy at $k = 0$ where they show a maximum in energy, corresponding to the highest point (the absolute maximum in valence band energy). Fourth, the conduction band shows a minimum directly above the valence band maximum but this is not the absolute minimum, which occurs towards the edge of the diagram along the (0 0 1) direction, that is, near $k = 2\pi/a$. Finally, the points $k = 0$ and $k = 2\pi/a$ are labelled with letters Γ and X, respectively. These notations originate

Box 3.1. Semiconductor band structure

The starting point for almost any discussion of semiconductor band structure is the behaviour of the totally free electron—by this we mean an electron in a vacuum which is subject to no other forces than that arising from the presence of a *uniform* electric field. An electron in a vacuum may be accelerated in the direction of a uniform electric field, gaining *kinetic* energy E in the process, where:

$$E = \tfrac{1}{2} mv^2 \tag{B3.1}$$

m being the free electron mass and v its linear velocity. The electron also acquires momentum p which is given by:

$$p = mv \tag{B3.2}$$

from which we readily obtain the relation between energy and momentum:

$$E = p^2/2m \tag{B3.3}$$

A plot of E vs p takes the form of a parabola with its minimum at the origin (see Figure 3.9). For future reference, we note that $dE/dp = p/m$ and $d^2E/dp^2 = 1/m$. In other words, the curvature of the parabola depends only on the electron mass m.

At this point, we have to take account of quantum mechanics in the form of the de Broglie wavelength λ which is to be associated with the electron momentum according to:

$$p = h/\lambda, \tag{B3.4}$$

h being Planck's constant and, from which, it follows that:

$$p = kh/2\pi \tag{B3.5}$$

where $k = 2\pi/\lambda$ is known as the propagation constant or 'wave vector' of the wave to be associated with the electron motion. In semiconductor parlance free carrier momentum is normally referred to as 'k'—the constant $h/2\pi$ being taken as read. Thus, instead of plotting E vs p, it is general practice to plot E vs k—the *shape* of the resulting curve being identical.

For an electron in the conduction band of a semiconductor, the shape of the E vs k curve is much more complex than for the free electron, as we discuss in the main body of the text. One important aspect of this difference concerns the

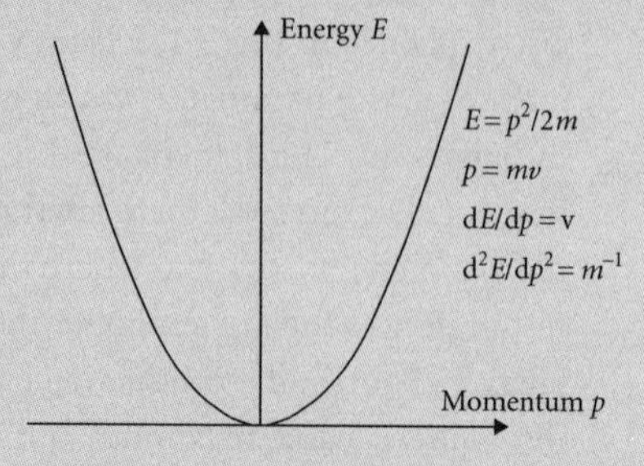

Figure 3.9. Relation between the energy E and momentum p of an electron *in vacuo*. The curve is a parabola, given by the equation $E = p^2/2m$, from which it follows that the slope of the curve at any point is equal to the electron velocity v. The curvature near the origin is a measure of the electron mass m. (*Note*: A similar relationship holds for the 'effective mass' in a semiconductor.)

Box 3.1. Continued

relation between the electron wavelength and the lattice spacing 'a' between atoms in the crystal. As k increases (and λ correspondingly decreases), the wavelength approaches 'a' from above and when $k = 2\pi/a$, a resonance occurs which results in 'standing waves' being set up and the E–k curve reaching a turning point. In fact, because of the periodic nature of the atomic arrangement in the crystal, it turns out that the E–k curve simply repeats for values of k greater than $2\pi/a$ so all the necessary information is contained in this first 'zone'. The point $k = 2\pi/a$ is referred to as the 'zone boundary'.

Finally, by comparison with the parabolic relation between E and p (or E and k) we can use the curvature of the E–k diagram near the valence band maximum (or the conduction band minimum) to define an 'effective mass' for the holes m_h (or electrons m_e) as follows:

$$m_h, m_e = \{d^2E/dp^2\}^{-1} = (h^2/4\pi^2)\{d^2E/dk^2\}^{-1} \quad \text{(B3.6)}$$

In an applied electric field F, the force acting on an electron is eF which produces an acceleration $a = F/m_e$, suggesting that the electron mobility (which is a measure of how readily the electron moves under the influence of an applied electric field) will be proportional to m_e^{-1}.

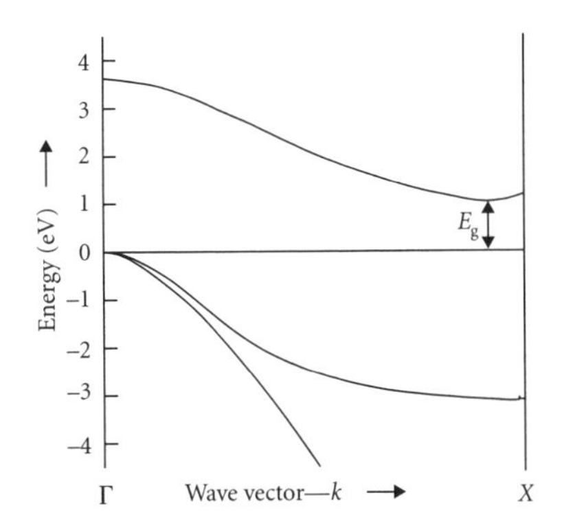

Figure 3.10. A simplified band structure (energy E vs wave vector k) for a silicon single crystal in the (0 0 1) crystal direction. 'a' is the lattice parameter of the silicon diamond structure. The maximum in the valence band appears at the Γ point (at the centre of the Brillouin zone, $k = 0$), while the lowest minimum in the conduction band lies near the X point, that is, near the edge of the zone along the (0 0 1) direction where $k = 2\pi/a$ (a being the lattice parameter of silicon). There are six equivalent conduction band minima, corresponding to (00 ± 1), (0 ± 10), and (±100).

from Group Theory—their precise significance need not trouble us—they can be regarded as convenient labels, nothing more.

Now we can begin to examine the significance of the structure shown in Figure 3.10. Note that the band gap of Si is the energy separation of the highest point in the valence band and the lowest point in the conduction band, and the band-edge absorption of a photon must take an electron from the valence band maximum at Γ to the conduction band minimum near X which occur at very different k-values. This implies that optical absorption involves not only a change in energy of the electron but also a change in momentum—the electron must gain both energy and momentum in making the transition from valence to conduction band. The energy gain is provided by the energy $h\nu$ of the photon but the momentum must come from another source—photons, being massless particles, carry very little momentum. In practice, the necessary momentum is acquired from interaction with a lattice vibration, or phonon (phonons exist with k-values covering the whole range from 0 to 2π/a) which means that the optical transition involves three particles (electron, photon, and phonon) interacting simultaneously. Such a process is inherently less probable than the simple two-particle interaction appropriate to an optical transition taking a valence electron into the higher energy conduction band minimum at the Γ point (which requires no change in momentum). The latter transition is known as a 'direct' transition while the former is known as 'indirect'. Clearly, the indirect optical transition at the Si band edge is significantly weaker (i.e. less probable) than that in a semiconductor which has its

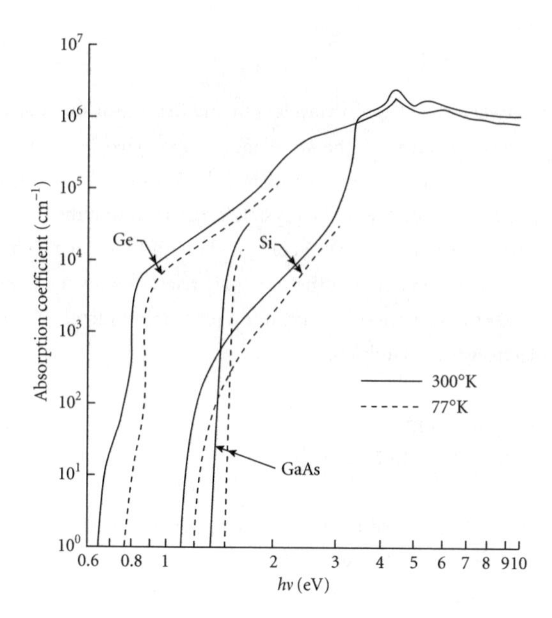

Figure 3.11. Optical absorption coefficients α for pure semiconductor crystals in the vicinity of the absorption edge (i.e. when the optical photon energy is close to that of the band gap). The fact that silicon and germanium are 'indirect gap' semiconductors means that α rises relatively slowly as the photon energy increases. This is illustrated by comparative data for the direct gap material GaAs. For silicon, α achieves a value of $10^4\ \mathrm{cm}^{-1}$ (corresponding to an absorption length of $1\ \mu\mathrm{m}$) only for photon energies of 2 eV or greater (roughly twice the band gap). Contrast this with the rapid rise to $10^4\ \mathrm{cm}^{-1}$ for GaAs. (From Sze, S. M. (1969) *Physics of Semiconductor Devices*, Wiley, New York, p. 54.) This material is used by permission of John Wiley & Sons. Inc

conduction band minimum at the same k-value as its valence band maximum (usually the Γ point) and this is reflected in the optical absorption coefficient at the band edge. In Figure 3.11, we show experimental data for the absorption coefficient in the region of the Si band edge which proves the point. The absorption edge is far from steep and the absorption coefficient is relatively small (compare this with the corresponding data for GaAs which is a direct gap semiconductor).

The second feature of the Si band structure to which we draw attention is the curvature of the bands near the 'critical points'—for example, the valence band maximum and the conduction band minimum. Take, for example, the X minimum in the conduction band where electrons reside under thermal equilibrium conditions. Close to the actual minimum, the band has a shape which is approximately parabolic, and comparison with the corresponding parabola for the free electron (see Box 3.1) allows us to define what is called an 'effective mass' for the electrons in the X minimum which are moving along the (0 0 1) direction.

Because the E–k curve in the semiconductor is quite different from that of the free electron, we must expect m_e to be different from the free electron mass, m. In fact, m_e is simply a convenient fiction to represent the fact that electrons in a semiconductor behave differently from free electrons *in vacuo*. More precisely, it says that electrons in the semiconductor are accelerated by an applied electric field *as though* they had masses equal to m_e. A mass greater than the free electron mass m implies that the electrons in the crystal are more sluggish than their counterparts *in vacuo*, a mass smaller than m, the opposite. What is, at first sight, somewhat baffling is the experimental fact that in many cases effective masses are, indeed smaller than m, that is, the effective mass ratio m_e/m (or m_h/m) < 1. How is it possible that electrons in a crystal, being obliged to negotiate the complex internal fields which exist there, can be more mobile than their free counterparts? This seeming *impasse* is a result of the periodic nature of atomic arrangement in a perfect crystal—as far as the crystal is concerned, an electron at a point within a particular Si tetrahedron would see an identical environment at the corresponding point within any other Si tetrahedron. In other words, the electron wavefunction is a periodic function which extends throughout the crystal and it is this which gives the electron its enhanced mobility.

Effective masses in Si and Ge were measured in the 1950s using a technique known as cyclotron resonance—a large magnetic field B is applied to the crystal which splits the bottom of the conduction band into so-called 'Landau levels' with energy separation $\Delta E = (h/2\pi)eB/m_e$. Transitions can be induced between these levels by an oscillating electric field of angular frequency ω when the quantum of energy $h\omega/2\pi$ is equal to ΔE, under which condition energy is absorbed from the field. (This is akin to the transitions between atomic energy levels which we referred to in Chapter 1, equation 1.1, though the energy quanta in cyclotron resonance are very much smaller.) By observing the values of B and ω for which this occurs, m_e is readily obtained from the relation:

$$m_e = eB/\omega. \tag{3.2}$$

Table 3.1 lists the masses measured for Si and Ge and requires some explanatory comment. The conduction band minimum shown in Figure 3.10 refers to the (0 0 1) crystal direction—the mass measured in this direction is known as the 'longitudinal mass' m_L—in the direction normal to the plane of Figure 3.10 the mass is different and is known as the 'transverse mass' m_T. Note that there are three equivalent (0 0 1)-type directions, (0 0 1), (0 1 0), and (1 0 0). In the case of Ge, the lowest conduction band minima occur along (1 1 1) crystal directions (of which there are four)—they are labelled as L minima—similar remarks apply to the masses. A suitable average of these basic masses must be taken when describing conductivity—this is m_c—a different average m_d is necessary to describe the density of conduction band states. Note that both m_c and m_d are much smaller than the free electron mass. For

Table 3.1. Effective masses for Ge and Si

	Conduction band				Valence band			
	m_L/m	m_T/m	m_c/m	m_d/m	m_{lh}/m	m_{hh}/m	m_{so}/m	m_d/m
Ge	1.64	0.082	0.12	0.22	0.044	0.28	0.077	0.39
Si	0.98	0.19	0.26	0.33	0.16	0.49	0.245	0.55

the valence band there are three isotropic masses, one for each of the bands—these are referred to as 'light hole' m_{lh}, 'heavy hole' m_{hh}, and 'split-off band' m_{so}, respectively.

Not surprisingly, the effective mass is an important parameter in determining the mobility of holes and electrons. We now explore this in a little more detail, considering electrons in the conduction band. Free electrons in thermal equilibrium with the crystal possess thermal energy of order kT in the form of kinetic energy $K = (1/2)m_e v_T^2$ (v_T being the thermal velocity). Thus:

$$v_T \sim \{kT/m_e\}^{1/2} \tag{3.3}$$

from which we find $v_T \sim 10^5$ m s^{-1} at room temperature—that is, electrons are buzzing about like a swarm of bees with random velocities of about 10^5 m s^{-1}, but during their thermal motion, they regularly collide with defects in the crystal (such as impurities or lattice vibrations), collisions which randomize their motion, that is, they change direction in random fashion so that, in the absence of an applied field, there is no net motion in any direction. If an electric field F is applied, this will superimpose a 'drift velocity' v on the thermal motion, giving rise to electrical conductivity. To calculate v, we need to know the average time between collisions τ. Thus:

$$v = e\tau F/m_e \tag{3.4}$$

and, using a typical value of $\tau = 10^{-12}$ s and a field of 10 V cm^{-1}, we obtain $v \sim 5$ m s^{-1} which is a factor of roughly 10^4 times smaller than v_T. The mean free path between collisions $1 = v_T\tau \sim 10^{-7} m = 100$ nm. Thus, the electrons travel several hundred lattice spacings between collisions (which justifies the concept of the periodic wavefunction spreading over a large number of atoms, which, in turn, justifies the effective mass description).

The electron mobility μ_e is related to v by the equation:

$$\mu = v/F = e\tau/m_e, \tag{3.5}$$

which confirms the dependence of mobility on reciprocal effective mass and allows us to estimate a typical value for $\mu \sim 0.5$ m^2 V^{-1}s^{-1}. The actual value to expect in any specific example will depend on the effective mass of the carrier and the precise value of scattering time τ. To learn more about τ we now consider scattering mechanisms in more detail.

In a perfect single crystal at absolute zero, electrons would have essentially infinite mobility—there would be nothing to inhibit their free motion through the regular lattice of Si (or Ge) atoms. In practice, a number of 'defects' in the crystal cause the scattering which we have already alluded to. These can be thought of in two groups: in the first group are lattice vibrations, due to thermal motion of the atoms, which result in their being displaced from their ideal positions; in the second group are impurity atoms, vacant lattice sites, interstitial atoms, dislocations, etc. which again represent departures from the perfect crystal. Lattice vibrational scattering (usually abbreviated to 'lattice scattering') is insignificant at low temperatures but, as the amplitude of the vibrations increases at higher temperatures, it becomes more and more important and is often the dominant scattering mechanism at, and above room temperature. This implies that carrier mobilities can be expected to decrease with increasing temperature, above room temperature. In fact, for the dominant acoustic phonon scattering, $\mu \propto T^{-3/2}$. The other important mechanism is that arising from charged impurity atoms in the crystal, which usually implies ionized donor or acceptor atoms. In an *n*-doped crystal, the free electrons in the conduction band originate as very lightly bound *extra* electrons, that is, those electrons not needed for atomic bonding. They are easily detached from their donor parents and are free to move through the crystal, leaving behind donor atoms (occupying normal lattice sites) which are positively charged. Free electrons, moving past such charged donors, are subject to a Coulomb attraction which deflects them from their original trajectories, that is, scatters them. The higher the doping level, the more scattering centres there are and the greater is the effect on electron mobility. Thus, $\mu \propto N_I^{-1}$ where N_I is the total density of ionized impurity atoms. This is relatively easy to understand—the dependence on temperature is rather less so. However, a little careful thought suggests that as T increases, the thermal velocity of the electrons also increases, resulting in a reduced interaction time, the electron spends less and less time within the force field of the charged donor atom, so the scattering effect becomes correspondingly smaller and the mobility increases. In fact, as a first approximation, $\mu \propto T^{3/2}$. Thus, the overall picture is of an electron mobility which increases from a relatively small value at low temperatures (say 4 K, the temperature of boiling liquid helium), reaching a maximum value at an intermediate temperature (of order 50–100 K), then decreases again above this temperature. The detailed shape will depend on the doping density, at high doping levels, ionized impurity scattering can be important even at room temperature, but we may anticipate a mobility–temperature curve of the general form:

$$\mu^{-1} = aT^{-3/2} + bT^{3/2}, \tag{3.6}$$

where the parameter 'a' depends on N_I. Note that equation (3.6) is written in terms of μ^{-1}, rather than μ because impurity scattering and lattice scattering must be combined according to:

$$\mu^{-1} = \mu_I^{-1} + \mu_L^{-1}. \tag{3.7}$$

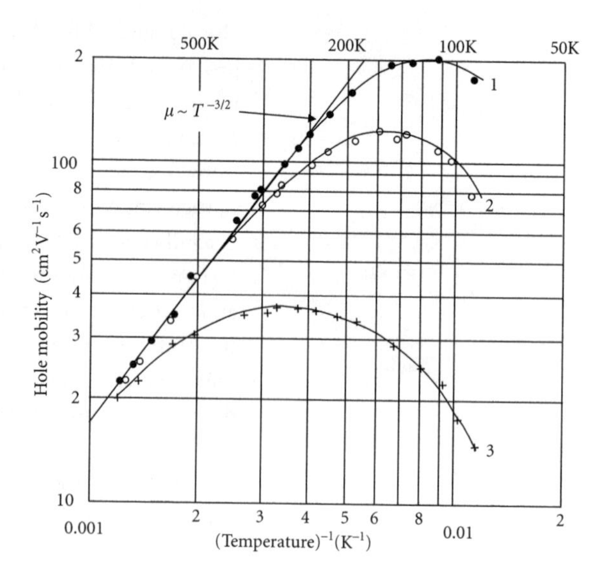

Figure 3.12. Measured hole mobilities for three boron-doped *p*-type silicon samples, plotted as a function of reciprocal temperature for a temperature range between 77 and 600K. The data show very clearly the $T^{-3/2}$ variation appropriate to acoustic phonon lattice scattering at high temperatures and also suggest the $T^{3/2}$ law expected for ionized impurity scattering at low temperatures. Doping levels were as follows: sample 1—$1.3 \times 10^{24}\,m^{-3}$, sample 2—$2.7 \times 10^{24}\,m^{-3}$, sample 3—$6.7 \times 10^{24}\,m^{-3}$. (Selected samples from Pearson, G. L. and Bardeen, J. (1949) *Phys. Rev.* 75, 865, figure 8A). Reprinted with permission from the American Physical Society.

As an illustration, we show in Figure 3.12 the temperature dependence of hole mobility in B-doped Si (taken from a paper by Pearson and Bardeen published in 1949—see Shockley 1950) which demonstrates the general correctness of equation (3.6). Since this time, many more detailed investigations have been undertaken but it is striking to see the excellent degree of understanding of carrier mobility which existed, as early as 1949. There can be no better justification for the hard work devoted to improving crystal quality.

The mobility of free carriers in Ge and Si is of importance for transistor operation in so far as it has an important bearing on the distance that injected minority carriers may travel in order to reach the collector contact, though it is worth emphasizing again here that minority carriers had no part in the original concept of the field effect which stimulated the Bell workers to look for a solid state amplifier. Their role was discovered somewhat by accident and it took some little time before their behaviour was properly understood. For our part, too, we have referred to them several times without making any attempt to explain their physical significance—it is time we made honest creatures of them.

In an intrinsic semiconductor, as we explained in Chapter 1, thermal excitation of electrons across the forbidden energy gap results in equal densities n_i $(=p_i)$ of free electrons and free holes, where n_i depends strongly on temperature through the relation $n_i = \sqrt{(N_C N_V)}\exp\{-E_g/2kT\}$. On the other hand, if we dope a semiconductor (let us say *n*-type), we generate a density of free electrons $n = N_D$ (where N_D is the density of donors), without introducing any holes at all. Similarly, $p = N_A$ in a *p*-type material containing N_A acceptors. If the temperature is low enough that n_i is negligible, the implication of these arguments is that, in *n*-type material, $p = 0$ and, in *p*-type material, $n = 0$. In fact, this is not quite true—there is a thermodynamic argument based on the Chemical Law of Mass Action which links the density of electrons and holes through the relation:

$$np = n_i^2. \tag{3.8}$$

Thus, in n-type material, there exists a hole density $p_n = n_i^2/n$, which we can evaluate when we know the value of n_i. For example, in Chapter 1, we estimated that for Si at room temperature, $n_i = 4.53 \times 10^{15}$ m^{-3}, so the hole density in Si doped with 10^{22} m^{-3} donors will be $p = (4.53 \times 10^{15})^2/10^{22} = 2.05 \times 10^9$ m^{-3}, orders of magnitude less than n, but not zero. These holes are therefore known as 'minority carriers', their density being very much smaller than that of the 'majority carriers', the electrons.

Note that equation (3.8) applies strictly to a system which is in thermal equilibrium—no voltage being applied, no current flowing, no light being absorbed, etc. If one or other of these stimuli is applied it may be possible to change the density of minority carriers considerably, as is the case during transistor action, for example. But the alternative example of a semiconductor illuminated by light having a photon energy greater than the band gap is probably rather easier to understand. We have already discussed this for the case of a p-type semiconductor under the heading of 'Photoconductivity' in Box 2.3. Electron–hole pairs, generated at a rate g (m^{-3} s^{-1}) by the light, recombine at a rate n/τ (where τ is the recombination lifetime) and, when a steady state is reached:

$$g = n/\tau \tag{3.9}$$

so the density of minority carriers (electrons in this case) becomes $n = g\tau$ and depends, therefore, on the light intensity and the recombination lifetime. If the lifetime is long and the light intensity is high, it may be possible to generate hole and electron densities which are even larger than the background doping level p_0 (this is known as the 'high injection' condition). In other words, the minority electrons are present with a density closely comparable to that of the majority carriers!

This example of optical carrier injection is relatively straightforward and easy to understand though it differs from the situation appropriate to transistor action. Absorption of light generates electrons and holes in equal numbers, implying that the injected minority carrier density can never exceed that of the majority carriers, whereas, in the base region of a transistor, minority carriers alone are injected from the emitter contact. It might, therefore, seem that, in this case, minority carriers could exist in even greater numbers than majority carriers—but that would be to overlook a basic natural law! It is widely known that Nature abhors a vacuum—she also hates any departure from strict charge neutrality. No sooner are positive holes injected into the n-type base region, than a corresponding density of negative charge, in the form of electrons, is drawn from the base contact to neutralize the injected positive charge, and this happens extremely rapidly. The time involved is known as the 'dielectric relaxation time' t_d:

$$t_d = \varepsilon\varepsilon_0\rho \qquad (3.10)$$

where ε is the relative dielectric constant of the semiconductor (a number close to 10 for most semiconductors), ε_0 the permitivity of free space, and ρ is the electrical resistivity of the base region. Assuming $\rho = 10^{-2}\,\Omega\,\text{m}$ (appropriate to the fairly low doping level in the base), it is easy to calculate a value of $t_d \sim 10^{-12}$ s for this process. So, once again, perhaps a little surprisingly, we find that the density of minority carriers never exceeds that of the majorities.

The point contact transistor is remarkable in that it depends for its operation on the fact that pressing a metal point into the surface of a piece of *n*-type Ge makes a small area of that surface *p*-type. The surface is said to be 'inverted', that is, converted from *n*-type to *p*-type, though the precise mechanism is not always clear. In some contacts which are 'formed' by passing pulses of current through them, minority carrier injection may be much improved and it has been argued that this is due to impurities from metal doping the surface region of the semiconductor. However, forming is not an absolute necessity for transistor action to occur, which serves only to confuse the interpretation—we shall not pursue the matter further but simply accept that inversion does occur and results in minority carrier injection when a forward bias is applied to the emitter contact. The collector contact (similar in physical form) is reverse biased which has the effect of sweeping minority carriers out of the semiconductor, to form a collector current, controlled by the emitter current.

The question remaining is: how do the minority carriers (holes in a *p-n-p* device) cross the base region? The answer is: they diffuse across! The process of injection results in a high concentration of holes close to the emitter, thus creating a concentration gradient which drives the diffusion process (this is rather like the expansion of gas through a leak into an evacuated vessel—gas molecules tend to spread out so as to fill the whole space available—Nature also abhors a concentration gradient!). Meanwhile, holes and electrons, being present in close proximity, tend to recombine, en route, thus reducing the number of holes reaching the collector, which is clearly a bad thing! How serious this is depends on the relative magnitudes of the quantities 'base width' W and 'diffusion length' L. The first of these is simply the emitter–collector separation which is defined by the positions of the point contacts, the second is a property of the base material which depends on minority carrier lifetime τ and on hole mobility μ (see Box 3.2). Thus:

$$L = \{\tau\mu kT/\text{e}\}^{1/2} \qquad (3.11)$$

Box 3.2. Minority carrier diffusion

In order to understand better, the diffusion of minority carriers across the base region of a transistor, we can represent it by a simple one-dimensional mathematical model. In Figure 3.13 we suppose that minority electrons are injected across the plane $x = 0$ into *p*-type material which extends to infinity on the right side of figure. Figure 3.13 represents a plot of electron density against distance x. Because $n(x)$ decreases as x increases, electrons diffuse to the right across unit area at a rate:

$$dn/dt = -D dn/dx \quad \text{(B3.7)}$$

where D is known as the diffusion constant which depends on the electron mobility. If we consider an element of material of thickness dx at distance x from the origin, we can write the rate of electron diffusion into the element as $-D[dn/dx]_x$, while the rate of diffusion out of the element is $-D[dn/dx]_{x+dx}$. Within the element, electrons are recombining with majority holes at a rate $n(x)dx/\tau$ and, equating 'rate in' to 'rate out' plus 'rate of recombination', we obtain:

$$-D[dn/dx]_x = -D[dn/dx]_{x+dx} + n(x)dx/\tau, \quad \text{(B3.8)}$$

which can be rearranged to give:

$$\{[dn/dx]_{x+dx} - [dn/dx]_x\}/dx = n(x)/D\tau,$$

that is,

$$d^2n/dx^2 = n(x)/D\tau \quad \text{(B3.9)}$$

This simple differential equation can then be solved for $n(x)$—it is easy to check that the complete solution is:

$$n(x) = A \exp\{-x/L\} + B \exp\{x/L\} \quad \text{(B3.10)}$$

where A and B are constants yet to be identified and L is the 'diffusion length' which is given by

$$L = \{D\tau\}^{1/2} \quad \text{(B3.11)}$$

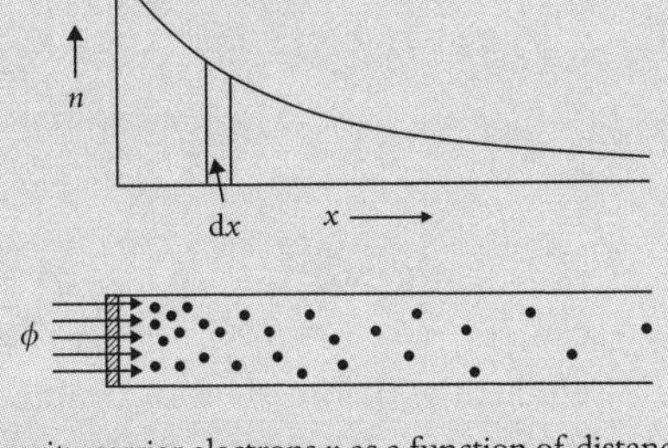

Figure 3.13. Variation of the density of minority carrier electrons n as a function of distance for diffusion into an infinite 'one-dimensional' *p*-type space. We assume that electrons ϕ are injected at the plane $x = 0$ so as to produce a density of $n(0) = n_0$. The resulting concentration gradient implies that electrons diffuse to the right, while recombining with majority holes. This recombination is characterized by a lifetime τ. (From Orton, J. W. and Blood, P. (1990) *The Electrical Characterization of Semiconductors: Measurement of Minority Carrier Properties*, Academic Press, London, p. 24). Reprinted with permission from Elsevier.

Box 3.2. Continued

As mentioned above, D depends on the mobility μ, the precise relation being $D = (\mu kT)/e$, so we obtain equation (3.11) in the main text.

The constants A and B can be determined by examining the boundary conditions—that is, what happens at the end points $x = 0$ and $x = \infty$. Thus, as x approaches infinity, $n(x)$ approaches zero and it follows that B must be equal to zero, otherwise $n(\infty)$ would become very large. Finally, at $x = 0$, we have $n(0) = n_0$, from which it follows that $A = n_0$ and the correct solution for $n(x)$ is:

$$n(x) = n_0 \exp\{-x/L\} \tag{B3.12}$$

This equation represents the minority carrier 'diffusion profile' for diffusion into an infinitely thick base region. The result of recombination is to reduce $n(x)$ by a factor $1/e$ for each increment of $x = L$.

This result serves to introduce the diffusion equation and one of its possible solutions but is not appropriate to the real situation in a transistor—in practice, the base width is finite (and, indeed, relatively small) so we need to modify the maths a little to take account of this. Clearly, it is the first of the boundary conditions which must be changed—instead of $n(\infty) = 0$, we should use $n(W) = 0$, where W is the width of the base region. This condition arises because the electric field at the collector sweeps electrons across the plane $x = W$ at a very high velocity. After a certain amount of manipulation, it then transpires that, for this case:

$$n(x) = \frac{n_0 \sinh[(W - x)/L]}{\sinh[W/L]} \tag{B3.13}$$

In a transistor, we are concerned with the minority carrier current at the collector, or, more precisely, the ratio α = (collector current)/(emitter current). The flux of electrons at any point x is given by:

$$[\mathrm{d}n/\mathrm{d}t]_x = -D[\mathrm{d}n/\mathrm{d}x]_x \tag{B3.14}$$

so the above ratio is simply given by $([\mathrm{d}n/\mathrm{d}x]_{x=W})/([\mathrm{d}n/\mathrm{d}x]_{x=0})$, that is,

$$\alpha = \mathrm{sech}\,[W/L]. \tag{B3.15}$$

From this expression, we see that, if $W \ll L$, then $\alpha = 1$ (all the injected current is collected), while, if $W = L/2$, $\alpha = 0.89$ or, if $W = L$, $\alpha = 0.65$. Clearly, a good transistor must have a base width significantly less than the minority carrier diffusion length.

and, if we take $\tau = 10^{-6}$ s, $\mu = 0.1\ \mathrm{m^2\,V^{-1}\,s^{-1}}$, and $T = 300$K, we can estimate $L = 5 \times 10^{-5}$ m, that is, 50 μm. In order to collect holes efficiently, before they recombine, the collector contact must be less than 50 μm (about the thickness of a sheet of paper!) from the emitter. Box 3.2 quantifies this rather vague statement.

We shall return to the question of transistor action in the next section but, first, we take the opportunity to look a little more carefully into the nature and properties of surface states. It is worth remembering that the original impetus towards finding an all-solid-state amplifier was fuelled by the idea that current through a piece of semiconducting material might be controlled by application of a large electric field at right angles to the direction of current flow (see Figure 3.1). It was the presence of surface states which confounded this idea and the eventual discovery of transistor action emerged from what were rather fundamental studies of these states. As we shall see, field effect transistors were later to be of enormous significance, once ways of minimizing the contribution of surface states were discovered, their importance was also quickly realized by Bardeen in relation to the theory of Schottky barrier heights (metal–semiconductor contacts). It is well worthwhile, therefore, for us to acquire at least some degree of understanding of their properties.

The existence of surface states was anticipated a long time before their effect became apparent in transistor studies. In 1932, Tamm predicted their existence on theoretical grounds and in 1939 Shockley, himself, also published a theoretical paper on the subject but it was not realized how important they might be in practice. The first question we might ask is: what is the density of surface states on a typical semiconductor? To obtain a rough feel for the answer, we note that there are approximately 10^{19} surface atoms per m^2 of surface, and, though it may be possible that each of these can account for one electronic state, this probably represents an upper limit to their density. We must also bear in mind that 'real' surfaces will normally be oxidized and thus contain some form of oxide layer which can be expected to modify the surface state density. As a result of such considerations, we should not, perhaps, be too surprised to find values of order $N_s = 10^{18}\ m^{-2}$. However, there is an important feature which cannot be guessed at from such simple ideas—surface states occur at energies which are distributed throughout the semiconductor band gap. In specifying their density, it is therefore necessary to refer to a density per square metre per electron volt, typical experimental values for chemically prepared surfaces being in the range $1\text{–}5 \times 10^{17}\ m^{-2}\ eV^{-1}$. This distribution in energy has many important consequences, one of which is its influence on field effect measurements.

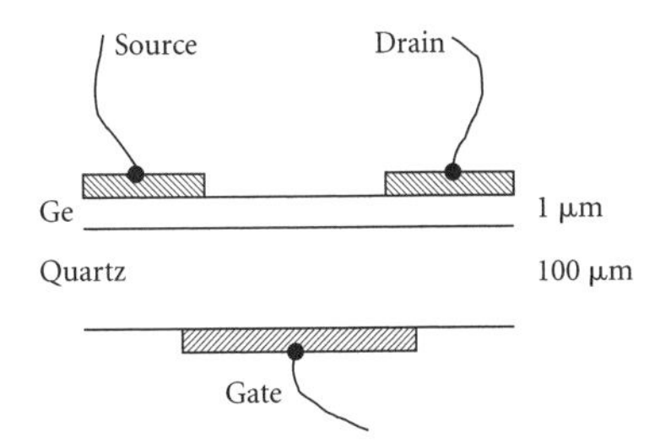

Figure 3.14. Schematic diagram to illustrate the field effect measurements made at Bell Labs while studying surface state phenomena. A thin film of germanium was deposited on a 100 μm thick quartz plate, together with a metal gate electrode on the opposite surface. Conduction was measured between electrodes on the Ge surface as a function of voltage applied to the gate.

Let us consider a typical field effect measurement such as made at Bell Labs in the 1940s. The experiment might consist basically of a thin film of Ge, thickness $t = 1$ μm, deposited on a 100 μm thick slice of quartz, having a metal plate electrode evaporated on the other side (see Figure 3.14). Suppose the Ge is n-type, it is characterized by a free electron density of $10^{22}\ m^{-3}$, corresponding to a sheet electron density of $n_s = nt = 10^{16}\ m^{-2}$. Application of a positive voltage to the metal field plate would be expected to induce further electrons in the Ge, thus increasing its conductance. If we suppose that a voltage of 100 V is applied, what change in electron density would be produced?

This is easily calculated making use of the relation between the charge Q induced on a capacitor C by an applied voltage V:

$$Q = CV \tag{3.12}$$

For an area of surface A, we note that $C = \varepsilon_i \varepsilon_0 A/d$ and therefore the charge *per unit area* induced is $\delta Q_s = e\delta n_s = \varepsilon_i \varepsilon_0 V/d$. Thus, finally, we have:

$$\delta n_s = \varepsilon_i \varepsilon_0 V/ed \tag{3.13}$$

which can be evaluated as $\delta n_s = 2 \times 8.85 \times 10^{-12} \times 100/1.6 \times 10^{-19} \times 10^{-4} = 1.1 \times 10^{14}\ \text{m}^{-2}$. Thus, the fractional change in conductivity of the semiconductor slice is $\delta n_s/n_s = 1.1 \times 10^{14}/1 \times 10^{16} = 1.1 \times 10^{-2}$, or approximately 1%, which would be readily measurable. However, real experiments showed very much smaller effects than this, the explanation (by Bardeen) being that most of the electrons induced in the Ge are trapped in surface states. In Box 3.3 we illustrate a calculation of the field effect for an identical experiment but with a density of $N_s = 3 \times 10^{17}\ \text{m}^{-2}\ \text{eV}^{-1}$ surface states included. As we see there, for every free electron induced by the field, 85 are trapped in surface states and the field effect is roughly 50 times smaller than might be anticipated. Similar arguments may explain the failure of Shockley's early attempts to observe a field effect in silicon, though there the screening effect appears to have been even larger—of order of a thousand times.

The high density of states on Ge and Si surfaces was soon seen to have a second important consequence in respect of metal–semiconductor contacts, or Schottky barriers. In Chapter 2 we described the use of these contacts for practical rectifiers and pointed out the significance of the energy barrier, which separates the metal from the bulk of the semiconductor, in realizing the rectifying characteristic. In fact, Mott's and Schottky's theories of rectification represented an important advance in the understanding of semiconductor behaviour. However, careful experiments to measure the height of the barrier began to throw doubt on the detailed application of their work, in which it had been postulated that the barrier height φ_b was determined by the work function φ_m of the metal concerned, according to:

$$\varphi_b = \varphi_m - \chi_s, \tag{3.14}$$

where χ_s is the electron affinity of the semiconductor, which takes values of 4.0 eV and 4.05 eV for Ge and Si, respectively. With the advent of high quality single crystals of Si, it gradually became apparent that the measured values of φ_b did not accord with the predictions of equation (3.14) and later work has only emphasized the discrepancies. Figure 3.16 shows data from the book by Rhoderick in which φ_b is plotted against φ_m for a range of metal contacts on etched surfaces of n-type silicon samples. It is clear that barrier heights vary much less strongly with metal work function than predicted by equation (3.14). In fact, a closer approximation might be to say that φ_b is, to first order, independent of φ_m.

Box 3.3. Surface states and the field effect

In order to illustrate the importance of surface states in a field effect measurement, we shall set up a model and calculate the field effect for reasonable values of the parameters. We assume that the n-type semiconductor sample has a thickness t, length l, and width w and that current flows along the length between suitable electrical contacts. The doping level is N_D (m^{-3}) and the density of surface states at top and bottom surfaces is N_s (m^{-2} eV^{-1}). The field effect is measured by placing a metal field plate a small distance d above the top surface of the semiconductor (thus forming a parallel plate capacitor) and applying a voltage to it. The resulting change in conductance along the sample length is monitored.

The presence of surface states in the forbidden energy gap which are occupied by electrons results in band-bending within the semiconductor, as shown in Figure 3.15. To maintain electrical neutrality in the semiconductor, the negative charge in surface states is balanced by an equal positive charge due to ionized donors within the band-bending (or depletion) region. The magnitude of the band-bending eV_b is related to the depletion width x_d by the equation:

$$x_d = \{2\varepsilon\varepsilon_0 V_b / eN_D\}^{1/2}. \quad \text{(B3.16)}$$

Suppose that a positive voltage V is applied to the field plate. This induces a net negative charge on the semiconductor given by:

$$Q = CV = (\varepsilon_i \varepsilon_0 A/d)V, \quad \text{(B3.17)}$$

where ε_i is the relative dielectric constant of the insulator which separates the field plate from the semiconductor and $A = lw$ is the area of the capacitor. By measuring V, we can determine the induced charge Q.

The induced negative charge is divided between an increase in electron charge in surface states and a decrease in positive charge in the space–charge (depletion) region. Filling up more surface states reduces the band-bending which, in turn, reduces x_d. The extra charge in surface states is $eN_s\delta V_b$ and the reduction in space charge is $eN_d\delta x_d$ (per unit area) so we can write:

$$Q = eA[N_s\delta V_b + N_d\delta x_d]. \quad \text{(B3.18)}$$

Because of the relation between x_d and V_b, we can write $\delta V_b = (dV_b/dx_d)\delta x_d$ and obtain:

$$Q = eAN_D\delta x_d[eN_s x_d/\varepsilon\varepsilon_0 + 1]. \quad \text{(B3.19)}$$

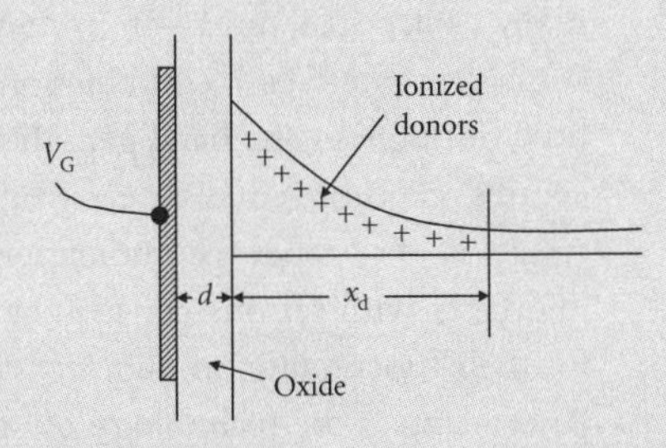

Figure 3.15. Band-bending diagram at the interface between n-type silicon and SiO_2. Positively charged ionized donors exist within the silicon depletion region, while interface states contain negative charge which balances this. When a positive gate voltage is applied, this induces negative charge at the interface and reduces the positive charge in the depletion region—that is, the band-bending and depletion width are reduced. This results in a small increase in the conductance of the silicon in a direction parallel to the gate.

Box 3.3. Continued

The term $eN_D\delta x_d$ represents free electron charge at the edge of the depletion region and is responsible for the measured increase in conductance. The first term within the square bracket represents the ratio of charge in surface states to free electron charge. Note that it is δx_d which is measured as a change in conductance because it represents an increase in the effective thickness of the conducting layer. From equations (B3.17) and (B3.19) we can obtain an expression for δx_d as follows:

$$\delta x_d = \varepsilon_i\varepsilon_0 V / \{edN_D[eN_s x_d/\varepsilon\varepsilon_0 + 1]\}. \qquad \text{(B3.20)}$$

Using the following values for the density of surface states $N_s = 3 \times 10^{17}\,\text{m}^{-2}\,\text{eV}^{-1}$, the semiconductor doping level $N_D = 10^{22}\,\text{m}^{-3}$ and band-bending $V_b = 0.5$ V, insulator thickness $d = 100$ μm and dielectric constant $\varepsilon_i = 2$, we can evaluate all these terms: thus, we obtain $x_d = 2.5 \times 10^{-7}$ m (0.25 μm), $eN_s x_d/\varepsilon\varepsilon_0 = 85$ (using $\varepsilon = 16$ for Ge). To find δx_d we suppose that $V = 100$ V and obtain $\delta x_d = 1.3 \times 10^{-10}$ m. If we now suppose that the semiconductor film is 1 μm thick, which implies a conducting region $(1 - 2x_d) = 0.5$ μm, the fractional change in conductance will be:

$$\delta\sigma/\sigma = \delta x_d/(1 - 2x_d), \qquad \text{(B3.21)}$$

which we can evaluate as $1.3 \times 10^{-10}/5 \times 10^{-7} = 2.6 \times 10^{-4}$, rather a small effect! This contrasts with the calculated effect (see text) for the surface-state free case where $\delta\sigma/\sigma \sim 1\%$. The explanation lies in the fact that, for every free electron induced in the conducting channel, no fewer than 85 are trapped in surface states. The surface charge effectively screens the interior of the semiconductor from the capacitor field.

The presence of a high density of surface states can account for these results in so far as the semiconductor bands are bent towards the surface, even in the absence of the metal. Thus, there exists a barrier whose height is determined simply by the surface states and which, if the surface is prepared in the same manner, will always have the same height. The application of the metal can do no more than modify this initial band-bending, resulting in the rather small variation of barrier height with metal work function seen in Figure 3.16. As we shall see later, surface and 'interface states' (states occurring at the conjunction of two different materials in intimate contact) play a major role in many device structures—an important practical problem being that of surface or interface recombination. Because surface states are distributed throughout the band gap, they can act as efficient recombination centres, capturing first an electron (say), then a hole. This frequently results in very rapid recombination of minority carriers at surfaces or interfaces with serious consequences for many devices which depend for their functioning on minority carriers—for example, the transistor. A vital step in the technology of such devices is the discovery of techniques for 'passivating' the surface and thereby minimizing its effect.

That brings us to the end of our first serious foray into the complexities of semiconductor physics. It is instructive to realize just how rapidly

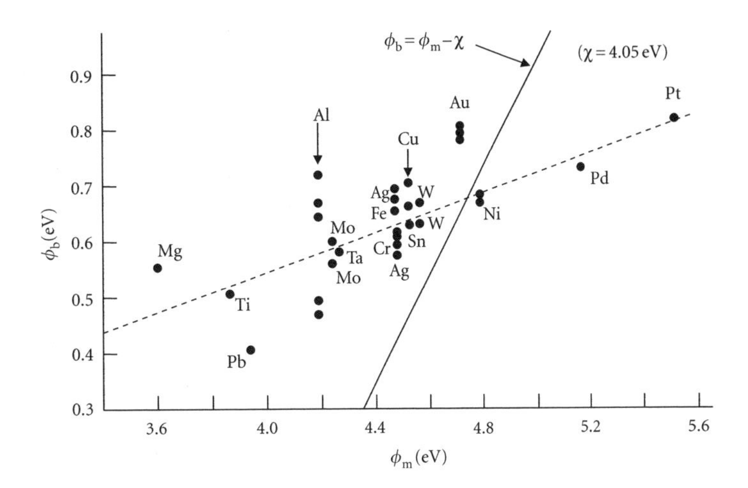

Figure 3.16. Plot of metal–semiconductor barrier heights for n-type silicon vs the metal work function. The dashed line represents a best fit to the experimental data, while the solid line is a plot of the relation $\varphi_b = \varphi_m - \chi$ (where φ_b is the barrier height, φ_m is the metal work function, and $\chi = 4.05$ eV is the electron affinity of silicon). The large disparity between theory and experiment demonstrates the role of surface states in determining barrier heights. (From Rhoderick 1978: 53). By permission of Oxford University Press.

the subject advanced in the wake of the transistor. The availability of high quality crystals of Ge and Si may have been essential for the appropriate device work but it also made possible enormously improved measurements of semiconductor properties—which, in turn, had a major influence on further device work. As we emphasized at the beginning of this section, in the field of semiconductor research, physics, devices, and materials form an inseparable triumvirate—just as do physicists, chemists, and device engineers in the relevant research laboratories.

It is now time to return to the device scene which we left in suspended animation at the point where point contacts had achieved the initial breakthrough. However euphoric may have been the initial reaction to the first transistor, there were people at Bell and, indeed, elsewhere who were all too keen to see further advances. We need to examine some of the consequences.

3.4 The junction transistor

While there can be no doubt that the invention of the point contact Ge transistor represented a critical turning point in electronics, it is also true that, in technological terms, it can now be seen as little more (!) than the proof of an existence theorem. Though the long sought solid state amplifier was now a reality, there was still a considerable step to be made before it could safely be allowed outside the research laboratory. Two immediate problems were apparent: first, reliable point contacts were notoriously difficult to make (though wartime experience could be called into play to assist in this aspect) and, second, placing two of them within 50 μm of each other in a controlled and reproducible fashion was never going to be easy. In particular, any commercial device would demand reliability and consistency in its operating characteristics which were certain to be compromised by variability in the character of the individual contacts or by inadequate control of their separation. Further to this, the point contact device suffered from high noise levels, undesirable in an amplifier to be used in a telephone repeater unit. It was at this point that Bell management took the significant decision to set up a completely new team to tackle the problems of industrialization, while encouraging the original inventors to concentrate on their fundamental research. Remember that their principal concern had, all along, been with semiconductor surface properties which were seen as having important fundamental significance. In this context, too, it is interesting to note that the papers announcing the vital breakthrough, demonstrating the first ever successful solid state amplifier, were published not in an engineering journal, nor even in the obvious (?) *Journal of Applied Physics* but in the *Physical Review* which is reserved for results of fundamental *scientific* importance. There was, in any case, much that needed elucidation with regard to the *modus operandi* of the point contact device.

Given the circumstances, this decision can be seen as an enlightened one, but, sadly, there were other forces at work which split the research effort at Murray Hill onto two diverging tracks. Shockley was less than happy to find himself playing second fiddle to Bardeen and Brattain, his own perfectly plausible ideas for a field effect device having met with total frustration, and, to give appropriate credit, he also had good reason to doubt that the point contact transistor represented an ideal commercial solution to Bell's needs. Motivation apart, the consequence soon became clear—Bardeen and Brattain were rapidly sidelined while Shockley, taking to himself a number of other collaborators, determined to develop what he saw as a more commercially viable transistor (and one which he could claim as his own!). The outcome was the junction transistor which not only became the model for all future single devices but, more importantly, opened the way to the integrated circuit and the much greater electronic revolution with which that has been associated. Shockley's man management would not win any plaudits from the schools of business management which have mushroomed during the latter part of the twentieth century but his single-minded determination and technical inventiveness did win a Nobel Prize! It must be said, too, that it steered not only Bell but the rest of the western world onto the right track. While it is true that Bell did manufacture a 'sanitized' version of the point contact device (referred to as 'the type A transistor'), sales were rather modest and, as we shall see, were rapidly challenged by later modifications which were more easily manufacturable.

Returning to mere technical issues, the first point of clarification concerned the nature of the minority-carrier injection process which made the transistor possible. Perhaps because of the obsession (?) with surface effects, it was, at first, thought that this too was a surface phenomenon and that the carriers were confined to the Ge surface between the two contacts. Shockley was convinced that this was not the case, but that injected carriers could also diffuse through the bulk of the Ge, a conviction which was vindicated by an interesting experiment performed by John Shive, another member of the semiconductor group. Shive made a new version of the point contact transistor having contacts on opposite sides of a thin sliver of Ge—the fact that it still gave satisfactory gain, Shockley realized, was clear proof that injected carriers could, indeed, diffuse through the bulk semiconductor. More importantly, it gave strong support to the idea he had been incubating for some months that a transistor could be fashioned from two *p-n* junctions, one forward biased to act as an injector (or emitter), the other reverse biased to act as collector. This would lay to rest once and for all the problems of non-reproducibility associated with metal point contacts—the junction transistor was essentially a bulk device, all its parts being (at least in principle) reliably manufacturable. It was also a 'one-dimensional' structure (in the mathematical sense) which was much easier to model and Shockley wasted no time in taking advantage of that.

The junction transistor being based on the *p-n* junction diode, we must first devote some attention to this fundamental building block (which, in any case, has other roles of its own). We saw earlier that the first recognized *p-n* junction came about rather by accident, as a result of impurity segregation in a molten Si rod, but it soon became possible to make such junctions deliberately by controlled doping during crystal growth. That they acted as rectifiers was soon appreciated but there was much else to be learned about them, including their ability to act as photodetectors, voltage-controlled capacitors, and (crucially) emitters and collectors of minority carriers. All these aspects can be understood qualitatively by thinking carefully about the behaviour of electrons and holes in the vicinity of the junction—a more quantitative understanding requires the application of mathematics, an example of which we illustrate in Box 3.4 for the case of minority carrier injection.

Let us first perform a 'thought experiment' and consider what happens when *n*-type and *p*-type samples (of the same semiconductor) are brought together (Figure 3.17(a)). As the contact is first made, near the interface in the *n*-type sample there exists a strong gradient in hole density, with a corresponding gradient in electron density in the *p*-type material. Thus, holes and electrons diffuse in opposite directions, holes from the *p*-side into the *n*-side, electrons from *n* to *p*. Concentrate attention first on the *p*-side where the in-flowing electrons rapidly recombine with majority holes, effectively removing free holes from the material close to the junction and leaving a region of negatively charged acceptor atoms. Similarly, on the *n*-side, there exists a layer of positively charged donors (Figure 3.17(b)). An electric dipole field is therefore created which opposes the diffusion currents due to the concentration gradients. Very quickly, an equilibrium is established where the electron diffusion current is exactly balanced by an opposing electron drift current (driven by the dipole field) and similarly for the holes. No *net*

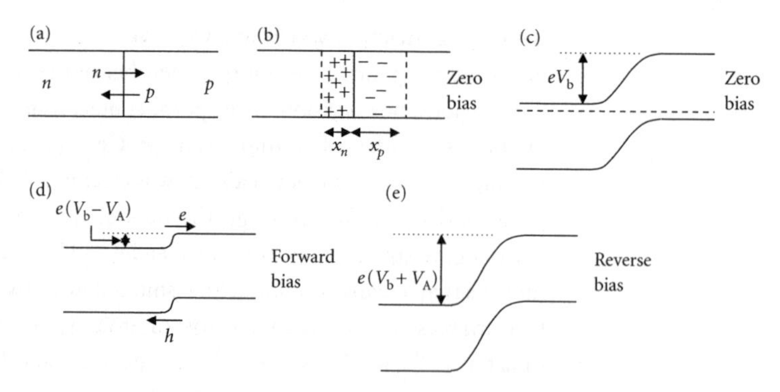

Figure 3.17. Diagrams to illustrate the properties of a typical *p-n* junction: (a) represents a physical picture of an *n*-type and *p*-type sample being joined together, showing electrons diffusing from left to right and holes from right to left; (b) the diffusion results in a space–charge region (depleted of free carriers) on either side of the physical junction, which sets up a dipole field opposing the diffusion currents. When the applied voltage is zero, the diffusion and drift currents are exactly balanced; (c) shows a band diagram of the junction region, demonstrating the presence of a 'built-in' voltage V_b due to the dipole field; (d) A forward bias has been applied (*p*-side positive, *n*-side negative) which reduces the voltage step across the junction so that the diffusion currents are no longer balanced by drift currents and a net current flows from right to left; (e) A reverse bias has enhanced the voltage step so that the diffusion currents are effectively quenched and the only current which flows is due to thermal generation of minority carriers (n_p and p_n) on either side of the junction. Reverse bias also leads to a widening of the depletion layer which reduces the depletion capacitance associated with the junction.

current flows across the junction, which is entirely consistent with the fact that no external voltage has been applied to it!—but individual drift and diffusion currents of both holes and electrons are happily pursuing their natural instincts. One important consequence of all this is a region of material close to the interface which is depleted of free carriers (and therefore known as a 'depletion region') but characterized by two space charges, positive on the *n*-side, negative on the *p*-side, somewhat similar, in fact, to the situation in a parallel plate capacitor C in which the two plates carry a charge Q when there exists a voltage difference V between them. As we saw above (equation 3.12), Q is directly proportional to the applied voltage V—but, in the case of the *p-n* junction with no external voltage applied, V appears to be zero so we should expect $Q = 0$ too! How can we reconcile this apparent *impasse*? Quite simply!—there *is* a voltage in the *p-n* junction, $V_J = V_b$ (the so-called 'built-in voltage'—see Figure 3.17(c)) which is associated with the dipole field. V_b is approximately equal to the band gap voltage E_g/e. However, because the charge is distributed throughout the depletion region, it turns out that, in this case, $Q \propto V_J^{1/2}$, (rather than $Q \propto V_J$).

Now we may ask what happens when an external voltage is applied. It matters significantly which way round we apply it. Suppose it is applied in the 'forward' direction—that is, so as to make the *n*-side negative and the *p*-side positive. In this sense the applied voltage V_A acts so as to reduce the voltage across the junction—that is, $V_J = (V_b - V_A)$ (Figure 3.17(d)). The drift current is correspondingly reduced but the diffusion current remains unaffected so we have a net current flowing across the junction in the direction of the diffusion component—remember this was made up of a hole current (i.e. a positive current) from p-side to n-side and an electron current from n-side to p-side (which also represents a positive current from p-side to n-side). Thus, applying a positive voltage to the p-side results in a positive current through the junction in the appropriate sense—but, surprisingly, perhaps, it is actually a diffusion current, not a drift current and, what is more, it takes the form of *minority carrier injection* across the junction plane. The field in the junction is actually *smaller* as a result of applying the forward voltage which also implies that the charge in the depletion region must be reduced and this comes about as a result of a narrowing of the depletion region (note that $Q = e(N_D x_n + N_A x_p)$ where N_D and N_A are the doping levels on either side and x_n and x_p are the depletion widths on either side, so Q is reduced if x_n and x_p are reduced). An important consequence of the reduction in depletion width is that the junction capacitance is correspondingly increased, illustrating the fact that C is a function of the applied voltage.

If the voltage is applied in the 'reverse' direction (Figure 3.17(e)), the junction voltage becomes $V_J = (V_b + V_A)$ and the depletion region expands, making the capacitance smaller. In fact, for most junctions, $C \propto V_J^{-1/2}$. At the same time, the increased junction field becomes so large as to reduce diffusion currents effectively to zero, suggesting that the reverse current should also be zero. This is not quite true—there is a small residual current which arises from the presence of thermally generated minority carriers in the bulk n- and p-materials. Such carriers (e.g. electrons on the p-side), far from being inhibited by the junction field, are accelerated across the junction to join their majority companions on the other side. The resulting current is small because, as we saw earlier (equation 3.8), the density of minority carriers in bulk material is also small and, what is more, it becomes rapidly smaller as the semiconductor band gap increases. Finally, we note that this reverse current is independent of the junction voltage (it depends only on the density of minority carriers because the junction field is large enough to sweep all minority carriers across the junction) so the reverse current saturates at a value I_s, where the diode current is given by:

$$I = I_s\{\exp(eV_A/k\mathrm{T}) - 1\} \tag{3.15}$$

very similar to the corresponding equation for a Schottky barrier diode (equation 2.19).

Let us try to summarize the properties of a p-n junction, as we now appreciate them:

1. The junction has an associated depletion layer which acts as a voltage-dependent capacitor $C \propto V_J^{-1/2}$.
2. The forward current increases rapidly (in fact, exponentially) with increasing applied voltage.
3. The forward current takes the form of a diffusion current of carriers injected across the junction which become minority carriers on the other side (see Box 3.4).

Box 3.4. The (n^+-p) emitter junction

The forward biased p-n junction is an essential part of the bipolar transistor, acting as the emitter of minority carriers into the base region. In this box, therefore, we analyse the injection behaviour of a specific case, that is, that of the asymmetric n^+-p junction which injects minority electrons into a p-type base. We shall assume that the donor level N_D on the n-side is so much larger than the acceptor level N_A on the p-side that the depletion region is effectively confined wholly to the p-side of the junction. This implies that the band-bending at zero bias is that shown in Figure 3.18(a) where the free electron density at the edge of the depletion region on the p-side n_{p0} is given by the expression:

$$n_{p0} = n_0 \exp\{-eV_b/kT\}, \tag{B3.22}$$

where n_0 is the equilibrium free carrier density on the n-side ($n_0 = N_D$) and V_b is the built-in voltage (or total band-bending across the junction). This result is based on the Boltzmann statistical probability for finding an electron at an energy eV_b above the conduction band on the n-side. When a forward bias V_f is applied across the junction, the asymmetric doping implies that V_f will appear entirely on the p-side so the band-bending is reduced to $e(V_b - V_f)$, as shown in Figure 3.18(b). It then follows that the free electron density at the edge of the depletion layer is:

$$\begin{aligned} n_p &= n_0 \exp\{-e(V_b - V_f)/kT\} \\ &= n_{p0} \exp\{eV_f/kT\}. \end{aligned} \tag{B3.23}$$

If we define $\Delta n(x) = [n(x) - n_{p0}]$, we see that at the edge of the depletion layer where $x = 0$ and $n = n_p$ we can write:

$$\begin{aligned} \Delta n(0) &= (n_p - n_{p0}) \\ &= n_{p0}[\exp\{eV_f/kT\} - 1]. \end{aligned} \tag{B3.24}$$

$\Delta n(0)$ is the excess electron concentration at the edge of the depletion layer ($x = 0$), compared with the equilibrium density of electrons in the bulk of the p-region n_{p0}. This excess implies there is a concentration gradient to the right of the plane $x = 0$ which results in a diffusion current of minority electrons into the p-side. The diffusion current across the plane $x = 0$ is given by:

$$\begin{aligned} J_D &= -eD[d\Delta n/dx]_{x=0} \\ &= -eD[-(\Delta n(x)/L)\exp\{-x/L\}]_{x=0} \\ &= (eD/L)\Delta n(0) \\ &= (eD/L)n_{p0}[\exp\{eV_f/kT\} - 1]. \end{aligned} \tag{B3.25}$$

In deriving this result we have made use of the diffusion profile calculated in Box 3.2 $\Delta n(x) = \Delta n(0)\exp\{-x/L\}$.

To obtain our result in its final form we note that:

$$\begin{aligned} n_{p0} &= n_i^2/p_0 \\ &= (N_C N_V/N_A)\exp\{-E_g/kT\}, \end{aligned} \tag{B3.26}$$

Box 3.4. Continued

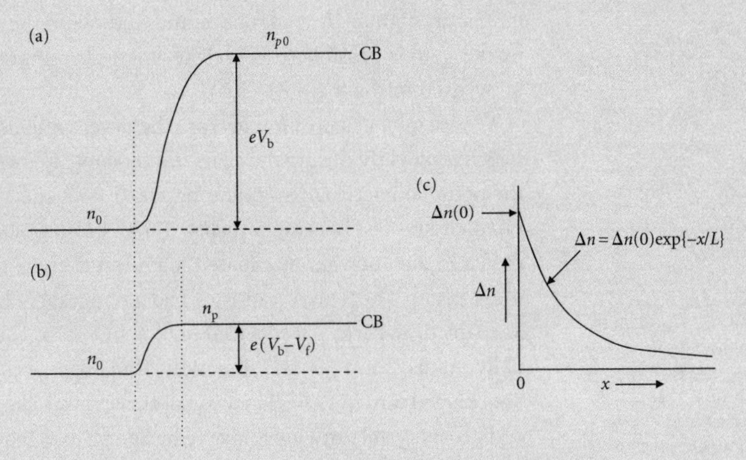

Figure 3.18. Band diagrams to illustrate the injection of minority electrons into the p-side of a (n^+-p) junction; (a) shows the conduction band-bending (CB) at zero bias which is assumed to lie entirely on the p-side of the junction. The free electron density n_{p0} at the edge of the depletion region is the equilibrium density in the bulk p-type material; (b) shows the corresponding picture when a forward bias V_f is applied; (c) shows the diffusion profile of minority electrons within the p-region.

so we can write the injected current density as:

$$J_D = J_s[\exp\{eV_f/kT\} - 1], \quad \text{(B3.27)}$$

where the saturation current density J_s is given by:

$$J_s = (eD/L)(N_C N_V/N_A)\exp\{-E_g/kT\}. \quad \text{(B3.28)}$$

Equation (B3.27) is the well-known diode equation for the special case of an n^+-p junction.

4. The reverse current saturates at a value I_s which depends on the density of thermally generated minority carriers in the bulk material on either side of the junction. I_s decreases rapidly with increasing semiconductor band gap.
5. There is an electric field in the depletion layer which increases under reverse bias and decreases under forward bias. It acts so as to oppose the diffusion currents referred to in point 3 but sweeps minority carriers across the junction.

How does all this relate to transistors? Briefly stated, the emitter is a forward biased junction which injects minority carriers into the base; these then diffuse across the base to reach the collector junction which is reverse biased so that its junction field sweeps them into the collector to form the collector current. (Figure 3.19 shows an appropriate band diagram which illustrates this behaviour.) If the base is narrow, compared

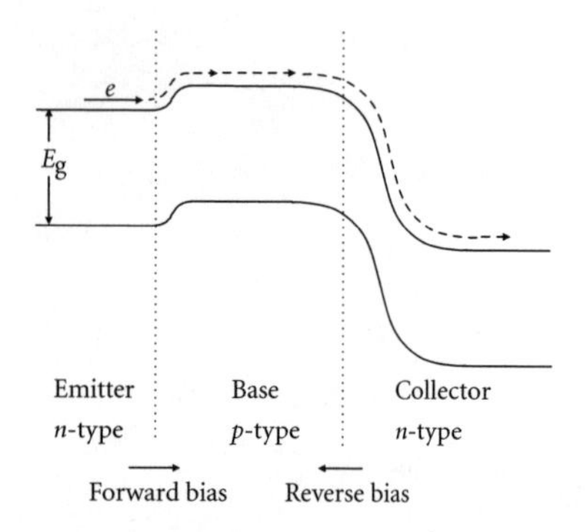

Figure 3.19. Band diagram of an *n-p-n* transistor, showing, on the left, a forward biased emitter junction which injects minority electrons into the base region (cf. Figure 3.18(b)) and a reverse biased collector junction whose high field sweeps these minority electrons into the collector contact to form the collector current.

with the minority carrier diffusion length L (see Box 3.2), the collector current is approximately equal to the emitter current but the collector circuit is high in impedance compared with the emitter circuit, so power gain is possible (this is clear when we remember that power can be written in the form $P = RI^2$).

We are now in a position to trace the developments in transistor design which took place during the years immediately following the invention of the point contact device—that is between 1948 and 1959 when the integrated circuit first became a reality. Table 3.2 summarizes these developments in a somewhat attenuated form but there is an obvious need for more detail. The type A transistor had a remarkably long life, considering its many drawbacks—it survived until about 1959—but it was doomed to oblivion as soon as the junction transistor was demonstrated by Shockley's team in 1950. Their initial success was largely the product of a dedicated crystal growing initiative by Sparks and Teal who developed the technique of 'double doping' during Czochralski growth, first introducing Ga to form a *p*-type base, followed by Sb to make the *n*-type collector. (As an aside, it is worth noting that Shockley, himself, was not in favour of any effort being devoted to the growth of single crystals, holding the opinion that polycrystalline material would suffice, provided it was pure enough. This blind spot concerning the importance of materials work has been widely demonstrated by other device-oriented scientists and development engineers, even in the face of overwhelming evidence—Shockley is not alone in this particular error!) The double-doped Ge transistor was of major significance in so far as it demonstrated an important principle—junction transistors worked!—but was of limited practical application due to its poor frequency response. The base width was difficult to control and was generally far too large to permit amplification of high frequency signals, though the use of slow crystal growth and thorough stirring did improve matters to the point that later devices showed

Table 3.2. Development of the junction transistor

Transistor type	Base width (μm)	Frequency response	Year
Ge point contact	25–50	10 MHz	1948
Ge grown junction	500–750	10–20 kHz	1950
Ge alloy junction	10–20	5–10 MHz	1951
Ge graded base	10–20	10–30 MHz	1952
Ge surface barrier	5	60 MHz	1953
Si grown junction	10–15	10 MHz	1954
Ge mesa diffused	5	150 MHz	1954
Si mesa diffused	5	120 MHz	1955
Ge mesa diffused	1	1 GHz	1959
Si planar	0.7–2	150–350 MHz	1959

cut-off frequencies as high as 1 MHz. This was just acceptable for the first transistor radios which came on the market during 1954 though totally inadequate to meet the requirement for FM radio which used the frequency band around 100 MHz.

We might pause here to consider this matter of frequency response which is generally limited by the base transit time t_D, this being the time taken for injected minority carriers to diffuse across the base. Strictly, it is not the time t_D itself which determines the cut-off frequency but the effect of pulse broadening due to the thermal motion of the carriers. However, for the case where the base width W is approximately equal to the diffusion length L, it turns out that the time spread Δt is actually equal to t_D (which is also equal to the lifetime τ of the minority carriers). More generally, $\Delta t = \{t_D \tau\}^{1/2}$ which makes it clear that a good high frequency response demands a short transit time and therefore a short base length.

The double-doped transistor was far from easy to make and there were further problems in making contact to the thin base region and this prompted the development of the alloy junction transistor by General Electric and RCA in 1951 (see Figure 3.20(a)). This was basically a cruder structure but was certainly easier to manufacture. It consisted of a 50 μm thick bar of n-type Ge to which was applied a pair of In dots, one on either side. The In was melted at a temperature of

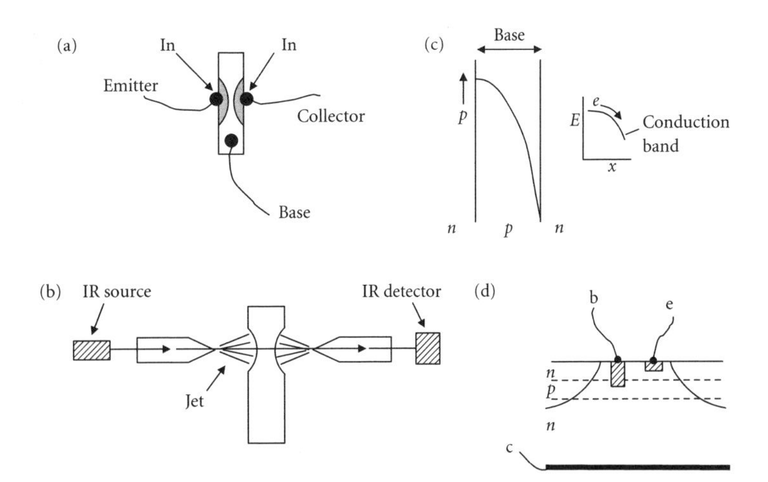

Figure 3.20. Early developments in transistor technology: (a) shows a schematic of an alloy junction transistor—a sliver of n-type germanium has In beads alloyed into it from both sides to form the emitter and collector regions of a p-n-p transistor. Accurate control of the base width was difficult; (b) shows the modified technology introduced by Philco. The germanium was etched from both sides by a jet etch and the base thickness monitored by measuring the transmission of an infrared (IR) beam. Alloy regions were then introduced as before; (c) illustrates the idea of grading the doping level in the base region of the transistor so as to set up a drift field across the base and speed the minority carriers on their way to the collector; (d) shows a mesa diffused n-p-n structure which allowed greater control of base width but suffered from the difficulty of making contact to the base layer—the contact had to be made through the emitter region.

about 500°C and alloyed into the Ge to produce *p*-type regions which acted as emitter and collector respectively. The alloying depth was such that the resulting base width was approximately 10–20 μm, though it was difficult to control accurately. However, considerably better control was obtained in the Philco surface barrier transistor which was an improved version of the alloy device. The Ge wafer was jet-etched from both sides by an electrolytic etch and the thickness of the base monitored by shining an IR beam through it and measuring the transmission intensity (Figure 3.20(b)). This enabled much thinner base regions to be made with correspondingly higher cut-off frequencies (of order 50 MHz). The emitter and collector were formed by plating In into the etched holes, followed by a micro-alloying process. This resulted in a more sophisticated device, though it was also considerably more fragile.

It was at this time that Herb Kroemer made the suggestion that the base region should not be uniformly doped, as had been standard practice, but should be graded in concentration (high at the emitter side, low at the collector side) so as to set up an internal electric field to assist minority carrier transport across the base (Figure 3.20(c)). This amounts to drifting carriers across, rather than relying on diffusion, thus reducing their transit time and improving the frequency response. The proposal had a further advantage in preventing 'punch-through'. In a fairly lightly doped base, the base–collector junction depletion region spreads into the base and, when base widths become very small, this depletion zone may reach right through to the emitter, effectively shorting out the base. By doping the base low at the collector side, it was still possible to apply an adequate collector voltage (to achieve the necessary collector sweep-out field) while the high doping at the emitter side prevented punch-through. Kroemer's idea is applicable to any junction transistor, however fabricated, provided, of course, that appropriate control of the base doping is possible.

The next important step was taken at Texas Instruments (TI) in 1954. Gordon Teal had moved from Bell Labs to TI in 1952 with a brief to develop Si as a transistor material, the chief motivation being to improve temperature stability for military applications. The rather modest band gap of Ge could often lead to thermal instability in power devices which tended to run hot. The increase of temperature generated further carriers which led to more current, which led to further rise in temperature, which led to more carriers, etc., a phenomenon known as 'thermal runaway'. Military specifications always tend to be more strict than those acceptable in civilian applications and, military funding being vital, at the time, to successful commercial transistor development, there was considerable urgency to employ the larger band gap of Si in military devices. Teal had been largely responsible for the Ge double-doped transistor when at Bell—it was, therefore, natural for him to pursue a similar line with Si, bringing an important success to this relatively young company and giving them a lead of, perhaps, 2 years on their competitors—a very significant advantage in a rapidly changing business. TI made electronic hay with their new-found devices between 1954 and 1958 but, once again, the rest of the world did not stand still. Bell rapidly came back into sharp focus with the application of impurity diffusion to transistor manufacture. First, in 1954 they demonstrated the mesa diffused Ge *n-p-n* transistor (Figure 3.20(d)), then in 1955, its Si equivalent, both important steps forward which were soon to lead to the planar diffused transistor, opening the way to rapid exploitation of the integrated circuit in the 1960s.

Diffusion doping was introduced in the search for better control of base width. The process involved heating a wafer of Ge or Si in a closed tube containing a suitable vapour pressure of donor or acceptor atoms. By controlling the temperature and vapour pressure, it was possible to tailor the diffusion depth and doping concentration with considerable accuracy. For example, to make a Si *n-p-n* transistor, a wafer of high resistivity *n*-type Si was simultaneously diffused with both Al and Sb, the lighter Al diffusing faster to form the *p*-type base, while the heavier Sb diffused less rapidly and, therefore, to a smaller depth, counter doping the surface region to form the *n*-type emitter. The base width was determined by the difference between the Al and the Sb penetration into the Si. Widths of 5 μm were readily produced, resulting in cutoff frequencies greater than the important 100 MHz target. There were, however, some drawbacks: first, it was necessary to etch a mesa structure to reduce the cross-sectional area (and therefore the capacitance of the collector junction); second, it was difficult to contact the base region; third, the relatively thick and lightly doped collector represented a large series resistance which, together with the collector capacitance constituted an unwelcome RC time constant. In a final tour de force, Theurer therefore introduced an epitaxial collector layer which was suitably lightly doped but deposited on a highly doped Si substrate, thus satisfying the requirement for a wide depletion region on the collector side of the junction, together with a low resistivity contact layer. This was the process on which Bell standardized—until the coming of planar technology in 1959.

Planar technology has now been so widely used for so long, it is difficult to realize that at one time it still needed to be invented! But invented it was, at a new start-up company called Fairchild, one of the early Silicon Valley ventures set up by a group of people who originally worked for Shockley after he left Bell Labs to start his own company, Shockley Semiconductor Laboratory in 1956. The inventor appears to have been a Swiss scientist named Jean Hoerni but the crucial material on which the process is based SiO_2 was discovered much earlier (in 1955), at Bell Labs by Carl Frosch—once again an accidental discovery

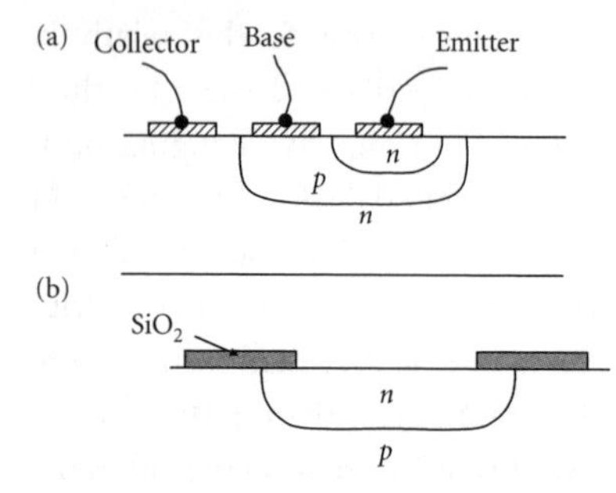

Figure 3.21. The ultimate junction transistor structure, employing planar technology: (a) shows a typical *n-p-n* structure fabricated using two diffusion steps through oxide masks. The *p*-type base was diffused first, followed by counter doping through a smaller mask to make the emitter. As can be seen, all metal contacts can be made to the top surface—hence 'planar' technology; (b) the effect of sideways diffusion under the diffusion mask—the junction meets the semiconductor surface underneath the oxide which acts as a surface passivant.

when water vapour was allowed to contaminate a diffusion capsule! Unlike germanium oxide which is water soluble and readily removed from the Ge surface, silicon oxide proved to be a highly stable and tenacious coating when formed on silicon, a property which, even more than its larger band gap, made silicon the ideal material for solid state devices. The planar structure is illustrated in Figure 3.21 which shows how an *n-p-n* transistor might be made. We shall leave discussion of the details until the next chapter which is concerned with integrated circuit development but merely note one or two important features. As in the diffused mesa transistor, the collector is made from the bulk Si wafer (lightly doped *n*-type) but the base and emitter regions are formed by diffusion through holes in an oxide mask. The diffusion moves sideways under the mask as well as downwards into the bulk Si which means that the junction, where it meets the Si surface, is covered by oxide which passivates it against atmospheric contamination (Figure 3.21(b)). This is of vital importance—all previous transistor designs had been plagued by the lack of such a passivating film and tended to become unstable as a result of moisture or dust landing on the junction and partially short-circuiting it. SiO_2 proved to be an ideal passivant. The emitter region was formed by a second diffusion within the base region and was similarly passivated. A further important feature is the possibility of placing all the metallic contacts on the top surface of the structure, which is the reason for the name 'planar'. This, as we shall see in our next chapter, is of crucial importance in integrated circuit technology.

Transistor technology had come a long way in this hectic decade and new companies had come into being to exploit it, in competition with the giants, such as RCA, GE, and Philco, already present. In the United States, Bell's policy of licensing the new technology to those prepared to pay the 'entrance fee' certainly encouraged this rapid spread of expertise, as did the American mobility of personnel but Europe, too, was not slow to climb on the band wagon. Philips, for example, made its first working transistor within a week of the first announcement from Bell in 1948 and its English subsidiary, Mullard, was manufacturing Ge alloy transistors as early as 1955. So too was GEC. In both companies, the important contribution of their already existing research laboratories was vital to success. The GEC laboratory in Wembley was opened in 1919, the Philips Nat. Lab. in Eindhoven in 1914, and the Mullard Research Laboratories in Redhill during 1946. Perhaps the principal lesson to be learned from these developments concerns the considerable step from the demonstration of an effect in a research laboratory to the realization of a reliable, commercially viable product. Ease of manufacture is as important as scientific inventiveness in an industry such as this and the Americans were probably rather quicker to develop it, coming, as they did, from an environment in which the practical and entrepreneurial had been all important prior to the Second World War.

They had two other advantages: relatively lavish funding from the US military and a tradition of personal mobility which contrasted starkly with the European (and, indeed, Japanese) culture of 'a job for life' which did little to encourage the spread of ideas. The nascent Japanese semiconductor industry (together with some European companies) suffered a serious setback towards the end of the 1950s when their investment in Ge was bypassed by the sudden emergence of Si, though Japan was to come into its own in the next decade, the decade of the integrated circuit. However, that constitutes another chapter.

Bibliography

Bardeen, J. and Brattain, W. H. (1948) *Phys. Rev.*, 74, 230.

Bernal, J. D. (1953) *Science and Industry in the Nineteenth Century*, Routledge & Kegan Paul Ltd, London.

Braun, E. and Macdonald, S. (1982) *Revolution in Miniature*, Cambridge University Press, Cambridge.

Cressell, I. G. and Powell, J. A. (1957) *Progress in Semiconductors*, 2, 138 (Heywood Co.).

James, F. (ed.) (1986) *The Development of the Laboratory the Place of Experiments*, Macmillan, London.

Many, A., Goldstein, Y., and Grover, N. B. (1971) *Semiconductor Surfaces*, North Holland, Amsterdam.

Pfann, W. G. (1957) *Solid State Physics: Advances in Research and Applications*, Vol. 14, p. 423 (eds. F. Seitz and D. Turnbull) Academic Press, New York.

Rhoderick, E. H. (1978) *Metal Semiconductor Contacts*, Clarendon Press, Oxford.

Riordan, M. and Hoddeson, L. (1997) *Crystal Fire*, W W Norton & Co, New York.

Seitz, S. and Einspruch, N. G. (1998) *Electronic Genie—The Tangled History of Silicon*, University of Illinois Press, Urbana.

Shockley, W. (1950) *Electrons and Holes in Semiconductors*, Van Nostrand, New York.

Smith, R. A. (1968) *Semiconductors*, Cambridge University Press, Cambridge.

Wilson, A. H. (1931) *Proc. Roy. Soc. A.*, 133, 458; 134, 277.

CHAPTER 4

Silicon, Silicon, and yet more Silicon

4.1 Precursor to the revolution

With the crucial advantage of hindsight, we are very well aware of the sea change consequent upon the invention of the transistor but it should not really surprise us to learn that those struggling to come to terms with it at the time were less readily persuaded. Yes, it was small and yes, it used far less power than the incumbent device (the thermionic valve) but there were disadvantages too. There was the problem of excess noise and the difficulty in producing devices which could amplify at high frequencies. Needless to say, in its early days, the transistor was seen essentially as a possible replacement for the valve—many of the companies taking part in its development were primarily valve (or, since they were mainly American companies such as RCA, GE, Sylvania, and Philco, *tube*) companies whose main business was, and continued to be for some considerable time, either valves or tubes. It is important to recognize that, though solid state circuitry was eventually to dominate the market, sales of valves did not even reach their peak until 1957 and showed little sign of serious decline until the late 1960s—the transistor might be an exciting technical advance but it was not at all obvious that it represented a major commercial investment. The possible exceptions were the small start-up companies, such as Texas Instruments (TI), Fairchild, Hughes, or Transitron, who carried none of the tube or valve baggage which encumbered the larger companies but they were, by definition, small and insignificant! They were, however, flexible and enterprising and it was from them that many of the important innovations in semiconductor technology were to come.

Technical innovation might be exciting and full to the brim with promise but, during the 1950s, the chief problem in transistor *manufacture* was one of reproducibility. We have already touched on the difficulty of controlling the base width in double-doped and alloyed structures which had a direct and crucial effect on cut-off frequency but there was the additional problem of encapsulation which frequently failed to stabilize the device against atmospheric pollution. Many manufacturers were obliged to divide their product into 'bins' containing high-grade devices which might sell for $20 apiece down to run of the mill (crumby?) specimens which they were lucky enough to offload for 75 cents! Only with the emergence of planar technology could these problems be overcome—and this process was not even invented until 12 years after the point contact transistor. And, needless to say, it took several more years to become widely accepted. Nevertheless, early transistors did find applications, first in hearing aids where the low power requirement, low weight, and small volume were obvious bonuses (though the excess noise associated with many devices could hardly have been welcomed by users!) and in small portable radios—the ubiquitous 'Transistor' which did more than anything to bring the word into common usage. Again, it was one of the small firms (TI) which saw the opportunity and forged an arrangement with the Industrial Development Engineering Associates (IDEA) to produce the 'Regency TR1' radio in October 1954. It was challenged, in the following year, by Raytheon with its own model and subsequently by numerous others, including, significantly, Sony who later contributed to the delight of youth (and the chagrin of the elderly!) with its highly successful 'Walkman' personal tape player. Applications in car radios followed soon afterwards and, in spite of various sticky patches, by the year 1960 there were some 30 US companies making transistors to a total value of over $300 million (see Braun and Macdonald 1982: 76–77).

Another area of application which attracted immediate attention was that of computers. These were still in a very primitive state of development during the 1950s—analogue computers had been used in radar systems as early as 1943 but the first general purpose digital electronic computer (ENIAC—Electronic Numerical Integrator and Calculator) was not built (at Penn State University) until 1946. It filled a large room, used 18,000 valves and dissipated 150 kW! British computing skills had been honed by code-breaking endeavours with the Colossus machine during the Second World War (Colossus was first introduced in 1943 and by the end of the war there were no less than 10 machines in use) and this experience was probably vital to the development in Cambridge of a rival to ENIAC, known as EDSAC (Electronic Digital Storage Automatic Calculator). This appeared towards the end of the 1940s, while the first transistorized computer was probably the TRADIC developed by Bell for the US Army in 1954, employing 700 transistors and 10,000 Ge diodes (all hand-wired!), followed by a commercial computer from IBM, containing over 2000 transistors, in the following year. The low dissipation and small physical size of the transistor gave it an immediate advantage and its solid state construction offered hope of much improved reliability—however, it was initially limited in speed by its relatively poorly controlled base width and, once the

Box 4.1. Digital transmission of information

As an example of a digital technique for information processing, we choose the everyday case of voice transmission along telephone lines. As the reader will be aware, the range of audio frequencies necessary to make voice transmission acceptably clear to the listener is about 100 Hz to 3 kHz, requiring a bandwidth of 3 kHz in the case of direct audio transmission. However, for digital transmission of the same information, a much wider bandwidth is necessary, demanding correspondingly faster electronic device performance. In this box we therefore consider the origin of this phenomenon and make a rough estimate of the switching speed necessary in a typical digital transmission system.

If we examine the nature of an audio signal, it can be represented in terms of an amplitude (or magnitude) which is changing with time in a fairly complex fashion (see Figure 4.1(a)), but, if we 'freeze' the signal at a specific instant (time t_1) it is characterized by a corresponding amplitude $A(t_1)$ which we can measure and represent in digital form—that is, in the shape of a set of binary digits. Thus, for example, the binary number $(10011011) = 1 \times 2^0 + 1 \times 2^1 + 0 \times 2^2 + 1 \times 2^3 + 1 \times 2^4 + 0 \times 2^5 + 0 \times 2^6 + 1 \times 2^7 = 1 + 2 + 8 + 16 + 128 = 155$. (We also need some additional information concerning the sign and the position of the decimal point but the details need not worry us here—we can leave this to the electrical engineer—suffice it to say that it introduces some further complexity into our digital signal.) Again, if we freeze the signal at a slightly later time t_2, the amplitude will take on a different value $A(t_2)$ and so on—in fact we can represent the audio signal as a sequence of these amplitudes, corresponding to a sequence of equally spaced times t_1, t_2, t_3, t_4, etc. (This overall process is known as 'analogue-to-digital conversion', ADC.) But we must accept that it is, at best, an *approximate* representation and therefore ask ourselves just how accurate it can be. Is it an adequate representation for the purpose in mind? In other words, will it introduce an unacceptable level of distortion into the signal? Will the listener find difficulty in recognizing the sound of the voice and the sense of its message? Two factors are relevant; the number of digits we employ to represent the amplitudes and the frequency with which we sample the audio signal—that is, the time differences $(t_2 - t_1)$, $(t_3 - t_2)$, etc. The more digits, the better accuracy which is possible (this is no different from when we are using decimal numbers, of course)—this is relatively simple but, more subtly, the sampling frequency is also vital. To quantify this, we refer to a standard theorem, known as the 'Nyquist sampling theorem' which tells us that, for adequate digital representation of a time-varying signal, we must sample at a frequency which is at least *double* that of the *maximum* frequency present in the analogue signal. In other words, we must measure the amplitude at least twice during each *period* corresponding to the maximum frequency present (see Figure 4.1(b)). In our case, this frequency is 3 kHz,

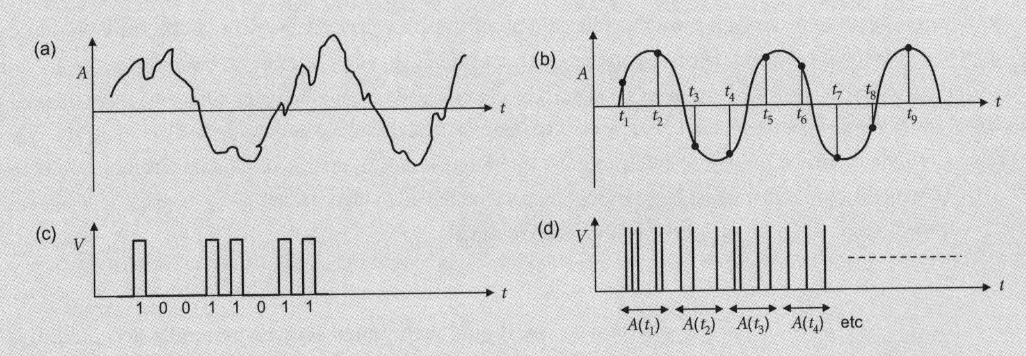

Figure 4.1. Stages in the digitalization of an audio signal: (a) shows the analogue signal; (b) illustrates the method of sampling a sine wave at a regular sequence of times t_1, t_2, t_3, etc; (c) provides an example of an eight bit digital signal which represents a single amplitude 155 mV; while (d) demonstrates the form of the digital representation of the sequence of sampled amplitudes which make up the final signal.

Box 4.1. Continued

so we must sample at least twice every 3×10^{-4} s—say once every 10^{-5} s. Finally, we shall choose eight digits (or eight 'bits') to represent each amplitude.

What, then, will our *digital* signal look like? Each digit will be represented by a voltage pulse in an electronic circuit (or along the telephone line) so, for example, the decimal amplitude 155 mV will take the form shown in Figure 4.1(c), a set of ones and zeroes in strict time sequence. It follows, then, that the overall signal will consist of an extended sequence of pulses such as shown in Figure 4.1(d), groups of eight bits (each group representing one amplitude) following one another in time. As we are sampling once every 10^{-5} s, the timescale for a single group must be no more than 10^{-5} s and the corresponding length of each individual pulse approximately 10^{-6} s. Immediately, we can see that the speed of an electronic device involved in processing digital information must be considerably faster than when handling the corresponding analogue signals—roughly 3 MHz compared with 3 kHz, in the case we consider here (3 MHz, rather than 1 MHz, because the device must respond to the rise and fall times of a digital pulse which are generally less than the pulse width, in order to form pulses with reasonably rectangular shapes).

Finally, let us briefly consider the problem of TDM switching (time division multiplexing) of digital signals. The idea here is to use only a minimum number of transmission lines when transmitting large numbers of simultaneous telephone conversations and, in order to do this, the individual digital signals are interleaved (in time sequence). We might, for instance, send the eight bits representing $A(t_1)$ for conversation '*a*', that is, $A_a(t_1)$, but before sending the next amplitude for conversation '*a*', we send, instead, the first amplitude for conversation '*b*', $A_b(t_1)$, then $A_c(t_1)$, $A_d(t_1)$, etc., before returning to $A_a(t_2)$, $A_b(t_2)$, etc. Various such sequences are possible but, whichever is to be chosen, there is an inevitable consequence—the individual pulses must be reduced in width by a factor equal to the number of simultaneous conversations to be transmitted. Thus, if we wish to send a hundred calls together down one line, the individual pulses need to be no longer than 10^{-8} s and the associated digital circuitry must then operate at speeds approaching 1 GHz (10^9 Hz). This makes clear why the British Post Office could not rely on solid state circuitry for TDM switching in 1956—the fastest Si transistors, still only in development, had cut-off frequencies in the region of 100 MHz.

The thoughtful reader may be wondering at this point why there is any need to use digital techniques at all when the consequences are so complicated! The reasons are two. First, digital signals are far more compatible with electronic computer circuits and, second they are far less prone to degradation by background (electronic) noise. In analogue computing, for example, the absolute magnitude of signal amplitude is significant (it is this which represents a number) and it is notoriously difficult to maintain it constant—the gain of an amplifier may change with temperature, for instance. Using digital representation, it is only necessary to be sure that a voltage (or current) pulse (a 'one') is present—its amplitude is of no consequence, provided it is large enough to be distinguished from a 'nought'. Similarly, if we think of an audio signal travelling down a telephone line, it is easy to imagine that the signal amplitude will gradually decay with distance and become more and more susceptible to noise interference. Again, the use of digital methods, requires only that the signal-to-noise ratio remains just large enough that the individual pulses can be recognized by a 'regenerating' circuit. Noise then has no effect on the clarity of the received signal.

decision to use digital techniques became generally accepted, this took on a more serious aspect because of the extra speed required for digital processing (see Box 4.1). Indeed, there was relatively little enthusiasm for the long-term future of such machines—a US survey at the end of the 1940s suggested that the likely *national* need might be satisfied by about a hundred digital computers! Such are the perils of technological forecasting! In mitigation, one must accept that, at that time, they were relatively expensive and ponderous instruments.

While the commercial and consumer markets for transistors and transistorized equipment were still in an uncertain state, there could be little doubt of the seriousness of US military interest. Much military electronic equipment had either to be portable, to be airborne, or to be attached to missiles where size, weight, and ruggedness were at a premium. The transistor therefore came as a heaven-sent opportunity to the military purchasing arm and, right from the word 'go', Government finance for transistor development was widely available—indeed, there was more than a hint to suggest that military backing kept the youthful transistor industry afloat during a large part of the 1950s. Something between 35% and 50% of all US annual semiconductor production was destined for military use during the period 1955–63 (Braun and Macdonald 1982: 80). (It should be remembered, too, that it was largely pressure from the military that led to the early demise of Ge in favour of Si as the preferred transistor material on the grounds of its much better resistance to thermal runaway.) Added to this came the decision by President Kennedy in 1961 to mount an intensive space programme, with the intention to 'put a man on the moon by 1970'. Once again, given the modest lifting capability of current US rockets, weight was a vital factor and all electronics must therefore be transistorized. Ruggedness and reliability, too, were better served by solid state devices than by the older, relatively fragile vacuum tubes. The European industry, though technically well advanced, received only a fraction of this level of support, and with inevitable consequences—competition with America was, at best, patchy and generally ineffective.

Nor was this state of affairs helped by some unfortunate technical planning. An unhappy example lies at the door of the British Post Office (then responsible for telecommunications as well as mail delivery, see Fransman 1995: 89–97). When it became clear, after the Second World War, that domestic and industrial demand for telephone services was soon likely to escalate, the Post Office, in 1956, took the bold decision dramatically to upgrade its telephone switching capabilities by leap-frogging from the rather ancient mechanical switching technology then in use to an advanced, *digital* 'time division multiplexed' (TDM) system, employing fast electronic switches. This was designed to bypass the more modest technology then being contemplated by most of their rivals, the cross-bar switching system and to give the United Kingdom an almost unassailable lead in this important field. It failed on account of the inadequacy of the components then available—a complete exchange was installed in Highgate Wood in 1962, only for it to succumb to excess heat from the 3000 thermionic valves employed (see Chapuis and Joel 1990: 62). At the time when the decision was made to go ahead, the transistor itself was far too uncertain a prospect (Ge devices were liable to thermal breakdown and Si had scarcely had time to assert itself—it was, in any case, rather slow for digital applications—see Box 4.1) so the choice of an old, well tried component technology was probably inevitable. (Even though this did contrast with the boldness of the overall project aims!) Success with similar TDM switching systems had, in fact, to wait until 1970 when suitable integrated circuits (IC) became available. What was worse from the UK industry viewpoint was the resulting attempt to salvage something from the ruins by reverting to the original mechanical switching technology, thus robbing the Post Office suppliers of the opportunity to develop intermediate switch technology, based on transistors and (as they became available) integrated circuits. It was a body blow for UK solid state device technology from which it never quite recovered.

These references to integrated circuits (ICs) serve to bring us back to our mainstream discussion of the development of solid state active devices, for it was the invention of the integrated circuit in 1958–9 which provided the jumping-off point for the real electronic revolution which still shows no sign of slowing. It was clear to many 'wizz kids' of the 1950s that the transistor had the potential for the development of large-scale, though compact, electronic circuits, and several attempts were made to facilitate progress in this direction. However, it soon became apparent that there was a limitation set by the necessary interconnections—all of which required individual attention with bonder or soldering iron—and several people began thinking of ways to overcome this. The first public proposal for integration has been credited to an Englishman, Geoffrey Dummer of the Royal Radar Establishment (RRE, as it then was) who presented a conference paper in Washington in May 1952, and who, by 1957, had persuaded the RRE management to fund a contract with the Plessey Company to build a flip-flop circuit based on his ideas. This resulted in a scale model which seems to have created considerable interest among American scientists but very little excitement within the United Kingdom! In fact, it was at TI in September 1958 that Jack Kilby first built an actual circuit in the form of a phase-shift oscillator. It used Ge, rather than Si because, at the time, Kilby could not lay hands on a suitable Si crystal and it employed external connecting wires individually bonded to the components but it demonstrated the use of the bulk Ge resistance to form resistors and a diffused *p-n* junction diode to provide capacitance—there was no need to add these functions by hanging discrete components onto the semiconductor circuit. As a demonstration of the integration principle, it may be likened to the point contact transistor—a huge step forward but some way from commercial viability.

The practical breakthrough came from Fairchild Semiconductors in the following year, in the form of a patent application by Robert Noyce claiming a method of making an integrated circuit using the Si planar process and forming the necessary interconnections by evaporating metallic films and defining them by photolithography. This was surely the practical way to go but it was nearly 2 years (March 1961) before Fairchild made their first working circuits based on these principles, closely followed by Texas in October 1961. These two companies were serious rivals, not only with regard to IC manufacture—a titanic patent battle also ensued over the question of priority in the basic invention (see the stimulating account given in Reid 2001). It took nearly 11 years of legal jousting before the Court of Customs and Patents Appeals finally adjudicated in favour of Fairchild—Robert Noyce was officially declared the inventor of the microchip! Not that it mattered very much—by that time the world of chips had moved on to such a degree that the issue had become of little more than academic interest and, in any case, the two protagonists Kilby and Noyce were, on a personal basis, more than happy to share the credit. In the year 2000, Kilby was awarded a half share in the Nobel prize and, doubtless Noyce would have joined him had he not died some 10 years earlier. That it should have taken the Nobel Committee more than 40 years to acknowledge a technical development of this magnitude must be seen as both remarkable in itself and sad in the extreme in that it prevented Noyce from receiving his rightful share of the honour.

So prodigious have been the ramifications of their invention that one is somewhat taken aback to learn of the initial lack of interest shown by equipment manufacturers in these early circuits. The problem was that they were too expensive—it was actually cheaper to build the same circuit from individual components, hard-wired together, than to buy the appropriate integrated version from TI or Fairchild. Sales were minimal. Stalemate! That was until May 1961 when President Kennedy threw down his famous challenge that America should put a man on the moon by the end of the decade. Almost immediately it became clear that the required rocket guidance would demand highly sophisticated computer technology and that such advanced circuitry could only be realized in integrated form. Hang the expense–this was the only way to go! Such a dramatic kick-start to a technological revolution smacked of divine intervention by a Higher Being with an unfair interest in the fledgling US chip industry—certainly no other country ever received a comparable boost. The result was demonstrated by the number of ICs sold: in 1963 the number was a mere 500,000, by 1966 it had risen to 32 million.

Government spending may have been the vital stimulus but the importance of diversification was quickly appreciated. Jack Kilby was put to work at TI to develop a revolutionary consumer product in the shape of a pocket calculator which appeared in 1971. No fewer than 5 million calculators were sold in 1972. At the same time the digital watch made its appearance and took the consumer market by storm.

Ted Hoff of Intel developed the first microprocessor also in 1971 and the first personal computer (PC) followed in 1975 in the form of a Popular Electronics kit! The revolution was well and truly launched and the industry has hardly cast a backward glance. Progress in increasing complexity of integrated circuits has shown a quite remarkable steadiness—in 1965 Gordon Moore (a physical chemist working in Noyce's group at Fairchild) made his famous pronouncement which came to be known as 'Moore's Law', that the number of components on an IC would continue to double every year and such has almost been the case. A careful examination of the data up to 1997 suggests that the annual increase is actually closer to a factor of about 1.6 but the really striking feature is its long-term consistency, encouraging confident prediction for future increases, at least as far as the end of the first decade of the new millennium.

Solid state circuitry has gone from 'small scale integration' (SSI, up to 20 'gates') in the 1960s to 'medium scale integration' (MSI, 20–200 gates) at the end of the 1960s through 'large scale integration' (LSI, 200–5000 gates) in the 1970s to 'very large scale integration (VLSI, 5000–100,000 gates) in the 1980s and what might be called 'ultra large scale integration' (ULSI, 100,000–10 million gates) by the end of the 1990s. Moore, himself, continued to play a role in these developments—together with Robert Noyce, he left Fairchild in 1966 to found Intel whose sales rose from $2700 in 1968 to $60 million in 1973 and in the year 2000 to $32 billion! The basis of this performance has, of course, been the steady *decrease* in size of the component transistors and we shall look in more detail at this anon. However, we must first backtrack to examine another important breakthrough, the development, at last, of a real field effect transistor (FET).

4.2 The Metal Oxide Silicon transistor

The Metal Oxide Silicon (MOS) transistor was yet another product of the fertile ground cultivated by Bell Telephone Laboratories and, once again, it involved just a small element of good fortune. The critical step in its invention was the (accidental!) discovery that the Si surface can be oxidized to form a highly stable insulating film which possesses excellent interface qualities (i.e. the interface between the oxide layer and the underlying silicon). We have already commented on the importance of this interface in passivating Si planar transistors which, in turn, led to the *practical* realization of integrated circuits. The further application in the metal oxide silicon field effect transistor (MOSFET) turned out to be a singularly important bonus.

We saw in the previous chapter that the quest for a FET (which would function in a manner closely parallel to that of the thermionic valve) had already acquired something of a history. It was a patent awarded in 1930 to a Polish physicist, Julius Lilienfeld (who emigrated to America in 1926), that thwarted William Shockley in his original attempt to patent such a device but, even more frustratingly, the existence of high densities of surface states on Ge and Si which prevented the Bell scientists from actually making one. Even though the application of a voltage to a 'gate' electrode may have been successful in inducing a high density of electrons in the semiconductor region beneath it, these electrons were not free to influence the semiconductor's conductivity because they were trapped in surface (or interface) states. What was needed was a surface (or, more probably, an interface) characterized by a low density of these trapping states (of order 10^{15} m^{-2} or less) but, at the time, no one knew how to produce it. Brattain and Bardeen had continued to study the problem of surface states until 1955, 8 years after their invention of the point contact transistor, but it was not until 1958 that another Bell Group under 'John' Atalla discovered the low density of states associated with a suitably oxidized Si surface. It was necessary to clean the Si carefully before oxidizing it in a stream of oxygen at a temperature in the region of 1000°C, the addition of a small amount of water vapour apparently making the process more readily controllable. The films were amorphous (i.e. non-crystalline), uniform, free of pinholes and appeared to consist of silicon dioxide, SiO_2. They behaved as good dielectrics (resistivity of order 10^{14} Ω m) with a dielectric constant of about 4 and breakdown fields close to 10^9 Vm^{-1} (meaning that voltages up to about 100 V could be applied across an oxide film 100 nm thick before significant current flowed through the oxide). They used two methods for characterizing the interface, first, performing field effect' measurements (semiconductor conductivity as a function of gate voltage) and second measuring the current–voltage behaviour of a *p-n* junction passivated by the oxide. These both confirmed the much reduced density of interface states, though the field effect allowed more detailed analysis and showed the existence of several different intergap states, having both donor and acceptor character. However, the key result was that their densities were below the above limit of 10^{15} m^{-2}, thus making possible the development of a practical FET. This was finally achieved at Murray Hill in 1960. Within a very few years RCA pioneered the introduction of MOS devices into integrated circuits and this technology rapidly came to dominate that of bipolar (e.g. *n-p-n*) devices in many applications.

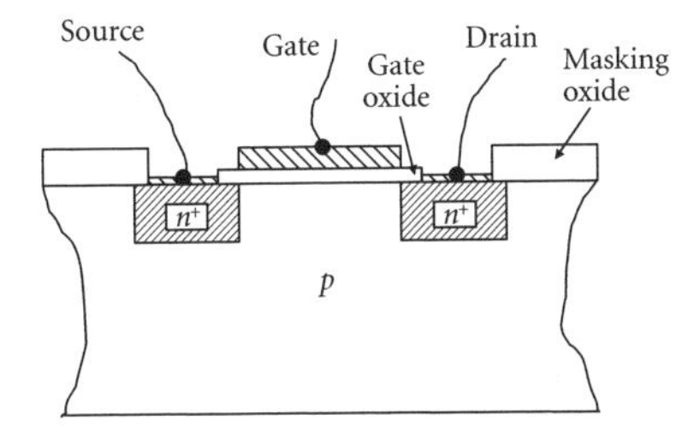

Figure 4.2. Schematic diagram of an N-MOSFET structure. A wafer of *p*-type silicon has two n^+ regions diffused into it, through an oxide mask, to form the source and drain of the MOSFET. These regions are contacted by evaporated aluminium films, again defined by oxide masks. The gate electrode is deposited on top of a thin gate oxide layer.

Figure 4.2 shows a schematic diagram of the structure of such a planar MOSFET. It is made by diffusing two *n*-type contact regions into a sample of *p*-type Si, using oxide masking to define their positions. A gate oxide layer about 100 nm thick is deposited over the region between the contacts and appropriate gate and contact metals are evaporated, again using oxide masks for their definition. (We shall examine planar technology in greater detail in the next section.) In the absence of a gate voltage, the connection between the two contacts, known as 'source' and 'drain' contacts, consists of two *p-n* junctions, arranged in a back-to-back configuration so that, when a voltage is applied between source and drain, one of these must inevitably be reverse biased and therefore very little source–drain current can flow. If, on the other hand, a suitable positive gate voltage is applied (between gate and source), this will forward bias the source n^+-p junction and allow electrons to flow into the region beneath the gate, thus inducing a sheet of electrons beneath the gate which forms a conducting *n*-type 'channel' between source and drain. For this reason, the device is called an NMOS transistor (it is also possible to make a PMOS device using *p*-type contacts diffused into an *n*-type substrate). As a result of applying the gate voltage, the channel region has effectively been converted from *p*-type to *n*-type conduction, a process which is known as 'inversion', the channel itself often being referred to as an 'inversion layer'. We shall examine the details of this inversion process later but, for the moment, it is sufficient to appreciate that the source–drain conductance can be controlled by varying the gate voltage V_G—the more positive we make the gate, the greater the conductance and, therefore, the larger the source–drain current I_{SD} which flows for any particular source–drain voltage V_{SD}. As we suggested above, this control is similar to that effected by applying a voltage to the 'grid' electrode in a thermionic triode valve and certain parallels can therefore be drawn between valve and MOSFET circuits.

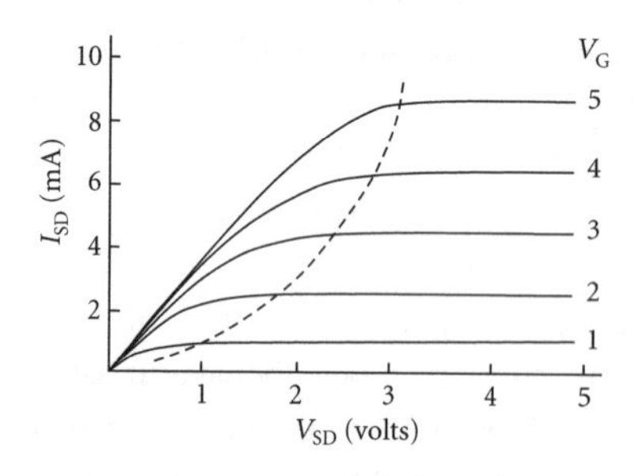

Figure 4.3. Example of MOSFET output characteristics, that is, source–drain current vs source–drain voltage, with gate voltage as parameter. Current saturation occurs at gradually increasing source–drain voltage as the gate voltage increases (indicated by the dashed curve). Note that the increments of saturation current are not linear in gate voltage.

Extending these ideas a little allows us to draw the so-called 'output characteristics' of an *n*-channel MOSFET, that is the I_{SD} vs V_{SD} curves with V_G as parameter. An example is shown in Figure 4.3. As we should expect, the source–drain current increases approximately linearly with source–drain voltage at first but, perhaps surprisingly, shows saturation behaviour at larger voltages. This effect arises because both drain voltage and gate voltage are applied with respect to the source contact so that, in the region of the channel close to the drain, the drain voltage effectively opposes the gate voltage. Thus, at large values of V_{SD}, the effective gate voltage is sufficiently reduced that the channel region is no longer inverted and this region forms a reverse biased junction which effectively prevents any further increase in current. As will be clear from this argument, the larger the gate voltage, the larger the value of V_{SD} at which saturation sets in, an obvious feature of the curves in Figure 4.3, as indicated by the dashed line.

A virtue of these curves is that we can use them to learn how the MOSFET may be used in electronic circuits and the first point to appreciate is that these may take the form of either linear 'analogue' circuits or digital 'switching' circuits. It is probably easier to understand the latter application so we shall begin with this. As already explained in

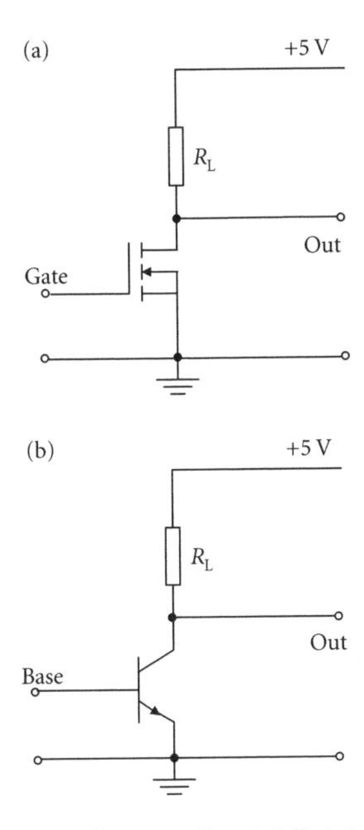

Figure 4.4. Schematic of typical digital switches, using (a) a MOSFET and (b) a bipolar transistor. The output voltage switches between 0 and +5 Volts according to whether the input is at +5 Volts or 0 V. These circuits have a superficial similarity, though the one is voltage controlled while the other is current controlled.

Box 4.1, digital signal processing (which is fundamental to present-day-computing and information transfer) depends on the use of short voltage (or current) pulses which are generated and moved around an array of electronic circuits in incredibly complicated fashion but the basis is, nevertheless, simple. At any point in the circuit, 'information' is represented by the presence (digital '1') or absence (digital '0') of a pulse voltage. Typically its amplitude is about 5 V but the exact value is less important than the ability of monitoring circuitry to determine, *with a high degree of certainty*, that the pulse is either present or absent. The circuit element which controls this condition is nothing more complicated than an on/off switch—though it must be operated electronically and be capable of switching on in a very short timescale, typically 1 ns or less. The MOSFET is ideal for this, as illustrated in the circuit of Figure 4.4(a) where the switching operation is controlled by the voltage applied to the transistor gate electrode. If $V_G = 0$, the MOSFET channel is non-conducting (i.e. 'off') so no current flows through it from the 5 V supply to 'ground'. This implies that no voltage is lost across the load resistor and the output terminal is at a potential of 5 V. On the other hand, if the gate voltage is raised to +5 V, the transistor channel is switched fully 'on', current flows through the load resistor, nearly all the supply voltage is dropped across R_L and the output terminal is at a potential very close to zero. (Another way of looking at it is to regard the conducting channel as short-circuiting the output so that the output potential is very close to ground potential.) The switching time is determined by the time taken to charge the gate capacitance C_G through the load resistor R_L ($t = R_L C_G$) which we estimate in Box 4.2 for a modern transistor as about 10^{-11} s, short enough for most purposes. In practice, the load resistor in an integrated circuit takes the form of a second MOSFET with its gate connected directly to its drain which means it will always be in the 'on' state so the channel acts as a pure resistor. This technique is convenient because it is easy to make this second MOSFET using the same processing steps as for the first and it takes up only the same amount of room or, as the jargon has it, 'real estate'.

For comparison, Figure 4.4(b) shows a similar circuit using a bipolar transistor as switching element. In this circuit, when the input is zero, the emitter junction is reverse biased by the voltage V_B, the collector current is therefore very small and the output voltage is close to 5 V. If an input voltage of 5 V is applied, the emitter junction is now forward biased, minority carriers are injected across the base and current flows in the collector circuit. This causes voltage to be dropped across R_L so the output voltage drops to a value close to zero. We shall not discuss these circuits in any more detail—our purpose is merely to indicate how transistors may be used as basic switching elements in digital circuitry. How these are combined to provide logic or memory functions is

Box 4.2. Switching speed of MOSFET circuits

In estimating the speed with which a MOSFET circuit may be switched from the off state to the on state we must consider the physical difference between the two states. In the off state, there are no electrons in the channel—in the on state there is a sheet density n_s (sheet charge density $Q_s = en_s$). These electrons are attracted to the channel by the application of a voltage V_G to the gate but they must be supplied from the drain voltage supply and must therefore flow through the load resistor R_L (see Figure 4.4(a)). There are two possible limitations to the time required, first the transit time t_T for electrons to travel the length of the channel region and, second the RC charging time t_C. We can estimate these times from a knowledge of the transistor dimensions and electron mobility in the channel.

The transit time is given by:

$$t_T = L/v = L/\mu E = L^2/\mu V_{SD} \tag{B4.1}$$

where L is the channel length, v the electron velocity, E the electric field in the channel, μ the electron mobility and V_{SD} the source–drain voltage. For a transistor with gate length $L = 1$ μm, $V_{SD} = 10$ V and electron mobility $\mu = 0.1\ m^2 V^{-1} s^{-1}$ we obtain $t_T = 10^{-12}$ s.

The charging time t_C is given by:

$$t_C = R_L C_G = R_L(\varepsilon\varepsilon_0/d)LW \tag{B4.2}$$

where ε is the relative dielectric constant of the oxide, ε_0 the permitivity of free space, d the thickness of the oxide film, and W the width of the gate electrode. Taking $W = 10$ μm, $d = 100$ nm, $\varepsilon = 4$, and $R_L = 3$ kΩ, we obtain $t_C = 10^{-11}$ s, showing that this is the limiting factor, rather than the transit time. It, nevertheless, represents a very acceptable switching time for a practical switch.

Let us now consider the more usual case in an integrated circuit when the load resistor is provided by a second MOSFET which is permanently in the on state. We shall assume that both transistors, in the on state, have the same effective gate voltage and therefore the same sheet charge density Q_s. The load resistance R_L, in this case, is given by:

$$R_L = \rho L/Wt = L/ne\mu Wt = L/n_s e\mu W = L/Q_s\mu W, \tag{B4.3}$$

where ρ is the effective resistivity in the channel and t the effective thickness of the channel.

Noting that the gate capacitance C_G is given by $Q/V_G = Q_s LW/V_G$, we can write the expression for the charging time as:

$$t_C = R_L C_G = (L/Q_s\mu W).(Q_s LW/V_G) = L^2/\mu V_G \tag{B4.4}$$

Evaluating this again for a gate length of 1 μm with $V_G = 5$ V, gives $t_C = 2 \times 10^{-12}$ s, implying a somewhat smaller value of load resistance than we assumed above. Note that t_C varies as the square of the gate length which illustrates the importance of reducing L in order to achieve fast switching times.

discussed in many textbooks but it would take us too far from our dedicated purpose to repeat these arguments here. What we *are* intent upon, however, is to complete our brief discussion of transistor applications by referring to their use in analogue circuits.

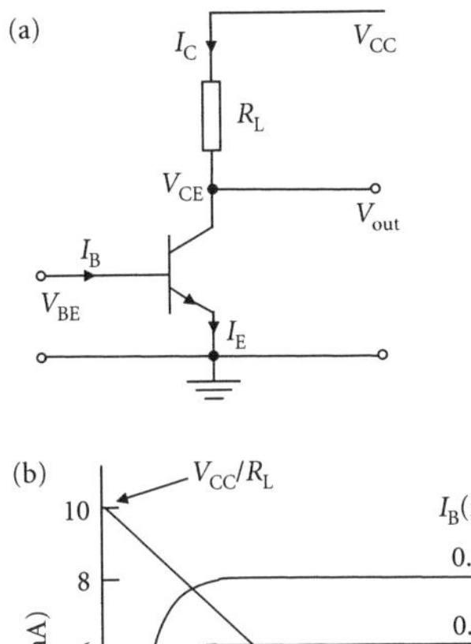

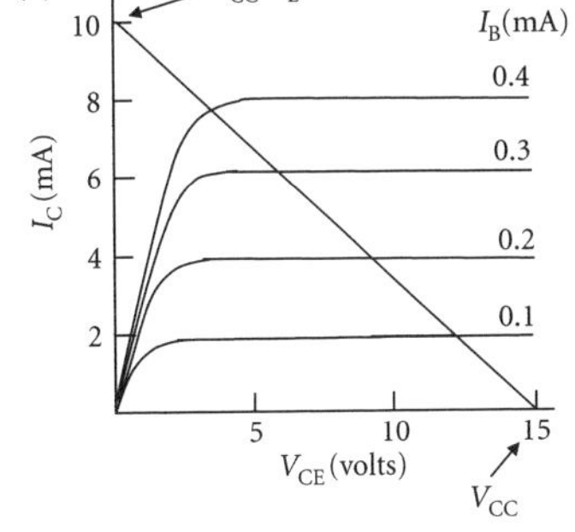

Figure 4.5. Circuit to illustrate the use of a bipolar transistor as a linear amplifier. The output voltage follows changes in base voltage with a gain in amplitude of approximately 300 times. In (a) the circuit is shown, while (b) represents typical output characteristics on which have been superimposed a load line to represent the effect of the load resistor R_L ($R_L = 1.5$ kΩ).

First we consider the use of the bipolar transistor as a linear amplifier. Figure 4.5(a) illustrates a typical circuit, while Figure 4.5(b) introduces the concept of the so-called 'load line', a line representing the effect of a load resistor R_L superimposed on a set of output characteristics (in this case, collector current I_C vs collector voltage V_{CE} with base current I_B as parameter). Two points should be noted concerning these curves: first, they show a saturation behaviour similar to that found in the MOS characteristics, though for a quite different reason—collector current is limited by the minority carrier current injected across the emitter–base junction—and second, the values of saturation current scale directly with base current—this is a consequence of the diffusion process (see Box 3.2). A typical value for the ratio $\beta = I_C/I_B$ is about 40. Now, it follows from the fact that the collector current flows through R_L that we can write a simple equation which relates the collector voltage to the supply voltage V_{CC}:

$$V_{CE} = V_{CC} - I_C R_L \tag{4.1}$$

and this equation can be represented by the load line drawn in Figure 4.5(b) (in this case $R_L = 1.5$ kΩ). Because V_{CC} and R_L are constant, it is clear from this that, if the base current changes as a result of applying a signal at the base of amplitude ΔV_{BE}, this will cause a change ΔI_C in collector current and a corresponding change in collector voltage ΔV_{CE}, which is given by:

$$\Delta V_{CE} = -\Delta I_C R_L = -\beta \Delta I_B R_L. \tag{4.2}$$

This equation expresses mathematically the relation between ΔV_{CE} and ΔI_C which can be seen intuitively by examination of the load line in Figure 4.5(b).

Finally, in order to arrive at an expression for the voltage gain of the circuit we need to examine the relation between base current and base voltage—in other words, we need to derive a value for the ratio $\Delta I_B/\Delta V_{BE}$. This is straightforward when we remember that it depends simply on the current–voltage relation for the emitter–base junction:

$$I_E = I_0 \exp\{eV_{BE}/kT\} \tag{4.3}$$

from which it follows that:

$$\Delta I_E = (e/kT) I_E \Delta V_{BE} \tag{4.4}$$

and, similarly:

$$\Delta I_B = (e/kT) I_B \Delta V_{BE}. \tag{4.5}$$

We can now substitute equation (4.5) into (4.2) and obtain:

$$\Delta V_{CE} = -R_L(e/kT)I_C\Delta V_{BE}. \tag{4.6}$$

So the voltage gain can be seen to be:

$$G = (e/kT)R_L I_C \sim 40R_L I_C. \tag{4.7}$$

(Note that the factor 40 which appears in equation (4.7) represents the value of (e/kT) at room temperature and should not be confused with the current gain factor β which coincidentally happens to take a similar value.)

This relation shows that, once the transistor has been biased at a specific value of collector current, the gain is simply proportional to the value of load resistor selected. Typically we may have $I_C = 5$ mA and $R_L = 1.5$ kΩ, giving $G = 300$. It is important to appreciate that the linear gain provided by such a circuit follows from the linear relation between I_C and I_B—in Figure 4.5(b) the fact that the curves for various values of I_B are equally spaced indicates this.

To see how a MOSFET may be used as a linear amplifier, we refer again to Figures 4.3 and 4.4(a). Suppose the transistor is biased into its saturation regime by an appropriate gate voltage. If, then, a small alternating signal voltage is superimposed on top of the gate bias, it will have the effect of swinging the drain current up and down by a small amount and this will result in a corresponding voltage drop across the load resistor. Thus, the voltage at the output terminal will also vary in sympathy, representing an amplified version of the input signal. It is straightforward to use a load line to describe the behaviour of the MOS amplifier but there is an important difference when compared with the bipolar amplifier described above. As is apparent in Figure 4.3, the characteristic curves are *not* equally spaced for equal increments in gate voltage. In fact, the relationship between drain current and gate voltage takes the form (see Sze 1969: ch. 10):

$$I_{SD} = (W/2L)\mu_n C_{ox} V_G{}^2 \tag{4.8}$$

where W and L are the width and length of the gate, μ_n is the electron mobility in the channel (assuming an NMOS device), and C_{ox} is the oxide capacitance per unit area. This square law dependence of saturation drain current on gate voltage means that the device is essentially non-linear—it follows that the variation of output voltage is also non-linear. However, if the swing in gate voltage is small with respect to its steady bias level, that is, $\Delta V_G \ll V_G$, it is easy to see that an approximately linear response may be obtained. Thus, if we write $V_{out} = KV_G{}^2$:

$$\begin{aligned} V_{out} + \Delta V_{out} &= K(V_G + \Delta V_G)^2 \\ &\approx K(V_G{}^2 + 2V_G\Delta V_G). \end{aligned} \tag{4.9}$$

And, since $V_{out} = KV_G^2$, we have:

$$\Delta V_{out} \approx 2KV_G \Delta V_G. \tag{4.10}$$

The MOSFET, in common with other types of FET may, therefore, be used as a linear amplifier only under very small signal conditions such as, for example, in a microwave receiver or when detecting weak light signals with a photodetector.

In summary, then, we see that, by the early 1960s, the two principal active devices, bipolar and MOS transistors had become available to the electronic engineer and the story from this point is one of continuing miniaturization to improve speed and packing density in IC design and, on the other hand, the development of large-scale devices with large voltage-handling capability for use in high power applications. Which device to use in which application depended, of course, on the specification required. In general, MOSFETs have an advantage in IC design on account of their lower power dissipation and modest demands on silicon area, though bipolar devices are capable of faster switching speeds at the expense of more power dissipation and greater demand on space. The dissipation advantage inherent in the use of MOS devices was further enhanced in the late 1960s by the development of Complementary MOS (CMOS) circuitry in which each switching element takes the form of a pair of transistors, one NMOS and one PMOS, the important feature being that power is dissipated only when the switch operates—in the quiescent state (whether storing a digital 0 or 1), no current flows.

We have seen already that the technology for making devices having better performance and greater ease of manufacture can be crucial to commercial development and this became more, rather than less, significant as time wore on and competition became fiercer. It is time for us now to look in more detail at the semiconductor technology which has played, and continues to play such a vital role.

4.3 Semiconductor technology

In one sense (the commercial sense) this section is the most important in the book! The reader should already be persuaded of the important part played by well controlled semiconductor materials in the development of transistors and integrated circuits. Without high-quality germanium the transistor could never have been discovered and without high-quality silicon the integrated circuit would still be a mere concept. However, even greater importance attaches to the role played by technology. Without the amazing skills built up by semiconductor technologists we might still be trying to wire together crude individual transistors on printed circuit boards, rather than linking powerful integrated circuit chips to build fast computers with almost unimaginable amounts of memory. Needless to say, the subject of semiconductor technology is a vast one and it would be presumptuous of me to suppose that this short section could do more than present the sketchiest of outlines—nevertheless, *some* impression is vital to the appreciation of what has been achieved during the past half-century, let alone what may yet be achieved during the next! So I am certainly obliged to try. This section, then, is concerned with the techniques employed in making silicon integrated circuits and with the development of those techniques in response to the never ending quest for faster processing of ever increasing quantities of information. It started with kilobytes at sub-megahertz frequencies, has now reached gigabytes at gigahertz, and will certainly progress into yet unknown territories of prodigality. It would be a rash prophet who ventured an opinion concerning the ultimate limit, though there is every likelihood of some major perturbation in method—the present technologies certainly *appear* to have limits. Here, however, (as elsewhere) I am concerned rather with history than futurology.

Perhaps the best place to start is with an attempt to list the requirements. What do we actually need in order to make a typical 'chip'? In the sense that it represents an electronic circuit, we clearly need gain, resistance, capacitance, and inductance—transistors, resistors, capacitors, and inductors. To such ends, we need high quality single crystal silicon, doped with a suitable level of donor or acceptor impurities, together with means for modifying its conductivity in specific, well-defined regions and for interconnecting these regions in intelligent patterns, while isolating them from undesired random connections. It turns out that resistance can be provided by, for example, a MOSFET channel, as described above, that capacitance is available in the form of the depletion capacitance of a *p-n* diode, and that inductance can be cobbled together by suitable combinations of these other elements so the most complicated individual device is the transistor, itself, together with appropriate isolation and interconnection. It also follows from the interconnection requirement that a planar technology, where all components are adjacent to the wafer surface, yields important practical advantages. In particular, it becomes possible to make connections in the form of evaporated metal films, defined as thin wires, certain of which may be taken to the edge of the chip and attached to contact pads to facilitate connection with the outside world. All this begs the question of how the individual transistors are fabricated and placed in the desired position on the slice. We have already seen (Figure 3.21) how diffusion of donors and acceptors may be used to form base and collector regions of a bipolar transistor and similar (though somewhat simpler) processing can obviously be applied to MOSFET fabrication but this still leaves unanswered the question of how to define their precise positions. This step, known as 'photolithography', probably representing the most important single contribution to the technology, originated in the printing industry and was adapted for microelectronic applications by a number of American companies such as Bell, TI, and Fairchild at the beginning of the 1960s. As an aid to understanding it, we refer to the earlier process for making mesa transistors, depending on selective etching to form the local bumps on the semiconductor surface which defined the active device area. First, a circular film of wax would be deposited through a metal mask and allowed to harden. This was then used as a barrier to prevent chemical etching of the mesa while the rest of the semiconductor surface was etched away. Photolithography (see Box 4.3 for more detail) employs the more refined technique of depositing a thick film of 'photoresist' which can be locally modified by shining ultraviolet (UV) light on it through a photographic mask to define the desired shapes. The illuminated regions (or, sometimes the unilluminated regions, depending on the nature of the photoresist used) are then washed away in a suitable solvent to leave an appropriate pattern of hardened resist on the semiconductor surface which may be used to define any subsequent process. These are the bare bones—to add the necessary flesh we shall now outline the overall process of making a typical integrated circuit. If this achieves nothing else, it should certainly make clear to the reader what an involved business it is and perhaps generate an appropriate degree of amazement at the efficiency with which present day practitioners thread their way through its complexities.

The starting point is a uniform cylindrical boule of Czockralski-grown silicon, 6 in. in diameter (say) which is first 'cropped' to remove atypical end material, then ground to the correct diameter and provided with 'flats' to mark appropriate crystallographic planes and etched to remove grinding damage. (Damage penetrates a few tens of microns below the ground surface and this material must be removed because its electrical properties are seriously degraded.) From this prepared boule, 'wafers' (i.e. slices normal to the boule axis) are cut with an annular diamond saw, some 100 μm thick, a process referred to as 'wafering'. The slices are then heated to a temperature in the range 500–800°C and rapidly cooled to room temperature for the purpose of annihilating oxygen donors. Oxygen is usually present at levels of about 10^{22} m^{-3} in Czochralski-grown material and it is vital that it does not form donors—the background doping level must be of order $p = 10^{21}$ m^{-3} for a typical NMOS circuit so the donor density should be at least an order less than this. Next, the wafers are 'edge rounded' on a diamond wheel to facilitate the spreading of photoresist (at a later stage) and then 'lapped' on a polishing machine to ensure wafer flatness (which is of vital importance for accurate photolithography). They must then be etched to remove damage, polished, and cleaned (mechanically and chemically) before being marked for future identification. They are

Box 4.3. Photolithography

Photolithography is the process whereby a desired pattern is transferred from a mask to the semiconductor slice. In particular, it is necessary, in fabricating a circuit, to make holes in various barrier layers through which will be diffused donor or acceptor impurities or via which electrical contact will be effected by the evaporation of a suitable metal. Depending on the application, the barrier may be SiO_2, Si_3N_4, photoresist, polysilicon, or metal but, in order to be specific, we shall assume it to be SiO_2 and we shall suppose that the immediate objective is to diffuse, within a bulk *p*-type region, a rectangular stripe of *n*-type Si suitable for the source or drain of an NMOS transistor. Figure 4.6 shows the various process steps, Figure 4.6(a) representing the oxidized silicon preparatory to starting.

The first step consists of 'spinning' on a film of photoresist to a thickness in the range 0.5–2.0 μm. The silicon slice is mounted on a vacuum chuck and spun at high speed (typically 1000–5000 rpm) and a drop of resist, dissolved in a suitable solvent, is allowed to fall on it, being spread to a thin layer by centrifugal force. Final thickness depends on the spin time, the viscosity of the resist solution and on spinning speed. The film is lightly baked (at about 90°C) to drive off solvent and enhance its adhesion to the SiO_2 and it is then ready for exposure. Intense UV light is incident on the surface through an appropriate photomask which defines the shape of the exposed region—in Figure 4.6(c) this is a simple rectangle but it could obviously take much more complex shapes if required. Assuming that we are using a 'positive' resist, the region which has been irradiated can readily be washed away with a 'resist developer' (a proprietary solvent supplied with the resist), leaving behind the unirradiated region. As can be seen in Figure 4.6(d) this results in the formation of a rectangular hole in the resist and it is now necessary to continue this hole through to the silicon surface by dissolving the oxide in hydrofluoric acid (HF). However, before this can be done, the resist must be hardened by a second baking step (typically 30 min at 150°C) which makes it resistant to the etch, thus preventing the HF from dissolving the oxide elsewhere but within the hole. Having thus defined the hole in the oxide, the resist is removed, either with a suitable chemical (again supplied with the resist) or by burning it off in a 'plasma asher'. The final step in the process is to diffuse phosphorus into the silicon at a temperature of about 1000°C so as to form a rectangular region of *n*-type material within the bulk *p*-type Si. (Clearly, in making a practical transistor, both source and drain regions would be made together.)

To appreciate one further subtlety in the photolithographic process, we now observe that it will be necessary to perform similar operations to define the MOS gate and the reader will appreciate that this requires not only a different shape of mask but also that the new shape must be accurately aligned with respect to source and drain. In the case of the fine geometries used in a modern IC, where the source–drain separation may be less than a micron, this alignment makes considerable demands on the skill of the operator. It is achieved by using 'alignment markers' on the photomasks which are viewed through a high resolution optical microscope. The first mask defines suitable marks on the wafer which can then

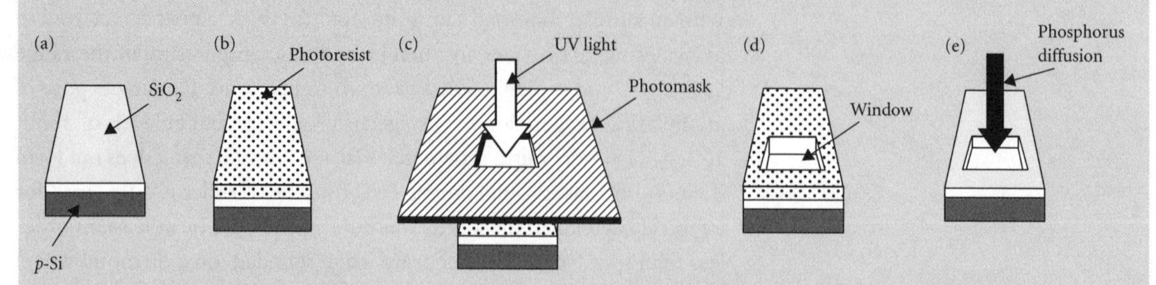

Figure 4.6. Sequence of steps to produce the source region in a typical NMOS transistor. In (a) the *p*-type silicon wafer has been covered with a thick film of SiO_2. In (b) this is further covered with a spun layer of photoresist. In (c) the photoresist is irradiated with an intense beam of UV light through a photographic mask. In (d) the resist has been dissolved away to expose the oxide which has then been etched away in hydrofluoric acid (HF). In (e) the resist has been removed and the oxide mask is being used to define the source region which is made by diffusing phosphorus through the opening at high temperature.

Box 4.3. Continued

be aligned with corresponding marks on each subsequent mask. In practice, a typical MOS technology may involve perhaps five masks (and bipolar transistors even more) so the alignment must be maintained at all stages to an accuracy significantly better than the minimum feature size in the circuit. It also goes without saying that dust particles cannot be tolerated in the processing area as just one such particle can ruin the function of any particular chip on which it may choose to alight. Ultraclean conditions are therefore essential and make considerable demands on the design and implementation of processing facilities (not to mention the budget which must provide them!).

This, then, represents a brief outline of the process but one major aspect has yet to be considered. It is clear that the viability of the whole business depends crucially on the accuracy of the photomasks available so we should not close this particular box without some reference to their manufacture. In the early days of integrated circuits, masks were made by photographic reduction from a large-scale model cut mechanically from a plastic sheet. The reduced mask was then replicated photographically to produce the multiple mask set needed for a whole silicon slice (remember there may be 500 chips on a single slice) by a 'step-and-repeat' process. Nowadays, the ever-decreasing minimum feature size has resulted in the basic mask being made by an electron beam lithography process which provides sub-micron resolution, the mask, in this case being made in a thin film of chromium metal, rather than photographic film. Its application follows the same pattern as before, except that it is no longer possible to use the photographic step-and-repeat method and the mask must be stepped mechanically over the wafer to satisfy the multi-chip requirement.

now ready for use. There follows a further sequence of processes designed to incorporate the active and passive components which go to make up a large number of identical integrated circuit chips. A single chip will typically be a few millimetres square, there being roughly 500 chips on a 6-in. wafer.

One final point to emphasize is that, here at least, size *is* important—because a silicon wafer is processed *in toto*, the more chips there are on the wafer, the more cost effective is their production, the appreciation of which has stimulated a steady increase in wafer diameter from the early value of 2 in. to the present 6 or 8 in., with a truly macho 12 in. in development. It should not be thought, however, that such enhancements can be lightly regarded—each quantum jump in wafer size requires a corresponding increase in handling capacity at each processing step and the inherent jump in investment level can only be afforded by a relatively few global companies.

Subsequent processes are designed to produce planar structures which perform the complex functions of memory, logic, or analogue signal processing—they are many and various but consist essentially of repetitious use of a number of basic steps. Those most commonly employed are: oxidation, chemical vapour deposition (CVD), thermal evaporation, sputtering, photolithography, diffusion, ion implantation, and etching, intercalated with frequent cleaning sequences. In certain cases, epitaxy (see Box 1.1) is also employed to deposit a crystalline silicon film. Apart from silicon, the commonly used materials are SiO_2, Si_3N_4, polyimide which act as barriers and insulators, polysilicon, various silicides, and metals (mainly aluminium) which act as conductors. It would be quite impossible for us to examine the intricate detail associated with real-life circuits—instead, we shall compromise by looking at a simplified process for fabricating a pair of MOS transistors for use in a CMOS circuit. This example, taken from McCanny and White (1987: 71), should be adequate to give a flavour of what is involved—the reader interested in greater detail can consult the references provided in the bibliography.

The essential features of the process are illustrated in Figure 4.7. We start with an oxidized *n*-type silicon wafer, the oxide to be used as a diffusion mask (Figure 4.7(a)). The first step involves defining a *p*-type region, within which will be formed the NMOS device, so this requires that a window be opened in the oxide by use of photoresist (see Box 4.3) and boron diffused through it at a temperature of about 1000°C in a quartz tube. In practice B is supplied at the Si surface in the form of boron trioxide B_2O_3 which is deposited from the reaction of diborane B_2H_6 with oxygen in a stream of an inert gas such as argon. (Diborane is highly poisonous and, for good measure, is also explosive so is almost always diluted with argon or nitrogen.) Following this initial diffusion, a second, thicker oxide layer is grown, during which the boron acceptors diffuse more deeply into the wafer to form a *p*-type well which isolates the NMOS device from the *n*-type substrate (Figure 4.7(b)). This new oxide is then used to define the source and drain regions for the two MOS transistors, but, because they involve opposite doping type (*n*-type for the NMOS and *p*-type for the PMOS), these diffusions must be performed separately. First, small windows are opened and B diffused to make the PMOS contact regions (Figure 4.7(c)), then the oxide is regrown and similar windows are opened for the *n*-type diffusions which define the NMOS source and drain regions. This diffusion follows similar procedures to the boron diffusion but uses phosphorus pentoxide as source of phosphorus donors, the P_2O_5 being, in turn, obtained from the reaction of phosphine gas (PH_3) with oxygen. Again, phosphine is highly toxic and explosive so similar dilution with argon is a vital necessity. Having defined the sources and drains, we now wish to form the gate oxide which is much thinner than the masking oxides used previously (typically 100 nm, as opposed to 1 μm). This oxide is crucial to the operation of the transistors and must be grown under carefully controlled conditions, typically using pure dry oxygen at a temperature near 900°C, whereas masking oxides which require a faster growth rate use wet oxygen at temperatures between 1000°C and 1200°C. The wafer has now reached the state represented in Figure 4.7(d). Next, contact windows are opened and metal contacts formed by thermal evaporation of aluminium in a vacuum chamber at background pressure of 10^{-6} torr ($\sim 10^{-4}$ Pa). Al is melted on an electrically heated filament and then evaporated to deposit a thin film on the adjacent

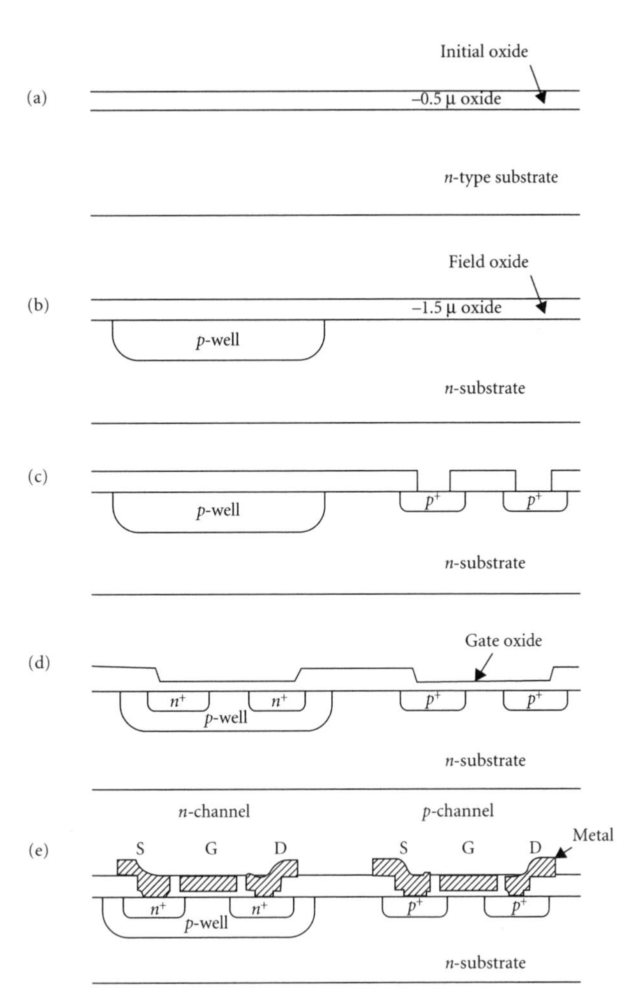

Figure 4.7. Sequence of steps for making a CMOS pair. A *p*-type well is formed as base for the NMOS device, while the PMOS structure is formed alongside by diffusing boron through an oxide mask to make the source and drain. Then phosphorus diffusion is used to make the NMOS source and drain. A gate oxide is grown at 1000°C for both devices and aluminium gate electrodes are defined by further masking. Finally, source and drain contacts are defined in similar manner. The overall process employs six photolithographic masks. (McCanny and White 1987: 71). Reprinted with permission from Elsevier.

silicon wafer. Finally, the metal has to be patterned so as to make separate contacts to each source, drain, and gate region, as shown in Figure 4.7(e). Note the convenience of using Al for all three metallizations, thus saving further process steps. Nevertheless, even this simple process is surely complicated enough!

One final comment concerns the mask set required for all these operations. No less than six are needed, to define: (1) the *p*-type isolation region, (2) the PMOS source and drain, (3) the NMOS source and drain, (4) the gate oxide, (5) the metal contact windows, and (6) the separate metal contacts. The number of masks required to complete any particular process is generally regarded as a measure of its complexity.

The CMOS process just described could well have been employed at the end of the 1960s but would be found totally inadequate for modern circuitry. While we cannot hope even to list all the changes and improvements that have occurred since then, it is certainly worthwhile looking at a few of them—again, with the object of obtaining a feel for recent trends, rather than any thought of being encyclopaedic. We may begin by looking again at Figure 4.7(e). It is apparent that the gate electrodes overlap the respective source and drain regions in both transistors, though less immediately obvious why this might be undesirable. That it *is* undesirable is due to the associated stray capacitance at the edges of the gate electrodes which effectively couple the gate to both source and drain, thus introducing undesirable feedback of the signal voltage (which becomes more and more serious as the operating frequency increases). Clearly, the gate must cover the whole of the channel length in order to switch it on completely but, ideally, no more than this. In other words, it should be perfectly aligned with the channel. Why, you may ask, can it not be? The answer lies in the inevitable alignment tolerances associated with the above technology—because it is impossible perfectly to position the gate defining mask, a small overlap (of the order of the alignment error) is essential and, while that may be of little significance in a large device, the need to miniaturize individual transistors, in order to increase packing density, ensures that it is significant today. When the channel length is less than a micron, an alignment error of, say, a third of a micron implies, perhaps, a doubling of the gate capacitance, no longer just a small perturbation! Something had to be done, and that 'something' was the development of a 'self-aligned' gate technology.

To achieve this, two innovations were necessary—the first and relatively minor one was the use of doped polysilicon instead of aluminium for the gate electrode, the second and much more significant was the introduction of ion implantation as a replacement for diffusion in forming the heavily doped source and drain regions. Ion implantation came on the scene towards the end of the 1960s, following considerable debate as to its likely viability. It involved the use of a high voltage accelerator to produce a beam of energetic ions of the required dopant atoms (20–200 keV) which was scanned over the surface of the semiconductor slice. At these energies, ions penetrate to a depth of order 0.1–1.0 μm before coming to rest and being neutralized by electrons flowing into the sample. The advantages to be set against the investment of some $3 million in the necessary equipment were three: it was a low temperature process which allowed photoresist to act as a barrier material, it provided a very precisely controlled dose of donor or acceptor atoms, and it positioned those atoms very accurately within the slice. The one technical drawback was that these same high energy ions tended to knock Si atoms out of their lattice sites, implying the need for a thermal annealing stage to repair the damage, following implantation. Figure 4.8 illustrates how the precise positional control can be used to achieve a self-aligned gate structure, the dopant ions being masked by the gate electrode itself which is defined before, rather than after the source and drain implant. It is true that the implanted ions do scatter to

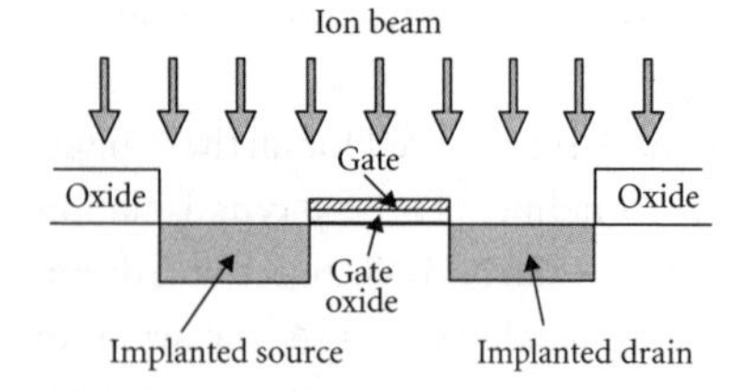

Figure 4.8. Schematic diagram illustrating the use of ion implantation to form a self-aligned MOS gate structure. The gate is formed first and the source and drain are implanted, using the gate metal as a mask. This process ensures a minimum overlap between gate and source/drain regions, thus minimizing the gate-to-source and gate-to-drain capacitances.

a very small extent underneath the gate electrode but far, far less than would occur in a diffusion process. To complete the picture, we need simply note that polysilicon (deposited by CVD) is used to make the gate contact because it will withstand any high temperature processing steps which follow the gate definition. Aluminium would certainly not.

Another innovation which grew in importance during the 1980s was the use of 'dry' etching for the controlled removal of silicon oxide, silicon nitride, polysilicon, etc., rather than the wet chemical etching which had previously been employed. The reason for its introduction bears a superficial similarity to that behind self-aligned gate technology, that is, the need for better positional control. The problem with wet etching is that it is an isotropic process, implying that etching proceeds at the same rate in all directions. This means that, as shown in Figure 4.9(a), when etching through a mask, material is removed under the mask to a distance roughly equal to the etch depth—this is referred to as 'undercutting' and, when it occurs in very fine geometries, may introduce an error at least as large as the intended window size. Ideally, a perfectly anisotropic etch is called for, as shown in Figure 4.9(b), where sideways etching is effectively zero. First attempts to achieve this employed ion beam etching where substrate atoms were physically knocked out by high energy argon ions but the damage inherent in such a process represented a major drawback. It also suffered from poor selectivity. Integrated circuit processing usually requires that one material be completely removed without significant attack on the layer beneath (e.g. SiO_2 being stripped from silicon) and with only minimal attack on the masking material. This can best be achieved with a chemical etch which etches the one material while having no reaction with the other (rather than a purely mechanical process which is essentially non-selective), and this led to the introduction of plasma etching, a chemical process using 'active' gaseous atoms, rather than ions in liquid solution. For example, a gas containing chlorine or fluorine atoms might be used to etch SiO_2, the sample being introduced into a vacuum system containing the gas, and a plasma formed by excitation with radio frequency (RF) power at 13.5 MHz. While excellent selectivity can be obtained, the ideal combination of selectivity and anisotropy is best achieved by combining chemical and physical attack, the active species from the plasma being accelerated (electrostatically) towards the sample surface. The resulting process is known as chemically assisted ion beam etching (CAIBE) and is now widely used throughout the industry. The investment level, of order \$100,000, compares somewhat unfavourably with the hundreds of dollars (euros?) demanded by wet chemical processing, though it is perhaps insignificant in comparison to the millions required for ion implantation equipment.

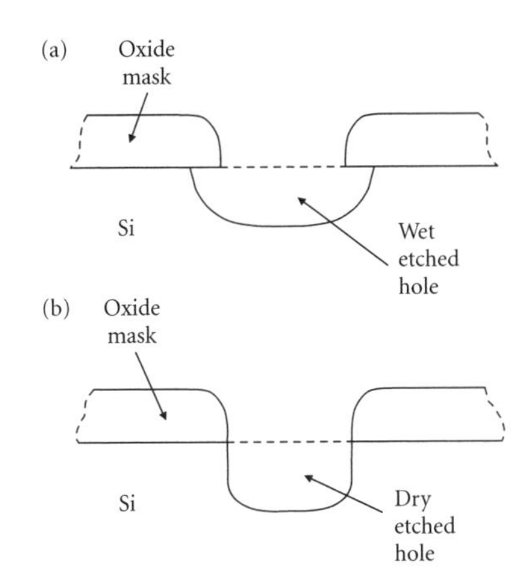

Figure 4.9. Comparison of wet and dry etching of holes in silicon. Wet (chemical) etching tends to undercut the mask to roughly the same extent as it etches downwards, whereas dry etching is a far more anisotropic process which results in minimal undercutting and produces etch pits with vertical sides.

At this point we should, perhaps, remind ourselves that this considerable investment relates directly to the quest for increased packing density and faster signal processing, both of which imply reduced feature size in the individual transistors. It is to this theme, therefore, that we now return. It is obvious enough that reduction in the overall size of the transistors will result in greater packing density and we have shown in Box 4.2 that the operating speed of MOS circuitry improves like $1/L^2$ (where L is the transistor gate length). It is not quite so obvious what other changes are inherent and what problems may accompany them.

The first question to be asked, when considering reduction of transistor size, concerns the basis on which to apply the scaling—what, for example, should remain invariant? Let us consider an MOS circuit in which the linear dimension of each transistor is reduced by a factor k. If we simply reduce size while leaving everything else unchanged, it is easy to see that device currents will decrease like $1/k$ because they scale with gate width. However, this implies an increase in total current per unit area of the chip by a factor k and, because we are keeping device voltage constant, the same factor applies to chip power density which is highly undesirable—before we can go very far, our chips will rapidly overheat. A better approach, and the one most frequently adopted, is to keep all electric fields constant, while scaling dimensions. This implies a reduction in all applied voltages, together with a similar reduction in gate oxide thickness. We now see that, though the current density per chip will increase by a factor k, the power density will remain constant, a much more satisfactory outcome.

The second important aspect of scaling is the effect on switching speed. Referring to Box 4.2 and, in particular, equation (B4.4), we see that, under the constant field assumption, delay time is reduced by a factor k (both L and V_G are reduced by k times) which is the origin of the ever faster speeds being quoted for PC processors. Connected with this, is a figure of merit often used in connection with switching circuits, namely the product of switching power and delay time (which is equal to the switching energy). The switching current scales like $1/k$, switching power like $1/k^2$, and delay time like $1/k$, so the power–delay product scales like $1/k^3$. This implies that switching energy per unit area of chip scales like $1/k$.

Clearly, there are many advantages to be gained from reducing the size of individual devices but this inevitably raises the question of whether we may eventually run into fundamental (or even practical) limits and a number of these are not difficult to predict. For example, scaling of the oxide thickness, so as to maintain a constant field as the gate voltage decreases and therefore maintain the channel conductivity, must run into difficulty when it reaches a value of about 5 nm. At this point, electrons are able to tunnel through the oxide and its insulating properties are rapidly lost. Another problem is that associated with the small depletion layers which exist between source and drain and the substrate (remember that the substrate is of opposite type from the source and drain regions). If the sum of these two depletion widths becomes equal to the channel length, the device ceases to function. While it can be minimized by increasing substrate doping, this implies the need for a larger gate voltage to turn the channel on and there is then a limit set by the breakdown field of the gate oxide which translates into an ultimate limit on gate length $L = 0.02$ μm. Another practical limit to scaling is implied by the need for circuit voltages to be reliably detected above background noise levels and an unlikely sounding factor here may be the incidence of errors due to naturally occurring ionizing radiation. Yet another problem may arise with interconnects between devices. Assuming that these are scaled in all three dimensions in proportion to the devices themselves, suggests that their intrinsic RC time constants will remain unaffected—however, this has to be seen in the context of rapidly decreasing device switching times which implies an eventual limit on speed set by the interconnects, rather than the transistors. Considerable research is therefore being undertaken into the possible use of optical interconnects (i.e. using light beams rather than electrons in wires), particularly for the more distant link-ups.

This is probably as far as we should attempt to go in discussing what is an increasingly complex subject, except for the very important matter of lithographic limitations. As device dimensions decrease below the micron barrier (psychological if not physical), the fundamental limitation set by the wavelength of light looms large on the lithographic horizon. As is well known (see any textbook on light!) the smallest spot to which a light beam may be focused approximates to the wavelength of the light. Current UV lithography is therefore limited to dimensions of about 0.2 μm (which is roughly where current technology has arrived). Further advances appear to demand a fundamental change of approach and two such have been very seriously explored during the past decade, based on the use of either X-rays or high energy electron beams instead of light. We have already referred to the use of E-beam (EB) lithography in mask-making—it depends on the fact that high-energy electrons have an associated de Broglie wavelength λ given by:

$$\lambda = (1.226/V^{1/2})\ \text{nm} \tag{4.11}$$

where V is the potential (in volts) through which they have been accelerated. Thus, 10 keV electrons have an associated wavelength of 1.226×10^{-2} nm, roughly four orders of magnitude shorter than typical UV light. This gives EB lithography a potential resolution very much greater than possible with UV light, though the highly sophisticated electron optics required makes the system considerably more expensive and there are also difficulties with back-scattered electrons which expose the special EB resist outside the intended area (often referred to as the 'proximity effect' because it becomes particularly serious when two closely spaced features are exposed, the scattered electron dose then being doubled). However its greatest drawback as a technique for 'direct slice writing' (i.e. defining patterns directly on the silicon slice), is its inherent slowness—this is because the beam must be scanned somewhat laboriously over the area to be 'written'. While it may be acceptable to use this approach for patterning expensive specialist devices for sophisticated microwave applications, this can hardly be true for a consumer product to be sold in a highly competitive market. The alternative of using ion beams, rather than electron beams, shows some promise. It suffers far less from proximity effects and resists are very much more sensitive to ions than to electrons (which promises considerably faster scanning) but there are serious difficulties in producing sufficiently intense ion beams for commercial application.

Soft X-rays ($\lambda = 0.5$–1.0 nm) at present appear to be the most promising alternative to UV light, well defined patterns with feature sizes of less than 100 nm having already been demonstrated. (This should be seen in relation to projections for a 64 Gbit dynamic random access memory DRAM, to be manufactured by the year 2010, based on a minimum feature size of 0.07 μm–that is, 70 nm.) The investment in both capital equipment and research manpower to achieve this objective is of such impressive magnitude as to make it worth our while examining some of the problems in detail so we shall bring this section to an end on a note of considerable technical achievement. For example, we might first consider the mechanical challenge of positioning a 3 cm $\times$ 3 cm mask to an accuracy of about 20 nm, that is, to something like 1 part in 10^6 which is significantly less than the thermal expansion of a typical solid when its temperature changes by 1°C! It also represents the distortion produced by placing a mass of only 100 g (3 oz) on top of the mask holder! Not only must the mask be positioned to this accuracy, it must also be easily moved to a new position when required, a demand met (typically) by the use of frictionless, air-lubricated bearings.

Designing the mask itself also makes considerable demands—it must be extremely thin and made from the lightest elements available in order to minimize absorption of the X-rays, while supporting areas of heavy elements which control the dark regions in the pattern. A typical mask consists of a membrane of CVD-deposited SiC or Si_3N_4 some 2 μm thick, stretched across a 3 cm $\times$ 3 cm aperture in a silicon wafer and it should obviously be as uniform as possible. This latter condition is far from trivial, bearing in mind that the tension in the membrane is sufficient to distort the silicon frame, while the non-uniform distribution of the absorbing material is sufficient, in turn, to distort the stress pattern. Another essential requirement is for minimal degradation of the film under intense X-ray fluxes—Si_3N_4 is not entirely satisfactory in this respect, SiC and hopefully diamond being preferable. The absorber material may be the heavy metal tantalum which must be about 0.5 μm thick in order to reduce the X-ray flux sufficiently. It is deposited by RF sputtering. Ta has an important advantage over Au, for example, in that it can be readily etched using the CAIB process, the patterning being done by EB lithography, using high-energy electrons.

Writing the necessary patterns by EB lithography presents its own difficulties when the aim is to reduce minimum feature sizes below 0.1 μm. To begin with, the masks inevitably contain very large numbers of patterns and this tends to increase the time required for mask making to an unacceptable degree. To combat this problem, it has been necessary to develop EB machines with variable beam shape, so as to expose significant areas of the EB resist without the need for scanning, though it is important to be able to reduce this when particularly fine geometries are required. Second, the problems due to back-scattered electrons become extremely serious when patterns include very closely spaced features (the proximity effect), demanding careful design of both machine and resist technology. In particular, it is becoming necessary to increase the beam energy from the values of order 10–30 keV common in current usage to the significantly higher value of 100 keV. This has the effect of spreading the back-scattered electrons over a wider area, thereby increasing the ratio of peak exposure to background. Finally, as is not difficult to appreciate, both the beam focus and beam placement electronics have to be designed to an unprecedented accuracy. Machines under development at NTT, for example, will provide patterns as fine as 50 nm with a placement accuracy of 10 nm.

Last, but very definitely not least, we come to the question of a suitable X-ray source. This, without doubt, represents the 64-million dollar question. The problem stems from the fact that photoresists are not very sensitive to X-rays and this demands a more than usually powerful source, the kind of source which is only available from a high voltage electron synchrotron. The synchrotron (an extension of the basic cyclotron accelerator) was first proposed in 1945 and various machines have been built for studies of high energy particle physics but these are generally large and cumbersome (typically of the order of 100 m in diameter) and, therefore, not convenient for use in semiconductor processing. The challenge, taken up towards the end of the 1980s in a number of processing laboratories was that of developing compact, high current electron synchrotron storage rings in which electrons are accelerated to energies of order 500 MeV, the beam then being bent by a superconducting magnet ($B = 1$–3 T) into a semicircular orbit. The bending process results in the emission of a wide spectrum of radiation (from IR to soft X-rays) but, provided the electron velocities exceed 90% of the velocity of light, with a highly collimated output beam (width of about 1 milliradian). Selection of wavelengths in the region of 1 nm is effected by focusing mirrors and by using selective windows, the resulting radiation beam being approximately parallel with a square cross-section of 3 cm × 3 cm.

This beam is transmitted into the semiconductor clean room where it can be used to expose a section of a silicon slice, located immediately behind the X-ray mask, and the whole slice exposed using a step-and-repeat sequence. Alignment of each mask with the pattern previously imposed on the silicon is effected using sophisticated optical interference methods, together with the ultra-fine positioning system referred to above. The electron storage ring may be arranged to provide several X-ray beams which allow for multiple lithographic facilities should they be required, the overall system representing a truly remarkable engineering achievement. As a result, Moore's Law (or, at least, an approximation to it) seems set to hold for another 10 years or so.

4.4 Wise men from the East

The explosive development of the microelectronics industry was remarkable enough, in itself, but the successful Japanese incursion which occurred during the 1970s shocked even the Americans, accustomed, as they were, to rapid and sometimes unpredictable change. In this section, therefore, we step briefly to one side to examine some aspects of these worldwide developments and outline possible reasons for their occurrence. I make no pretence, of course, at providing a comprehensive treatment—the reader with a specific interest in the business side of the industry should consult the considerable literature—but there is no doubt that these developments represent as fascinating a study for the dedicated scientist/technologist as for the business psychologist! I make no apology for their inclusion. Indeed, the implication that the scientist might have little interest in such matters probably points up one of the significant differences between the eastern and western approaches to the subject.

We have already made brief reference to the disparity between the 'take-up' of microelectronics in Europe compared with United States which, at least in part, depended on the relative lack of European government and military spending. Though there can be no doubt that serious attempts have been made to rectify these early omissions, through both National and EC programmes (such as ESPRIT and BRITE), it is nevertheless true that Europe has struggled to compete with the Americans, perhaps largely due to differences in industrial attitudes. To quantify the early 'failure' of the European Industry, we may simply note that, though the *demand* for semiconductors in Europe and Japan was slow to take-off, by the mid-1970s it had actually surpassed that from the American market—nevertheless, this was largely satisfied from American *production*—only 14% was actually supplied by European companies. This is in spite of the fact that in 1978 Philips took a creditable third place in the league table of semiconductor sales, with

7.4% of world sales and Siemens was tenth with 3.3% (TI was first with 10.5%). However, this was achieved, in part, by their considerable investments in American companies—Philips had bought Signetics, and Siemens held significant shares in Litronix and Advanced Microelectronic Devices. France, too, came into the picture largely by the takeover of Fairchild by Schlumberger in 1979. Native European *production* of semiconductors was extremely weak when compared with the demand for *applications*. In addition, European companies had been obliged to take out licensing agreements with American firms in order to keep up with the rapid advances in semiconductor technology, while TI, for example, had established a subsidiary in England (at Bedford) from which it could supply European demand. (It was, too, the only American company to establish a manufacturing activity in Japan.) To analyse exactly why this state of affairs came into being would take us well beyond the scope of the present book (not to mention its author's competence!) though it has been analysed in numerous other books and articles. Here we may simply note a number of relevant factors and take the opportunity to contrast the European with the Japanese experience—which has, undoubtedly, been a much happier one.

A startling feature of American success was the frequent and rapid rise of small start-up companies such as Texas, Fairchild, Hewlett Packard, and Intel. It was these companies which carried the momentum of microelectronics forward, rather than the large, well-established monoliths but there has been a noteworthy absence of such entrepreneurial success within Europe. This is surely connected with another feature of the American industrial scene, the remarkable mobility of personnel between companies. Staff from Bell were responsible for setting up the semiconductor activity at TI, Shockley, himself, left to found Shockley Semiconductors, then members of his staff departed to form Fairchild and, finally Moore and Noyce moved on to start Intel. These were the high profile moves which determined the culture—they were to be emulated by hundreds of others, moves which ensured a rapid diffusion of expertise and technological know-how throughout the growing industry. The contrast with Europe and Japan could scarcely have been sharper—job security, pension rights, and company loyalty, in various measures took precedence over entrepreneurial initiative and, in any case, venture capital within Europe, at least, was a particularly scarce commodity. Other cultural differences were probably of equal significance. We noted, earlier, the very high regard within Europe for *pure* scientific achievement, based, as it was, on the remarkable developments in 'modern' physics during the 1920s and 1930s. This resulted in a lingering preference for pure research which meant that *mere technology* failed to attract a sufficient following to spearhead the necessary thrust into new industrial developments. Certainly within the United Kingdom, research in University and Government laboratories acquired considerably greater 'kudos' than that associated with the immediate requirements of industry, and many of the country's brightest researchers were effectively 'lost' to the industrial cause. As an American commentator put it: 'The English took a scientist's route and the Americans took a technologist's route.' Oddly enough, this lukewarm attitude towards new technology seemed to permeate management even as thoroughly as it did the scientific community. Those who *were* concerned to push forward the frontiers of technology usually found it frustratingly difficult to gain the backing of their seniors, as exemplified rather poignantly by the failure of the RRE scientist Geoffrey Dummer to obtain a positive response to his early proposal of the concept of the integrated circuit—that was in 1952, some 6 years before Jack Kilby's demonstration of the first experimental circuit!

Another, and, possibly related, aspect of European scientific culture concerns the *coupling* between research and production, the importance of which has been demonstrated by many Japanese companies. Within Europe there was a tendency to keep research and production very much at arms length. As an extreme example of the 'European' approach one may cite the 'Casimir doctrine', as practised within Philips Research during the 1950s—crudely paraphrased as: 'Recruit the best scientists, give them a well founded laboratory, containing the best available equipment—and leave them to it! Something useful is bound to emerge.' This is not without theoretical merit—scientists, like artists, work most effectively when pursuing ideas to which they, themselves have given birth—but it fails to address the surprisingly difficult problem of translating good ideas into saleable products. Two things tend to happen in practice: the production staff (who have no vested interest in the new idea!) work a bit harder and find ways to improve the present manner of doing things (the well-known 'sailing ship' effect) and quietly forget about it altogether or they completely re-invent the idea in a guise 'better fitted to manufacture' (thereby giving themselves a personal stake in it!). This may not be serious in a 'traditional' industry which innovates at a modest rate but the waste of time involved can be catastrophic in an industry like electronics which has such viciously short timescales and rapid price reductions. Even Philips was obliged to accept the truth of this argument in their 1980s restructuring.

While musing on such matters, we should now devote a few moments to examining the contrasting topic of Japanese experience. During the 1970s there was much agonizing in the West with the aim of discovering the hidden secret of their success, many observers feeling there was something unfair, even sinister about Japanese competitive achievements. More recently, this has been the subject of numerous business studies and several books have been written which tend to play down the hysteria and offer quite rational explanation. In this short section, we make no attempt to compete with them but simply try to summarize a few of the relevant factors. The first thing to be said about the development of the microelectronics industry in Japan is that it grew from the most unpromising beginnings. Japan was devastated by the Second World War, many of its cities in ruins and its government subject to an occupying force—that, by the 1970s, Japan was also a force in microelectronics represents a remarkable recovery. It is also noteworthy that this was not in any way a consequence of military support, Japan not being allowed any military investment during the period of interest. It is equally clear that the phenomenon of personnel mobility and frequent small company start-ups was not a contributory factor, either—the Japanese culture of company loyalty was probably stronger by far than that pertaining in any other part of the globe. What, then, was the 'secret'?

To say that personnel mobility was negligible should not lead to the conclusion that technological know-how was characterized by an equally small 'diffusion coefficient'—the unique method of control exercised by Japanese government over the innovation process ensured that those companies which worked together within a research contract effectively shared new technology. The subtleties of this process have been well described by Fransman in discussing the rise of the Japanese computer and telecommunications industries—he uses the phrase: 'controlled competition'—an example being the development of electronic switching systems for the Japanese telephone system. NTT, the ultimate user, would supervise research undertaken by the suppliers, NEC, Hitachi, Fujitsu, and Oki while making its own contribution to the overall project and, of course, assessing the final product. In return for a fair guarantee of future sales of the resulting equipment, each company would accept a considerable degree of sharing of its own new technology with its rivals, to the mutual benefit of all concerned. This contrasts rather favourably with the preferred American or European system which is probably closer to 'uncontrolled warfare'! While it is certainly true that the Japanese government did, indeed, provide strong support for most sections of the microelectronics industry, this was only made available after very careful planning, resulting in well thought-out targets which could be agreed by all participants. Largely a matter of common sense! Perhaps the West should have recognized this earlier, rather than conducting a sometimes wild search for the 'magic' which might explain Japanese commercial success.

But there was more to Japanese organization than government planning. Another major factor in their success concerned the planning of product development within a company, quite irrespective of government support. The essential philosophy was one of starting with a product which was perceived to be marketable and thinking backwards to define a research and development programme designed to realize precisely this. With this in view, Japanese companies made sure their research, development, and production and marketing functions could work easily together—they therefore housed them together, as far as possible so they could readily talk to one another, rather than keeping them at arms length according to the western pattern. Transfer of personnel between functions is another aid to effective product innovation. Simply more common sense! But, perhaps the single most significant aspect of the Japanese approach has been their concern to achieve the highest possible quality in manufacture. This more than any other factor accounted for their success in winning large market shares and, ironically enough (according to Reid 2001), it was all based on ideas promulgated by an *American* management consultant, W. Edwards Deming. Deming's attempts to persuade his fellow countrymen apparently fell on deaf ears whereas, in Japan, his word was accorded the status of gospel truth and his ideas were followed with almost religious zeal, so much so that in 1980, Richard Anderson, a Hewlett Packard divisional manager, was able to produce statistical evidence for the marked superiority in quality and reliability of Japanese memory circuits over their American counterparts. It took nearly 10 years and a great deal of heart searching for the US industry to fight back to a state of parity. Once again, the 'magic' was nothing more than careful attention to detail.

Without doubt, a significant factor in the very complex equation which defines commercial success was the existence of a highly buoyant home market for electronic equipment. Japanese houses are generally small (even among the professional classes) on account of the severe shortage of level ground on which to build. They are also, by long tradition, somewhat sparsely furnished so what can be more tempting for an increasingly affluent society than to equip itself to the full with the best available automobiles and in-home entertainment facilities, both representing industries in which Japan has made major incursions. Adding to this a deeply felt need to develop its own telecommunications capability, gave Japan an impressively powerful motive for building a strong microelectronics industry. Nor should we forget that the policy of political isolation which characterized Japanese foreign policy in the two centuries prior to the year 1868, inevitably left, in what, below the surface, is a strongly conservative society (see, for example, the excellent social history of modern Japan by Professor J. E. Thomas) an in-built need to be self-sufficient. It was hardly surprising that a country still deeply affected by the war should feel suspicious of allowing foreign intervention in this vital postwar recovery activity—it had to be 'home made' as far as this was technically possible.

That the new industry could not be built entirely without foreign assistance was well illustrated by the deal Sony was obliged to make with Western Electric to licence its transistor technology in 1953 for the huge sum (to Sony) of $25,000, some 10% of its total assets! The outcome was Sony's very successful venture into transistor radios which began the Japanese incursion into the US microelectronics market. By 1968, 90% of all Japanese-made transistor radios were being exported, largely to America! Sony, by the way, represents the odd-man-out in the Japanese electronics industry—it really was a small start-up company in 1946 and grew to maturity as a result of aggressive entrepreneurial policies, always aiming to be technical leader in its chosen fields, rather than a careful follower. (An interesting history of Sony has recently appeared in *Sony: The Private Life*, by John Nathan 1999.) Almost all the others were well-established companies before the war, taking up solid state electronics as a natural development of their earlier activities, which may possibly explain their more conservative policies. In business circles there is currently much discussion of what are called the 'core competences' of an enterprise, it being seen as important that full use is made of these in planning company strategy, and Japanese companies appear to have been uniquely aware of such considerations. Looking, too, at Company Japan, in the early stages of microelectronics, it is, perhaps, significant that high quality optics (developed during the war) represented an essential competence in the business of integrated circuit manufacture.

Having said all this, it would be misleading to suggest that Japan could beat the West hands- down under all circumstances. The Japanese success has, in fact, been patchy. In some areas, such as TV and hi-fi, they have nearly swept the board, in telecommunications they have kept their market pretty well to themselves but have failed to penetrate the major American market, in computers they have struggled, in ICs they have had remarkable success in memories but made rather little penetration into the world microprocessor market, and so on. It is interesting to consider some reasons for this lopsided success, one of the principal being that of market standards. For example, the world standard for the microprocessor which is at the heart of the now dominant PC was set when IBM was obliged to change its philosophy from mainframe to PC. Because of its large size it was able to dictate the standards for operating system and main processor which it subcontracted to Microsoft and Intel, respectively. This effectively excluded competition in these essential areas. It was not that Japanese companies were lacking in the necessary competences to design either software or microprocessors—NEC marketed its own PCs (the 9800 series) very successfully in Japan (taking more than half the Japanese market) but neither its hardware nor software was IBM-compatible so it was impossible to build sales outside Japan. A similar situation holds in the telecommunications field. Each major country tends to set its own standards for transmission and switching equipment which are incompatible with all others. ATT sets the US standards, BT the UK standards, and NTT those in Japan with the result that any equipment supplier must design for a specific market. The natural tendency is for US suppliers to aim for the US market, Japanese suppliers for the Japanese market, etc., so we are not surprised to learn that Japanese electronics companies supply virtually all the equipment required by NTT, while having an almost negligible share of the US market.

This contrasts, for example, with the situation in respect of computer games—Nintendo was able to dictate the appropriate standards here and has a correspondingly large share in the world market. On the other hand, the standards for random access memory chips have not been circumscribed and Japanese companies have dominated the world market in DRAMs to a degree which has caused considerable concern in the West. In the mid-1990s, Japan held more than 50% of the market for computer components, one reason undoubtedly being the fact of Japanese companies' ability to produce high quality components at very competitive prices. It is a noteworthy feature of their business strengths that they operate with much smaller profit margins than those considered necessary in the West, a consequence of the Japanese philosophy of long-term investment with minimal demand from shareholders for fat dividends. Furthermore, this is an appropriate opportunity to emphasize another important aspect of the IC market. Since the 1970s, the nature of the business has changed fundamentally, in so far as the extremely large and complex chips now being developed require huge capital investment. It is no longer possible for a small entrepreneur to enter the basic memory business—only the giants can afford the development costs. The fact that many of these are Japanese giants seems to be a reflection of their devotion to detailed planning and their tough commercial outlook (allied with a bit of assistance from MITI!).

Perhaps the surprising strength of the Japanese incursion into *some* areas of the world microelectronics market has led to a perception of competition in terms of Japan vs the West but this tends to overlook at least two important points. First, there are numerous examples of severe competition between Japanese companies and, second, other eastern countries have recently made their presences felt. Some observers, for example, saw the famous battle of the video cassette recorder (VCR) as one being fought between Philips and Japan. Far from it—the real commercial battle was that between Sony who, in 1975 made a pre-emptive strike with their Betamax system, and JVC who threw down their challenge a year later with the video home system (VHS). These two slugged it out in the world markets until, in 1988, Sony finally had to accept defeat and switch to their rival's system. Outside Continental Europe, the Philips V2000 system was never a serious contender. The idea that Japanese companies are so cushioned by government support that they need fear no competition is hardly supportable. Similarly, it is now quite clear that competition from other eastern countries, Korea and Taiwan, in particular, is making serious inroads into Japanese-dominated markets. During the 1990s, Samsung has become one of the world's largest manufacturers of computer memories. Again, this seems to refute the idea, inherent in some western thinking, that there was something 'unfair' about Japanese success. In fact, it was largely the result of careful attention to detail in planning appropriate strategies, allied to high quality production standards. It was also supported by remarkably low interest rates, too low by far, as it emerged, a feature brought starkly to the fore by the sudden economic crash which hit Japan towards the end of the 1990s. Like everyone else, Japan has come to realize that it must abide by the rules which govern sound economic management. But, no matter, the new millennium promises to be an exciting time in electronics, with new countries such as China entering the field to create even greater competition. However, the reader will recall that the subject we are concerned with here is history, not futurology.

4.5 Power and energy—sometimes size *is* important

Having followed this chapter so far, the reader may have reached the conclusion that the future of silicon is to be simply 'more and more of less and less, and so ad infinitum' but he would be wrong! A great deal of silicon is actually used for quite other applications than integrated circuits. In fact, such is the popularity of space exploration, that most people are familiar with silicon solar panels, those strange appendages which appear to hang rather tenuously from a space craft in the interest of providing power for the crazy complex of electronic gadgetry that controls (or sometimes—the times we tend to hear most about—*doesn't* control) its every twist and turn. The silicon solar cell has a history almost as long as that of the transistor—it was invented in 1954 at (believe it or not!) Bell Labs, as an offshoot of transistor development. Indeed, if the *Handbook of Semiconductor Silicon Technology* is to be believed, no less than 11,000 tonnes of silicon per annum is devoted to photovoltaic applications, compared with only half that amount to integrated circuits! (The reason for the disparity is, of course, that the solar panel depends on large areas of silicon, whereas the integrated circuit tends, in the interest of yield, to stay relatively small.) What is, perhaps, less widely appreciated is the comparable importance of silicon devices in connection with the electric power, aerospace, and automobile industries, where, again, devices tend to be relatively large. We shall leave discussion of solar cells until a later chapter—in this section, we look, albeit briefly, at the subject of power semiconductor devices where the emphasis is on high voltage and high current carrying capacity, rather than high speed and high packing density.

A convenient starting place for the discussion of power devices, an application familiar to most readers, is that of the electric light dimmer switch. The desirability of smooth, continuous control of light intensity

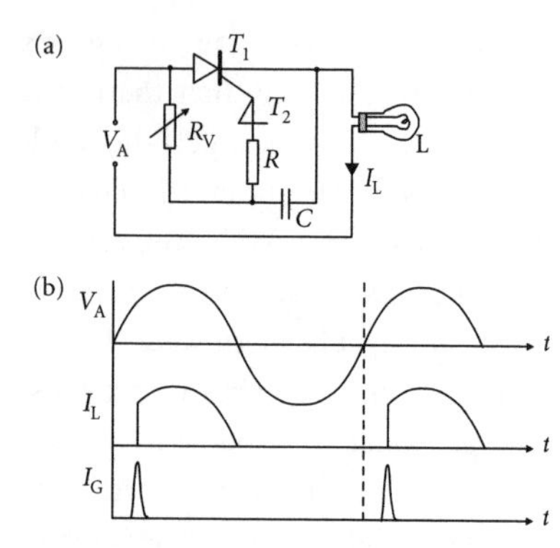

Figure 4.10. (a) Circuit used as a dimmer switch for controllable electric lighting. It employs two thyristors, T_1 being gated, T_2 not. Current flows to the lamp only when T_1 is conducting, determined by the switching action of T_2 when the capacitor C has charged to a sufficient voltage through the variable resistor R_V. The time constant CR_V controls the time at which T_1 switches on and therefore the fraction of time for which current flows through the lamp L. (b) Typical waveforms are shown here.

is self-evident enough to need no discussion but how to achieve it in practice is less obvious. The simplest method would be to incorporate a variable resistor (or rheostat) in series with the lamp but this has the obvious disadvantage of wasting power. A better alternative might be to employ a variable tapped transformer to step down the mains voltage but such a solution would be both bulky and relatively expensive—far easier, therefore, to do the trick with a couple of tiny electronic devices. The principal of the method is to use an electronic switch to control the *average* power delivered to the lamp by varying the *time* during which the supply voltage is actually connected to it. If we do this on the timescale of the mains frequency, the overall brightness can be modulated, without detectable flicker (a consequence of the light integrating facility of the human eye). The circuit shown in Figure 4.10 represents one such approach, but it requires some explanation, not least because it incorporates a new device, the 'thyristor' (also known as the semiconductor controlled rectifier or SCR).

We have already referred to the fact of William Shockley's leaving Bell Labs to form his own company, Shockley Semiconductor Laboratory, in 1956, one of many moves involving the principal semiconductor scientists during the late 1950s. Sadly, it was not a commercial success, ending in the early departure of most of the newly recruited staff members to set up Fairchild. One of the main reasons was Shockley's obsession with another of his own inventions, the four-layer semiconductor device, which possessed bi-stable characteristics and which he saw might therefore be used as a digital switching device. Though this was, without doubt, an intriguing idea, the four-layer structure proved too difficult to manufacture with the technology then available to him, and, instead of developing the junction transistor as a commercial product (which probably stood a far greater chance of success), the company was crucially divided and failed to make adequate progress with either. Once again Shockley demonstrated his considerable visionary skill while failing to appreciate the significance of the inherent technological difficulties. As it happened, the four-layer device was destined for glory as a *power* device, the thyristor, where its bi-stable characteristics were ideal for switching large currents and voltages, rather than the tiny signal currents used in digital circuitry. It was ironic that both the FET and the thyristor had to await the availability of the necessary technologies which emerged only after Shockley had been obliged to abandon them! Not that, in the case of the thyristor, there was long to wait—it was as early as 1957, in the GE laboratories, Schenectady, that commercial power devices were first demonstrated. They were based on 5 mm silicon chips using a mesa/alloy technology and showed 300 V blocking with 25 A current capacity.

Figure 4.11(a) shows the structure of a modern thyristor, with appropriate doping levels and layer thicknesses. One method of making it

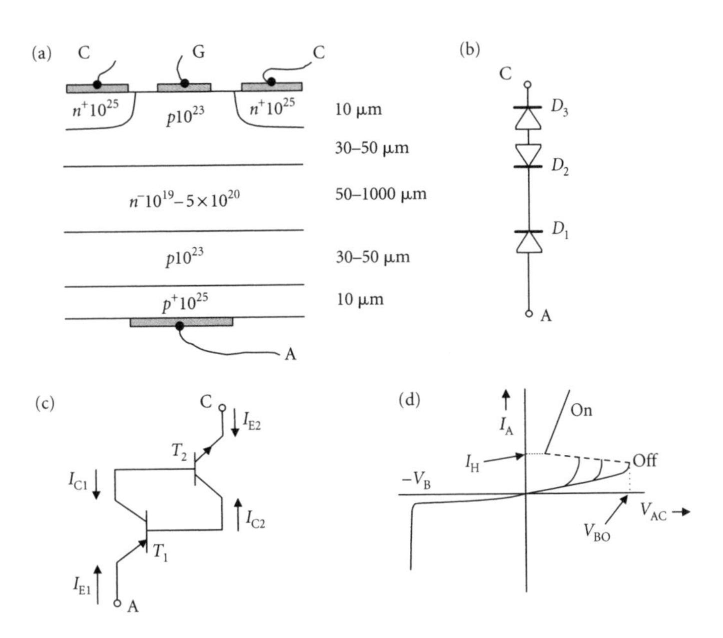

Figure 4.11. In (a) is shown the structure of a typical gated thyristor (SCR), while in (b), is shown the equivalent circuit when the device is reverse biased. It consists of three diodes, two being reverse biased but, as D_1 is a *p*-n^- structure, it has the larger breakdown voltage and therefore controls the reverse blocking capability of the thyristor. In (c) the effective circuit when the thyristor is under forward bias is represented, the *p-n-p* and *n-p-n* transistors being interconnected via the structure. In (d) typical forward switching characteristics are illustrated, showing the break-over voltage V_{BO} at which the thyristor switches from high impedence to low impedence behaviour. The various values of V_{BO} are determined by the gate current I_G, the larger I_G, the smaller is V_{BO}.

would be to select a suitable n^- substrate, diffuse in the two *p*-type layers, followed by the shallow p^+ contact layer (while the top surface is protected by an oxide mask) and, finally diffuse the n^+ region, using a patterned mask to define its annular shape (with the bottom surface protected). An example of a current–voltage characteristic appears in Figure 4.11(d), showing the critical switching behaviour under forward bias (positive cathode). On first applying bias, the forward current remains relatively small (the device presents a high impedance) until V_F exceeds the 'break-over' value V_{BO}, at which point the impedance drops abruptly, the 'on' state being characterized by high current and low voltage ($V \sim 1$ V), so long as the forward current is maintained above the holding level I_H. One final feature to note is the possibility of obtaining a range of break-over voltages by varying the gate current I_G. This allows the thyristor to be switched on by applying a gate potential and offers considerably greater flexibility in operation. Box 4.4 provides a simple analysis of thyristor operation in terms of its constituent *n-p-n* and *p-n-p* transistors (see Figure 4.11(c)). Now (at last!) we are in a position to understand the working of the dimmer switch shown in Figure 4.10.

Initially, the gate-controlled thyristor T_1 is in its 'off' state so no current can flow to the lamp and the full voltage V_A appears across the series combination of R_V and C. Thus, C charges through R_V (in a time $\tau \sim R_VC$) until the voltage V_C is sufficient to exceed the break-over voltage of T_2, at which point ($t = t_1$) a gate current can flow to T_1 which causes it to switch to its low impedance state. Current now flows from the source to the lamp until the applied voltage falls below the holding value for T_1. T_1 then switches off and blocks all current flow until the corresponding point in the next cycle $t = t_1 + T$ (where T is the period

Box 4.4. Thyristor switching characteristics

In order to understand the switching behaviour of a (four-layer) thyristor we refer to Figure 4.11. First consider the reverse characteristic, that is with the anode negative (i.e. V_{AC} negative). As is clear from Figure 4.11(b), this polarity implies that the two outer diodes D_1 and D_3 are reverse biased so only a very small current flows through the thyristor until both diodes run into avalanche breakdown ($V_{AC} = -V_B$). In practice, diode D_1 has a much larger breakdown voltage than D_3 and therefore it is this which determines the overall breakdown behaviour. To understand the forward (switching) characteristic, it is convenient to regard the thyristor as made up of two transistors T_1 (*p-n-p*) and T_2 (*n-p-n*) which are intimately connected, as shown in Figure 4.11(c). Initially, we assume that the gate is unbiased ($I_G = 0$). The collector currents of the two transistors are given by:

$$I_{C1} = I_{C01} - \alpha_1 I_{E1} = I_{C01} - \alpha_1 I_A \tag{B4.5}$$

$$\text{and} \quad I_{C2} = I_{C02} - \alpha_2 I_{E2} = I_{C02} + \alpha_2 I_A, \tag{B4.6}$$

where I_{C0} represents the collector current in the absence of injected minority carrier emitter–base current and α is the transistor current gain.

If we now sum the currents into T_1, we have:

$$I_A + I_{C1} - I_{C2} = I_A + I_{C01} - \alpha_1 I_A - I_{C02} - \alpha_2 I_A = 0 \tag{B4.7}$$

Solving for I_A, and taking account of the fact that the cut-off currents I_{C01} and I_{C02} are of opposite sign (because the transistors are of opposite type):

$$I_A = I_{C0}/[1 - (\alpha_1 + \alpha_2)], \tag{B4.8}$$

where $I_{C0} = I_{C02} - I_{C01}$ (an essentially positive quantity).

For very small applied voltages V_{AC}, the current gains are small, but as V_{AC} is increased, $(\alpha_1 + \alpha_2)$ approaches unity and the anode current I_A tends to increase without limit. In practice, it is limited by the impedance of the external circuit, the impedance of the thyristor being very small in the 'on' state. Reducing the anode current to a low value ($I < I_H$) returns the transistors to their low gain condition so that the thyristor switches back to its high impedence state. Note that, if the gate electrode is biased positively with respect to the cathode, this will cause an increase in I_{E2} which implies an increase in I_A (for a constant V_{AC}) which causes the thyristor to switch into its 'on' state at smaller values of V_{AC}. Thus, a pulse of gate current can be used to turn the thyristor on at any chosen time.

of the AC supply). The net effect is an average current into the lamp which is controlled by the time t_1—the longer t_1, the smaller the average current—and, because t_1 is determined by the time constant $R_V C$, varying R_V controls the power level at the lamp, as required. This, of course, represents a low power application, the current level being a fraction of an ampere, and the corresponding device area typically 1 mm^2. As we shall see, there exist many high power applications with correspondingly larger devices, 3 in. diameter being not uncommon.

Though our ambition here runs no higher than outlining a few representative examples, the reader may find it interesting to glance at Table 4.1 which lists a number of applications for power devices, both consumer and professional. While making no claim to completeness, it helps to indicate the wide range of applications in which electronic control plays an important role, including the large number of applications associated with electric motors, all the way from the 1 kW domestic motor to the 5 MW motors to be found in high speed electric trains. There can be no doubt that motor control represents a large part of power electronics so it makes sense to look in more detail at motors and their characteristics.

Broadly speaking, we need to consider three types of motor, DC, induction, and synchronous, all three being widely used in industry and in the home. (In fact, one of the more noteworthy developments in home economics during the second half of the twentieth century is the

Table 4.1. Areas of application for semiconductor power devices

1. Power supplies
2. Battery chargers
3. Power converters
4. TV deflection coils
5. Computer peripheral drives
6. Electric light dimmer switch
7. Food mixer speed control
8. Vacuum cleaners
9. Washing machines
10. Electric shavers
11. Power tool speed control
12. Electric trains
13. Electric cars
14. Ship propulsion
15. Forklift trucks
16. Wind and solar cell power control
17. Refrigerators
18. Elevators
19. Blowers, fans, pumps
20. Induction furnace control
21. Electric welding
22. Steel rolling mills
23. Cement mills
24. Textile and paper mills
25. Electrochemical processing
26. Aircraft flight controls
27. Automobile ignition and lighting

tremendous increase in the number of domestic electric motors—most modern households contain more than 20.) Each motor type possesses its own characteristics and any discussion goes well beyond the scope of this account. We note, simply, that DC machines are ideal in applications where a high starting torque is desirable but suffer the disadvantage of needing brushes and a commutator which require more or less frequent maintenance, induction machines function without the need for electrical connection to the rotor and can be widely used where severe loading is avoided and in situations where highly flammable gases are present, synchronous machines are ideal in applications where constant speed is a prime consideration and where load fluctuations are minimal. However, in all cases, there are usually requirements for control of speed, torque, or motor current which can best be satisfied by the introduction of power electronic devices. Often, as in food mixers, power tools, or electric traction, for example, it is important that control be maintained over a wide range of speeds. In addition, considerably greater operational flexibility is provided by the use of electronic power converters, when, for example, a DC motor is to be driven from an AC supply or an induction machine is to be run from a variable frequency source. The range of applications is enormous and we can do no more than hint at its complexity by describing a very few examples.

Electric trains have been in existence for a surprisingly long time, the first electric locomotive having been demonstrated as long ago as 1879, though the real expansion in their application was largely a post-Second World War phenomenon. Some electrification of main line train services occurred in the early years of the twentieth century but its widespread use in France, Japan, Germany, and England had to wait until the 1950s, partly because of the huge capital cost involved (and, partly, we may imagine, by the lack of convenient electronic controls!). Nonetheless, in the 'western world' today (though, interestingly, excluding the United States!), typically 30–60% of track has been electrified, the motivation having been one of increased flexibility and much reduced maintenance costs, electric power sets being extremely reliable, and requiring only minimal attention. What is more, by 1970, steam locomotives had been completely replaced by diesel-electrics, even on non-electrified lines, in the interests of reducing maintenance and pollution so electricity rules the tracks almost everywhere. However, this having been said, it is far from obvious to the uninitiated which electric traction system should be favoured. DC motors have several advantages, particularly in providing high starting torque, but require a relatively low voltage supply (typically ~1 kV) which implies heavy gauge supply lines. (A well-known example, is the use of a 600 V power rail for London Underground trains, carrying currents of thousands of amperes!) The alternative (and generally preferred) mode of supply is high voltage, overhead wire AC which can be transformed down *in loco*, the European standard being

25 kV, 50 Hz. This system demands both rectification to provide the necessary DC and some form of speed control—its use is, nevertheless, widespread because of the high torque at low speed, low torque at high speed characteristic appropriate to series-connected DC motors. (Further justification is provided by the relatively modest cost of the necessary electronics!) Even less predictable, *ab initio*, is the *modus operandi* of the typical diesel-electric locomotive—this (would you believe?) employs an alternator to generate an AC supply which is then rectified to drive a DC motor. There are, of course, good reasons for this but our concern here is with the essential electronics so we shall not comment further. A second example we might consider relates to the ever-increasing use of small power tools in and around the home, again a relatively recent development. Take, for example, the well-known electric router which makes light of the task of cutting accurate rebates and mouldings in home carpentry. An essential feature here is to maintain the peripheral speed of the bit at an appropriate constant value to optimize efficiency, while avoiding the burning which results from too high a cutting rate. As router bits can vary considerably in diameter, a wide range of control over angular velocity (i.e. motor speed) is essential for satisfactory performance. Finally, there are numerous applications where it is desirable to maintain constant motor speed in the face of widely varying loads. A typical commercial situation occurs in the use of electric power for crushing stones for motorway construction. How, then, are semiconductor devices to be utilized?

We have already seen how thyristors may be used to control the illumination level in our living rooms by varying the average current supplied to a lamp. Similar methods are applicable to DC motor control. It is a characteristic of a DC machine that its speed depends on the voltage applied to the rotor winding, so all that is required to vary the speed is a corresponding change in 'firing angle' α (the angular equivalent of the time t_1 in Figure 4.10(b)) of one or more thyristors in a converter circuit. If the supply is AC (as it often will be), a typical circuit is that shown in Figure 4.12(a) which represents a bridge rectifier, employing gated thyristors. When the source voltage is positive, the motor current flows through T_1 and T_2, when negative, through T_3 and T_4. Figure 4.12(b) illustrates a typical sequence of voltage pulses which appear across the load. The bridge circuit is used so as to utilize both half-cycles of the supply voltage and, thereby, reduce the ripple in the output. A similar effect can be achieved from a DC source by using a single thyristor in series with the motor to generate a rectangular waveform (this is known as a 'chopper' circuit). While larger motors will generally be driven from a three-phase AC supply, speed control is achieved in a similar manner.

As a second example of electronic motor control, we consider the interesting case of an induction motor driven by high frequency AC. The induction motor runs at a speed which is usually somewhat lower

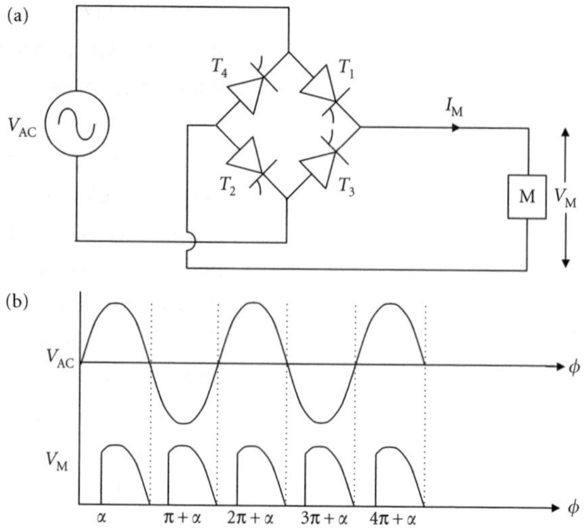

Figure 4.12. (a) Bridge rectifier circuit used to drive a DC motor from an AC source. The use of gated thyristors facilitates control of the average voltage applied to the rotor which, in turn, controls the motor speed; (b) shows the typical waveforms. Note that the timing circuit is not shown.

than the so-called synchronous speed but is, nonetheless determined by the supply frequency. The use of the mains frequency of 50 or 60 Hz is often perfectly satisfactory but suffers from the generation of low frequency acoustic noise which can be highly undesirable in many instances. The solution is to use much higher frequencies, of order 20 kHz or more, which are inaudible to the human ear but these frequencies must, of course, be generated by a suitable circuit, known as an 'inverter'. This takes a DC input and converts it to AC. We shall not delve into details of the circuits used—suffice it to make two points: the switching sequence employed requires that the switch must be capable of being turned *off* as well as on and the need to use high frequencies implies rather short switching times (20 kHz corresponds to a time $\tau = 1/\omega \sim 10\ \mu s$). By varying the output frequency, it is also possible to vary the motor speed but this implies even shorter times, typically down to 1 μs or less, the pursuit of which leads us naturally to examine the development of power electronic devices in rather more detail.

The first power devices were, unsurprisingly, diodes—as early as 1952, Hall reported on power germanium diodes made using a mesa/alloy process which demonstrated 200 V blocking and 35 A current carrying capacity. As mentioned already, the first power thyristors made their appearance in 1957 but the demands of a wide range of applications have led to the development of an equally wide range of new devices. The list is impressive: diodes, transistors, thyristors, (SCRs), gate-turn-off-thyristors (GTOs), reverse conducting thyristors (RCTs), light-activated thyristors (LATs), diode AC switches (DIACs), triode AC switches (TRIACs), power MOSFETs, insulated gate bipolar transistors (IGBTs), static induction transistors (SITs), static induction thyristors, MOS controlled thyristors, etc. As new challenges have appeared, new device structures have emerged to meet them, and the search is by no means

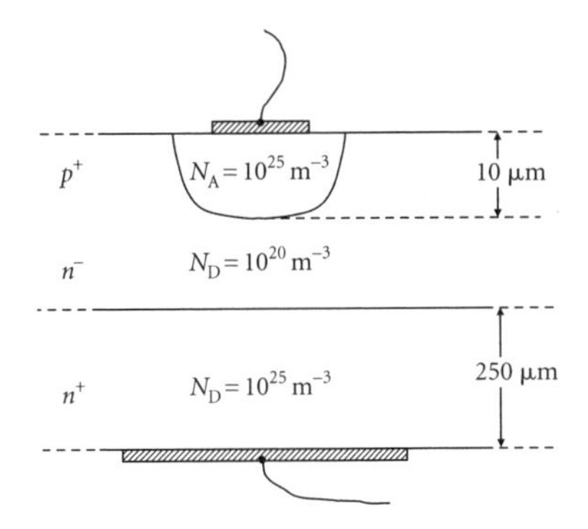

Figure 4.13. Structure of a p^+-n^- power diode, based on an n^+ silicon substrate. The width of the n^- drift region depends on whether the device is designed for maximum reverse breakdown (i.e. blocking) voltage or for punch-through operation which allows faster switching at the expense of reduced breakdown voltage.

completed, as exemplified by recent developments of so-called 'smart power' devices which combine both control circuitry and power devices on the same chip, a kind of power integrated circuit. Whole books are devoted to the description of these developments (e.g. Mohan et al. 1995; Bose 1997; Benda et al. 1999)—we can do no more than touch briefly on a few of them. As representative of the choice available we consider the diode, the GTO, the power MOSFET, and the IGBT and, following our earlier discussion, we examine how well they perform in respect of reverse blocking, forward loss, and switching time.

Much can be learned about power device principles by considering the simple rectifier diode, shown in Figure 4.13. To make it, one might start with an n^+ silicon wafer (doped with 10^{25} phosphorus atoms per m^3), then grow an n^- layer epitaxially upon it, and complete the structure by diffusing the p^+ top layer with boron or aluminium atoms. An alternative approach might use an n^- doped substrate into which two diffused layers are introduced, antimony for the n^+ contact and aluminium for the p^+ layer, as before. (Note that nearly all power devices are vertical structures, rather than planar—in this they differ fundamentally from their IC counterparts.) The thickness of the n^- 'drift region' is determined by the need for a large reverse breakdown voltage, as discussed in Box 4.5. In words: the breakdown voltage increases as the doping level of the drift region is reduced but, at the same time, the depletion width expands, requiring a thicker drift region to accommodate it. For example, to good approximation, we need a doping level of $N_D = 10^{20}\ m^{-3}$ and a thickness of 100 μm to obtain a blocking voltage of 1000 V.

The importance of achieving high blocking voltages for some applications is illustrated by the case of an electric train with a 5 MW motor. If the motor runs at 1 kV, this implies a current of 5000 A which, in turn, implies a device area of $5 \times 10^{-3}\ m^2$ (50 cm^2) (based on a safe current density of $10^6\ A\,m^{-2}$), corresponding to a 3-in. diameter device wafer. Clearly, the rectifier diodes need to block reverse voltages of at least 1000 V, though we can envisage some degree of trade-off between voltage and current, some present-day diodes being capable of 10 kV blocking voltages. However, the reader should appreciate that, in order to approach the ideal blocking performance represented by the calculation of Box 4.5, it is important for the device designer to incorporate geometrical modifications in the form of surface bevelling or the addition of 'guard rings' (sometimes referred to as 'Kao's rings'), concentric diffused p^+ rings surrounding the p^+ diode top contact region in order to minimize those regions where the field is higher than average. In either case, this tends to reduce the effective device area below its nominal value. Even more important is the need for the silicon itself to be highly uniform in quality, an essential aspect being uniformity of doping level. In fact, one of the most beautiful examples of device technology was the introduction in 1976 of neutron irradiation doping to obtain

Box 4.5. Diode breakdown voltage

In order to sustain a large blocking voltage under reverse bias, the drift region of a power diode must contain a low doping level. Thus, in Figure 4.13, we suggest a value of order 10^{20} m^{-3}. Mathematically speaking, this is expressed by the fact that, for a p^+-n^- asymmetric junction where we can neglect the voltage dropped on the p-side, the breakdown voltage V_{BD} is given by:

$$V_{BD} = \varepsilon\varepsilon_0 F_{BD}^2/2eN_D, \tag{B4.9}$$

where N_D is the doping level and F_{BD} is the breakdown field which is characteristic for each semiconductor material. In the case of silicon, $F_{BD} = 2 \times 10^5$ V cm^{-1}, so we can write:

$$V_{BD} = 1.3 \times 10^{23}/N_D \text{ volts } (N_D \text{ in m}^{-3}). \tag{B4.10}$$

It follows that, to achieve a blocking voltage of 1000 V, we need $N_D < 1.3 \times 10^{20}$ m^{-3}.

The corresponding depletion layer width W_{BD} increases as N_D decreases according to:

$$W_{BD} = 2V_{BD}/F_{BD}, \tag{B4.11}$$

which, for silicon, means:

$$W_{BD} = 0.1V_{BD}\ \mu\text{m } (V_{BD} \text{ in volts}) \tag{B4.12}$$

and, for $V_{BD} = 1000$ V, $W_{BD} = 100$ μm.

In the above we have assumed that the depletion region does not punch-through into the highly doped n^+ contact region. Sometimes, it is desirable to make the drift region thinner than the value calculated above (see the discussion of switching speed in the text) so punch-through occurs and we find ourselves confronted by a slightly more complex argument. Some part of the reverse voltage V_1 is dropped across the drift region, some part V_2 within the contact region but, because the doping in the contact region is so much larger than in the drift region, $V_1 \gg V_2$ and we obtain another simple (approximate) result:

$$V_{BD} = V_1 = F_{BD}W_d, \tag{B4.13}$$

where W_d is the thickness of the drift region.

The drift region behaves rather like the dielectric material in a parallel plate capacitor, having a constant electric field within it. Suppose, for example, we choose $N_D = 10^{20}$ m^{-3}, corresponding to $V_{BD} = 1300$ V for a non-punch-through diode, but we make $W_d = 10$ μm, in the interest of rapid switching, we obtain $V_{BD} = 200$ V, a very much smaller value, illustrating the trade-off between blocking voltage and switching speed.

n-type doping, uniform to within ±1% across a 3-in. diameter slice. Remarkably enough, the effect of thermal neutrons is to transmute Si^{30} atoms into P^{31} atoms which act, of course, as n-type dopants. One might be forgiven for thinking that Nature had a yen to see high blocking voltage devices made from n-type silicon!

A second important property of the diode is its 'on-state resistance' which results in undesirable power loss under forward conduction. Obviously, this should be kept as small as possible and a useful criterion is the comparison with the inevitable forward loss due to the nature of the forward current. In the drift region, forward current is carried by minority carriers (holes in the case of Figure 4.13) which recombine with electrons injected from the n^+ contact and, in recombining, they generate phonons (i.e. heat). The power loss per unit area incurred is given approximately by the product of the diode current density and the band gap energy (in volts)—i.e. $P_0/A = JV_g \sim J$ (since $E_g \sim 1$ eV for Si). For a typical value of $J = 10^6$ A m^{-2}, the minimum forward loss is therefore about 10^6 W m^{-2}. This is unavoidable—but, in addition, there is avoidable loss due to resistive dissipation which can be estimated if we know the series resistance of the diode under forward bias. This can be calculated as $P_1/A = \rho W_d J^2$, where ρ is the resistivity of the drift region and W_d its thickness (100 μm from Figure 4.13) but we must be careful how we calculate ρ—if we were to use the bulk resistivity appropriate to $N_D = 10^{20}$ m^{-3}, we should arrive at $\rho = 1$ Ω m and $P_1/A = 10^8$ W m^{-2}, a hundred times greater than P_0/A. However, this overlooks the fact that, in forward bias, the drift region contains a high density of injected carriers—a simple calculation proceeds as follows: $J = eW_d n/\tau$ (where τ is the recombination lifetime) so $n = J\tau/eW_d$. Therefore, if $\tau = 1$ μs, we obtain $n = 10^{23}$ m^{-3} and $P_1/A = 10^5$ W m^{-2}, some ten times *less* than P_0/A. This probably underestimates P_1/A (carrier mobilities and lifetimes are both reduced at high carrier densities) so we may expect P_1/A to be roughly comparable with P_0/A in practical silicon devices. Another way to express this is to note that the diode forward voltage takes a value of about 2 V, rather than the 1 V appropriate to an ideal device. While it is possible to reduce the voltage drop associated with the drift region by reducing its thickness W_d, this also results in reverse 'punch-through' (see Box 4.5) with consequent reduction in the reverse blocking voltage, yet another example of the kind of trade-off facing the designer of power circuits. It should also be clear that care in designing low resistance contacts is essential—it is all too easy for further losses to creep in as a result of voltage drop at contacts.

Finally, we must consider the question of switching times, the times taken for the diode to switch from its off state to its on state and vice versa, which are particularly important in inverter circuits designed to drive induction motors with high frequency alternating voltages. We shall not discuss this rather complex behaviour in detail—suffice it to say that switching is limited by the time taken to establish (or remove) the minority carrier charge stored within the drift region and, in consequence, both switching times are of the same order as the recombination lifetime τ, typically of order 10–20 μs for lightly doped silicon. However, as we pointed out earlier, to achieve switching speeds faster than the audio range (i.e. 20 kHz or more) implies switching times of less than 10 μs so, in practice, it is usually necessary to take action to reduce τ. This can be achieved by introducing certain impurity atoms such as Au or Pt into the silicon or by irradiating the silicon with a flux of high energy electrons, to generate 'damage centres'. In either case, the effect is to produce 'recombination centres' which actively increase the recombination rate. These procedures work very effectively but tend to result in increased forward losses, as referred to above, so in many power devices it is necessary to strike a compromise between these conflicting requirements, a compromise usually determined by the application. So much for the simple diode—there are many more sophisticated power devices and we end this section by summarizing the properties of some of them.

By the end of the 1950s, power transistors (bipolar junction transistors, BJTs) and thyristors were well established and the addition of a gate electrode (to form the SCR) allowed external control of thyristor switching, though their ratings were extremely modest by comparison with present-day devices. Thus, in 1960, a typical thyristor specification was 200 V reverse blocking voltage, combined with 100 A current carrying capacity. Since then, both these parameters have improved steadily, leading to today's impressive values of 10,000 V and 5000 A in a single device, largely as a result of the introduction of larger and more uniformly doped silicon wafers (crystal growth continues to play a pivotal role!), together with improved thermal design and packaging. However, in addition, several important new devices have emerged. First, in the early 1960s came the so-called GTO, a modified version of the SCR. Not only could the GTO be turned on by a positive gate pulse, but the arrangement of interdigitated gate and cathode regions provided sufficient coupling between them that a negative pulse on the gate could turn the device off as well, thus achieving much greater flexibility. (Recall the need for turn-off in inverter circuits which drive induction motors at high frequency.) Then, after considerable development effort, the power MOSFET appeared in the late 1970s, providing the first switch having a high input impedance. Being a voltage controlled device, it has the advantage over the BJT of requiring very low switching power. (Because the power BJT has a large base width, it tends to have rather low gain, implying the need for large base switching current.) In addition, being a majority carrier device, it also had the advantage over the bipolar transistor of much enhanced speed and has replaced the BJT in many applications where switching frequencies up to 1 MHz are required. However, it does suffer the disadvantages of having much smaller current and blocking capabilities and being considerably more expensive. Finally, in the early 1980s the combination of MOS - and bipolar technologies led to the introduction of the IGBT which is a versatile switching device with speeds in the range 10 kHz–1 MHz combined with low switching power and low on-state loss. Development continues, of course, to improve details of performance and to meet specific challenges but it is clear that the range of devices now available is capable of satisfying most demands, and applications are growing steadily. As the scale of integrated circuits has become ever smaller, that of power devices has grown larger and its success is almost equally impressive (if, perhaps, not so widely heralded!). Interestingly, recent developments point to the combination of power and logic circuits on one and the same semiconductor chip leading to improved cost-effectiveness and to even greater flexibility in application. A noteworthy example is the control of all the electrical functions in an automobile by running a high current bus round the body with switching elements at each point of use (e.g. each headlamp, sidelight, indicator, etc.), but one may be sure that this is only one of many such applications which will materialize in the coming years. Electric power is clean and convenient and its efficient use and control will continue to be of paramount importance to civilised man—'smart power' is certainly here to stay.

4.6 Silicon is good for physics, too

It will not be lost on the reader who has followed our story thus far that the burgeoning use of silicon in our daily lives began with physics. It was the urge to comprehend the very nature of electrical conduction in these new materials (which came to be known as semiconductors) that led, eventually to their application in solid state devices. The discovery of transistor action, itself, was an outcome of attempts to understand the physical nature of surface states on silicon and germanium. The physics of the *p-n* junction came before its application in high power silicon rectifiers. Even when new devices appear (as they sometimes do) before the physics of their operation is fully understood, a proper understanding is usually essential to their ultimate optimization. Physics and technology are clearly inseparable in any high-tech developments such as those described here, and we shall find numerous further examples in subsequent chapters. Physics obviously stimulates technological advance—what is, perhaps, less immediately obvious is the reverse process in which technological advance stimulates physics. In this section, we shall examine a small handful of ways in which physics has benefited from the advance of silicon as a material of technological importance, looking first at some aspects of silicon itself, then an exciting development which grew from the study of MOS structures.

In the case of both silicon and germanium two important features resulted from the considerable effort devoted to their technological development: first, in the early 1950s, they became available in the form of large, high quality single crystals and second, they were prepared with controlled doping, down to levels of the order 10^{19} m^{-3} (roughly 1 part in 10^{10}). This probably made them the purest materials available to man, presenting both a challenge and an opportunity to the semiconductor physicist. An important aspect of such high purity is the relatively large separation of dopant atoms (of order 10^{-6} m). Compare this distance, for example, with the size of the Bohr orbit of the donor electron described in Box 1.4 (approximately 2×10^{-9} m). This implies that donor atoms can be regarded as strictly isolated and there is no need to consider interactions between them, a distinct advantage in studying their physical properties. We shall look at two examples of studies which gain from this, both dating from the early years of silicon physics, one from the mid-1950s and the second from the early 1960s.

In his recent book *How the Laser Happened*, Charles Townes (1999) describes the exciting developments which preceded the first laser: these were concerned with a radically new microwave amplifier known by the acronym MASER (Microwave Amplification by Stimulated Emission of Radiation) which eventually found application in the first satellite communication experiments and made a vital contribution to radio astronomy. In fact, it was a maser which detected the first microwave radiation to be recognized as coming from the cosmologically famous Big Bang. Townes and others spent much time during the 1950s searching for suitable atomic systems with which to demonstrate maser action, a principal requirement being an 'inverted population'. Einstein had provided the necessary physical understanding as early as 1917. If we think of any atomic system which can be described in terms of just two energy levels, interacting with a 'radiation field' we can define three processes involving exchange of energy between the atom and the field: (1) absorption—a photon may be absorbed from the field by an atom in its 'ground state' (the lower of its two energy levels), the atom thereby being excited into its upper level, (2) spontaneous emission—an atom in its excited state may spontaneously emit a photon into the field, thus losing energy and falling back to its ground state, (3) stimulated emission—an atom in its excited state may be stimulated to emit a photon by interacting with the radiation field—this is the inverse of the absorption process. It is this third process which is important in maser or laser action because it effectively generates photons which are coherent with the stimulating photons, thereby increasing the intensity of the radiation, that is, amplifying it. The difficulty in utilizing the process to make a practical amplifier lies in the inevitable coexistence of absorption which has the opposite effect of removing photons from the field. It is easy to see that, if there were more atoms in the upper state than in the lower, then stimulated emission would win over absorption and lead to net gain in radiation intensity, the problem being that, in thermodynamic equilibrium, there are always more ground state atoms than excited atoms. The secret of achieving maser action is to trick the atomic system into reversing this order of things, thereby producing an inverted population. This is not a trivial exercise and various approaches were tried before the ruby maser demonstrated practical success. One of these approaches utilized the energy levels of donor electrons bound to phosphorus impurity atoms in silicon.

Before we can understand the nature of the energy states involved, we must take note of the relevant photon energies. To make a microwave amplifier at a frequency of 10 GHz (10 kMcs sec^{-1}, as it was then known!) we need a pair of energy levels separated by $h\nu = 4.13 \times 10^{-5}$ eV (see equation 1.2), very much smaller than any energy difference we have considered in connection with semiconductor properties, to date. In practice, it turns out that the most convenient method of obtaining such small differences is to apply a magnetic field H to the silicon which produces an appropriate splitting of the ground state levels of the donor electron in its hydrogen-like orbit round the P atom. This comes about as a result of the electron 'spin'. In the absence of the magnetic field, each electron energy level is twofold degenerate in spin—application of the field splits the degenerate states apart by an amount $\Delta E = g\beta H$, where g is the electron g-factor (approximately equal to 2) and β is the Bohr magneton which takes the value 5.79×10^{-5} eV T^{-1}. It follows that a (conveniently generated) magnetic field of $H = 0.36$ T results in a splitting which corresponds to the required frequency of 10 GHz. All that is necessary in order to set up the experiment is to cool the silicon down to a sufficiently low temperature to ensure that all the donor electrons are trapped on P atoms ($kT \ll 50$ meV) and, ideally and much more restrictively, that most of the electrons are in their spin ground states ($kT \sim 10^{-5}$ eV). This demands $T \sim 0.1$ K. In practice, temperatures of about 1 K are used because they are obtainable using liquid helium—to obtain lower temperatures would make the experiment unduly complicated. This limits the amount of population inversion obtained but does not invalidate the principle.

To confirm the validity of the above analysis, experimental work was undertaken, first in France, at L' Ecole Normale Superieure, then at Bell Labs (where the high quality silicon came from, in the first place). First, an electron spin resonance study was made—this is essentially a measurement of absorption of microwave photons by the spins and is detected by placing the sample in a microwave cavity (essentially a metal box which is resonant at a frequency $\nu = 10$ GHz) and varying the applied magnetic field H until the condition defined in equation (4.12) is satisfied:

$$h\nu = g\beta H. \tag{4.12}$$

The first surprise was that not one but two absorption lines were seen, separated by an energy of order 10^{-6} eV, due to so-called 'hyperfine' interaction between the electron spin and the nuclear spin of the P^{31} nuclei ($I = 1/2$). Note that this result immediately tells us that the wavefunction of the loosely bound electron has a finite amplitude at the central nucleus—in other words, the electron penetrates the outer screen of bonding electrons to interact with the P nucleus within. In fact, this implies that the original assumption on which the hydrogen model of the donor is based cannot be quite accurate—this assumption was that the core of P nucleus and bonding electrons together would act like the nucleus of a hydrogen atom and the fact that the electron penetrates into the centre of the core contradicts this. Indeed, it is consistent with the observation that the binding energies of donor electrons do depend slightly on the nature of the donor (see Box 1.4).

The next stage in the attempt to achieve maser action required the population distribution between the two electronic energy levels to be inverted and this might be achieved by sweeping the magnetic field rapidly through the absorption condition (a result which was first demonstrated in nuclear resonance experiments). This experiment was shown to work and net gain was obtained (though barely sufficient to overcome the losses inherent in the microwave cavity), but even this degree of success could only be achieved by resorting to an amazing experimental initiative. George Feher at Bell Labs found that the line width of the spin resonance absorption was considerably increased by the interaction between the orbiting electron and the nuclear spins of Si^{29} nuclei which are present in naturally occurring silicon at a level of about 5%. By growing a silicon crystal using material from which most of the Si^{29} had been removed, he was able to reduce the line width by a factor of 12 and obtain a corresponding improvement in gain. Having done this, however, the next question was: how long would the inverted population last before thermal equilibrium was re-established? In most spin resonance experiments the so-called spin-lattice relaxation time is found to be of order milliseconds (or even microseconds). The second surprise, in this case, therefore, was the observation of very much longer times, amounting to as much as a whole hour with the sample cooled to 1 K, though decreasing fairly rapidly if the temperature was increased. Such amazingly long relaxation times spoke volumes for the extremely high quality of the silicon crystals used.

A completely different approach to studying the properties of donor impurities uses the technique of photoluminescence in which a silicon sample is excited by means of a light source with a photon energy greater than the silicon band gap, while the experimenter monitors (with a monochromator and suitable light detector) the nature of light emitted from the sample at energies below the band edge. The exciting light generates electrons and holes in the silicon which rapidly thermalize to the bottom of the conduction band (top of the valence band) and then recombine by one or more processes. Silicon, being an indirect gap semiconductor, radiative recombination (i.e. recombination which generates photons) tends to be very inefficient because the recombination process must conserve both energy and momentum and therefore requires the involvement of a phonon to provide the necessary momentum. Nevertheless, at low temperatures, a rich spectrum of emission lines can often be observed. Chief among these are lines associated with 'exciton' decay so we must first come to terms with the nature of excitons.

An exciton is essentially a hydrogen-like 'atom' made up of a hole and an electron, held together by their Coulomb attraction, the binding energy E_X being calculated along the same lines as used in Box 1.4 to determine the donor binding energy E_D. The only change involved is the use of a reduced effective mass $m_r = \{m_e^{-1} + m_h^{-1}\}^{-1}$, rather than the simple electron mass. For a semiconductor like silicon, in which the electron and hole masses are approximately equal, this implies an exciton binding energy roughly half that of E_D, that is $E_X \approx 15$ meV (see Box 1.4). Such an electrically neutral species is free to wander through the silicon crystal as a consequence of its having thermal energy and this thermal motion is reflected in the line width of the luminescence spectra. It will eventually decay as a result of electron–hole recombination and we note that, in silicon, the process must involve an optical phonon to conserve overall momentum. As a consequence, the photon energy $h\nu$ is given by the relation:

$$h\nu = E_g - E_X - E_p, \tag{4.13}$$

where E_p is the appropriate phonon energy. This is the so-called 'free exciton' which is not associated with any impurity or defect in the crystal. However, it turns out that excitons may become localized by being bound to impurities, in which case they lose their momentum to the lattice. In particular, they can form complexes with either neutral donors in *n*-type silicon or with neutral acceptors in *p*-type material. The neutral donor complex which is our current interest can be regarded as consisting of a positive donor core which binds three particles, two electrons, and one hole. Overall, it is electrically neutral. Because the complex is strongly coupled to the lattice, recombination may occur without phonon involvement, the necessary momentum being given directly to the lattice, so the emission energy is now given by the equation:

$$h\nu = E_g - E_X - E_B, \tag{4.14}$$

where E_B is the binding energy associated with the exciton localization. In a moment we shall see the relevance of all this to photoluminescence measurements on silicon crystals.

The confident statement above, about excitons being bound to neutral donors or acceptors can now be supported by perhaps hundreds of measurements on a wide range of semiconductors but in the late 1950s

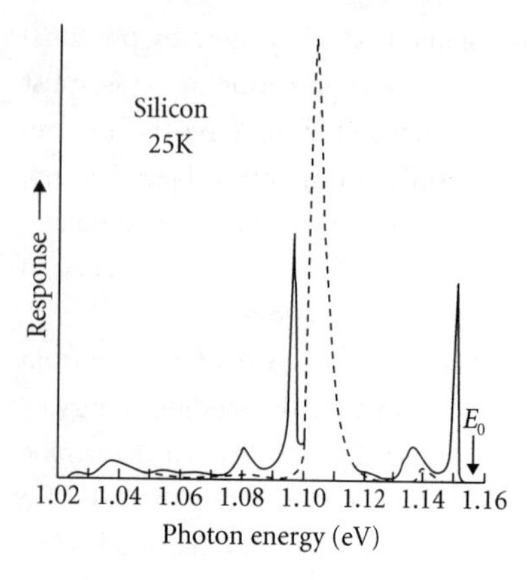

Figure 4.14. Photoluminescence spectra of two silicon samples measured at a temperature of 25 K. The broader line at 1.1 eV represents free exciton recombination in a high purity crystal, the sharper lines are attributed to the recombination of excitons bound to neutral As donor atoms in a sample doped with 8×10^{22} As atoms per m^3. (From Haynes, J. R. (1960) *Phys. Rev. Lett.* 4, 361, fig. 1). Reprinted with permission from the American Physical Society.

no such experience was available. It was only in 1960 that Richard Haynes discovered and interpreted some new luminescence spectra from *n*-type silicon crystals in terms of donor-bound exciton recombination. Haynes had earlier worked with Shockley on the junction transistor and is probably best known for the now famous Haynes–Shockley experiment (1949) which demonstrated the diffusion of minority carriers along a narrow filament of germanium. Following this tour de force, Haynes spent much of the next decade studying the emission spectra of germanium and silicon and elucidating the nature of near band edge optical transitions so it was a natural development for him to recognize these new lines. The experiments were made possible by the availability of high quality silicon crystals with, on the one hand, very low impurity content and, on the other, well-controlled levels of known impurities. Bell Labs was the right location for any scientist needing such a facility in 1960.

Luminescence measurements were made at a temperature of 25 K on (1) a pure silicon crystal and (2) a crystal containing $8 \times 10^{22}\ m^{-3}$ arsenic atoms, added to the melt. The two spectra are shown in the same plot in Figure 4.14 and clearly differ in various ways. The pure crystal shows a strong, somewhat broadened emission line which peaks at a photon energy of 1.099 eV. If we take the band gap of silicon at low temperatures to be $E_g = 1.165$ eV, the exciton binding energy $E_X = 14$ meV and the transverse optical (TO) phonon energy $E_{TO} = 58$ meV we can calculate the expected position of this line (from equation 4.13) to be $h\nu = 1.093$ eV, in rather good agreement. The width of the line (7 meV) and its asymmetric shape arise from the thermal motion of the exciton through the crystal lattice. On the other hand, the lines from the As doped sample are extremely sharp (Haynes measured a value less than 0.5 meV), showing that these lines are to be associated with bound, rather than free excitons. The measured photon energies of 1.091 and 1.149 eV are exactly one TO phonon energy apart, suggesting that emission occurs both with and without phonon participation. Note that the lower energy line appears just 8 meV below the corresponding free exciton peak, suggesting that the binding energy of the exciton to the neutral donor is $E_B = 8$ meV. This value is, in fact, a little too high—the comparison to be made is with the free exciton energy corresponding to those excitons which have zero kinetic energy and, making this adjustment, results in a more accurate value of $E_B = 6.5$ meV.

Haynes repeated these measurements for a range of donors and acceptors and found there was a simple approximate relationship between the measured values of E_B and the donor or acceptor ionization energies: $E_B = 0.1E_i$. This relationship has become known as Haynes' Rule and (with some qualification) is observed to hold for many other materials.

The development of the MOS transistor was, as we have emphasized earlier, a vital one for the long-term future of the integrated circuit.

It depended not only on the near perfection of silicon as a semiconductor but also on the technologist's ability to grow a high quality oxide on its surface. The oxide performs as the dielectric in what is effectively a silicon-oxide–metal capacitor and the device depends on the fact that applying a voltage to the metal gate produces an inversion layer close to the oxide–silicon interface, in the case of an NMOS transistor, a thin sheet of free electrons. The technological importance of the structure led to its being the subject of in-depth study during the 1960s, particularly with regard to the nature of the electron states existing in the oxide and in the region of the interface. It was the low density of such states in the Si–SiO_2 system which made the MOS transistor possible in the first place but the response of these states to rapid changes in gate potential also plays an important role in device operation. In addition, the presence of fixed or relatively immobile charge in the oxide can influence the static characteristics of the transistor and result in undesirable drift. Studies of capacitance and conductance at frequencies in the 1–100 kHz range demonstrated the existence of: (1) 'slow states' in the oxide associated with sodium impurity atoms which were finally eliminated by attention to good housekeeping during oxide growth, (2) fixed charge near the oxide–silicon interface due to the presence of excess silicon ions which can be minimized by suitable choice of growth conditions, and (3) 'fast states' at the interface which were shown to be distributed in energy within the silicon energy gap, their density being roughly 10^{15} $m^{-2}eV^{-1}$ near mid-gap, rising to 10^{16} $m^{-2}eV^{-1}$ near the band edges (qualitatively similar to the deep states in amorphous silicon which we shall meet in Chapter 10).

Apart from the details of the capacitor response, the mobility of free carriers in the MOS channel is also of importance in determining device performance and this led to yet another branch of MOS studies which grew to particular prominence in 1980 with the discovery of the quantum Hall effect. The electron mobility in an inversion layer may be measured in several different ways but one important method is based on the Hall effect. As explained in Section 2.4 (see Box 2.2), by measuring both Hall coefficient and resistivity, it is a straightforward matter to derive values for the free carrier density and mobility. This had been done over a range of temperatures in order to understand the behaviour of electrons in bulk silicon and the results interpreted in terms of appropriate scattering mechanisms. Broadly speaking, the mobility was found to increase as temperature was reduced below room temperature (being limited by lattice scattering), reach a maximum, then decrease again (as a result of ionized impurity scattering—the donors which supply the free electrons are ionized as a result of losing an electron and are therefore positively charged). Lattice scattering becomes less important as the temperature is decreased because fewer phonons are excited at lower temperatures, while scattering by ionized impurities becomes stronger at low temperatures because electrons are moving more slowly (they have less thermal energy) and experience the Coulomb potential of the impurity for a longer time. Similar studies of electrons in the MOS channel revealed them to be less mobile than their bulk counterparts at room temperature, due to scattering by structural imperfections at the silicon–oxide interface, known as 'roughness scattering'. The mobility showed a similar increase as the temperature was lowered but showed no evidence for the down-turn at very low temperatures, this being due to the different way in which the electrons are produced—the fact that they are induced by the gate voltage means there is no necessity for donors, and ionized impurity scattering is therefore minimal. In fact, it was the availability of relatively high concentrations of high mobility electrons at low temperature ($\mu_e \sim 30{,}000\ \mathrm{cm^2 V^{-1} s^{-1}}$) that led to the discovery of the quantum Hall effect which we now go on to describe.

As explained in Chapter 2, the Hall voltage developed across a Hall bar under the influence of a magnetic field applied normal to the sample surface increases linearly with both sample current I_x and magnetic field B_z. This is the classical Hall effect and allows us to determine the free carrier density n in the bar. Thus, from equation (B2.5), we have:

$$V_H = R_H B_z I_x / W, \tag{4.15}$$

where the Hall coefficient $R_H = -1/en$ and W is the thickness of the conducting channel ($W \sim 10$ nm in a MOS channel). We can rewrite this equation in terms of a so-called 'Hall resistance' V_H/I_x as:

$$V_H/I_x = R_H B_z / W, \tag{4.16}$$

which suggests that the Hall resistance should increase linearly with magnetic field or with the Hall constant (in a MOS device R_H may be varied by varying the gate voltage). Experiments at low magnetic fields showed precisely this behaviour but when the field was increased to the much larger values obtainable with superconducting magnets (typically 10 T or more) a new and totally unexpected result was observed, as reported in a joint publication from Klaus von Klitzing, Gerhard Dorda, and Mike Pepper in the 1980 volume of *Physical Review Letters*—a European success based on three separate institutions, illustrating the way in which much science is now conducted.

Their sample design is illustrated in the inset to Figure 4.15. It consisted of a MOS bar of *p*-type Si, resistivity 0.1 Ω m, length (typically) 400 μm, width 50 μm with an aluminium gate electrode, and having metal contacts at either end and probe contacts along its side which allowed four point measurement of longitudinal resistance and Hall resistance. The oxide thickness differed between samples from 100 to

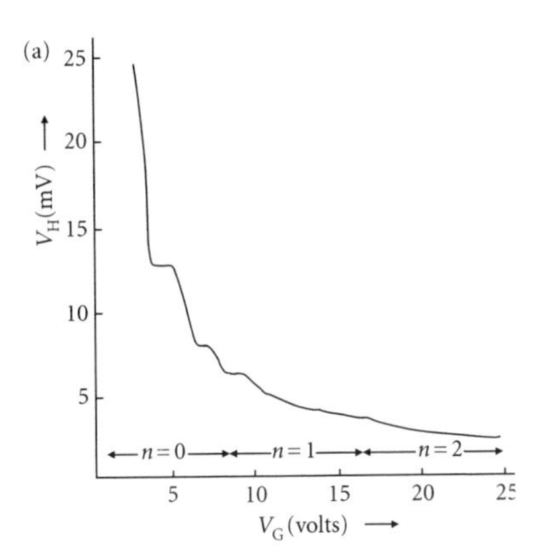

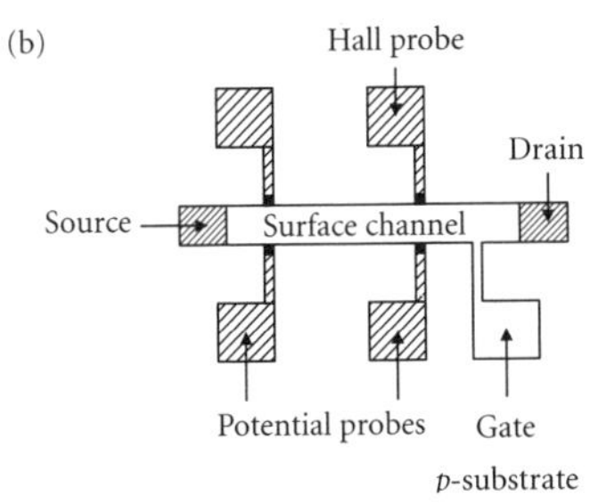

Figure 4.15. (a) Hall voltage as a function of gate voltage measured at 1.5 K on a MOS structure. The magnetic field was held constant at 18 T and the source–drain current at 1 μA. The feature to note is the existence of well-defined plateau in the curve which occur for values of the Hall resistance $V_H/I_x = h/ie^2$ (where $i = 1, 2, 3, \ldots$). Note that $n = 0, 1, 2$ refer to the lowest, first excited and second excited Landau levels, respectively.
(b) Illustration of the MOS structure. (From von Klitzing, K., Dorda, G. and Pepper, M. (1980) *Phys. Rev. Lett.* 45, 494, fig. 1). Reprinted with permission from the American Physical Society.

400 nm, the length/width ratio was similarly varied from 25 down to 0.65. Measurements were made at a temperature of 1.5 K in magnetic fields up to 18 T. The unusual feature of their results is illustrated in Figure 4.15 which shows the Hall resistance plotted against gate voltage, for a constant magnetic field of 18 T—the Hall resistance no longer shows a simple linear behaviour but contains flat 'plateau' regions at specific values of resistance which are given by:

$$[V_H/I_x]_{\text{plateau}} = h/ie^2, \qquad (4.17)$$

where h is Planck's constant and i is an integer ($i = 1,2,3,\ldots$). This remarkable result was accurately reproduced in all their samples, demonstrating that it was not dependent on sample shape or oxide thickness. It also occurred in a plot of Hall resistance against magnetic field with carrier density kept constant.

Since then, many other workers have confirmed the universality of the effect which has now been adopted as a standard of resistance, the value of plateau resistance (for $i = 1$) being 25812.807 Ω, to an accuracy approaching one part per billion. (Readers wishing to pursue this aspect can find further details in a series of papers which appeared in IEEE transactions IM-34 1985: 301 et seq.) We shall discuss the physics of the effect in greater detail in Chapter 6 with reference to AlGaAs/GaAs heterostructures but there can be no doubt of its considerable importance both practically and theoretically. The Hall effect had been regarded as fully understood for almost a hundred years before quantum effects made their appearance—another triumph for semiconductor technology/physics (or physics/technology, depending on one's point of view!).

Bibliography

Baden-Fuller, C. (ed.) (1996) *Strategic Innovation*, Routledge, London.

Benda, V., Gowar, J., and Grant, D. A. (1999) *Power Semiconductor Devices: Theory and Applications*, John Wiley & Sons, Chichester.

Bose, B. K. (ed.) (1997) *Power Electronics and Variable Frequency Drives*, IEEE Press, New York.

Braun, E. and Macdonald, S. (1982) *Revolution in Miniature*, Cambridge University Press.

Chapuis, R. J. and Joel, A. E. (1990) *Electronics, Computers and Telephone Switching*, North-Holland, Amsterdam.

Fransman, M. (1995) *Japan's Computer and Communications Industry*, Oxford University Press.

Jaeger, R. C. (1990) *Introduction to Microelectronic Fabrication*, Addison-Wesley, Reading, MA.

Matsuda, T. and Deguchi, K. (1998) 'Microfabrication technologies using synchrotron radiation', *NTT Rev.*, 10, 40 (see also several related articles in the same issue).

McCanny, J. V. and White, J. C. (eds.) (1987) *VLSI Technology and Design*, Academic Press, London.

Mohan, N., Undeland, T. M., and Robbins, W. P. (1995) *Power Electronics—Converters, Applications and Design*, John Wiley & Sons, New York.

Morris, P. R. (1990) *A History of the World Semiconductor Industry*, Peter Peregrinus Ltd (IEE), London.

Nathan, J. (1999) *Sony: The Private Life* Houghton Mifflin Co, Boston, MA

O'Mara, W. C., Herring, R. B., and Hunt, L. P. (eds.) (1990) *Handbook of Semiconductor Silicon Technology*, Noyes Publications, Westwood, NJ.

Prahalad, C. K. and Hamel, G. (1990) 'The core competence of the corporation', *Harvard Bus. Rev.*, May–June 1990, 79–90.

Reid, T. R. (2001) *The Chip*, Random House Trade Paperbacks, New York.

Riordan, M. and Hoddeson, L. (1997) *Crystal Fire—The Birth of the Information Age*, W W Norton and Co, New York.

Seitz, F. and Einspruch, N. G. (1998) *Electronic Genie—The Tangled History of Silicon*, University of Illinois Press, Urbana, IL.

Sze, S. M. (1969) *Physics of Semiconductor Devices*, John Wiley & Sons, New York.

Thomas, J. E. (1996) *Modern Japan—A Social History Since 1886*, Addison-Wesley Longman Ltd, Harlow.

Townes, C. H. (1999) *How the Laser Happened*, Oxford University Press, New York, Oxford.

CHAPTER 5

The compound challenge

5.1 Why bother?

The dramatic commercial success of silicon in digital and power electronics has tended to leave its rivals trailing sadly behind. Was there really any need to 'bother' with an alternative material? The current world market for silicon devices is roughly $200 billion, while its nearest challenger GaAs can boast device sales of little more than $5 billion, a tiny enough fraction when viewed against silicon, though far from negligible in absolute measure—clearly, this size of market made it well worth someone's while to bother! However, from our point of view, it is interesting to examine some of the reasons for, on the one hand, the dominance of silicon and, on the other, the necessity for other semiconductors to be involved at all.

There are, as we noted in Chapter 1, upwards of 600 known semiconductors, so why should just one of them assume such a dominant position? The answer is, of course, a commercial one. Silicon (obtained from sand) is not only one of the commonest elements on earth but is also, technologically speaking, the *simplest* semiconductor material with an appropriate band gap, making it significantly less expensive than all its rivals. As we have seen, germanium was the first semiconductor to be purified and made available in the form of high quality single crystals but its commercial promise was blighted by the problem of thermal runaway, inherent in its small band gap. Silicon was marginally more difficult to tame but came with a significantly larger gap and, as an unforeseen bonus, a stable oxide, which provided a low density of interface states. These advantages have allowed silicon to beat off all challengers in the area of what we might call (to beg most of the questions!) 'conventional' electronic devices, which may stimulate the reader into wondering whether it has *any* disadvantages. The answer is definitely 'yes'; silicon is not quite an *ideal* semiconductor for commercial application. Briefly, it possesses only modest electron and hole mobilities and its band gap is indirect. The first of these defects renders it less than ideal for devices, which must operate at very high frequencies, while the second precludes its application to lasers and one or two other optoelectronic devices which demand a steep band edge and strong electron–photon coupling (i.e. a strong probability that electron–hole recombination will result in light emission, rather than heat generation). In the context of light emission, it is also important that silicon's band gap (1.12 eV) is far too small to permit *visible* emission, which demands gaps in the range 1.6–2.8 eV (or greater). On the other hand, it is also far too large to absorb the long wavelength radiation (photon energies in the range 0.1–0.4 eV) employed in thermal imaging systems. Clearly, no one semiconductor can be expected to fulfil all such applications—a wide range of band gaps is essential and, in consequence, we shall devote several later chapters to these various aspects. The important point to grasp is that such applications tend to be more specialist than the broad span of conventional semiconductor electronics for which silicon caters and they therefore represent significantly smaller commercial markets. No one material can hope to achieve the same success as silicon in purely numerical terms, though this should certainly not be interpreted to imply lack of importance—imagine, for example, the world of communications without fibre-optics or the audio industry without the compact disc! Nor should we overlook the rich variety of physics associated with the wide range of semiconductor materials already developed—to risk being accused of tedious repetition, let me emphasize again the importance of the interaction between pure and applied research in furthering both intellectual and commercial advance.

It is in this context, therefore, that we approach the subject of compound semiconductors. Why compounds? Because the list of elemental semiconductors is strictly limited. Having taken account of germanium and silicon, there are rather few others—diamond is sometimes regarded as a semiconductor inspite of its band gap of 5.5 eV while selenium and tellurium also show semiconducting properties (remember the early selenium rectifier, described in Chapter 2) but there are few others, and none at all which can boast of being technologically developed. The extension to compounds immediately increases the number of options dramatically. We need look no further than the group III-V and II-VI materials to realize another 20-odd possibilities, many of which have received serious attention from the technologist. Add to these the possibility of forming ternary and quaternary alloys (e.g. InGaAs, CdHgTe, AlGaInP, ZnCdSSe) and the range of semiconducting properties becomes almost infinite! Not that such infinite variety comes without considerable toil and sweat—a correspondingly near-infinite amount of time and effort has been required to develop these multifarious materials to a state of practical usefulness—and it is a point of some interest that there should have been appropriate commercial pull to justify the necessary financial investment. At the time of writing, we can look back on no less than 50 years of developmental effort devoted to these compounds alone. I am not sure whether anyone has attempted to estimate the overall investment, which this represents, but it undoubtedly runs to several billion dollars. One must recognize that the compounds are inevitably more difficult to control than their elemental predecessors and bring with them considerably steeper (and correspondingly more costly) learning curves. Much of this vast development effort has therefore been based on quite remarkably far-sighted acts of commercial faith, stimulated by a small army of enthusiasts in industrial, university, and government research laboratories who had the vision to pursue the physics and device applications of compounds, frequently in the face of considerable scepticism. One such enthusiast, Cyril Hilsum has written a fascinating account of the British experience in III-V materials, which I strongly recommend as further reading (Hilsum 1995).

Given that there has, indeed, been significant success, both commercial and scientific, in the development of these materials, any attempt to present a coherent account of such wide ranging activity is faced with a fundamental dilemma. Work on the various compounds has taken place in parallel strands, each material having its own development programme, so a strictly chronological presentation is clearly impractical. The approach usually adopted in the past has been to treat the III-V and II-VI compounds as separate groups, which is convenient from a purely pedagogic viewpoint. However, it overlooks the undoubted fact that different materials have been developed in distinctly different contexts, usually (though not exclusively) associated with their band gaps. Thus the III-V compound InSb was the subject of a large development effort in the 1950s because it was seen as a serious candidate for the post of far-infrared detector in chief. This puts it in a quite different category from GaAs (which owes its commercial development to the interest in microwave transistors and semiconductor lasers) while suggesting a close relationship with the II-VI alloy CdHgTe (CMT), which has also been studied as a far-infrared material. Similarly, the wide gap III-V nitride semiconductors InN, GaN, and AlN are of prime interest as visible light emitters, which links them with wide gap II-VI materials such as ZnS, ZnSe, CdS, MgS, etc. against whom they are in direct commercial rivalry. A contrary argument might run along the lines that semiconductor physics is common to all such materials and their different applications are largely irrelevant—from this viewpoint, indeed, it probably makes better sense to discuss the different groups of materials in the conventional manner—hence the dilemma. One must make a choice and choices never please everyone! Mine is based on the premise that the *history* of semiconductors has been driven by applications and I shall therefore depart from the standard approach and treat each material in its commercial context. You may see this as the *Engineer* in me is taking precedence over the *Physicist*. (They have always been rivals, however friendly!) It does not, of course, imply any less respect for the importance of the appropriate physics—I have already nailed firmly to the mast my belief in the vitality of their mutual interdependence.

It merely (!) remains to decide how to order the different groups of materials. Once again, chronology hardly helps—work on examples of most of these semiconductors can be traced back to the 1950s or even earlier (for interesting accounts of the early work on III-Vs see Welker 1976 and the article by Hilsum 1995 referred to above). However, if one examines the degree of commercial interest pertaining to their development, a somewhat clearer picture emerges—apart from InSb, which we consider in Chapter 9, the first material to be taken up with a seriousness rivalling that of silicon was GaAs, followed, with somewhat less intensity, by InP and it is their fortunes which will dominate the present chapter. We follow this, in Chapter 6, with an account of so-called 'low-dimensional structures' (LDS) which were, in the first instance, largely based on GaAs and which strongly influenced the later development of GaAs devices. Two other materials which benefited from rapid technological development were GaP and InSb, the former for visible light emitters, the latter for infrared imaging systems, and their stories will be taken up in Chapters 7 (wide gap materials), and 9 (narrow gap materials), while the fascinating and vitally important subject of optical communications is dealt with in Chapter 8. As we shall see, there is much exciting subject matter which made it worthwhile for both device technologist and pure phycisist to 'bother'. The worldwide success of hundreds of commercial companies selling products based on compound semiconductors indicates that it was also worth the while of entrepreneurs and investors to follow suit.

5.2 Gallium arsenide

Investigation of many compound semiconductors had made remarkable progress by the end of the 1950s, an observation made quantitative by the equation given by Madelung (1964) in the preface to his book '*The Physics of III-V Compounds*'. This states that the number of papers published per year between 1952 and 1964 is given by:

$$N = 125[\exp\{0.1(t - 1952)\} - 1] \quad t \leq 1961$$
$$N = \text{const.} \qquad t \geq 1961 \tag{5.1}$$

In words, the number increased from 0 in 1952 to 300 in 1961, then remained sensibly constant for the next 3 years. According to Hollan et al. (1980), equation (5.1) somewhat exaggerates the initial steepness of the curve though they agree with the figure of 300 for the mid-1960s. After 1965, they suggest that, in fact, the number continues to rise to a value of about 2000 per annum in 1980. Whichever account one accepts, this represents an impressive rate of increase and sets compound semiconductor development only a matter of a few years behind that of silicon. However, we must be careful to compare like with like—perhaps a better measure of this time-lag would be to compare the years in which high quality single crystals first became available. This, for silicon, was 1952, when Teal and Buehler pulled crystals from the melt, while the first successful application of this Czochralski method to GaAs was reported only a few years later, in 1956, when Gremmelmaier described the use of a magnetic puller. However, the development of the now favoured LEC (Liquid Encapsulation Czochralski) technique had to wait until 1965, which simply emphasizes the difficulty in making meaningful comparisons of this kind! What *is* undoubtedly true is the fact that, in publishing their books in 1961 and 1964, respectively, Hilsum and Rose-Innes, and Madelung were able to summarize the physical properties of a wide range of III-V compounds, many valid measurements having been made on small crystals, which would hardly be viable as the basis for a commercial activity. While scientific knowledge lagged no more than a couple of years behind that of silicon, the lag in commercial development was probably nearer 10 years. But we are running on too quickly—first it is necessary to understand the potential importance of GaAs as a commercial rival to silicon and that requires some familiarity with its electronic properties.

GaAs crystallizes in the cubic, zinc blende form, each gallium atom being surrounded by a regular tetrahedron of four arsenic atoms and each arsenic atom by four gallium atoms. Both Ga and As appear in the same row of the periodic table as Ge, lying immediately to the left and right of it, an observation which prompted Gunn (1976) to describe GaAs as 'only germanium with a misplaced proton'! The similarity to Ge is also reflected in the bond lengths (i.e. Ga—As and Ge—Ge), which differ by no more than one part in a thousand, the respective lattice constants a_0 being 0.5658 nm (Ge) and 0.5653 nm (GaAs). Also, because the average (72.3) of the atomic weights of Ga (69.7) and As (74.9) is close to that of Ge (72.6), their densities are correspondingly close at 5323 and 5318 kg m^{-3}. Here, however, the similarities cease—GaAs is characterized by a direct energy gap of 1.43 eV (at room temperature), compared with the indirect gap of 0.664 eV for Ge (a consequence of the ionic component of the GaAs chemical bond) and most other properties show significant differences, too. Consequent upon the difference in band gap, the intrinsic free carrier densities differ very considerably, being $2.3 \times 10^{19}\,m^{-3}$ for Ge and $2.1 \times 10^{12}\,m^{-3}$ for GaAs (at room temperature), the corresponding resistivities being 0.5 and $10^6\ \Omega\,m$. This large value of resistivity for intrinsic GaAs can be put to practical use in the guise of so-called 'semi-insulating' substrate material on which conducting epitaxial films may be grown to fashion, for example, field effect transistors (FET). The substrate, having an extremely high resistance, represents a quite negligible parallel conducting path. How to achieve intrinsic conduction in practice may not be immediately obvious. A moment's thought makes it clear that using shallow donors or acceptors is impossibly difficult—it would be necessary either to reduce such dopant levels below the intrinsic free-carrier level of 2×10^{12} m^{-3} (i.e. roughly one atom in 10^{17}!) or to balance the densities of donors and acceptors at individual levels of 10^{20} m^{-3} to an accuracy of about 1 part in 10^7! Practical realization depends upon using impurities or defects, which provide so-called 'deep levels', lying close to the centre of the energy gap. These effectively trap any free electrons or holes resulting from unintentional shallow dopant atoms and prevent them playing any part in electrical conduction. It is only necessary that these deep impurities are present in densities greater than those of the shallow impurities. We shall see how this can be achieved when we consider the practicalities of doping GaAs.

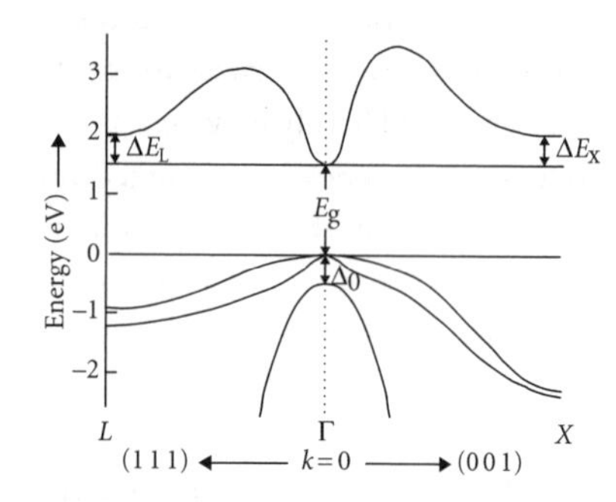

Figure 5.1. Simplified representation of the band structure of GaAs, plotted as energy vs momentum (E vs k) for the two crystallographic directions (0 0 1) and (1 1 1). (This should be compared with the corresponding case of silicon shown in Figure 3.10). The lowest conduction band minimum lies at the centre of the Brillouin zone ($k = 0$), known as the Γ point. This corresponds with the highest valence band maximum, making GaAs a direct band gap material. The next minima occur at the L point, with $\Delta E_L = 0.33$ eV, the X minima being 0.19 eV higher, that is, $\Delta E_X = 0.52$ eV. The valence band consists of three distinct bands which are closely spaced at the zone centre. Two of them are degenerate at $k = 0$, while the third (known as the 'split-off' band) is separated by an energy $\Delta_0 = 0.34$ eV.

A simplified version of the GaAs band structure is shown in Figure 5.1 for the two principal crystal directions (0 0 1) and (1 1 1). The valence band is similar to those of Ge and Si, being composed of three distinct bands, referred to as the 'light' and 'heavy' hole bands, which are degenerate (i.e. have the same energy) at the centre of the Brillouin zone (the Γ point), together with the 'split-off' band which lies a distance $\Delta_0 = 0.34$ eV below the other two at the zone centre. (This splitting is a consequence of spin–orbit coupling but the details need not concern us.) It is the conduction band, which differs significantly from those of Ge and Si, that gives GaAs its practical importance in the device world. The lowest conduction band minimum occurs at the zone centre, while the X and L minima lie some 0.3–0.5 eV above (remember, it is the L minima which are lowest in Ge and the X minima in Si). These numerically rather small differences have two immediate and important consequences: first, as we have already said, GaAs is a direct gap material which means that optically induced transitions across the fundamental gap take place without the need for phonon assistance and are therefore very much stronger than in the elemental semiconductors; second, as is generally the case for minima at the Γ point, the curvature of the band near its minimum is considerably sharper than is the case for X or L minima—in other words, the electron effective mass is much smaller. The value measured by cyclotron resonance for GaAs is $m_e = 0.067$ m, which is essentially the same for all crystal directions (compare this with the strongly anisotropic and much larger values for Ge and Si in Table 3.1). Recalling, from equation (3.5), that the electron mobility μ_e is inversely proportional to m_e, (light electrons are more easily accelerated by an electric field) we see that this leads us to anticipate a larger mobility in GaAs, compared with silicon. In fact, the ratio of masses is close to four so, other things being equal, we might expect a similar ratio of electron mobilities, however, scattering mechanisms differ somewhat and the actual ratio is even greater, the values being 0.9 m^2 V^{-1} s^{-1} for pure GaAs and 0.15 m^2 V^{-1} s^{-1} for pure silicon, a ratio of six. As already indicated, this suggests that GaAs will have an advantage over silicon in *electron* devices required to operate at high frequencies. (Because of the similarity in their valence bands, hole mobilities in the two materials are roughly equal, and no corresponding advantage can be anticipated for holes.) In the case of an *n-p-n* bipolar transistor, the diffusion of minority electrons across the base should be much faster and, in an FET, the electron drift velocity in the channel should be larger. We discuss these points in detail later (see Section 5.6).

Returning to the other major difference between GaAs and silicon, the direct energy gap of GaAs, we note that this gives rise to strong optical absorption above the band edge (i.e. when the photon energy $h\nu$ exceeds the energy gap E_g) and a correspondingly large rate of radiative recombination. We discuss radiative recombination in more detail in Box 5.1 but the important conclusion is easily stated: the radiative recombination lifetime τ_r takes values of order nanoseconds rather than microseconds and this implies much reduced minority carrier diffusion lengths, compared with silicon. It raises important problems for the development of GaAs bipolar transistors, as we discuss in Section 5.6 but, more importantly, it has very positive consequences for light emitting devices such as light emitting diodes (LEDs) and lasers, which we examine in Section 5.5.

Yet another difference between GaAs and the elemental semiconductors results from the nature of their chemical bonds, being purely covalent in the latter case but partially ionic in the former. In other words, the Ga—As bond is polar—that is to say there is an electric dipole associated with each (Ga—As) pair of atoms and this has consequences for the crystal as a whole. GaAs is a piezoelectric material—if it is squeezed mechanically along certain crystal directions (e.g. along (1 1 1)), a macroscopic electric field results. Alternatively, if an electric field is applied to the crystal, the lattice distorts. This property is important in a less obvious manner in that it contributes an important scattering mechanism between free carriers and phonons. Lattice vibrations represent periodic distortions of the crystal lattice so, in a piezoelectric material, they carry with them oscillating electric fields, which interact with free carriers. The resulting scattering, known as 'polar optical-phonon scattering' serves to limit free carrier mobilities in GaAs at room temperature, while it has no counterpart in silicon.

It has proved straightforward to dope GaAs either *n*-type or *p*-type. In this respect it follows germanium and silicon, though the reader should appreciate that this is not true of all semiconductors. Some wide gap materials, in particular, demonstrate a marked reluctance to exist in both conductivity types. For *n*-type doping, it is convenient to choose an element from Group VI of the periodic table, having one more outer electron than As. Thus, S, Se, or Te have all been used effectively, as they tend to substitute for As in the GaAs lattice and donate their 'spare' electrons to

Box 5.1. Radiative recombination in GaAs

The most significant difference between GaAs and the elemental semiconductors is, without doubt, its direct energy gap which implies a steep optical absorption edge and strong radiative recombination of electrons in the conduction band with holes in the valence band. It is easily understood that the probability for this direct recombination depends on the product of the electron and hole densities so we can write an expression for the rate of recombination R as:

$$R = -dn/dt = Bnp, \tag{B5.1}$$

where B is a constant. (The negative sign represents the fact that the recombination process results in a *decrease* in electron density). If we now consider a p-type sample with background doping p_0 and assume that the density n of minority electrons is much smaller than p_0, we can write:

$$-dn/dt = Bp_0 n, \tag{B5.2}$$

and, noting that Bp_0 is a constant, we can readily solve this differential equation to obtain an expression for n as a function of time:

$$n(t) = n(0)\exp\{-t/\tau_r\}, \tag{B5.3}$$

where τ_r is the 'radiative lifetime' and is given by:

$$\tau_r = (Bp_0)^{-1}. \tag{B5.4}$$

For direct gap materials, the constant B takes a value of order $B = 10^{-16}\ \text{m}^3\,\text{s}^{-1}$ so, for a GaAs sample doped at a moderate level of $p_0 = 10^{24}\ \text{m}^{-3}$, we find $\tau_r = 10^{-8}$ s or 10 ns. Compare this with the values of microseconds typical of lifetimes in silicon.

This short radiative lifetime has an important consequence. When we discussed minority carrier diffusion in Chapter 3, we defined a diffusion length $L = (D\tau)^{1/2}$ where D is related to the free carrier mobility μ as $D = \mu kT/e$. Thus, we can express the diffusion length as:

$$L = (\mu\tau kT/e)^{1/2} \tag{B5.5}$$

and evaluate it using a typical GaAs electron mobility of $0.5\ \text{m}^2\text{V}^{-1}\text{s}^{-1}$ and $\tau = 10^{-8}$ s to obtain $L = 11\ \mu\text{m}$. This value should be compared with the 100 μm typical of diffusion lengths in silicon.

the conduction band. (Oxygen, on the other hand gives an impurity level much deeper in the forbidden band and therefore cannot be used as a donor). According to the hydrogen model for shallow donors, we should expect the donor ionization energy to be small and the effective Bohr orbit to be correspondingly large, as a result of the small electron effective mass. Using equation (B1.6), with $m_e/m = 0.067$ and $\varepsilon = 12.5$, we obtain a value of $E_D{}^H = 5.8$ meV, which agrees well with experimental measurements of ionization energy. The corresponding Bohr radius is roughly 10 nm (some 40 times the Ga—As bond length). For p-type doping, the common

Carl Ferdinand Braun who shared the 1909 Nobel Physics prize with Guglielmo Marconi in recognition of his work on crystal detectors. Courtesy of The Nobel Foundation.

Famous photograph of John Bardeen, William Shockley and Walter Brattain following the invention of the Ge point contact transistor in 1947. Brattain resented Shockley's taking centre stage, when his involvement had been only peripheral. Courtesy of Lucent Technologies and AIP Emilio Segre Visual Archives.

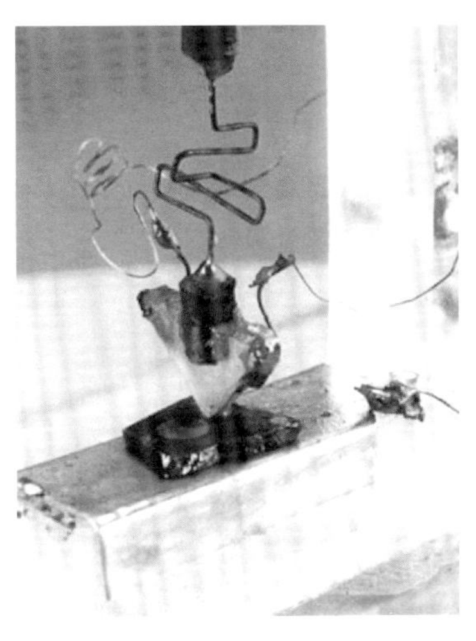

The first transistor – Brattain's innovative construction which allowed a pair of point contacts to be made to the Ge crystal with only 50 micron separation. Courtesy of Lucent Technologies Inc.

Photograph of Jack Kilby in middle life. Kilby demonstrated the first integrated circuit in 1958 shortly after joining Texas Instruments. Courtesy of Texas Instruments.

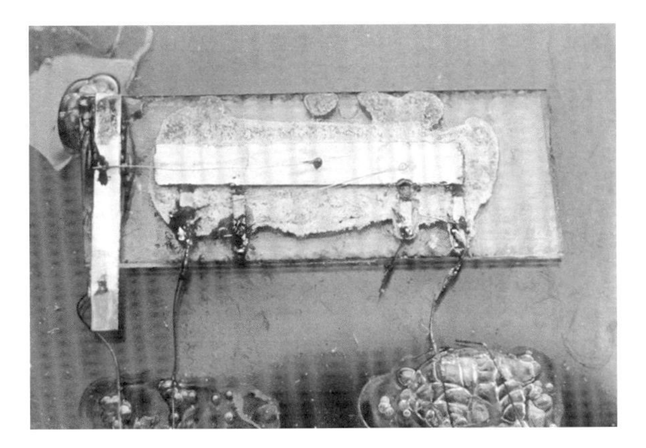

Kilby's innovative integrated circuit based on a single crystal of germanium. The functions: resistance, capacitance and gain were all provided by the Ge, without the need for external components. Courtesy of Texas Instruments.

Robert Noyes, who was responsible for the first planar integrated circuit which set the scene for the future explosion in integrated circuit technology. Noyes was employed by Fairchild Semiconductors at the time (1959). Courtesy of IEEE.

Robert Hall of GE Schenectady whose group won the race to demonstrate laser action in a forward biased p-n junction semiconductor diode toward the end of 1962. Courtesy of General Electric Laboratory and AIP Emilio Segre Visual Archives.

Marshall Nathan of IBM who led one of the four research groups to demonstrate laser action in GaAs or (in the case of Nick Holonyak, GaAsP) towards the end of 1962. Courtesy of Marshall Nathan.

Group photograph of attendees at the 1987 CLEO-LEOS conference which took place on the 25th anniversary of the first laser demonstrations. Courtesy of Nick Holonyak.

Nick Holonyak and Zhores Alferov (two of the leading semiconductor laser pioneers) meeting outside the Ioffe Institute in Leningrad in 1967. Courtesy Nick Holonyak.

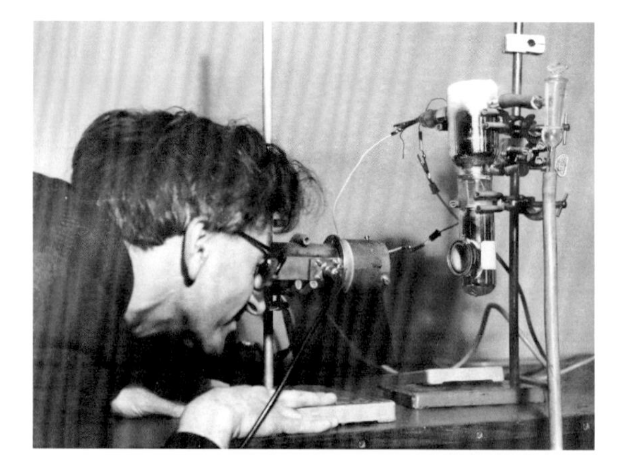

Cyril Hilsum using an image intensifier to examine the output from the first European laser diode in 1962. Courtesy Cyril Hilsum.

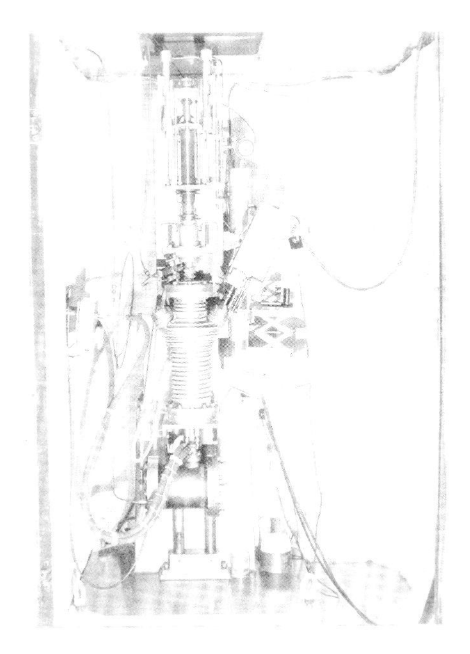

Photograph of one of the first high pressure Czochralski crystal pullers with which Brian Mullin and colleagues at RSRE Malvern demonstrated the virtues of the liquid encapsulation technique for growing bulk crystals of GaAs, GaP, InP, etc. Courtesy of Brian Mullin.

A recent photograph of Brian Mullin in his latter role as conference proceedings editor. This was the occasion of the ICCG-13 conference in Kyoto. Courtesy of Brian Mullin.

A 1970 photograph of Al Cho (with his technical assistant Charles Radice) in his Murray Hill laboratory during the pioneering days of III-V MBE growth. Courtesy of Al Cho.

Tom Foxon in the Philips Redhill Laboratory with the MBE team which in 1985 achieved GaAs two-dimensional electron gas structures, showing electron mobilities of over a million $cm^2V^{-1}s^{-1}$. Courtesy of Tom Foxon.

Bruce Joyce who played a major role in setting up the MBE activity at Philips, Redhill with the first MBE machine to be constructed there. It was later moved to Imperial College, London and remained active until very recently. Courtesy of Bruce Joyce.

Close-up view of a modern MOVPE machine, illustrating just how far it has acquired the appearance of its long-standing MBE rival. Stainless steel now abounds in both machines. Courtesy of Thomas Swann Scientific Equipment Ltd.

Photograph of Art Gossard (centre) and the team who first achieved modulation doping of an AlGaAs/GaAs heterostructure in 1978. The background is provided by yet another MBE machine. Courtesy of Art Gossard.

Sir Nevill Mott who shared the 1977 Nobel Physics Prize for his contribution to the understanding of amorphous semiconductors. Mott enjoyed a remarkably long scientific career, remaining active in research into his nineties. Courtesy of The Nobel Foundation.

Klaus von Klitzing discussing low dimensional structures with one of his research students. Von Klitzing won the Nobel Physics Prize in 1985 for his 1980 discovery of the Quantum Hall Effect. Courtesy of Klaus von Klitzing.

Philip Anderson of Bell Labs who shared the 1977 Nobel Prize with Mott. He first predicted the now famous 'mobility edge' in amorphous materials. Courtesy of Lucent Technologies Inc.

Daniel Tsui, who shared the 1998 Nobel Prize with Horst Stormer and Bob Laughlin for their discovery and interpretation of the Fractional Quantum Hall Effect in 1982, in his laboratory. Courtesy of Dan Tsui.

Horst Stormer, 1998 Nobel Prize winner, following his work at Bell Labs in 1982. Courtesy of Lucent Technologies Inc.

Bob Laughlin of Stanford University who was awarded a Nobel Prize for his theoretical interpretation of the Fractional Quantum Hall Effect, following its discovery at Bell Labs in 1982. Courtesy of Bob Laughlin.

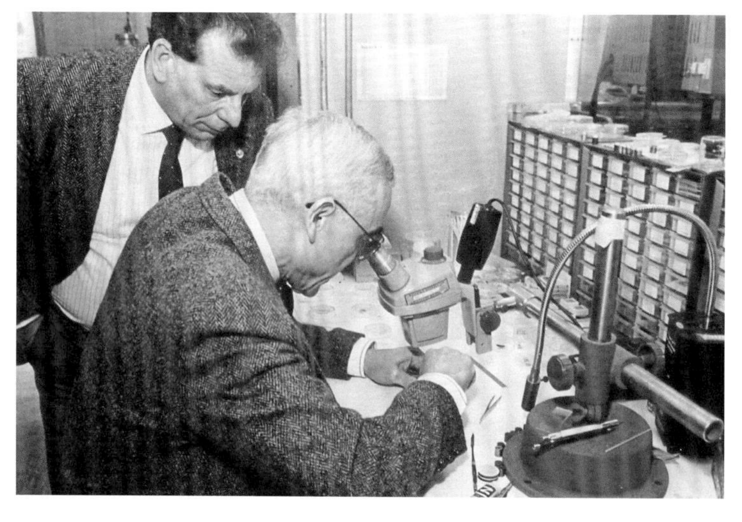

A 1988 photograph of Nick Holonyak and Zhores Alferov at work in Nick's laboratory at the University of Illinois, Urbana, examining a quantum well sample under the microscope. Alferov shared the Nobel Physics Prize in 2000. Courtesy of Nick Holonyak.

Herb Kroemer who shared the 2000 Nobel Physics Prize with Alferov and Kilby for his contribution to a wide range of semiconductor device problems. Courtesy of the Nobel Foundation.

Peter LeComber (left) and Walter Spear of Dundee University celebrating with their wives at a reception following the award of the Optoelectronics Prize for their work on amorphous silicon thin film transistors. Courtesy of Walter Spear.

Martin Powell who was responsible for the early development of amorphous silicon at the Philips Redhill laboratory. Martin went on to make a major contribution to the understanding of instabilities in a-Si:H. Courtesy of Martin Powell.

The first full colour liquid crystal TV display in Europe, which was developed at Philips, Redhill in 1987. The screen had a diagonal of six inches. Courtesy of Martin Powell.

Cyril Hilsum with a group of GEC colleagues looking at some of the first polysilicon TFTs in 1985. Courtesy Cyril Hilsum.

Martin Green holding a meeting of his Research Group at the University of New South Wales. This Group has been responsible for several important developments in the physics and technology of solar cells. Courtesy of Martin Green.

The Author with his Research Group at Philips Research Laboratories, Redhill in 1990.

Isamu Akasaki whose group first obtained p-type conduction in epitaxial GaN by the use of a low temperature AlN buffer layer, together with electron beam irradiation to activate the Mg acceptors. They later grew p-n junction diodes and obtained efficient blue electroluminescence for the first time. Courtesy of Isamu Akasaki.

Shuji Nakamura who, while at the Nichia Chemical Company, developed the first blue semiconductor laser diode, together with efficient blue and green LEDs which opened the door to full colour LED displays and to high brightness white emitters. Courtesy of Shuji Nakamura.

Until April, 1985

(a)

Since May, 1985

(b)

Persuasive evidence for the dramatic improvement in epitaxial film quality when GaN epitaxy was preceded by a low temperature buffer layer. The use of sapphire substrates allowed film quality to be monitored by the transmission of visible light. Courtesy of Isamu Akasaki.

Lester Eastman at work in his Cornell office. Eastman was a pioneer in the application of MBE growth to microwave semiconductor devices and has recently demonstrated high microwave powers from AlGaN/GaN HEMTs. Courtesy of Lester Eastman.

Al Cho of Bell Labs receiving the US National Medal of Science from President Clinton and Vice President Gore in 1993. Courtesy of Al Cho.

Enrico Capasso (left front) with the Bell Labs team which developed the quantum cascade laser using energy states in conduction band quantum wells in the AlInAs/GaInAs material system. Courtesy of Lucent Technologies Inc.

Leo Esaki (left) and Hiroyuki Sakaki (right) deep in discussion of the properties of advanced quantum well structures in 1976. Esaki won the Nobel Physics Prize in 1973. Courtesy of Hiroyuki Sakaki.

Just to prove that even the most dedicated scientists do not spend all their time in the laboratory – Hiroyuki Sakaki on the golf course at St Andrews in 1991. Courtesy of Hiroyuki Sakaki.

acceptors are those from Group II, including Be, Mg, Zn, and Cd all of which substitute for Ga. The valence band is characterized by a hole effective mass $m_h = 0.36$ m, which results in an acceptor ionization energy of $E_A{}^H = 31$ meV, which is in fairly good agreement with experiment, though there is some variation between the different dopants, the so-called 'chemical shifts'. This is not yet the whole story concerning doping. Elements from Group IV represent an interesting case of 'amphoteric' behaviour (i.e. in principle, they can act as either donors or acceptors). Thus, Si atoms entering Ga sites in the GaAs lattice act as donors, while Si atoms substituting for As act as acceptors. In practice Si normally behaves as a donor and has been widely used in molecular beam epitaxy (MBE) growth as the most convenient *n*-type impurity. However, under special circumstances such as epitaxial growth on the (1 1 1) crystal surface or at very high doping levels, it is also possible to observe acceptor behaviour. On the other hand, Ge and C atoms in GaAs generally tend to act only as acceptors, while Sn is always a donor.

The shallow levels associated with these impurities are reasonably well understood in terms of a slightly modified hydrogen model but there exists a wide range of impurities, which give deep levels within the energy gap. For example, many of the transition group metals such as Ag, Au, Ni, Cr, or Pt act as deep impurities and influence GaAs electronic properties

Table 5.1. Properties of Ge, Si, and GaAs

Property	Ge	Si	GaAs
Crystal type	Diamond	Diamond	Zinc blende
Lattice constant (nm)	0.5658	0.5431	0.5653
Density (kg m^{-3})	5323	2329	5318
Melting point (°C)	937	1412	1238
Band gap (eV)	0.667	1.12	1.43
CB minimum	L	X	Γ
CB effective mass ratio	$m_L = 1.64$	$m_L = 0.98$	$m_e = 0.067$
	$m_T = 0.082$	$m_T = 0.19$	
VB effective mass ratio	$m_l = 0.044$	$m_l = 0.16$	$m_l = 0.08$
	$m_h = 0.28$	$m_h = 0.49$	$m_h = 0.5$
	$m_{so} = 0.077$	$m_{so} = 0.25$	$m_{so} = 0.2$
CB effective density of states (m^{-3})	1.04×10^{25}	2.84×10^{25}	4.74×10^{23}
VB effective density of states (m^{-3})	6.14×10^{24}	1.044×10^{25}	8.04×10^{24}
Intrinsic carrier density (m^{-3})	2.34×10^{19}	1.00×10^{16}	2.10×10^{12}
Electron mobility (m^2 V^{-1} s^{-1})	0.39	0.15	0.90
Hole mobility (m^2 V^{-1} s^{-1})	0.19	0.05	0.04
Dielectric constant	16.2	11.7	12.5
Optic phonon energy (ev)	0.037	0.063	0.035
Thermal conductivity (wm^{-1} K^{-1})	60	130	46

in a more or less serious manner. Such impurity atoms may trap free carriers, thereby reducing the material conductivity and they may also act as non-radiative 'recombination centres', strongly reducing minority carrier lifetime. For example, a free electron may be trapped, then a free hole may recombine with this trapped electron, both processes giving energy to the lattice vibrations, rather than it being emitted in the form of light. The process has become known as 'Shockley–Read recombination' after the authors of a definitive paper published in 1952 and discussed in most semiconductor textbooks. (They were concerned with Ge and Si but the principle is completely general.) Two deep levels are of particular practical significance because they lie near the centre of the forbidden gap—these are Cr which can be incorporated in GaAs in a fairly well controlled manner and the so-called EL2 defect, which is associated with oxygen and appears in most bulk GaAs crystals. Either of these may be used to make GaAs semi-insulating, as described above.

For ready reference, Table 5.1 contains a brief summary of the main properties of GaAs where they are compared with the corresponding values for Ge and Si.

5.3 Crystal growth

To risk labouring a point made several times already, the key to success with any semiconductor material is the ability to prepare single crystals of high structural perfection with very low background impurity levels. Silicon came of age with the perfection of the Czochralski technique of pulling crystals from the melt and the steady increase in size and decrease in cost of integrated circuits (ICs) rides on the back of the ever-increasing diameter of Czochralski boules, from approximately 1 in. in 1952 to 10 in. today. One might therefore expect a similar progression for GaAs and, in a sense, that is what has happened, though with one important qualification—because of the difficulty in producing bulk crystals of adequate quality, the majority of GaAs commercial developments have depended on the use of epitaxial films. However, these films must be supported by a suitable substrate and this is universally taken to be a Bridgman or Czochralski-grown wafer of GaAs, which automatically provides a perfect lattice match with the film. The race to develop larger and larger substrates has therefore followed a similar pattern to that of silicon, though GaAs is currently limited to 6 in. diameter LEC wafers (most Bridgman wafers being only 3 in. in diameter).

But, having said this, it is important to emphasize two essential differences between the steady progression of silicon and the somewhat chaotic development of the III-V compounds. Because it was discovered relatively early that high quality bulk silicon crystals could be grown from the melt there has been little motivation to develop alternative growth methods, whereas the progress in III-V materials has been characterized by a plethora of growth techniques, both bulk and epitaxial. This is partly the result of the inherent difficulty in obtaining high quality III-V material, encouraging the hope that each new method might result in significant improvement over those already existing, but also stems from the impressively wide range of specialist applications for III-Vs, which have materialized over the years, each presenting its own challenges to the materials scientist. In practice, then, bulk single crystals of GaAs have been grown by both horizontal and vertical Bridgman methods in addition to the LEC process, while thin films have been produced by a variety of liquid epitaxial methods (tipping, dipping, and sliding, to mention but three), several vapour-phase processes (hydride, halogen, and metal-organic) and numerous modifications of the basic molecular beam epitaxial process (solid source, gas source, atomic layer, and migration-enhanced). While it is clearly beyond the scope of a global overview, such as we are engaged upon, to provide fine detail, some appreciation of the merits and demerits of these various techniques is essential to a proper understanding of compound semiconductor development—in this section we shall attempt to provide it.

Much of the GaAs used in early experiments (i.e. during the 1950s) was polycrystalline, grown from a solution of GaAs in liquid gallium (see, for example, the article by Cunnell in Willardson and Goering 1962). It was possible, nevertheless, to obtain small single crystal samples by carefully selecting appropriate areas from the otherwise polycrystal mass and these proved invaluable as a means to establish basic physical properties of the material. Indeed, some of the early laser work was performed with such samples, but it was clear that serious further development required larger, more uniform crystals which, in turn, demanded dedicated application of previously established growth methods. First to be tried were those generally known as Bridgman methods in which polycrystalline GaAs is melted in a suitable crucible, together with a small seed crystal and slowly cooled in such a way as to solidify the boule from one end (the seed end). Careful control of temperature profile and rate of cooling produced a single crystal boule. In the horizontal Bridgman process, illustrated in Figure 5.2, the charge is melted in a silica or PBN (pyrolitic boron nitride) boat within a horizontal furnace tube, the boat being slowly withdrawn so as to achieve the desired cooling profile. An alternative, which avoids mechanical movement (with concomitant vibration and slip problems), is to use a set of heater elements, which are programmed so as to move the temperature profile along the boat, this being known as the 'horizontal gradient freeze' method. In either case, special care is required to

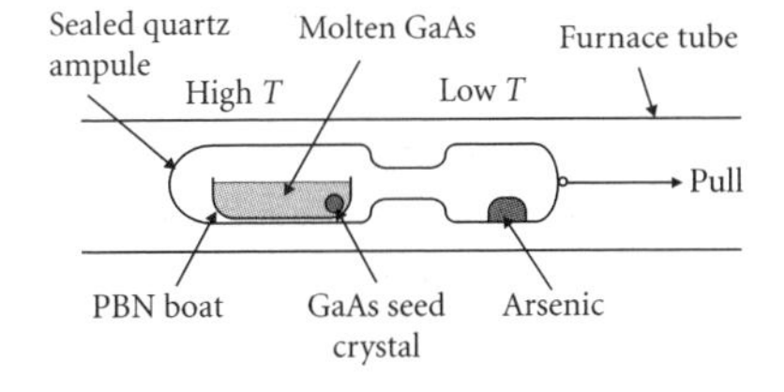

Figure 5.2. Schematic diagram of a Horizontal Bridgman crystal growth apparatus for the growth of GaAs. Molten GaAs is contained in a horizontal boat, sealed within a quartz ampule which is located within a horizontal furnace tube. The ampule is slowly withdrawn from the furnace so as to cool the GaAs boat from the seed end, leading to the formation of a single crystal boule. The arsenic, contained in a separate part of the ampule is maintained at a temperature of 617 °C to provide an arsenic vapour pressure of approximately 1 atm over the GaAs surface, thus preventing loss of arsenic from the boule.

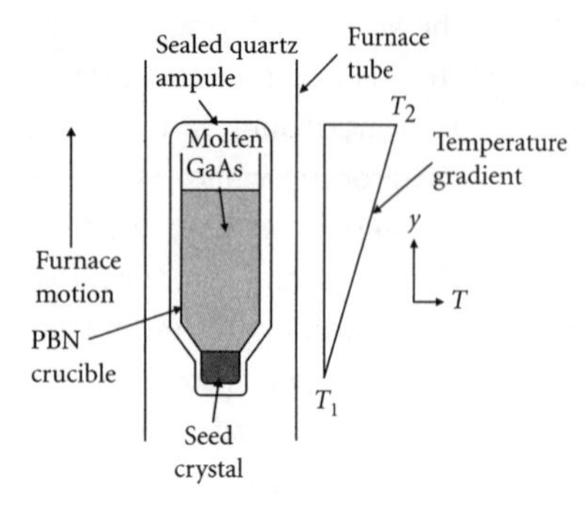

Figure 5.3. Schematic diagram of a VGF (vertical Bridgman) apparatus for the growth of GaAs crystals. A temperature gradient is established, as illustrated and moved slowly in the direction indicated to cool the GaAs from the seed end, as in the horizontal method.

prevent the loss of arsenic from the melt (As being a highly volatile element) and this is achieved by using two temperature zones in the furnace. In the second zone is a boat containing solid arsenic at a temperature of 617°C, such that the vapour pressure of As generated corresponds to the dissociation pressure of As over GaAs at the melting temperature of 1238°C (approximately 1 atm). The use of a horizontal boat minimizes contact between boat and melt, providing the twin advantages of reduced strain in the crystal and reduced impurity incorporation from the boat (particularly important in the case of silica boats, which act as a source of Si doping). It has the disadvantage of producing a crystal whose cross-section lacks axial symmetry and is therefore less convenient to handle in subsequent processing steps. It also suffers from a non-uniform impurity distribution, so the majority of Bridgman growth now uses the vertical configuration (introduced in 1986) shown in Figure 5.3. This results in a crystal with circular cross-section. It also allows the crystal to be rotated during growth so as to improve uniformity.

A second temperature zone, to provide an As over-pressure may still be used, though an alternative, made possible by the vertical arrangement, employs a liquid encapsulation of boric oxide (B_2O_3), which floats on the surface and seals the melt against possible As loss (an innovation borrowed some 24 years after its introduction into Czochralski growth of GaAs!). An advantage common to these Bridgman methods, and particulaly important in the development of GaAs lasers, is the relative ease with which low dislocation densities can be achieved. This, more than anything, has maintained the commercial importance of the vertical gradient freeze (VGF) method. The *n*-type crystals required for laser and LED production can be readily obtained by doping the melt with an appropriate amount of Si, Te, or Sn.

We have already met the Czochralski method of pulling crystals in Chapter 3 in connection with the growth of Ge and Si (Section 3.2), an example of the apparatus being shown in Figure 3.8. Following the considerable success in producing high quality crystals of these materials, it was natural that crystal growers should wish to apply it to the growth of GaAs too and Gremmelmaier's magnetic puller came on the emerging GaAs scene at a remarkably early date (1956). Because of the high vapour pressure of arsenic, he chose to enclose the melt in a sealed tube which, in turn, required the pulling and rotating mechanism to be magnetically coupled. Later work has made use of cleverly designed motion seals but these are always problematic in the presence of the corrosive arsenic vapour, given that the whole apparatus must be at a temperature of, typically, 630°C (sufficient to maintain the necessary arsenic pressure). The important breakthrough came with the development of the LEC method by Brian Mullin and co-workers at the Royal Radar Establishment, Malvern in 1965. By sealing the top surface of the melt

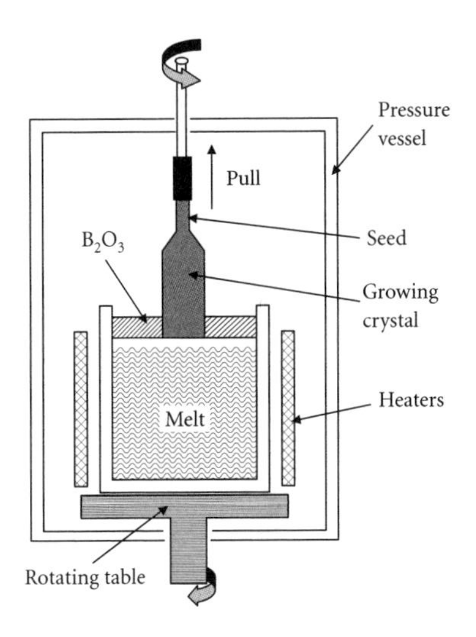

Figure 5.4. An example of a LEC system for the pulling of GaAs crystal boules from a melt which is enclosed below a layer of molten boron oxide which prevents arsenic loss from the melt. A GaAs seed crystal is introduced through the B_2O_3, then slowly pulled upwards into the cooler part of the furnace, while, at the same time being rotated about a vertical axis. This helps to achieve improved radial uniformity in the grown boule. An inert gas pressure is maintained within the pressure vessel to balance the arsenic vapour pressure over the melt.

with boric oxide, they were able to do away with the need for an As over-pressure and grow, instead, under a pressure of an inert gas such as nitrogen or argon. This neatly avoids the problems associated with corrosive atmospheres such as arsenic or (in the growth of GaP or InP) phosphorus. A typical apparatus is sketched in Figure 5.4. By loading the PBN crucible with arsenic and gallium, it is possible to synthesise GaAs *in situ* at a temperature of about 800°C, As loss being prevented by the boric oxide, which softens at the relatively low temperature of about 450°C and flows over the charge, to protect it. The temperature is then raised to the melting point (1238°C) and the seed crystal dipped through the boric oxide to make contact with the melt, the temperature stabilized and then the seed withdrawn, growing first of all a narrow neck to minimize dislocation density, followed by a gradual expansion to the final diameter of 2, 3, 4, or (now) 6 in. as required (for some reason, crystal diameters seem not to have been metricized!). This stage of the growth requires carefully programmed control of heat input and pulling rate which, in today's commercial pullers, has been taken over by the ubiquitous digital computer. This is also true of diameter control, using a sensor in the form of a laser beam reflected from the growth meniscus or, more frequently, the measured weight of the crystal. Growth rate for GaAs is approximately 1 cm/h with a rotation rate of 10 rpm. The inert gas pressure is some 20 atm. The LEC process has been widely used for producing semi-insulating wafers for FET processing, originally Cr-doped but more recently undoped—it was discovered in 1977 that synthesis of GaAs *in situ* in a PBN crucible, beneath a boric oxide film resulted in samples sufficiently pure that they were semi-insulating even without Cr doping. A critical aspect is the control of the water content of the B_2O_3, which results in efficient gettering of unintentional impurities.

Finally, it is well to remember that the grown boule is no more than the raw material. Nearly all applications call for a cut and polished wafer, whether it is to be used directly in device fabrication or simply as the substrate for epitaxial growth of the active layers. Thus, the crystallographically aligned boule must first be cut into slices using a diamond impregnated saw, then cutting damage is removed by chemical etching in (for example) sodium hypochlorite solution and, finally, both surfaces are polished using a combination of mechanical and chemical actions. This usually involves a fairly soft polishing cloth together with a solution of bromine in methanol, used on an automatic polishing machine. Immediately before use, the slice will usually be degreased, then given a light etch to remove surface oxides and any remaining contamination. Such preparation is often of crucial importance.

In the early 1960s, which saw the generation of efficient *p-n* junction luminescence, the demonstration of laser action and the discovery of the Gunn effect, the best GaAs available, grown either by the

Bridgman method or from solution in liquid gallium, had a total ionized impurity concentration ($N_D + N_A$) in the low 10^{22} m^{-3} range, with electron mobility at room temperature only slightly more than half the value of 0.9 $m^2 V^{-1} s^{-1}$ expected for pure material. This compared rather unfavourably with the situation obtaining for Ge and Si, where values of ($N_A + N_D$) in the region of 10^{17}–10^{18} m^{-3} could be achieved, and it became clear that alternative growth methods should be tried in the hope of realizing higher purity samples, there being a clear need for better material for the rapidly developing field of microwave Gunn oscillators. This provides the background to the development of vapour phase epitaxy (VPE) and liquid phase epitaxy (LPE) in the mid-1960s. LPE, as a development of the solution growth method, came first, in 1961, when Nelson of RCA described the so-called 'tipping' method of bringing a super-saturated solution of GaAs in liquid Ga into contact with a horizontal-Bridgman-grown GaAs substrate. VPE followed in 1965 when John Knight and co-workers at Plessey, Caswell in the United Kingdom and D. Effer also at Caswell (!) more or less simultaneously described the use of the $AsCl_3$–Ga–H_2 system to grow GaAs films on GaAs substrates. Both techniques were successful in producing material of improved purity and many other workers took up the challenge, developing ingenious variants on the basic methods.

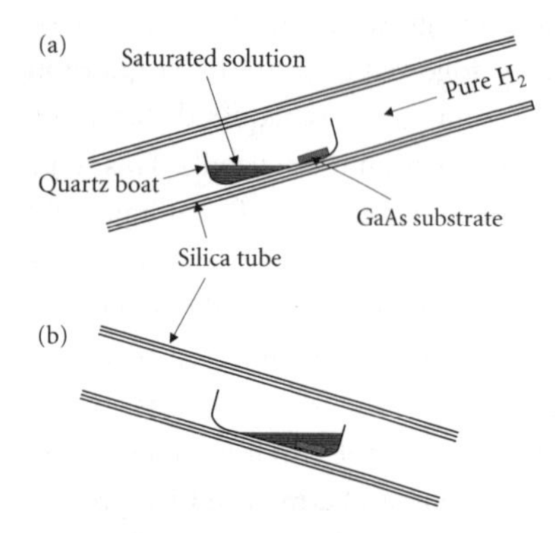

Figure 5.5. Schematic diagram of the 'tipping' method of LPE of GaAs. (a) GaAs is dissolved in liquid gallium at one end of the boat to form a saturated solution which is then tipped over the GaAs substrate located at the opposite end of the boat. A small reduction in temperature causes GaAs to be deposited from the solution to form an epitaxial layer on the substrate. (b) When the required thickness has been deposited, the solution is tipped back to its original position and the apparatus slowly cooled to room temperature.

The first LPE method, introduced by Nelson, is illustrated schematically in Figure 5.5. It featured a horizontal furnace tube mounted on a gimbal, which allowed it to be tipped in the vertical plane. The saturated solution of GaAs in liquid gallium was retained at one end of a silica boat, while, at the other end was held a GaAs substrate, highly purified hydrogen gas being flowed through the surrounding quartz tube. The solution was then allowed to flow over the substrate by tilting the furnace and the temperature slowly reduced (at a few degrees per minute) so as to grow a thin film of GaAs on the surface of the substrate. Finally, the solution was moved back to its starting position by reversing the angle of tilt. It was possible to grow films ranging in thickness from about 1–100 μm, doped films being produced by including suitable dopant elements in the solution or by introducing them into the hydrogen flow. Typical growth temperatures were in the range of 600–900°C. Since this early work, a great many modifications to the basic process have been introduced. In 1966, Ruprecht at IBM, Yorktown Heights reported on the so-called 'dipping' process in which a GaAs substrate is held above a saturated solution in a vertical furnace arrangement until a steady state is achieved, then the substrate is dipped into the solution to grow a desired thickness of GaAs, and, finally, removed to its original position. Slow cooling rates of less than 1°C min^{-1} were found to result in higher purity in the grown film. However, the most popular LPE

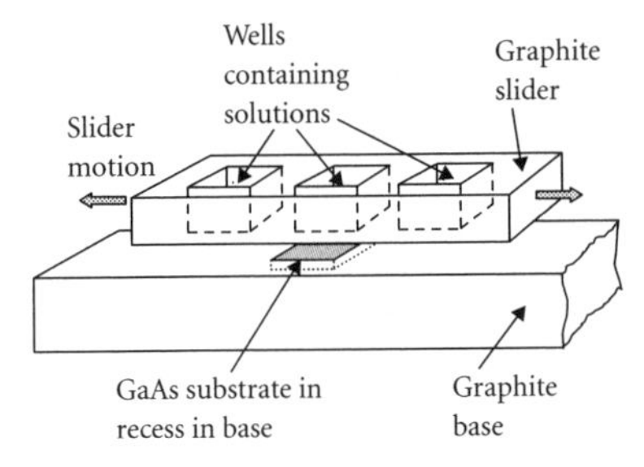

Figure 5.6. An example of a 'sliding' LPE apparatus for growth of GaAs and AlGaAs films on GaAs substrates. The substrate is contained in a small declivity in the graphite base, while various saturated solutions of GaAs or AlGaAs are held in wells in the slider. A small temperature gradient is maintained so that the substrate is at a slightly lower temperature than the melts. To deposit a film from a particular well, this well is moved over the substrate for an appropriate length of time, then moved away again and, if a second layer is to be grown, replaced by an other well.

method, which has the advantage of much greater flexibility, makes use of a sliding boat system similar to that illustrated in Figure 5.6. The basic idea was first put forward in 1969 by Mort Pannish and co-workers at Bell Labs, Murray Hill and has been much copied and improved upon. As clear from Figure 5.6, it makes use of several reservoirs in which different saturated solutions may be contained to allow growth of multilevel structures (e.g. *p-n* junctions or heterostructures such as AlGaAs/GaAs). The substrate (though, indeed, several may be used) is held in a shallow recess in the base plate to facilitate easy transfer of the slider over it and it is a common practice to add a suitable wafer at the top of each well in order to ensure the solution remains at saturation point. The component parts are generally made from graphite, rather than quartz, because the required precision demands that they be accurately machined. Purity of the grown films was found to improve when pyrolytic graphite was used—i.e. graphite treated at high temperature to densify it. Needless to say, the base and slider are mounted in a quartz tube through which pure hydrogen is flowed and the whole structure contained within a horizontal furnace tube. Quite large systems have been developed for application to device production and several manufacturers still offer LPE wafers for optoelectronic applications in the visible and infrared spectral regions.

That LPE realized a very important improvement in material quality is apparent from the state-of-the-art towards the end of the 1960s when the best samples of 'undoped' GaAs showed free electron densities of about $10^{19}\ \mathrm{m}^{-3}$ with total ionized impurity levels $(N_D + N_A) \sim 10^{20}\ \mathrm{m}^{-3}$ (cf $10^{22}\ \mathrm{m}^{-3}$ for bulk material). Though this is still some two orders of magnitude greater than the best silicon crystals and represents the effect of 'compensation levels' N_A/N_D in the region of 0.8, it nevertheless represented a dramatic step forward and, importantly, more than satisfied the requirements for Gunn oscillator material. In particular, room temperature mobilities were limited entirely by lattice scattering at a value of about $0.9\ \mathrm{m}^2\mathrm{V}^{-1}\mathrm{s}^{-1}$. The reader should appreciate, though, that such a favourable result was not achieved without much attention to detail—only the highest purity silica ware, gallium and GaAs starting material could be used and every item of apparatus had to be carefully cleaned and, in many cases, baked in pure hydrogen before use. Care was also necessary to minimize the take-up of Si from silicaware, a consideration based on thorough understanding of the chemical interaction between silica and hydrogen. The amphoteric nature of Si impurities in GaAs also plays a crucial role, depending, as it does, on both substrate orientation and on growth temperature and cooling rate.

It would be wrong, however, to give all the credit for these advances to LPE alone—VPE material very soon caught up with, and even surpassed, these LPE vital statistics therefore we must briefly examine the

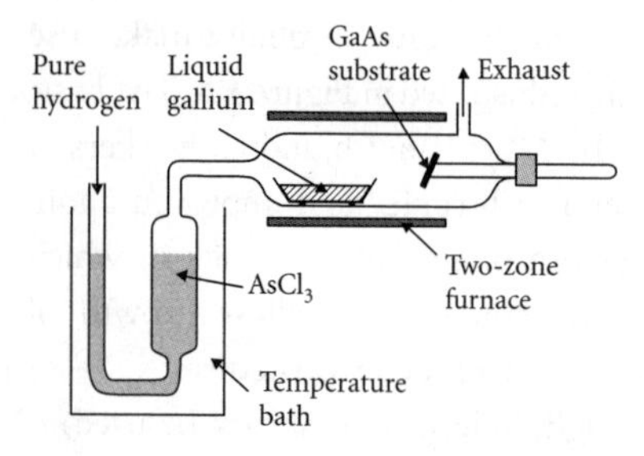

Figure 5.7. An outline of a VPE equipment for growing GaAs. Ultrapure hydrogen is bubbled through liquid $AsCl_3$ and the resulting gas mixture passed over a boat containing liquid gallium at a temperature of typically 800°C. This forms arsenic and gallium chloride which are transported to the GaAs substrate, held at 700°C, where GaAs is deposited and the waste gas pumped away, largely as HCl.

principles of epitaxial growth from the vapour phase. The first application to GaAs growth was based on the $AsCl_3$–Ga–H_2 system and, while being more complicated than the solution method, is also considerably more versatile. A typical apparatus is shown in Figure 5.7. A two-zone furnace contains, at one end a boat holding a charge of high purity gallium, at the other end a GaAs substrate. Pure hydrogen gas is bubbled through a silica bubbler containing (liquid) $AsCl_3$ and the saturated carrier gas passed over the Ga boat where it reacts to produce gaseous arsenic and gallium chloride. These gases are transported to the substrate end where a further reaction occurs, which deposits solid GaAs on the substrate and generates HCl as a waste product. Typical temperatures are 800–850°C for the Ga boat and 700–750°C for the substrate. Unsurprisingly, it is necessary to choose high purity gallium and arsenic trichloride but the purity of the epitaxial film also depends on the $AsCl_3/H_2$ ratio in the gas stream and on the precise orientation of the substrate. It is also of crucial importance that air leaks are minimized, as oxygen tends to make the deposited GaAs insulating. The best material produced by this halide transport technique at the end of the 1960s was characterized by (unintentional) doping levels of $N_D = 5 \times 10^{19}$ m^{-3} and $N_A = 2 \times 10^{19}$ m^{-3}, giving a free electron density $n = 3 \times 10^{19}$ m^{-3} and a compensation ratio $N_A/N_D = 0.4$. Doped films could readily be produced using a variety of methods. Addition of dopants such as Si, Sn (*n*-type) or Zn, Cd (*p*-type) to the gallium solution represents the easy approach but lacks flexibility. It is better to use a gaseous source such as H_2S or H_2Se for *n*-type doping or Zn vapour for *p*-type, which allows separate control of doping density and avoids contamination of the Ga melt. Several alternative VPE methods were subsequently developed based on the use of different starting chemicals—for example, the so-called 'arsine' or 'hydride' method using AsH_3–Ga–HCl, which was reported by Tietjen and Amick of RCA in 1966 but we shall not discuss them here. Suffice it is to say that each has its peculiar advantages and disadvantages, which make it suitable for specific applications—all that we need to appreciate is the considerable amount of work and ingenuity expended on their development.

It is interesting to examine the rapid spread of these epitaxial growth techniques through the esoteric world inhabited by those concerned with compound semiconductor development. A glance at the first four volumes of the Proceedings of the (then) newly established conference series '*GaAs and Related Compounds*' (which began in 1966 at Reading, UK simply as 'GaAs' and has now evolved into a very large international conference under the more general title of 'Compound Semiconductors') reveals no less than 14 companies (or institutes) with established skills in LPE and 12 with similar expertise in VPE. From this, it is clear that, up to 1972, these were mostly old friends with well-established programmes in silicon—names such as RCA, IBM, Fairchild, Texas

Instruments, Hewlett-Packard, Monsanto, GE, GTE, Bell Labs, and Lincoln Labs in the USA, Philips and Siemens in continental Europe and Plessey, STL, Mullard, and RRE Malvern in England all have a familiar ring. In Japan, NEC were first out of the starting blocks, soon to be followed by many others. At all events, the situation at the end of the 1960s was a great deal more satisfactory than it appeared at their beginning. Not only were both Bridgman and LEC, doped and semi-insulating substrates available in adequate quality but careful attention to epitaxial deposition had enabled the production of single crystal films of a purity appropriate to meet the demands of current device technology. Why should anyone ask for more? That anyone should, (and did!) was a simple consequence of the march of science (or was it technology?).

As we shall see in Chapter 6, LDS were to demand epitaxial techniques geared to the growth of films whose thickness would be measured in nanometres, rather than micrometres and this was to throw open the crystal growth doors to yet further contestants. Conventional LPE and VPE, as they existed in the early 1970s were ideally suited to growing films of 10–100 μm thickness at growth rates of order 1–10 μm/h but were ill-designed for the very much slower growth rates demanded by LDS—hence the emergence of two new protagonists Metal-Organic Vapour Phase Epitaxy (MOVPE) and Molecular Beam Epitaxy (MBE) both of which made an impressive impact in the burgeoning field of 'nanostructures'. However, the initial motivation for their development was somewhat less esoteric, springing, as it did, from the need for well controlled AlGaAs/GaAs heterostructures for lasers and solar cells. AlGaAs films could be grown from gallium solution by adding Al to the melt, though it was not easy to control the Al/Ga ratio in the film and there were serious difficulties caused by aluminium's marked tendency to oxidize and generate an oxide scum on the melt surface. VPE systems could also be modified to include two or more gas streams, one for GaAs, another for AlGaAs, though this required the GaAs substrate to be translated rapidly between them, thereby introducing undesirable complication into the growth process. But what was probably the greater difficulty was inherent in the nature of the film deposition processes—both LPE and VPE are sensitive to the precise control of growth temperature, a small change in temperature making the difference between deposition (i.e. film growth) and etching (film removal). While this is hardly a problem when growing a single layer on a thick substrate, it can be catastrophic when growing multilayer structures, a slight error in temperature resulting in the complete dissolution of a previously deposited film, and, with the soon-to-be-established demand for ultrathin films, this problem became particularly intransigent. It was perhaps fortunate that, for quite different reasons, both MOVPE and MBE suffered hardly at all from this defect and they were quickly able to demonstrate considerable flair in the demanding business of multilayer growth.

MOVPE first established its credentials in the scientific literature in 1968 when Manasevit of North American Rockwell Corporation reported, in an *Applied Physics Letter*, the growth of epitaxial films on a variety of oxide substrates, though, ironically enough, he made no mention of his new growth method until a later, and more comprehensive, paper published in the *Journal of the Electrochemical Society* in 1969. For the transport of the Group V element (As, P, or Sb) he used the hydrides (AsH_3, PH_3, or SbH_3), which were available in standard gas cylinders, diluted with hydrogen. For the Group III element he employed the appropriate tri-methyl or tri-ethyl organo-metallic compound, for example, to grow GaAs, tri-methyl gallium, (TMG), chemical formula $Ga(CH_3)_3$. These compounds were available as liquids (at room temperature) in sealed containers with control valves. Pure hydrogen was bubbled through the TMG, premixed with the (AsH_3–H_2) in a gas line and supplied to a heated GaAs substrate mounted on a graphite susceptor, which could be inductively heated by an RF power supply. The quartz reactor was thus at a much lower temperature than the substrate (in contrast to the usual arrangement in VPE where the whole reactor is contained within a resistively heated furnace tube) an obvious advantage from the viewpoint of layer purity. The reaction between AsH_3 and TMG to generate GaAs occurs *only on the substrate*, roughly according to:

$$(CH_3)Ga + AsH_3 \rightarrow GaAs + 3CH_4. \tag{5.2}$$

By growing under a large excess of AsH_3, growth rate was found to be independent of temperature between 600°C and 800°C and controlled by the TMG flow, while the reverse reaction which would result in etching of the substrate occurred at an entirely negligible rate. Here was an important difference from conventional VPE, which was to have vital significance when the demand for ultrathin multilayer structures emerged some 5 years later. Manasevit reported the growth of GaAs, GaP, GaAsP, and GaAsSb and was also able to demonstrate good control of both *n*- and *p*-type doping, using H_2Se (a convenient gas) or di-ethyl zinc (DEZ), respectively. DEZ was supplied by bubbling hydrogen through the liquid in exactly the same way as for TMG.

From the device point of view, the most immediate requirement was for well controlled AlGaAs/GaAs heterostructures with application to lasers, photodetectors, and solar cells so it was no surprise that, within a couple of years, MOVPE was demonstrating success in this sphere too. The obvious modification of adding a separate bottle of tri-methyl aluminium (TMA) was relatively straightforward and this, as might be anticipated from the observation that growth rate is directly proportional to the TMG (TMA) flow rate, enabled excellent control over the Al/Ga ratio. The only problem was the universal one where aluminium is concerned, that of oxidation. Extreme care to exclude oxygen at all points was essential—success in this, however, was usually sufficient to guaranteed overall success.

While MOVPE can be seen as a fairly natural development from earlier VPE activity, the emergence of its arch rival MBE has the appearance of being quite remarkably coincidental, developing, as it did, out of the purest of pure semiconductor surface studies. During 1968–9 John Arthur at Bell Labs was busy trying to understand the behaviour of Ga and As atoms on the surface of GaAs crystals, studies which required him to work under ultra-high vacuum conditions ($P < 10^{-9}$ Torr) in order to minimize surface contamination due to undesired background species. This involved the use of a range of esoteric techniques such as a stainless steel vacuum system, cryogenic pumping, high resolution mass spectroscopy, and reflection electron diffraction. One of his objectives was to measure the 'sticking coefficient' of Ga and As atoms on an ultraclean GaAs surface, which involved impinging a beam of atoms from a 'Knudsen cell' (a rather special thermal evaporation source) onto the surface and measuring the fraction, which bounced off with a mass spectrometer. It was then only a small step for him to apply both Ga and As beams together and demonstrate the growth of a molecular layer (i.e. a monolayer) of GaAs—thus was GaAs MBE born. True, there had been earlier attempts to grow GaAs by evaporation methods but most of this material was polycrystalline—it was only with the much greater understanding of the growth process provided by Arthur's work that MBE became a practical technique at the time it did. The importance of close interaction between the pure and the applied has been a recurring theme in our story and here we see yet another example. Arthur continued his work in surface science and appeared to take no direct role in the development of MBE but there can be little doubt that his work strongly influenced those colleagues at Bell who were motivated to finding practical solutions to the problems of making better lasers.

To turn these fundamental surface studies into a practical growth method required a number of modifications. The Knudsen cell, itself, was ideal for doing physics—it was possible to calculate the emitted flux of atoms (or molecules) simply from a knowledge of the cell temperature—but the fluxes available were far too small to allow realistic growth rates. It was therefore necessary to enlarge the aperture when designing a practical growth cell, though this carried with it the disadvantage that the flux could no longer be simply determined. Practical considerations demanded that the number of cells be increased to take account of the need for two dopant species (*n*- and *p*-type) and the likelihood that the system would be called upon to grow heterostructures such as AlGaAs/GaAs (which means the inclusion of both Ga and Al sources). It was also necessary to arrange for a vacuum interlock to allow substrates to be taken in and out of the apparatus without letting the system up to air. This reflects the fact that ultra-high vaccum (UHV) systems clean up very gradually over a significant time span, implying that the quality of the epitaxial films improves slowly from one run to the next—letting the system up to air immediately returns the process to square one. This change further required that some mechanical translational system had to be incorporated so as to move the sample through the vacuum interlock and place it accurately in position, in front of the source cells, thus ensuring that it would intercept the beams of As, Ga, Al, and dopant atoms. The geometry of the cell arrangement also required that the substrate should be rotated about its axis during growth to improve the uniformity of the deposited film and this proved not a little difficult, bearing in mind that any rotating shaft had to pass out of the vacuum system without the benefit of lubrication, which is incompatible with UHV technology. Finally, control of alloy composition or doping during a growth run required that a moveable shutter be placed in front of each source cell to switch off any particular flux when it was not wanted. Each MBE chamber, therefore, contained numerous motion drives, all of which must be leak-tight and mechanically reliable over a long period of time—any defect would almost certainly imply the abortion of the current growth run, and, as the time for making the repairs, the necessary bake-out and repumping could well be as long as a week, frequent breakdowns were clearly not acceptable.

Having in mind all these complications, it may strike the reader as remarkable that MBE ever did become a viable growth method but, within less than 2 years from John Arthur's pioneering work, MBE had already been used by Al Cho and co-workers at Bell Labs to grow GaAs, AlGaAs, and GaP and to dope them both *n*-and *p*-type. Very soon others followed and MBE rapidly became established as a serious rival to MOVPE for the controlled growth of complex heterostructures. A snapshot of the MBE world taken in the early 1970s would have revealed strenuous activity not only in Bell Labs, Murray Hill but also at IBM, Yorktown Heights, where Leo Esaki and his collaborators were exploring the properties of short period superlattices (more of this anon) and at Mullard (Philips) Research Laboratory at Redhill in England where Tom Foxon and Bruce Joyce were establishing the first European MBE facility. The period up to 1977 saw the entrance of at least two Japanese groups (Takahashi at the Tokyo Institute of Technology and Gonda at the Tokyo-based Electrotechnical Laboratory) and that of Klaus Ploog at the Max Planck Institute in Stuttgart. The emphasis of all this work was on GaAs, GaP, GaAsP, and AlGaAs, material of quality comparable with that produced by other methods being clearly demonstrated.

A typical MBE machine for growing AlGaAs is shown schematically in Figure 5.8. Within the stainless steel UHV chamber, pumped by a

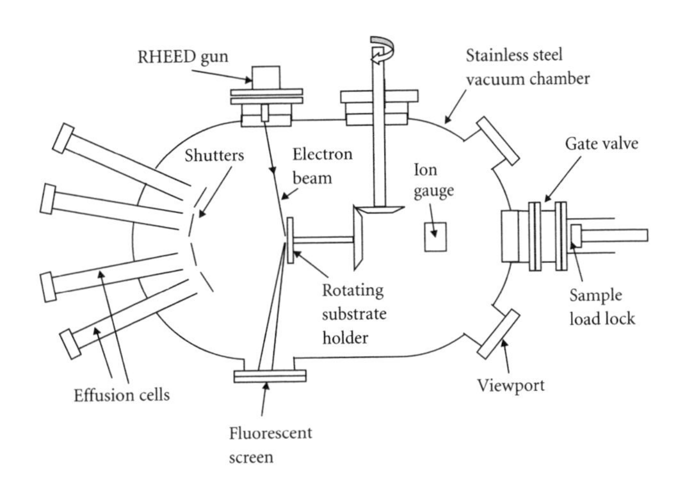

Figure 5.8. A typical MBE apparatus for the growth of AlGaAs. A GaAs substrate is held on a rotating specimen holder kept at a temperature of typically 650°C while beams of Al, Ga, As, and dopant atoms are directed at it from a series of thermal evaporation sources (effusion cells). The apparatus is contained in a stainless steel vessel evacuated to a base pressure of order 10^{-10} Torr, to minimize incorporation of background impurities in the grown film. The crystal structure of the growing surface can be monitored by means of RHEED using an electron beam at glancing incidence which is diffracted to form a RHEED pattern on the ZnS fluorescent screen.

combination of ion and titanium sublimation pumps, is a heated substrate holder, in line with a set of molecular beam sources, and a variety of sophisticated diagnostic techniques. The sources, containing samples of the pure elements Ga, Al, As, and dopants are electrically heated to generate appropriate atomic or molecular fluxes, which are collimated by the geometry of the cells to form beams directed towards the substrate. Beam intensities may be measured by an ion gauge, while a mass spectrometer is available for analyzing the background gases present before and during growth, and the atomic structure of the growing layer can be monitored by the method of reflection high energy electron diffraction (RHEED) (see Box 5.2). A 5-kV gun directs an electron beam at near glancing incidence onto the film and the diffracted intensity is displayed on a phosphor screen on the opposite side of the chamber. This facility, retained from earlier surface studies, has proved invaluable in establishing optimum growth conditions, providing information about 'surface reconstruction' (the surface atomic arrangement being slightly different from that in the bulk crystal), which can be correlated with the growth process. The substrate temperature during growth is usually in the range 500–800°C and a typical growth rate for this so-called 'solid source' MBE process is 1 μm/h (which, for GaAs, corresponds to one monolayer per second). The shutters that switch beams on and off can be operated in something less than a second, which gives MBE the capability for producing extremely sharp interfaces, and the relatively low growth temperature minimizes interdiffusion and ensures that this notional advantage actually obtains in practice. The use of 'RHEED oscillations' (see Box 5.2) provides a unique method of instantaneous monitoring of growth rate, literally counting monolayers as they are deposited. Exotic it may seem, but MBE has impeccable credentials for the growth of ultrathin epilayers, as will become apparent in Chapter 6. Successful *n*-type doping is usually achieved using a beam of Si atoms from a standard source cell and *p*-type doping by a beam of Be.

Box 5.2. The use of RHEED in MBE

A unique feature of MBE growth, which stems from its association with fundamental surface studies, is the use of RHEED as an *in situ* diagnostic technique. In this Box we shall briefly examine three aspects, the study of surface reconstruction, the determination of growth mode—whether two- or three-dimensional and the use of RHEED oscillations to 'count monolayers'.

The geometry of the experiment is clear from Figure 5.8—electrons from the RHEED gun are diffracted from the surface of the film and form a diffraction pattern on the phosphor screen and the nature of this pattern provides a surprising amount of detailed information concerning the surface structure, that is, the precise positions of the surface atoms. In the ideal situation of an atomically flat surface, the pattern consists of a set of *streaks* (not spots) whose spatial separation depends on the effective lattice parameter on the film surface and the observation of this streak pattern is a reliable indication that the film is growing in a two-dimensional fashion—that is, layer by layer. If, on the contrary, growth becomes three-dimensional, that is in random, spatially separate hillocks, the streaks change to spots. (Essentially, this arises because electrons penetrate through the hillocks and are diffracted as from a region of bulk material.) As high quality films are usually obtained only as a result of two-dimensional growth, the streaks represent vital evidence for optimum growth conditions. Furthermore, the symmetry of the streak pattern may change according to the nature of the crystal orientation and according to substrate temperature, a feature which is often used to confirm that the growth temperature is in the desired range. The different surface structures which correspond to each pattern also enable the expert to understand details of growth which could only be guessed at in the absence of such data.

This kind of information, in itself, was of considerable value in the early days of MBE but a welcome bonus came with the discovery by the Mullard (Philips) group in 1981 of the phenomenon of 'RHEED oscillations'. They noticed that the intensity of the specularly reflected spot in the RHEED pattern oscillated in time with a period which corresponded precisely to the time for deposition of a single monolayer of material, an observation of tremendous import for the controlled growth of very thin films, later applied to excellent effect in the growth of LDS (see Chapter 6). A complete understanding of the mechanism is difficult but a good qualitative picture emerges from a model which treats the electron beam in terms of an optical analogue. The wavelength of 5 kV electrons is approximately 0.02 nm (see equation (4.10)) which is small compared to a monolayer step (0.28 nm in GaAs) so step edges serve as efficient scattering centres and the greater their density on the film surface, the greater the scattering and the smaller the final intensity. Suppose growth is started from an atomically smooth surface—for the case of nucleated growth (i.e. nucleation of small, monolayer islands randomly sited on the surface), the density of steps increases monotonically up to approximately half-a-monolayer coverage, then decreases again as further nucleation fills in the spaces between islands, until, at one complete monolayer coverage, the original smooth surface is recovered. Electron scattering thus increases up to half-a-monolayer coverage, then decreases again to its initial value after a single monolayer has been deposited, accounting for the observed intensity oscillation.

This interpretation was verified in a classic experiment by the same group in 1985. The oscillation behaviour was monitored for growth on so-called 'vicinal' surfaces of GaAs (surfaces misoriented from the precise (0 0 1) crystallographic orientation by a few degrees so as to introduce a more or less regular series of steps with spatial separation, which depends linearly on the misorientation angle). Their experiment demonstrated that, as the growth temperature was increased, the oscillations disappeared due to a change from random nucleation to so-called 'step edge growth'. The reason is that Ga becomes more mobile on the surface as the temperature increases until, eventually, all the deposited Ga atoms can reach steps, where they are trapped and two-dimensional nucleation ceases altogether.

In conclusion, we should note that the measurement of RHEED intensities has also been central to the development of so-called 'atomic layer epitaxy', which proceeds by the deposition of a single layer of Ga atoms, followed by a single layer of As atoms, and so on. It is the ability to monitor the deposition of a single layer which makes such a technique practicable. Similar remarks apply to the development of 'migration enhanced epitaxy' in which there is a pause between deposition cycles to allow surface atoms to migrate to appropriate sites before starting the next layer. These techniques allow high quality material to be deposited at unusually low growth temperatures ($T < 500°C$).

The more usual acceptor atoms Zn and Cd have extremely short lifetimes on the GaAs surface (i.e. they desorb extremely rapidly) and are wholly ineffective. Mg has been used with partial success, though it, too, has a short surface lifetime and a strong tendency to diffuse at the growth temperature, making it difficult to control the doping profile.

Finally, before closing this extended section on crystal-growing methods, we must also note the development of two variants of MBE, which were developed from the basic technique during the latter half of the 1980s. The use of solid sources in conventional MBE suffers from one drawback, in so far as the cells inevitably require replenishment from time to time and this means taking the system up to air. Their replacement with gas sources removes this limitation and allows greater flexibility in operation. Perhaps ironically, these changes made use of experience gained from MOVPE in the use of metal-organic compounds to supply the Group III elements (TMG, TMA, etc., carried on a stream of pure hydrogen) and of hydrides to supply the Group V elements. Metal-organic MBE or MOMBE carried this approach into UHV practice, mass flow controllers providing the beam switching mechanism, rather than shutters, while gas source MBE took the process to its logical conclusion and used arsine or phosphine or, later, ammonia to grow arsenides, phosphides, or nitrides, respectively. (An in-joke within the MBE community in the United Kingdom noted that these developments completed the hierarchy of honours—MBE, OBE (organo-metallic beam epitaxy), and CBE (chemical beam epitaxy) though we should obviously treat such frivolity with the degree of respect it undoubtedly deserves.)

The reader who has struggled through this section on crystal growth methods may question the merit of such a lengthy account but he or she will surely now appreciate the enormous effort, which has been devoted to it in practice. While one's main interest may, indeed, relate to the physics of the resulting material or to the sophisticated devices, which have thereby been realized, it is important to recognize the essential role of materials research in making them possible and, not least, the huge financial investment involved (e.g. the installation of either an MOVPE or an MBE facility in practice costs well over a million dollars!). I therefore make no excuse for the length of my exposition (or should it be 'deposition'), nor for the fact that the following section is concerned not yet with devices but with the rise of characterization measurements. The opportunity to taste some of the mouth-watering fruits of the III-V tree will materialize (if that is the appropriate word?) in Section 5.5.

5.4 Material characterization

The considerable effort committed to crystal growth during the 1960s and 1970s demanded a corresponding investment in material characterization, it being meaningless to develop new techniques of epitaxial deposition in the absence of appropriate means for measuring the results of one's efforts. To say that LPE and VPE achieved considerable improvement in background doping levels and compensation ratios implies the existence of reliable methods for measuring these parameters. Furthermore, it is interesting to recognize that the considerably greater difficulties involved in developing high quality III-V materials (compared with those of silicon) led to a rapid expansion of characterization methods. While, for over a decade, silicon characterization had made do with little more than the measurement of resistivity, GaAs was treated to the luxury of Hall effect measurements over a wide temperature range, capacitance–voltage free carrier profiling measurements, a range of deep level studies, photoluminescence (PL) investigation at low temperatures and, often, much more. This may reflect, to some extent, the developing interest in semiconductor measurements per se but, nonetheless, also reflects a real need to gain control over a much less amenable material system. It also led to a plethora of books explaining the intricacies of these measurements and their interpretation (a representative selection includes: Look 1989; Stradling and Klipstein 1990; Orton and Blood 1990; Blood and Orton 1992; Perkowitz 1993; Schroder 1998; Runyan and Schaffner 1998) with the result that today's crystal growers are positively besieged with advice on how to characterize their precious samples. Not that characterization is a new phenomenon—even in the 1960s techniques were growing rapidly in sophistication and importance, a state of affairs which obliges us, at least, to comment on some of the more widely used techniques.

The reason why silicon could prosper with nothing more sophisticated than a knowledge of its resistivity lies in the fact of its properties being considerably more reproducible that those of the III-Vs. Specifically, we note that there is a well established relationship between the density of free carriers in silicon and their mobilities (one relationship for electrons, that is, another for holes). Remembering from our earlier discussion of the Hall effect in Box 2.2, that resistivity ρ is related to free carrier density n (or p) and the appropriate mobility μ_n (or μ_p) by equation (5.3):

$$\rho = (ne\mu_n)^{-1} \text{ or } (pe\mu_p)^{-1} \tag{5.3}$$

it follows that, if mobility and carrier density are precisely related, a measurement of resistivity is adequate to determine all three parameters. However, a difficulty arises with this simple approach in the case that the material is compensated to a significant (though unknown) degree. Consider, for example, an n-type sample containing N_D donors and N_A acceptors ($N_D > N_A$). A measurement of the free electron density at room temperature (using the Hall effect) provides a value for $(N_D - N_A)$ but still leaves N_D and N_A unknown. The electron mobility, on the other hand is related to the total ionized impurity density $(N_D + N_A)$, through ionized impurity scattering, but this implies that μ_n is no longer *simply* related to n. In order to derive values for all four parameters ρ, N_D, N_A, and μ_n, we need to make *two* measurements, those of ρ *and* of n. From these, using the relationship $\rho = (ne\mu_n)^{-1}$, we can obtain μ_n and this, in turn, provides a value for $(N_D + N_A)$. Finally, a knowledge of both $(N_D - N_A)$ and $(N_D + N_A)$, allows us to determine N_D and N_A separately. The essential difference between silicon and GaAs is that, broadly speaking, silicon corresponds to the (easy) uncompensated case (N_A being negligible) while GaAs belongs to the compensated category, which demands more sophisticated treatment. In practice, the mobility measured at a temperature of 77K is much more sensitive to variation in $(N_D + N_A)$ than is the room temperature value so it became a common practice to use this when characterizing GaAs. There was therefore a growing demand for Hall effect and resistivity measurements at both room temperature and at 77K.

The above discussion is based on the assumption that there exists an *established* relationship between mobility and total ionized impurity density and the inquisitive reader will wonder, no doubt, how such a relationship came to be so established. The answer is: by means of sophisticated analysis of measurements of resistivity and Hall effect over a wide temperature range, typically from 4K to room temperature. As the temperature is reduced below room temperature, free carriers gradually become trapped on donor atoms (we refer to 'carrier freezeout') so n decreases rather strongly and this, incidentally, allows one to measure the donor ionization energy E_D but the detailed nature of this freezeout is also related to the compensation ratio N_A/N_D so careful analysis of the experimental data for $n(T)$ can yield values of both N_D and N_A. It is then possible to measure $(N_D + N_A)$, together with μ_n(77K) for a range of samples and plot the desired curve of μ_n(77K) vs $(N_D + N_A)$. Altogether, this involved a considerable amount of very careful work, as will be apparent from the data plotted in Figure 5.9, taken from a 1970 paper by Wolfe and Stillman (then at Lincoln Labs, MIT). This shows the measured electron mobilities of four GaAs epitaxial samples over the temperature range 4–100K, compared with mobilities calculated from values of N_D and N_A derived from $n(T)$ data. The excellent agreement between calculated and experimental values, not only confirms the results of the $n(T)$ analysis but also demonstrates the depth of understanding of scattering mechanisms, which limit the mobility in GaAs. No less than five different scattering processes are included in the calculations used to obtain Figure 5.9.

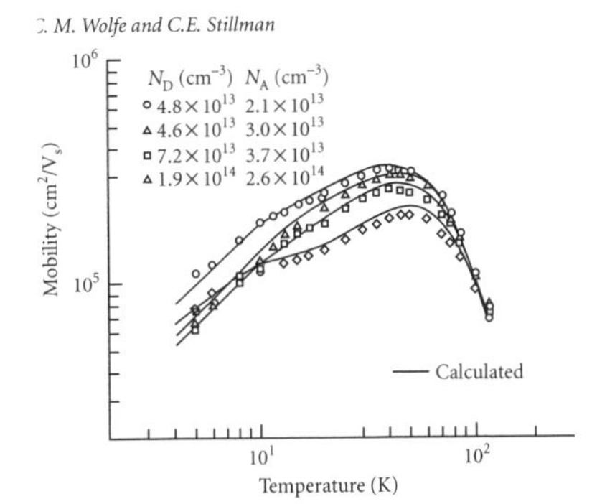

Figure 5.9. Plots of the electron mobilities as a function of temperature for a set of high purity epitaxial GaAs samples. The symbols represent experimental data from Hall effect measurements, the solid lines calculated values obtained by combining five different scattering processes. Note that the analysis of the data allows one to derive values for both N_D and N_A. (From C. M. Wolfe and G. E. Stillman (1970) 'Gallium Arsenide and related compounds' *Institute of Physics*, Conference series No. 9, p. 10.) Courtesy IOP Publishing Ltd.

One of the principal tasks of the characterization expert is that of simplifying measurement procedures as far as possible so they may be applied routinely to a large number of samples. This is seen at its best in the case of resistivity measurements on silicon where use is made of the so-called 'four point probe', a set of four metal point contacts arranged in line, spaced about 1 mm apart and brought down with controlled pressure on the surface of a silicon slice. A constant current is passed between the outer pair of probes and the resulting voltage measured between the inner pair. This, together with appropriate formulae (depending on whether the measurement is made on a bulk sample or a thin epitaxial film) is sufficient to determine the resistivity of a silicon area defined by the probe spacing. There is no need for special sample shapes or for making ohmic contacts. An even simpler procedure is available in the form of a single probe used to measure 'spreading resistance', the resistance associated with current which flows into a semiconductor from a metal point contact, though this is somewhat less reliable and, for good accuracy, requires careful calibration. Clearly, either method offers a rapid and easy-to-perform characterization tool and both have served silicon technology extremely well over the years.

GaAs and other compound materials present a greater challenge. Quite apart from the need for more detailed investigation, as explained above, attempts to apply metal probe contacts were met with an immediate rebuff—contact resistance was found to be extremely high and current—voltage relationships strongly non-ohmic (i.e. non-linear). An element of sample preparation became inevitable. This has usually taken the form of the 'clover leaf' shape shown in Figure 5.10, which may be cut from a bulk wafer by abrasive action (sand blasting! but with a mild abrasive such as SiO_2 powder) or similarly defined for an epitaxial film, simply by cutting through the film as far as the substrate (which must be semi-insulating to avoid short-circuiting the measurement). Four metal contacts are made as indicated, one on each leaf, to allow both resistivity and Hall effect measurements to be made using appropriate pairs of current and voltage contacts. The procedure depends on formulae derived in 1958 by L. J. van der Pauw, at the Philips Research Laboratories in Eindhoven, the method being known, appropriately enough, as the van der Pauw method. It simplifies sample preparation but is clearly much more complicated than the probe methods used to characterize silicon. It also uses considerably more material and is destructive of that particular piece of what may be a very precious sample.

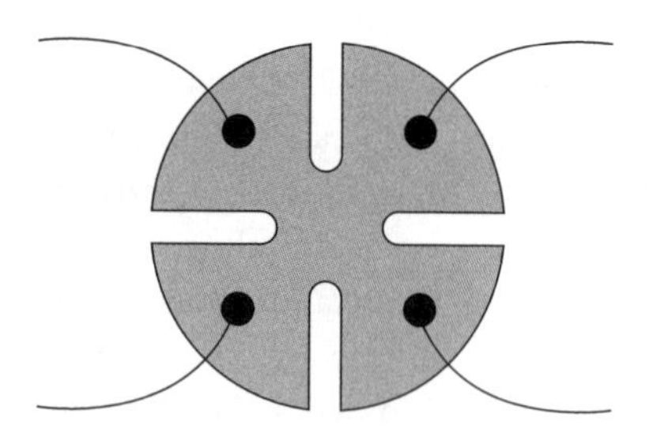

Figure 5.10. The clover leaf shape used for van der Pauw measurements of resistivity and Hall effect on semiconductor samples. The 'wings' effectively act as contacts for the small area in the centre of the sample.

Perhaps the second most important technique to be applied to GaAs was the *C–V* profiling measurement whose principal purpose was to provide a depth profile of free carrier density—that is, a measure of $n(x)$ (or $p(x)$), where x represents distance below the sample surface. It is all too easy to assume that epitaxial layers, for example, are uniform in their electrical properties but there are various reasons why this may not always be true and, in some applications, it may be of critical importance whether, or not, it *is* true. The *C–V* method yields such information in a reasonably straightforward fashion, from the measurement of the capacitance C of a Schottky barrier contact as a function of an applied reverse bias V. Increasing reverse bias expands the depletion region associated with the barrier (in identical fashion to the p-n junction discussed in Section 3.4) and correspondingly reduces the depletion layer capacitance. As is shown in several of the standard texts (e.g. Blood and Orton 1992, equation 5.49), $n(x)$ can then be obtained from the relation:

$$n(x) = -(C^3/e\varepsilon\varepsilon_0 A^2)(dC/dV) \tag{5.4}$$

where ε is the relative dielectric constant of GaAs and A is the area of the Schottky contact. The depth x at which the measurement $n(x)$ is made is simultaneously obtained from:

$$x = \varepsilon\varepsilon_0 A/C. \tag{5.5}$$

The required Schottky barrier may be formed by thermal evaporation of a gold film through a suitable mask onto the GaAs surface (which has been carefully cleaned beforehand in the best tradition of semiconductor practice), it then being necessary to bond a thin gold wire to the contact. It is relatively easy to automate the measurement and the calculation of $n(x)$ from equation (5.4) and of x from (5.5) to obtain a plot of the desired profile, the only limitation resulting from the reverse breakdown of the Schottky diode. At sufficiently large reverse bias, the electric field within the depletion region exceeds the breakdown field of GaAs and the reverse current increases very rapidly as a result of an avalanche process—the few free carriers within the depletion region acquire sufficient kinetic energy to create further electron–hole pairs on impact with lattice atoms and the process rapidly runs away, leading to destruction of the diode unless the current is limited by an external circuit element. At all events, it limits the amount of reverse bias, which can be applied and, therefore, the maximum depletion depth available. As though life were not complicated enough, this limiting depletion depth depends on the sample doping level, which it is the object of the measurement to measure, an approximate indication being: $x_{max} = 100\ \mu m$ at $N_D = 10^{20}\ m^{-3}$ falling to $x_{max} = 1\ \mu m$ at $N_D = 3 \times 10^{22}\ m^{-3}$.

The somewhat lengthy process of evaporating and contacting a suitable Schottky barrier can readily be seen as a serious drawback to routine C–V profiling so a good deal of attention was paid to alternatives. One such, which gained popularity in the 1970s made use of liquid mercury for the contact instead of gold and, apart from the care necessary to avoid mercury poisoning of the operator (!), this proved rather satisfactory. The contact area was defined by the diameter of the tube containing the mercury and it was straightforward to make and unmake the contact in a matter of seconds. In fairness, one should admit that the results were not always quite so accurate as with the more laborious method but it was undeniably advantageous when dealing with a multiplicity of samples. The fundamental limitation due to diode reverse breakdown was also circumnavigated in 1974 by a modified version of *C–V* profiling due to Ambridge and Faktor, working at the UK Post Office, in which the barrier took the form of an electrolyte. This had the important advantage of allowing material to be controllably removed by electrolytic action, the amount dissolved being monitored by measuring the total amount of electric charge which flowed. This undoubtedly represented a major step forward in terms of both flexibility and ease of use and it was found possible to apply it to a range of materials, with suitable choice of the electrolyte. Commercial versions of the equipment were developed and proved highly successful in many research and production environments around the world.

We have referred already to the significance of deep levels in semiconductors. They may result from deliberate introduction of impurities, such as that of Cr in GaAs to produce semi-insulating material or Au in Si to reduce the minority carrier lifetime, or they may be a consequence of uncontrolled (and usually undesired) impurities or point defects in the crystal. Once again, the coming of GaAs and other compound semiconductors led to a burgeoning interest in the measurement of deep level properties, largely because the compounds were generally less pure and more defective than elemental semiconductor crystals. The range of techniques brought to bear on the problem grew rapidly and is described in detail in the books already referred to (in Blood and Orton 1992, the discussion of deep level measurements runs to something over 350 pages!)—here, we can do no more than outline a few general principles. While photoconductivity measurements can be used to provide information on deep levels in high resistance samples, the large majority of work has made use of junction or Schottky barrier capacitance methods, which are more widely applicable. Being based on junction or barrier depletion regions, they effectively depend on the production of a high resistance region in an otherwise conducting sample.

The characterization of a deep level requires that we obtain four pieces of information about it—how deep is it (i.e. what is its energy location within the forbidden gap), how readily can it capture free carriers from the allowed bands, how readily does it emit carriers to the bands, and how many centres exist (i.e. what is their density). (We also wish to know whence it originates, of course, but this can often be surprisingly difficult.) A centre with energy close to the middle of the gap may interact with carriers from both bands (in so-doing it acts as a 'recombination centre') but more often than not a deep centre will be strongly associated with one band only and, for the sake of simplicity, we shall consider centres which are in the upper half of the gap and interact only with the conduction band.

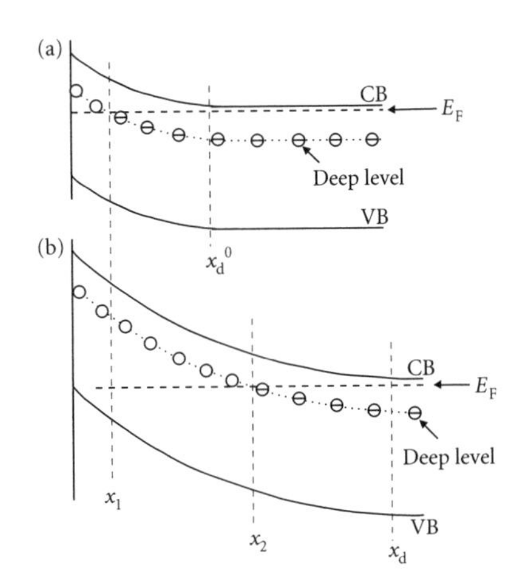

Figure 5.11. Band diagram to illustrate the change in occupancy of deep impurity levels in the band bending region of a semiconductor beneath a Schottky barrier contact. The dashed line represents the Fermi energy E_F. Deep levels which lie above E_F are essentially empty of electrons, those below are essentially full. (a) Represents the case of zero bias, and (b) that for a reverse-biased barrier. The application of a reverse bias enlarges the depletion width x_d and changes the occupancy of deep levels in the region between x_1 and x_2.

The principle of a deep level capacitance measurement can be understood with reference to the energy diagram shown in Figure 5.11. It represents the band bending region associated with a Schottky contact on the surface of a semiconductor. The essential feature concerning the deep levels is that the band bending implies that those nearer the surface are empty of electrons, while those nearer the edge of the depletion region are full. The point of demarcation is defined by the horizontal line which is labelled E_F and known as the 'Fermi level'. The concept of the Fermi level is discussed in greater detail in Box 5.3 but, for those not wishing to be embroiled in the mathematical complexities, it is sufficient to know that electron states which lie above the Fermi level are largely empty of electrons, those below being largely full. Figure 5.11(a) represents the degree of band bending associated with zero bias applied to the diode, (b) the situation under applied reverse bias. Comparison of the two pictures shows the depletion depth to be increased under bias from x_d^0 to x_d and that, as a result, there is a change in occupancy of some of the deep levels which lie in the vicinity of the Fermi energy, that is, those centres which lie in the space between x_1 and x_2 were originally full but are now empty. The electrons they contained have been emitted to the conduction band and swept into the bulk of the semiconductor to the right of the depletion edge at x_d. This process represents a change in electric charge within the depletion region, which makes a contribution to the capacitance of the diode (remember that $C = dQ/dV$—Q being charge and V the voltage across the capacitor). It is easy to see that, if there are N_t deep centres per unit volume, the change in charge will be $eN_t(x_2 - x_1)$ per unit area, that is, it depends on N_t and this gives us a way of measuring N_t.

Imagine, now, that, instead of changing the reverse bias in a quasi-static manner, we apply a rapidly alternating voltage across the diode so that the bias increases, then decreases, then increases again, etc. In principle, the occupancy of deep centres near the Fermi level will change in sympathy and we should be able to measure this in terms of an oscillating change in diode capacitance. However, we must remember that these changes in occupancy depend on the ability of the traps to emit electrons to, or to capture electrons from, the conduction band and these processes do not occur instantaneously but require a finite time—for instance, the emission time τ_e is related to the emission rate e_n according to:

$$\tau_e = e_n^{-1} \tag{5.6}$$

and this time can be measured by varying the frequency of the applied bias voltage. For angular frequencies ω less than e_n, the trap occupancy will follow the changes in voltage but for frequencies greater than e_n it can no longer do so. Observation of the frequency at which this

Box 5.3. The Fermi level and Fermi function

One of the concepts most frequently used in semiconductor physics is that of the Fermi energy level, concerned with the mathematical description of the statistical probability that an electron state will actually contain an electron. We have already met the idea, for example, that conduction band states in an intrinsic semiconductor are more likely to be filled at high temperatures than at low temperatures. This implies the existence of some temperature-dependent probability function which defines the occupancy and this function is known as the 'Fermi function' (more strictly, we should refer to the 'Fermi–Dirac distribution function because single electrons obey Fermi–Dirac statistics but the more concise form represents common usage). The Fermi energy E_F is inherent in its use.

Simply stated, the probability that, under thermal equilibrium conditions, a single electron state at energy E, in a semiconductor at temperature T, will contain an electron is given by the probability function:

$$P(E) = f(E) = [1 + \exp\{(E - E_F)/kT\}]^{-1}. \qquad \text{(B5.6)}$$

Another way to think about this is to say that, if there are N_t electron states at energy E_t, the number which will be occupied by electrons is n_t, where $(n_t/N_t) = f(E_t)$.

Note that equation (B5.6) is not immediately straightforward to apply because it contains not only the energy of the state E (which, in principle, we always know) but also the Fermi energy E_F which, a priori, we do not know. We clearly need some further information in order to determine E_F and we shall illustrate the point with an example. Consider an n-type semiconductor with a density n of free electrons in its conduction band. Using the Fermi level concept, we can write the following equation for n:

$$n = \int N(E) f(E)\, dE, \qquad \text{(B5.7)}$$

where the integral is taken from the bottom of the conduction band at $E = E_C$ to $E = \infty$, in other words, over all the states in the conduction band. $N(E)$ represents the distribution of conduction band states, that is, how many states are there with energy E. $N(E)$ can usually be written as:

$$N(E) = N_0(E - E_C)^{1/2} \qquad \text{(B5.8)}$$

Now, provided the Fermi level is more than a few kT in energy below the bottom of the conduction band, we can approximate $f(E)$ by the so-called 'Boltzmann function':

$$f(E) = \exp\{-(E - E_F)/kT\} \qquad \text{(B5.9)}$$

in which case, it is possible to perform the integral in equation (B5.7) and we can obtain the very useful relation:

$$n = N_C \exp\{-(E_C - E_F)/kT\}, \qquad \text{(B5.10)}$$

where N_C is known as the 'effective density of conduction band states'. Its use *effectively* represents the conduction band as a single energy state at energy $E = E_C$. Equation (B5.10) can be used to define E_F, according to:

$$E_F = E_C + kT \ln(n/N_C) \qquad \text{(B5.11)}$$

Box 5.3. Continued

Note that $n/N_C < 1$ so $\ln(n/N_C)$ is negative and E_F lies *below* E_C.

The thoughtful reader will, perhaps, be wondering how this result can really be useful. All that we have achieved is to relate n and E_F, exchanging one unknown for another! However, the Fermi level really *is* useful in that, once it is defined (in terms of n), we can use it to determine the occupancy of other electron states in the system. Two examples will serve to illustrate the point. Suppose we have an n-type semiconductor with N_D donors per unit volume and wish to know, at any particular temperature, how many electrons are in the conduction band and how many on donor atoms. We can use the Fermi function to write an expression for the latter:

$$n_D = N_D[1 + \exp\{(E_D - E_F)/kT\}]^{-1} \tag{B5.12}$$

and combine this with equation (B5.10) and the relation $N_D = (n + n_D)$ to obtain the following quadratic equation for n:

$$n^2 + N_C'n - N_C'N_D = 0 \tag{B5.13}$$

where $N_C' = N_C \exp\{-(E_C - E_D)/kT\}$. This can now be solved for n in terms of known parameters, N_C, N_D, and $(E_C - E_D)$, the donor ionization energy.

Our second example depends on an important property of the Fermi level that, *in thermal equilibrium*, the Fermi level is flat (i.e. it is represented by a horizontal line on an energy level diagram such as that shown in Figure 5.11. This allows us to define the occupancy of the deep levels within the band-bending region of the diagram with respect to the position of the Fermi level in the bulk of the material, as we do in the text. Note that, in equation (B5.6), if $E = E_F$, $f(E) = 1/2$, so, at the point where E_F intersects the line of the deep energy levels, half of the deep states are occupied. When $(E_t - E_F) > 3kT$, the deep states are effectively empty and when $(E_F - E_t) > 3kT$ they are effectively all filled. Thus, at finite temperature the Fermi function is not a step function but changes gradually over an energy range of a few kT. However, when we are dealing with an energy level several tenths of an electron volt below the conduction band and, at low temperatures, where kT may be no more than 0.01 eV, it is a reasonable approximation to think of it as an abrupt step.

cross-over occurs gives us a value for the emission rate e_n. What is more, because e_n depends on temperature according to the relation:

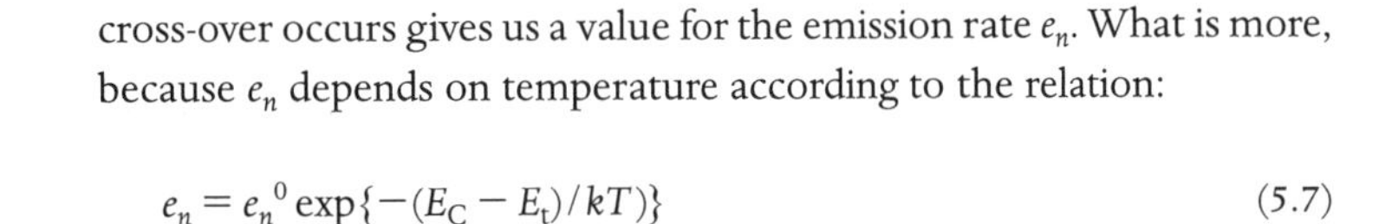

$$e_n = e_n{}^0 \exp\{-(E_C - E_t)/kT)\} \tag{5.7}$$

this means we can obtain a value for the trap energy E_t by measuring e_n as a function of temperature and plotting $\log_{10} e_n$ against $1/T$. From equation (5.7), we can easily derive:

$$\begin{aligned}\log_{10} e_n &= \log_{10} e_n{}^0 - [(E_C - E_t)/kT]\log_{10} e \\ &= \text{Const} - 0.4343(E_C - E_t)/kT.\end{aligned} \tag{5.8}$$

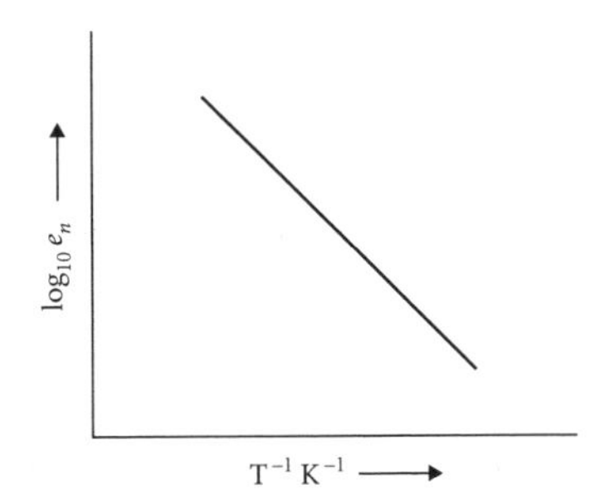

Figure 5.12. Typical plot of the logarithm of measured emission rate e_n for a deep centre against reciprocal temperature (known as an Arrhenius plot). The slope of the line is a measure of the depth of the centre below the semiconductor conduction band.

Such a plot is shown in Figure 5.12 and it is clear that the slope of the line is a measure of the trap energy. From equation (5.8):

$$\text{Slope} = -0.4343(E_C - E_t)/k. \tag{5.9}$$

(Note that '*e*' in these equations is the exponential—it must not be confused with e_n, which is a variable parameter, characteristic of the trap. Nor should either of them be confused with the electron charge *e* which does *not* occur in the equations at all! Note, too, that this account represents something of an over-simplification but my intention is to give the reader a feel for the principles of the measurement—more complete accounts will be found in the appropriate book references.)

The temperature-dependence of the emission rate is made use of in the classic method of 'deep level transient spectroscopy' (DLTS), proposed by D V Lang at Bell Labs in 1974, in which the sample temperature is slowly scanned (usually in a range between 77K and room temperature) while clever electronic wizardry is employed to locate the specific temperature at which the emission rate coincides with a reference rate set by the electronics. This reference rate is then changed and a second temperature scan made, then a third, etc. until a map of e_n vs T is obtained, from which the trap energy is found as described above. Finally, we should note that the electron capture rate can be related to the emission rate by thermodynamic arguments, so all the data required to characterize the trap are now determined.

All the above characterization techniques make use of electrical measurements—we now turn attention to the use of optical methods, which have the advantage of not requiring electrical contacts and of being non-destructive. The principal such method is that of PL in which minority carriers are generated within the sample by the absorption of laser light (having a photon energy greater than the band gap) while the resulting electron–hole recombination is monitored by way of the luminescence generated. Standard practice is to disperse the recombination light with a high resolution monochromator and detect it with a sensitive photomultiplier tube to present an output in the form of a plot of emission intensity against photon energy (or wavelength). This is the 'luminescence spectrum' which may be measured over a wide range of sample temperatures, though frequently concentrating on temperatures in the liquid helium range, 1.5–4.2K. While radiative recombination tends to be much more *efficient* in direct gap materials, like GaAs, it can generally be observed, even in indirect gap samples. We saw in Section 4.6 how luminescence measurements were used by Haynes to study the properties of silicon, an example where the recombination process involved excitons, bound to donor and acceptor atoms. More generally, recombination may occur via several different mechanisms, providing information about band structure and about impurities or point defects in a wide range of semiconductors. (An elementary application is to the measurement of the band gap of ternary systems such as AlGaAs where the alloy composition is often known only imprecisely.) Three parameters are generally of interest for characterization: the emission intensity, the photon energy at the peak of an emission line, and the width of the line.

An example of interest in the case of GaAs is that of identifying acceptor impurities. We have already referred to the problem of compensation in *n*-type GaAs, where unknown acceptors are often present at densities only slightly less than those of the dominant donors. From the crystal grower's viewpoint, it is important to discover which acceptor is concerned, the hope being that, once it has been identified, it may then be eliminated. The trouble is that at least seven acceptor species exist, C, Be, Mg, Zn, Cd, Si, and Ge, the question being how to identify the culprit! Careful doping studies have established that the various acceptors have slightly differing ionization energies, ranging from 26 meV for carbon up to 40 meV for Ge, so a measurement of acceptor energy provides a reliable method of identifying the particular acceptor involved and this can readily be done by low temperature PL. At a temperature of 4K, the majority of electrons are frozen out onto donors and effectively all the holes are on acceptors so recombination is dominated by so-called 'donor–acceptor transitions', which we shall discuss in some detail later. However, as the temperature is raised to about 15K, electrons are gradually released from their donor atoms (though holes remain trapped due to the much larger acceptor ionization energies—$E_A \sim 30$ meV, compared with $E_D \sim 6$ meV for donors) and the emission spectrum is dominated by conduction band—acceptor transitions (i.e. free electron-bound hole). It is easy to see from Figure 5.13 that the photon energy for such a transition is given by:

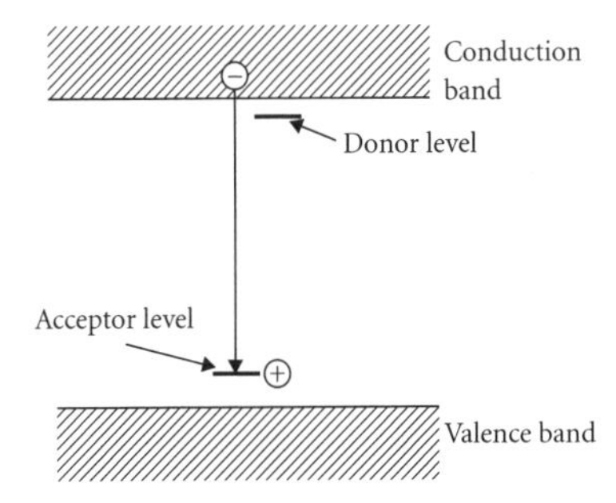

Figure 5.13. Energy diagram to illustrate the free electron-to-acceptor transition which gives rise to PL at moderate sample temperatures in a *p*-type sample of a direct gap semiconductor. Provided the band gap of the semiconductor is known, a measurement of the photon energy of the emission line, gives a value for the acceptor binding energy E_A.

$$h\nu = E_g - E_A, \tag{5.10}$$

where E_A is the ionization energy of the acceptor involved, so, provided we know the band gap at the appropriate temperature (which, in this case, we certainly do), a measurement of the photon energy yields a reliable value for E_A. Typical spectral line widths are 3–4 meV so it is possible to measure E_A to an accuracy of better than ± 1 meV, which is more than adequate to identify the acceptor. Its elimination from future crystals then depends on some inspired guesswork by the growers as to the mechanism of its incorporation—but knowing the relevant chemical species is, without doubt, a key piece of the jigsaw.

A somewhat different approach to the monitoring of unintentional impurities in GaAs makes use of exciton spectra. Electrons and holes created by the exciting laser light thermalize very rapidly to the conduction and valence band edges, respectively, (where they remain mobile) and may then come together to form excitons and these excitons may then become bound at donor or acceptor atoms as already discussed in connection with Haynes' work on silicon. The recombination of the constituent electron and hole gives a sharp emission line, which is characteristic of the specific donor or acceptor involved. However, if the impurity density in the sample is very low, the excitons may recombine before finding a suitable donor or acceptor at which to lodge, in which case the emission line is characteristic of free, rather than bound, exciton recombination and occurs at a photon energy greater than that of a bound exciton line, by an amount which corresponds to the exciton localization energy. It follows that the ratio of the intensities of bound to free exciton emission is a measure of the impurity density and can be used as a figure of merit in monitoring crystal growth. In the case of silicon, it has even been possible to establish a quantitative measure of the impurity density from such data, covering the range 10^{18} to 10^{23} m^{-3}.

Emission line energies are modified in many cases by the presence of strain in the crystal and a classical example of this is provided by the numerous attempts made to grow device quality GaAs on silicon substrates. Because silicon does not offer the possibility of efficient light emission, there would be many optoelectronic applications for GaAs if they could be combined with the use of silicon electronic circuitry. One such involves the use of optical coupling of signals between different parts of high speed silicon integrated circuits where the conventional metal interconnects suffer from significant time delay. However, it has proved extremely difficult to grow GaAs lasers on silicon because of the large lattice mismatch between the two materials. The lattice parameters of GaAs and Si are 0.5653, and 0.5431 nm, respectively, differing by approximately 4% which, in crystal growth terms, is highly significant. A very thin film of GaAs will tend to grow isomorphically on silicon (i.e. with the same lattice parameter as Si), which means that it is heavily strained but, as the film is made thicker, the strain energy becomes so great that it becomes more energy efficient for dislocations to form, relieving the strain and allowing the GaAs to adopt its natural lattice size. However, the high density of dislocations is anathema to the successful operation of GaAs lasers which explains the difficulty. In practice, strain relief by dislocation generation *may* be complete at the growth temperature but additional strain is induced as the sample cools to room temperature because the thermal expansion coefficients of film and substrate differ significantly ($\alpha_{GaAs} = 7 \times 10^{-6}$ K^{-1}, $\alpha_{Si} = 2.5 \times 10^{-6}$ K^{-1}) and this, we should note, results in strain of opposite sense from that due to lattice mismatch.

Measurements of PL have proved valuable in providing a simple method of determining the nature of the residual strain in the GaAs film. Because the silicon lattice is smaller than that of GaAs, the GaAs film is expected to be under compressive strain in the plane of the interface (as a result of lattice mismatch) and this can be represented in terms of a reduction in bulk lattice parameter a_0, together with a tensile axial strain along the normal to the interface (i.e. in the direction of growth). The reduction in a_0 has the effect of *increasing* the GaAs band

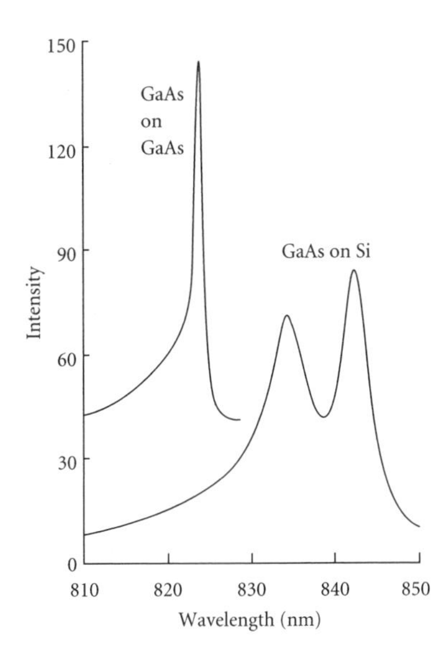

Figure 5.14. Experimental low temperature, near-band edge spectra from samples of epitaxial GaAs films grown on GaAs or silicon substrates. The GaAs-on-Si samples show spectra which are shifted to longer wavelength (lower photon energy), are split into a pair of lines and are significantly broader than the homoepitaxial sample. These effects all result from strain in the heteroepitaxial films. (From H. P. Lee, S. Wang, Y. Huang, and P. Yu (1988) *Appl. Phys. Lett.* 52, 215). Reprinted with permission from the American Institute of Physics.

gap which predicts an *upward* shift of emission energy, while the axial component of strain lifts the degeneracy of the light and heavy hole valence bands at the Γ point (see Figure 5.1) and causes a splitting between the two components of the emission spectrum (e-hh and e-lh). Both the shift and the splitting provide a measure of the residual strain in the GaAs film. On the other hand, the effect of thermal expansion mismatch will result in a *downward* shift of emission energy, together with a splitting of opposite sign. What actually happens in practice? Figure 5.14 shows an example of low temperature PL spectra taken from a standard sample of GaAs on GaAs, together with the spectrum from a sample grown on silicon. Two features are apparent: first, the spectra are shifted to *lower* energy, second the single sharp line is split into two peaks whose linewidths are considerably increased, compared with the standard sample. The increased line width is the result of a distribution in strain between different parts of the film which effectively gives slightly different spectra from different regions and spreads the lines out (referred to as 'inhomogeneous line broadening'). However, the significant result is, of course, the *reduction* in photon energy compared to the standard sample which implies that the residual strain is due to thermal expansion mismatch, rather than lattice mismatch. Strain relief by dislocations at the growth temperature is apparently rather efficient.

Various other techniques were brought to bear in order to characterize as many samples, as fully as possible. Infrared absorption was applied to the identification of unknown impurity atoms by way of their 'local mode' vibrational spectra, a technique able to provide information concerning the symmetry of the site as well as the nature of the impurity atom. Far infrared photoconductivity was used to excite electrons out of shallow donor states into the conduction band and thus measure donor ionization energies with remarkable accuracy, an accuracy which permitted differentiation between different donor atoms in much the same way as we described above for acceptors but, because donor energies in GaAs are as small as 6 meV, it was necessary to achieve a resolution of 0.01 meV in order to distinguish them, quite beyond the capability of PL. Photoreflectance and piezo-reflectance methods were used to study fine details in exciton spectra, as were various magneto-PL experiments. Time-resolved luminescence techniques have given invaluable information about the rates of recombination processes and a range of techniques has been developed for the measurement of minority carrier diffusion lengths. Raman spectroscopy was much used to identify the energies of lattice vibrational modes and, again, to measure strain, making use of the sensitivity of certain modes to small distortions of the lattice.

Structural quality (also of major significance) was monitored by X-ray diffraction (XRD), transmission electron microscopy (TEM), scanning electron microscopy (SEM), and, more recently, by atomic force microscopy (AFM), which allows atomic resolution in surface roughness measurement. Chemical composition has been confirmed (or not!) by Auger electron spectroscopy (AES), secondary ion mass spectroscopy (SIMS), electron probe micro-analysis (EPMA), X-ray photoemission spectroscopy (XPS), energy-dispersive X-ray analysis (EDX), extended X-ray absorption fine structure analysis (EXAFS), and Rutherford backscattering (RBS). Another interesting technique brought to bear in a few cases is that of electron–positron annihilation spectroscopy which has the capability of characterizing vacancies in semiconductor crystal lattices. The list seems endless—but we must, nevertheless, end this account in order to concentrate on some of the device developments which were to give GaAs its central importance in the scheme of things. Hopefully, however, the reader will now have some feel for the huge effort devoted to characterization measurements during the 1960s and continuing, indeed, to the present day. Over the years, many different materials have benefited but it was all sparked off by the need to gain control over GaAs.

5.5 Light emitting devices

The huge investment in crystal growth and characterization techniques which formed the subject matter of the two previous sections clearly had to be motivated by more than a simple interest in semiconductor physics. The hope of material gain likely to accrue from the development of GaAs depended on the essential differences between it and the already well-established material, silicon; differences, on the one hand, in band structure, resulting in qualitatively different behaviour and on the other, quantitative differences in electronic parameters which might give it an edge in device performance. In fact, it was the direct band gap, and the efficient radiative recombination consequent upon that, which dominated the initial commercial development of GaAs, leading, as it did, to the semiconductor laser, essential to the success of the compact disc audio system. Philips launched their first optical disc system (the Video Disc) in 1972 but it used a He–Ne gas laser and never quite caught the public imagination. It was the availability of the semiconductor laser in 1979 which contributed to the dramatic success of CD audio and its subsequent clones, the Interactive Video Disc, the CD ROM, more recently the Digital Versatile Disc. The world market for semiconductor lasers is now some $4 billion, of which more than $1 billion belongs to these applications alone—but no one could possibly have guessed at such riches in the 1960s when GaAs light-emitting diodes were first under development—the early pioneers certainly deserve our approbation for their tenacity.

The first reports of electroluminescence from GaAs appeared in 1955, based on point contact diodes, the 'light' (actually infrared) having a photon energy some 300 meV below the band gap. However, it was only with the development of well controlled *p-n* junction devices that true band-edge luminescence could be consistently generated (E_g = 1.52 eV at low temperatures, decreasing to 1.43 eV at room temperature, corresponding to infrared wavelengths in the range 800–900 nm). To understand the operation of such a diode (LED) we refer back to the discussion of diode behaviour in Section 3.4 (and Box 3.4). As we saw there, a *p-n* junction diode under forward bias draws a forward current which is carried by minority carriers, injected across the junction plane and recombining within a few diffusion lengths of the junction. In contrast to the case of silicon or germanium diodes, in a GaAs diode, a significant portion of the recombination results in the emission of 'light', resulting in a simple light emitter that operates at low voltage (approximately 1.3 V) and moderate current (~10 mA). Even though the external efficiencies of these early LEDs were rather poor (~0.01%) and the radiation was invisible, it was immediately apparent that they would provide highly convenient sources for applications such as optical couplers (a combination of an emitter with a photodiode to transmit electrical signals over short distances, with the advantage of complete electrical isolation) and many workers took up the challenge to develop practical devices. (What, of course, was not appreciated at this time was the potential for the LED to be developed into the commercially much more significant laser diode (LD).)

Fabrication was based on the diffusion of zinc acceptors into an *n*-type sample of GaAs, which initially took the form of a bulk single crystal, though, when epitaxial techniques became available in the 1960s, this was replaced by a thin *n*-type film. The *p-n* junction was formed at a depth of some 20 μm below the surface where the *p*-type doping profile reached a density equal to the background *n*-type level (see Figure 5.15) and the light emitted within a few microns of the junction plane. To make individual devices, the GaAs slice was diced into squares and each square alloyed onto a tin plated metal base to form the back contact to the *n*-region (remember that tin is a donor in GaAs). A small In–Zn sphere was then alloyed to the *p*-type layer to make the top contact and a thin wire attached to provide electrical connection to a current source. Finally, the *p*-region was etched to define a device area of about 10^{-3} cm^2.

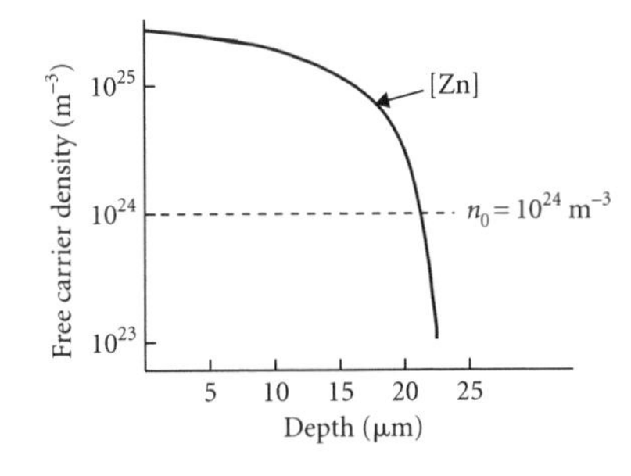

Figure 5.15. A profile of zinc concentration vs depth for a zinc-diffused GaAs *p-n* junction such as used to make a light-emitting diode. The junction is formed at the point where the zinc concentration [Zn] falls below the background *n*-type doping level $n_0 = 10^{24}$ m^{-3}, that is, at a depth of about 20 μm.

The performance of the resulting devices varied considerably and clearly depended in some way on material quality and device structure. Though room temperature performance was modest, there was evidence to suggest that high radiative efficiency might be achieved at low temperatures (77K and below). This (ill-understood) temperature dependence was still one of the principal problems being addressed at the time of the first Gallium Arsenide Conference in 1966 but it was not allowed to get in the way of what proved to be dramatic progress towards the development of the first semiconductor lasers only a little more than 2 years after the announcement of the first solid state (ruby) laser by Theodore Maiman at Hughes Research Laboratories in July 1960, itself closely followed by Ali Javan's He–Ne gas laser at Bell Labs.

The story of the laser has been splendidly told by Charles Townes (who shared the 1964 Physics Nobel Prize for his contribution to its invention) in his book 'How the Laser Happened' (Townes 1999). Very briefly, it began at Columbia University in 1954 with the demonstration of the first ammonia *maser*, a unique microwave oscillator which depended on some brilliant scientific wizardry to produce an 'inverted population' (see Box 5.4 and our earlier discusion in Section 4.6) between two energy levels in a beam of ammonia gas molecules. This provided the conditions for stimulated emission of radiation to dominate the more usual spontaneous emission and allow coherent amplification of a microwave frequency corresponding to the energy difference between the two levels. The same principle was later applied in developing a solid state maser amplifier (coincidentally, also using a ruby crystal) for detection of the first trans-Atlantic satellite communication signals via Telstar in 1963 and for similar applications in radio astronomy. The question of whether this same principle could be employed to amplify light signals was warmly debated but was finally answered by Maiman's success in 1960—the laser oscillator became a reality. It immediately led to speculation concerning the feasibility of making a laser based on a semiconductor *p-n* junction diode and, given the rather crude state of semiconductor technology at the time, the brief span of 2 years which elapsed can only be regarded as truly remarkable. Two other things are remarkable about the development of the first semiconductor lasers; first, the fact that no fewer than four different experimental research groups claimed success within a matter of weeks of one another (September–October 1962) and, second, the number of widely differing theoretical studies made between the years 1953 and 1962, leading up to these practical demonstrations (but which, for the most part, were largely irrelevant to the practical realization! The first, and one of the most detailed, by John von Neumann, though written in 1953, was not even published until 1963 and then only in summary form). The interested reader will find details, including the full text of von Neumann's paper, in a series of Special Issue Papers published in the IEEE Journal of Quantum Electronics, June 1987 (e.g. the introductory paper Dupuis 1987)—they make fascinating reading. With the familiar advantage accruing from hindsight, it is now clear that the key requirements for a successful LD were: (1) a semiconductor with direct band gap which favours radiative, rather than non-radiative recombination, (2) a semiconductor with at least one band characterized by a small effective

Box 5.4. The semiconductor laser

Whole books are dedicated to describing the theory and practice of semiconductor lasers (e.g. Casey and Panish 1978; Agrawal and Dutta 1993) so how can it be possible to condense all this understanding into one small box? Not at all, of course, but we must attempt a brief outline of some important features in order to clarify the discussion in the main text. The laser, in its usual format, takes the form of an oscillator—that is a source of radiation—characterized by coherence, narrow line width, and narrow (angular) beam width, characteristics resulting from the nature of the emission process (stimulated emission) and the structure employed.

First, we consider the process. The laser acronym, Light Amplification by Stimulated Emission of Radiation, implies that the semiconductor material acts as a 'gain medium'—in other words, it acts as an amplifier for light signals and it does so by virtue of the process of 'stimulated emission'. If we think of an atomic system having two energy states, separated by an energy $h\nu$, and situated in a radiation field of frequency ν, three radiative processes are possible: (1) absorption of a photon from the field by an atom which raises the atom from its ground state to its excited state, (2) spontaneous emission of a photon by the atom into the field, and (3) stimulated emission of a photon into the field due to the stimulating effect of the field. This latter process, which contains the essence of laser action, is effectively the exact reverse of the absorption process and it occurs with exactly the same probability. What is more, the emitted photon is in phase with the stimulating photon which gives rise to the property of coherence. However, in normal circumstances the vast majority of the atoms are in the lower energy state (the ground state) so absorption will dominate emission and stimulated emission is of only marginal importance. The magic of the laser lies in the stroke of ingenuity which produces an 'inversion' of normality, the upper state now being more highly occupied than the lower—we speak of an 'inverted population' in the atomic system. Stimulated emission now dominates and the semiconductor acts as an amplifying medium—photons already present in the radiation field stimulate further emission from the atoms and the density of photons is thereby increased. But how is this achieved in practice?

Let us first make it clear that different types of laser achieve inversion in various different ways but the semiconductor laser achieves it by injection of minority carriers across a *p-n* junction (hence the name 'laser diode'). If, for example, we think of a *p*-type sample containing a high hole density and we inject into it an equally high density of electrons, we generate a high probability for electron–hole recombination (with corresponding photon emission), which may exceed the probability for photon absorption. Can we quantify this? Yes, indeed, we can—it was done for us in 1961 by two Frenchmen from CNET, Bernard and Duraffourg. Refer to Figure 5.16 which represents an electronic transition between a state in the valence band and a corresponding state in the conduction band of our semiconducting medium. We can write a simple expression for the transition rate as follows:

$$r_{12} = B_{12} f_1 (1 - f_2) \rho(\nu_{12}), \tag{B5.14}$$

where B_{12} is the so-called Einstein coefficient, f_1 the probability of the valence band state being occupied by an electron, f_2 the probability of the conduction band state being similarly occupied and $\rho(\nu_{12})$ the density of photons in the radiation field. In identical fashion, we can write an expression for the rate of *stimulated* emission from state 2 down to state 1:

$$r_{21} = B_{21} f_2 (1 - f_1) \rho(\nu_{12}). \tag{B5.15}$$

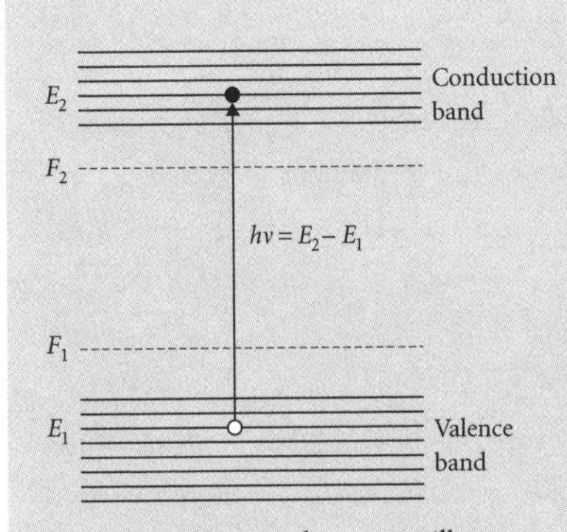

Figure 5.16. Energy diagram to illustrate the absorption of a single photon of energy $h\nu$ which excites an electron from a valence band state E_1 to an empty state in the conduction band E_2. F_1 and F_2 are the two quasi-Fermi levels which describe the occupation of valence and conduction band states, respectively.

Box 5.4. Continued

Bearing in mind that $B_{12} = B_{21}$, it is now easy to see that the condition for the rate of stimulated emission to exceed that of absorption is:

$$f_2(1-f_1) > f_1(1-f_2), \quad \text{(B5.16)}$$

which, for reasons which will become clear in a moment, may be written as:

$$[1/f_1 - 1] > [1/f_2 - 1]. \quad \text{(B5.17)}$$

All that now remains is to express this result in terms of the so-called quasi-Fermi levels F_1 and F_2 indicated in Figure 5.16. Note that in Box 5.3 we defined a single Fermi level, which characterize the occupancy of states in both bands—in the present case, where the system is very far from thermal equilibrium on account of the high injection level, we define *two* Fermi levels, one for each band. Thus, we can write:

$$f_1 = [1 + \exp\{(E_1 - F_1)/kT\}]^{-1} \quad \text{(B5.18)}$$

from which it follows that:

$$[1/f_1 - 1] = \exp\{(E_1 - F_1)/kT\} \quad \text{(B5.19)}$$

and, finally, the condition for stimulated emission to dominate absorption is simply written as:

$$\exp\{(E_1 - F_1)/kT\} > \exp\{(E_2 - F_2)/kT\}, \quad \text{(B5.20)}$$

which can be rearranged as:

$$(F_2 - F_1) > (E_2 - E_1) \quad \text{(B5.21)}$$

implying that the separation of the quasi-Fermi levels should be greater than the photon energy of the emitted radiation. The beauty of this very simple formulation lies in the fact that F_1 and F_2 provide a measure of the density of free carriers in their respective bands. For example:

$$n \sim N_C \exp\{-(E_C - F_1)/kT\}. \quad \text{(B5.22)}$$

N_C being the effective density of states in the conduction band, so we can readily express our condition in terms of the injected carrier density required to achieve inversion and this, in turn, allows us to estimate the appropriate injection current (known as the 'transparency current', the current at which net optical absorption in the semiconductor tends to zero). Strictly, equation (B5.22) is only valid when $(E_C - F_1) > 3kT$ which may not be true for high injection conditions so it should be taken as giving only a rough estimate of n but that alone may well be valuable. In practice a value of $n \sim 2 \times 10^{24}\ \mathrm{m}^{-3}$ is found appropriate for so-called 'transparency' to be reached (when inversion occurs, net absorption in the GaAs becomes zero and the material is completely transparent at the laser wavelength).

Box 5.4. Continued

So much for the process. What about the structure? This has the prime purpose of converting an amplifier into an oscillator and, as the reader will no doubt be aware, this depends upon the provision of positive feedback. The laser structure which performs this function is illustrated in Figure 5.17. It consists simply of a pair of (partially reflecting) parallel mirrors in the form of cleaved semiconductor crystal facets. As the stimulated emission process generates new photons within the material, these photons are reflected back at each mirror facet and return to stimulate further emission, the photon density (radiation intensity) increasing exponentially along a line normal to the mirror planes (which accounts for the narrow beam characteristic of the laser). At each mirror, a fraction of the radiation escapes from the 'laser cavity' and is therefore lost to the gain medium. In addition there is loss within the cavity which can be characterized by a loss coefficient α (having units of m^{-1}) and it is easy to show (see Figure 5.17(b)) that the condition for the gain to exceed the losses is:

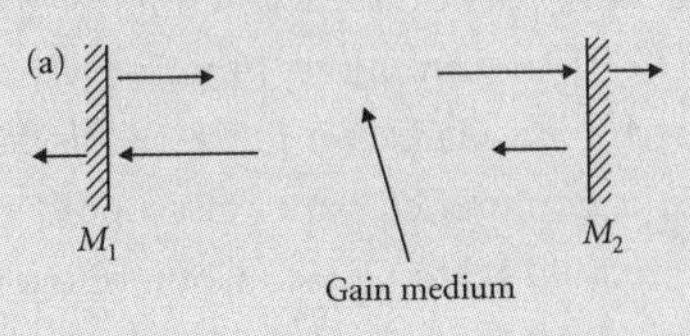

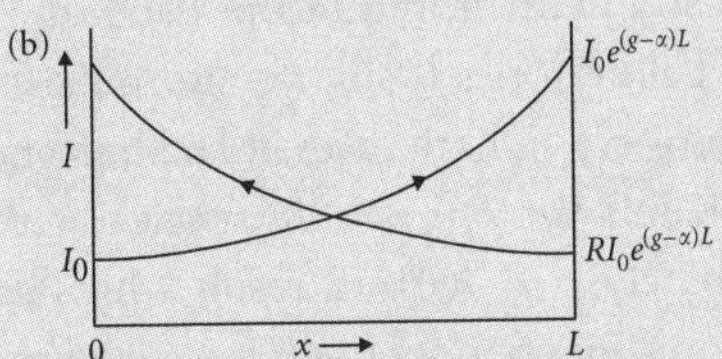

Figure 5.17. Schematic diagram demonstating the use of a pair of partially reflecting parallel mirrors to form an optical cavity in a semiconductor laser. Light is reflected backwards and forwards between them along the axis of the cavity, while gain is provided by stimulated emission of light within the semiconductor material. (a) Shows the physical arrangement, while (b) provides a plot of light intensity as a function of position within the cavity. The overall gain just balances cavity losses when $R\exp\{(g-\alpha)L\} = 1$.

$$(g-\alpha) > (1/L)\ln(1/R), \qquad \text{(B5.23)}$$

where g is the laser gain coefficient (i.e. gain over a length $x = \exp\{gx\}$), L the length of the cavity and R the reflection coefficient at each mirror (assumed equal). Highly reflecting mirrors and long cavities make the task of the gain medium less onerous but it is clear that g must, at least, exceed the cavity loss α. Note that equation (B5.23) defines a threshold condition for laser action which implies a well defined threshold current density J_{th}, somewhat greater than the transparency value. For current densities below this, spontaneous emission dominates, above it, stimulated emission takes over and light output increases much more rapidly, there being a sharp knee in the (light vs current) characteristic.

The cavity performs another important function in that it assists in determining the precise laser emission frequency ν (or wavelength λ). The effective 'round trip' gain within the cavity peaks sharply when an integral number of half-wavelengths corresponds precisely to the cavity length and this results in a set of so-called 'longitudinal cavity modes' separated in wavelength by $\Delta\lambda$, where:

$$\Delta\lambda = \lambda^2/2\mu L \qquad \text{(B5.24)}$$

μ being the refractive index of the semiconductor. For a GaAs laser emitting at $\lambda = 850$ nm, with a cavity length of 300 μm and refractive index $\mu = 3.5$, the mode separation turns out to be approximately 0.35 nm (corresponding to a frequency difference of 1.51×10^{11} Hz or, expressed in energy units, 0.62 meV). In practice, the gain curve is very much wider than the mode separation so a large number of modes fall within it and this often results in lasing at several wavelengths simultaneously—we speak (somewhat disdainfully!) of a 'multimode laser'.

Finally, in this brief summary of laser characteristics we should comment on the width of the laser emission line. It is an important feature of laser action that the line width becomes extremely narrow above the lasing threshold and this is often used as evidence for laser action. It is characteristic of any oscillator and applies equally to an LC-controlled RF oscillator such as used in radio circuits—when the cavity gain overcomes all losses, the cavity Q-factor tends to infinity and, in consequence, the emission line width tends towards zero. 'Infinity' and 'zero' are never obtained, of course—second-order processes serve to limit the effect—nevertheless extremely narrow lines are observed in practice. Whereas the laser gain curve may have a width of order 10^{13} Hz (corresponding to the width of spontaneous emission below threshold), the emission line can be as narrow as 10–100 MHz.

mass—this results in a small density of states and makes it easier to achieve the inversion condition discussed in Box 5.4, (3) the ability to achieve high doping densities, both *n*- and *p*-type, (4) the incorporation of a suitable optical cavity to provide the feedback necessary for laser action. In this context, therefore, it is interesting to note how many of the early papers referred to silicon or germanium, rather than GaAs and how few seemed to appreciate the importance of using an optical cavity. However, everything changed in the early part of 1962 when it became widely known (at least within the US III-V community) that very efficient luminescence could be obtained from GaAs *p-n* junction diodes. This was first reported in March in a conference paper presented by Sumner Mayburg of GTE, then at the Solid State Device Conference in July by two other groups, Keyes and Quist from Lincoln Laboratories and Jacques Pankove from RCA Laboratories (Pankove also published a Physical Review Letter on the subject). In both cases, it had been necessary to cool the diodes to 77K but the *internal* efficiencies at that temperature were estimated to be close to 100%, a result which sent several conferees back to their laboratories hotfoot to exploit such promise in the form of a first demonstration of a semiconductor laser. Chief among these were Nick Holonyak (then at GE Syracuse), Robert Hall (GE Schenectady), Marshall Nathan (IBM, Yorktown Heights), and all three, together with the Lincoln Labs Group were in at the kill, just 2 months later. Hall actually won the race by a short head—his paper was received on 24 September, just ahead of the IBM paper, received 6 October, with Holonyak's on the 17th and Quist's on the 23rd of the same month. Though work was certainly in progress elsewhere (notably in Soviet Russia), the Americans appear to have made a clean sweep of this particular contest! The key to success lay in the addition of a suitable optical cavity to the basic diode structure and this was most easily provided in the form of a pair of parallel mirror facets, made by polishing optical flats on the laser crystal. Holonyak chose to use the ternary alloy material, GaAsP which emitted red light, while the others used GaAs but the important requirement of a direct energy gap was common to all.

There can be no question that the events of September 1962 represented a major turning point in III-V semiconductor science but it is, nonetheless, salutary to examine what had been achieved in technological terms. Coherent light had been generated by a small, convenient, inexpensive, low-voltage device but only when immersed in the (far from convenient!) medium of liquid nitrogen. The required current densities were also somewhat alarming—thresholds were, in all cases, greater than 10^4 A cm^{-2}, which restricted operation to short pulses in order to minimize heat dissipation. So, once the fog of initial euphoria had cleared, the goal of room temperature, CW operation looked a very long way ahead. Indeed, so it was! Having reached first base in a mere 2 years, the laser diode was destined to wait another 8 years for its next major advance. What was the problem? There were, in fact, three problems: the material quality was still not good enough to allow efficient radiative recombination at room temperature, the structure had no means for confining the light close to the *p-n* junction nor was there any means for confining the injected carriers, which were free to diffuse away from the junction region and recombine where they would (rather than where they should—close to the junction). One is reminded of Dr Johnson's adage about dogs walking on their hind legs—one is not so much surprised to see it done badly, as to see it done at all! The remarkable fact is that the LD pioneers were as successful as they were.

Let us look at these three areas in slightly greater detail. First, material quality. The early lasers were made from bulk GaAs samples and, as we have already suggested in Section 5.3, the introduction of epitaxial growth methods was crucial to improvement here, the issue specific to laser development being one of radiative efficiency. Injected carriers may recombine radiatively to generate the photons required for stimulated emission to take control but, alternatively, they may recombine via so-called recombination centres, producing only phonons (i.e. heat), which represents an undesirable energy loss from the radiation field. In particular, the competition between the two processes tends to favour non-radiative recombination at higher operating temperatures, a property which mitigates against room temperature operation. The key to improving material quality was, therefore, to minimize the inclusion of unwanted impurity atoms and lattice defects which act as recombination centres. Compared with melt growth, the much reduced temperatures used in epitaxy (typically 700°C, rather than 1300°C) had the advantage of reducing both point defect concentrations (there is a thermodynamic argument in favour of this) and impurity incorporation (most impurities being more soluble at higher temperatures) and it was particularly fortunate, therefore, that LPE growth of GaAs should appear on the laser scene in the early 1960s. But probably an even more important aspect of LPE was its ability, demonstrated in 1967 by Jerry Woodall of IBM, to grow the ternary alloy AlGaAs. It was this which made possible the development of heterojunction lasers, leading, in turn, to the 'separate-confinement-heterostructure' (SCH) device, which bore the semiconductor laser into the all-important consumer market.

Several people had given thought to the problem of minority carrier diffusion which allowed precious carriers to disperse spatially, rather than remaining close to the junction region where they could contribute to stimulated emission, and in 1963 Kroemer put forward the idea of countering this tendency by using a heterostructure. He suggested, for example, using a thin layer of Ge sandwiched between two outer GaAs layers in order to confine free carriers in the germanium, the larger

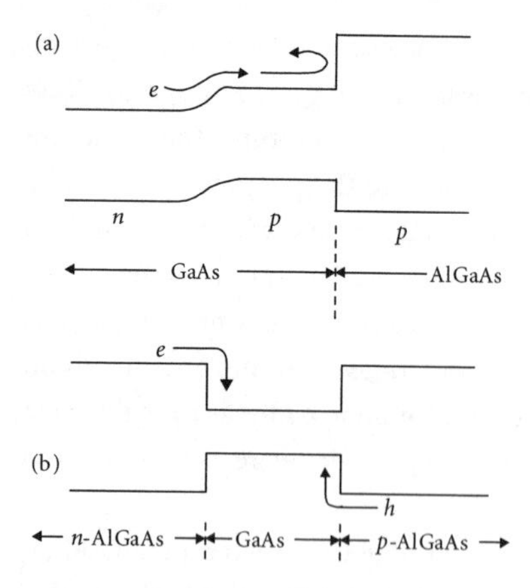

Figure 5.18. Schematic band diagrams of AlGaAs/GaAs heterostructures used to confine the recombining carriers in a semiconductor LD. (a) Represents a single heterostructure which prevents injected electrons diffusing away from the junction region. (b) Shows a double heterostructure which captures both holes and electrons so that they are constrained to recombine within the energy well, close to the junction.

energy gap of GaAs providing a barrier to their outward flow. This radical departure from then current practice was based on the fact that Ge and GaAs have closely similar lattice constants and could therefore be grown successfully together (though as we have already commented, the indirect gap of Ge makes it unsuitable for use in the active region of a laser). It was doubtless one of the ideas which won Kroemer his share of the 2000 Nobel Prize but, oddly enough, the people then responsible for making lasers appeared not to notice it! Alferov at the Ioffe Institute in Leningrad (now St Petersburg) had a similar idea for using a GaAsP/GaAs structure but did not publish it until 1967 so, again, no immediate stimulus resulted. The breakthrough had to wait until 1969 with the use of an AlGaAs/GaAs heterostructure (see Figure 5.18) and (as seems to happen in this field!) three separate groups came more or less simultaneously to the same solution. Nelson and Kressel at IBM and Hayashi and Panish at Bell Labs published accounts of what came to be known as 'single heterostructure lasers' (Figure 5.18(a)) in which electrons injected across a GaAs *p-n* junction were prevented from diffusing away by the presence of an energy barrier, due to an adjacent AlGaAs layer. Also in 1969 Alferov and co-workers reported an AlGaAs/GaAs 'double heterostructure laser' (DHL) in which carriers were confined in both directions (Figure 5.18(b)). While threshold currents for single heterostructures were very much improved as compared to homojunction devices (room temperature values of 8.6 kA cm^{-2}, compared to 50 kA cm^{-2}), Alferov's DH structure did even better, with a best result of 4 kA cm^{-2}. Room temperature CW operation followed soon afterwards, Hayashi and Panish and Alferov et al. both reporting success in the following year, with threshold currents now down to about 2 kA cm^{-2}. The long struggle to achieve a viable room temperature device had been crowned with success at last.

The use of AlGaAs for confining layers was based on the extremely convenient fact that the lattice parameter of AlAs differs by only about 0.5% from that of GaAs, making it possible to grow the desired heterostructures without significant mismatch, important in order to minimize strain and avoid the incorporation of dislocations which are commonly associated with lattice mismatched combinations. This was later recognized to be more than ever true when it was realized that a single dislocation could lead to catastrophic failure of a GaAs laser after relatively few hours of operation. The band gap of AlAs, too, is sufficiently larger than that of GaAs to permit adequate confinement with modest amounts of Al in the alloy. Very approximately, the direct energy gap difference is given by $1.25x$ (eV) (where x is the mole fraction of Al) and this difference is divided between the conduction and valence bands in the ratio of $\Delta E_C : \Delta E_V \approx 2:1$. In order to effect adequate confinement the step in energy must be of the order of a few times kT, which, at room temperature requires only that $x > 0.15$, an easy enough criterion to satisfy. It is also possible to dope AlGaAs alloys both *n*- and *p*-type using the same dopants as are effective in GaAs, convenient for ease of epitaxial deposition.

Typically the first DH laser structures employed cavity lengths of about 500 μm and carrier confinement regions *d* between 0.5 and 1.0 μm thick, with the Al fraction in the confining layers $x \sim 0.3$. It was soon recognized that a simple method to reduce threshold current lay in reducing *d*, which had the effect of reducing the total volume of material to be pumped (though, of course, it also reduced the output power) and a linear reduction was, indeed, observed experimentally down to a thickness of about 0.4 μm. Below this, however, a law of diminishing returns set in because of mismatch between the confinement of carriers and the distribution of the light and this led in 1973 to the introduction of the so-called 'separate confinement double heterostructure laser' (SCL), proposed independently by Thompson and Kirkby at STL Laboratories in England and by Hayashi at Bell Labs. This structure makes use of another feature of the $Al_xGa_{1-x}As$ alloy system, namely that the refractive index decreases with increasing x (roughly as $n = 3.59 - 0.72x$) and this facilitates optical confinement in the shape of a slab waveguide (confinement occurs when the outer layer of material has a lower refractive index than that of the central strip). The structure, shown in Figure 5.19, allows the carrier and light confining regions to be optimized independently and provides the ultimate semiconductor laser design. Threshold currents came down routinely to about 500 A cm^{-2}, some two orders of magnitude below those measured on the original homostructure lasers, 10 years earlier (the book by Gooch 1973: 5 provides an interesting graph of J_{th} vs Year for the decade 1962–72). As also indicated in Figure 5.19, practical lasers

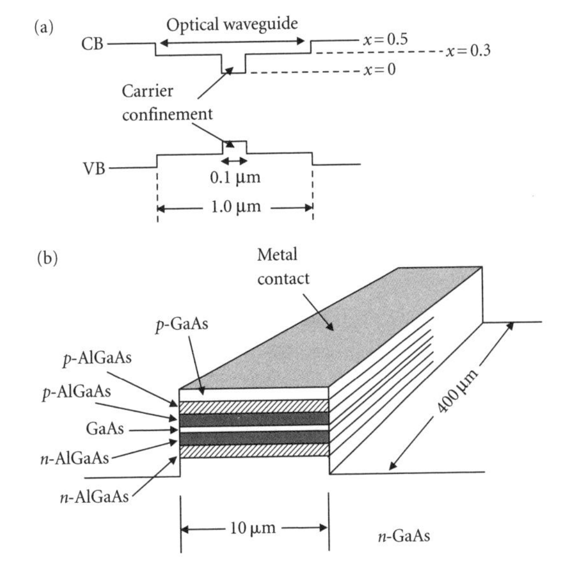

Figure 5.19. Energy band and physical structure of a separate confinement double heterostructure laser. The band diagram in (a) shows how two different $Al_xGa_{1-x}As$ layers ($x = 0.5$ and $x = 0.3$) are used to confine both carriers and light. This arrangement optimizes the interaction between the light and the recombining carriers, recombination taking place at the point where the optical intensity is maximum. A typical mesa stripe laser is shown in (b), light being emitted through the end facet—that is, normal to the plane of the diagram.

employed some kind of stripe geometry to confine the lasing region in the horizontal direction, stripe widths varying between about 50 and 5 μm. There was still much to be explored in the way of controlling the pattern of lasing modes, temperature dependence of threshold current, the character of the output beam and the improvement of total output power but the GaAs laser was now a viable commercial device.

This came at a convenient time for the introduction of the CD player by Philips and Sony in 1978 (the other essential ingredient being the availability of high speed integrated circuits (IC) to cope with the digital signals employed—see Box 4.1) but, even then, it should not be thought that all the technical problems had been overcome. In designing the first CD systems, the choice of laser wavelength was important as this limits the minimum spot size to which the laser beam can be focussed and, therefore, the density of information which can be stored on the disc—the shorter the wavelength, the more music per disc. The compromise of 780 nm, which was reached demanded yet further development of the fledgeling semiconductor laser. To reduce the wavelength from the value of 880 nm appropriate to GaAs it was necessary to include about 15% of Al in the active layer to increase the band gap by the required amount and then to increase the Al content of the two confining layers in sympathy, not a major step in scientific understanding but one which brought its own frustrations for the technologist. At the same time, other choices had to be made, concerning the method of epitaxial growth to be employed. LPE, VPE, MOVPE, and MBE were all available by then and making strong claims for preferment, but whichever of these routes was followed, growth had to be on low-dislocation-density substrates to minimize catastrophic failure due to 'dislocation climb' through the active region of the laser.

We shall leave the laser story for the time being but, before doing so, a final comment may be in order. The reader can no longer be in any doubt (if he or she ever was) concerning the enormous gulf between the demonstration of a scientific principle and the realization of a commercial device based upon that principle. Semiconductor laser action was clearly demonstrated in 1962 while the first major commercial application did not emerge until 1978, 16 years later! Should the sceptic be tempted to ask 'what kept you?', we need merely point out the number of technological developments incorporated in the narrow stripe, separate confinement, double heterostructure, red emitting laser which eventually found its way into the CD player in 1978, when compared with its 1962 predecessor. Not that any of this detracts one iota from the debt owed to the pioneers—the point to grasp is that both dramatic innovation and technological hard grind are essential to ultimate commercial success and neither can manage without the other.

5.6 Microwave devices

The early days of GaAs device research were characterized not only by the exploitation of its efficient light emission but by attempts to use the high electron mobility, values as high as 0.85 $m^2 V^{-1} s^{-1}$ having been measured on moderately pure material by 1960 (nearly six times greater than the corresponding value for silicon). It was clear that advantage might therefore accrue for both bipolar and field effect transistors, tunnel diodes, and microwave Schottky diodes to be employed as detectors (compare the silicon cat's whisker of wartime fame) and mixers. In all cases there was hope of faster operating speeds, in particular the application to the microwave region of the spectrum between 1 and 10 GHz. For comparison, we note from Table 3.2 that at the beginning of the 1960s Ge and Si bipolar transistors were struggling to reach 1 GHz, while both radar and satellite communication systems were functioning at frequencies of 3–10 GHz or more. There was considerable incentive, therefore, to move solid state electronics into this spectral region and GaAs appeared to offer the greatest promise of success.

Few predictions ever work out quite as anticipated and this one was no exception. The bipolar transistor made only a brief appearance (though, as we shall see, it was later resuscitated in the form of the heterojunction bipolar transistor HBT), being fairly abruptly killed by severe trapping of minority carriers in the base region and, though the tunnel diode may have made a considerable impact on semiconductor physics, it had no lasting influence on the device scene, while one of the most significant microwave devices, the Gunn diode, was not even on the original list. This section will therefore be concerned largely with the FET, the mixer diode, and the Gunn diode, all successful microwave devices of the 1960s and still performing well today.

It is often difficult to state with certainty which of the several devices actually came first but it is probably fair to say that the Gunn diode emerged with a slight lead in the early 1960s so we shall give it priority here. It should really be known as the Ridley, Watkins, Hilsum, Gunn (RWHG?) effect diode in honour of its various inventors and discoverer but because it depends for its operation on the so-called 'Transferred Electron Effect' it is also known as a transferred electron diode (TED). In its simplest format, it consists of an extremely basic structure, a single, uniform film of lightly *n*-doped semiconductor (usually GaAs or InP), a few microns thick, sandwiched between a pair of metal contacts and mounted in a suitable microwave package. The application of a DC voltage of just a few volts between the contacts results in the generation of milliwatts of microwave power at a precise frequency determined by a combination of film thickness and circuit characteristics. It has been, and is widely used as the local oscillator in a range of heterodyne microwave receivers and as the microwave source in mini-radar systems such as used by the police for monitoring automobile velocities in restricted areas of the road system (to mention but two of its many applications). It has been used to generate microwave frequencies all the way from about 3 to 300 GHz (wavelengths from 10 cm to 1 mm) and, given this considerable versatility, represents one of the simplest and cheapest semiconductor devices ever developed.

The story began in 1960 and, in marked contrast to the laser, it was an essentially British story. Brian Ridley, Tom Watkins, and Ron Pratt, at the Mullard Research Laboratories in Redhill, Surrey were enthusiastically hunting for so-called 'negative resistance effects' in semiconductors and in 1961 Ridley and Watkins published two papers in which they predicted that a negative resistance (i.e. a negative slope in a current–voltage characteristic) should be observable in suitable samples. They considered two mechanisms, one involving the field-assisted capture of electrons by a negatively charged deep impurity centre, the other the transferred electron effect. In the first case, the kinetic energy gained by electrons from the applied electric field would overcome the Coulomb repulsion between the electrons and the negatively charged centre and increase their rate of capture. Thus, an *increase* in field would lead to a *reduction* in the density of free electrons, that is, a *reduction* in conductivity. In the second case, they envisaged electrons (or holes) in one band minimum being accelerated by an applied electric field, the excess energy thereby gained allowing them to transfer to a different minimum at somewhat higher energy which was characterized by a larger effective mass (and correspondingly lower mobility). This reduced mobility implied that, on transfer, the carriers would slow down, thus reducing the average drift velocity of the overall carrier distribution and therefore the current flowing in the semiconductor. Thus, as the applied voltage was increased, the current would first increase, but then decrease and this negative slope (dI/dV) was the negative resistance they sought. In principle, it might be used to cancel the normal, positive resistance in an electrical circuit and result in amplification of an injected signal (or generate spontaneous oscillatory behaviour). They were aware that the germanium they were studying did not have the appropriate band structure for the transferred electron effect but proposed that the application of strain might shift the minima in such a manner as to produce it artificially.

Experimentally, they succeeded in measuring a negative resistance due to electron capture at negatively charged Au centres in germanium, though it was necessary to reduce the ambient temperature to 20K in order to see it. On the transferred electron front they were less successful but this particular torch was immediately taken up by Cyril Hilsum at the Services Electronics Laboratory, Baldock, Herts who, in a paper submitted just 4 months after theirs, pointed out that the III-V materials GaSb and GaAs already possessed the necessary structure in their conduction bands and reported calculations of the TE effect which encouraged further exploration. In particular, the GaAs CB structure was believed (at the time) to have a central (Γ) minimum roughly 0.36 eV below the first excited (X) minima, with the L minima perhaps 0.14 eV higher still. (More recent work suggests that the order of the X and L minima is reversed (see Figure 5.1) though this makes no difference to the gist of the argument.) On this basis, Hilsum estimated that, for a lattice temperature of 100°C, at applied electric fields greater than about 3 kV/cm ($3\times10^{5}\,\mathrm{V\,m^{-1}}$), a negative differential resistance would be found. He also suggested the need to use semi-insulating GaAs in order to minimize the dissipation due to large electric currents at such fields.

Once again, experimental success was frustratingly elusive and little further progress could be made at Baldock. In fact, the next step was actually made at the IBM Thomas J. Watson Research Center, Yorktown Heights, though by yet another Englishman, J. B. Gunn who had joined IBM in 1959 with an interest in applying short pulse techniques to semiconductor research. However, it was very much by chance that he discovered high frequency oscillations emanating from samples of GaAs and InP while studying 'hot electron effects' and, though he was aware of both Ridley and Watkins' and Hilsum's papers, he proved to his own satisfaction (and, incidentally, also to Watkins' during a visit to IBM) that their predictions did not provide a satisfactory explanation for his observations. He published his findings in 1963 in Solid State Communications—above a certain threshold field, he observed coherent oscillations at frequencies (in the range 0.47–6.5 GHz), which showed an inverse dependence on sample thickness (somewhat surprisingly, sample thickness seemed to play a dominant role in the determination of the oscillatory frequency) and represented microwave to DC conversion efficiencies of 1–2%, values which 'would appear to be high enough to make the phenomenon of some technological importance'. But he also concluded that 'The mechanism leading to the oscillations is not yet understood'. It *was*, nevertheless, as revealed by later experiments, the transferred electron effect—the RWHG effect was an experimental reality, even though it took a further 2 years for the proof to emerge.

The experiment which finally clinched the point was reported by A. R. Hutson and colleagues from Bell Labs in 1965. They studied the behaviour of the oscillations while applying hydrostatic pressure to their GaAs samples and observed, first of all, that the threshold field was reduced with increasing pressure, then that further increase caused the oscillations to die away altogether. Earlier work had shown that the principal effect of hydrostatic pressure was to reduce the energy separation between the Γ and X conduction band minima, so the implication of their results was that the oscillations resulted from the transfer of hot electrons from the Γ to the X minima. This gradually became easier as the separation decreased but, when the energy difference became small enough that thermal energy was sufficient for the transfer to occur, the negative resistance disappeared and the oscillations ceased.

At this point, the discerning reader can be forgiven for wondering exactly how the application of a steady DC voltage across a thin film of uniform semiconductor could possibly give rise to oscillations at microwave frequencies (or, for that matter, at any frequency!). It is very far from obvious. Interestingly, the explanation became available more or less coincidentally with, though quite independently of, the first experimental observation. In yet another paper, published in 1963, Ridley pointed out that, in the presence of a negative resistance of the transferred electron type, a uniform distribution of electric field would be unstable and would break up into regions of high and low field with which would be associated a non-uniform free carrier density. We discuss the precise form of these non-uniformities in Box 5.5—here let it suffice to say that a region of high field and modified carrier density (known as a 'domain') propagates through the device from one electrode to the other with a velocity $v_D \sim 10^5\,\mathrm{m\,s^{-1}}$, resulting in a series of current pulses in the external circuit which repeat with a frequency of $\nu = 1/\tau_t = v_D/L$, where τ_t is the domain transit time and L is the sample length (in quantitative agreement with Gunn's observations). For a device with thickness $L = 10\ \mu\mathrm{m}$, we therefore find $\nu = 10$ GHz (a frequency known in engineering jargon as 'X-band'). The presence of such travelling domains was first measured experimentally by Gunn himself and reported at a meeting in Paris in 1964 more or less coincidentally with the publication of an IEEE paper by our old friend Herb Kroemer who made probably the first convincing argument for the oscillations being of this type and resulting from the TE effect. As we have already noted, the final proof of his argument came just 12 months later in the paper by Hutson et al.

From these faltering beginnings grew a whole industry. Gunn himself, in reminiscences published as one of a series of such papers by the IEEE (Gunn 1976), describes how

> the number of labs and individual workers increased quite appreciably. At one point, the field seemed to be so crowded that, if one did not perform an experiment on the day that one thought of it, it would have been done elsewhere and the opportunity would have passed. The bandwaggon had now reached its terminal velocity, and the nature of the field was changing. The key to progress became, not fast pulse electronics and new experimental techniques but materials technology and device engineering.

Such is life when the sniff of future commercialization begins to turn into reality! By 1970, there was worldwide commercial activity—epitaxial growth had taken over from the rather crude bulk samples used

Box 5.5. The Gunn oscillator

The basis for all calculations on transferred electron devices is the electron 'velocity–field' characteristic which, for n-type GaAs, is shown in Figure 5.20. Let us be clear about its interpretation. Free electrons in the Γ conduction band minimum are accelerated by an applied electric field and gain kinetic energy. In lightly doped GaAs at room temperature, the dominant scattering mechanism which limits their velocity is polar optic phonon scattering and this has the property that, once electrons gain kinetic energy beyond a certain threshold value, their velocities tend to increase very rapidly, indeed, and they acquire sufficient energy to transfer to the higher X minima, some 0.36 eV above the Γ minimum. This transfer is mediated by interaction with 'zone boundary' phonons which provide the necessary momentum, the characteristic transfer time being of order 10^{-13} s. The negative differential mobility above threshold results from the fact that electrons in X minima have much smaller mobilities than those in the Γ minimum so that the average mobility suddenly decreases. The threshold field for electron transfer is 3.31×10^5 V m^{-1} and the value of the differential mobility dv/dF is approximately -0.2 m^{2}V^{-1}s^{-1}.

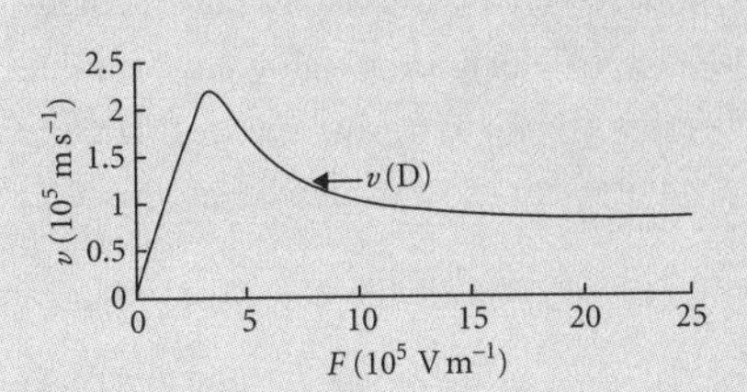

Figure 5.20. Experimental velocity–field curve for GaAs, v being the average drift velocity for electrons in an electric field F. The velocity increases with field until electrons begin to transfer to low mobility minima in the conduction band, when they are abruptly slowed down. Beyond the peak, the negative slope of the characteristic represents a negative resistance which can be used to generate microwave oscillations. The velocity v_D represents the velocity with which a dipole domain will travel between cathode and anode of a Gunn diode.

Once the device is biased above threshold, the negative resistivity leads to a space-charge instability in the form of a dipole domain which is illustrated in Figure 5.21. This domain is formed near the cathode electrode and travels through the device with a velocity v_D which is close to the 'valley' (or saturation) velocity shown in Figure 5.20—that is, $v_D \sim 10^5$ m s^{-1}. Note that the peak field in the domain is very high—typically of order 5×10^6 V m^{-1}—and this means that the field outside the domain is well below the average field $F_{av} = V/L$ so, once a domain is formed, there is no possibility for a second one to nucleate until the first leaves the device at the anodecontact.

The time constant τ_D associated with domain formation (the negative dielectric relaxation time—compare equation (3.10)) is:

$$\tau_D = \varepsilon\rho' = \varepsilon/(n_0 e\, dv/dF), \tag{B5.25}$$

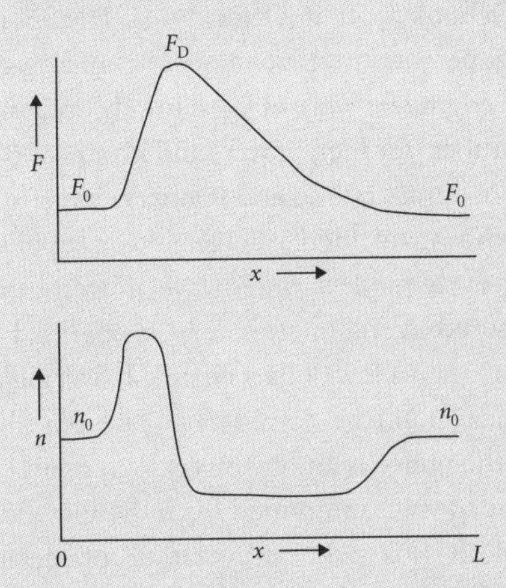

Figure 5.21. Schematic diagram to illustrate the properties of a dipole domain in a Gunn diode, showing the variation of electric field with position within the Gunn layer, together with the corresponding free electron concentration. The background doping level is n_0. The field outside the domain F_0 is too small to allow a second domain to be nucleated.

where ε is the dielectric constant of GaAs at microwave frequencies, ρ' is the (negative) differential resistivity and n_0 is the background free electron density in the layer. For a doping level of $n_0 = 10^{21}$ m^{-3}, $\tau_D \sim 3 \times 10^{-12}$ s, which suggests a time for domain formation of about 10^{-11} s. For comparison, we may note that, for an X-band (10 GHz) device ($L = 10$ μm) the transit time is $\tau_t = 10^{-10}$ s.

We now consider the nature of the current flowing from the device into the external circuit. During the time when the domain is fully formed the current density J_0 is given by:

$$J_0 = n_0 e v_D \tag{B5.26}$$

Box 5.5. Continued

but when the domain reaches the anode, the electron velocity rises to a value close to the peak value v_{pk} in Figure 5.20 ($v_{pk} \sim 2.21 \times 10^5$ m s^{-1}) so the current rises to a peak level of:

$$J_{pk} = n_0 e v_{pk} \quad \text{(B5.27)}$$

and the result is a sequence of approximately triangular current pulses, superimposed on a DC level. To maximize the efficiency of power generation it is necessary to make the pulse rise and fall times (which are roughly equal to one another) equal to half the transit time—that is, $\tau_D = \tau_t/2$ and, as τ_D is inversely proportional to n_0, this implies that there is a need for an optimum choice of n_0. Also, as τ_t is proportional to L, the condition can most readily be expressed in terms of the n_0L product, as follows:

$$n_0 L \sim 10^{16}\,\text{m}^{-2}. \quad \text{(B5.28)}$$

Another way of looking at this is to say that the domain length should be approximately half the sample length. Calculation of the DC-to-RF conversion efficiency under this condition gives a value of about 5%. An optimized X-band device, with $L = 10$ μm and doping level $n_0 = 1 \times 10^{21}$ m^{-3}, might have an area of 10^{-8} m^2, a DC input of 100 mA at 5 V (current density 10^7 A m^{-2} = 1 kA cm^{-2}) and generate an output of some 25 mW of RF power.

initially and emphasis had passed from basic understanding to exploitation. In the style characteristic of the fast moving electronics industry, new products and new applications were appearing at an alarming rate. Even Mills and Boon published a book about it (Orton 1971)! The reason was easy to see—the Gunn diode was the first *convenient* microwave source. Up to 1970, microwave engineers were obliged to rely on either magnetrons or travelling wave tubes (for high power) and klystrons (for low), all of which not only were cumbersome and relatively expensive in themselves, but required extravagant high voltage power supplies. The possibility of generating even modest microwave power using a device whose size was measured in millimetres, which operated at 10 V DC and which cost no more than a few dollars changed, overnight, the range of applications which could be considered. The portable Doppler radar system used by the police is an obvious case in point.

Probably the most important advances stemmed from the introduction of epitaxy. As explained in Box 5.5, optimization of device performance required that the doping level in and the thickness of the active material should be accurately controlled. For X-band operation this required $n_0 = 10^{21}$ m^{-3} and $L = 10$ μm, for Q-band (35 GHz) $n_0 = 3.5 \times 10^{21}$ m^{-3} and $L = 3$ μm—these being the most popular bands for many of the early applications—and bulk GaAs could scarcely hope to meet such demands on a routine basis. With epitaxy it was perfectly possible. The standard technique was to grow an active n-layer on an n^+ substrate which formed the anode, the top contact being either an evaporated metal film or a second n^+ layer grown epitaxially (followed by metallization). Two growth methods were used, liquid phase (from Ga solution) and vapour phase using chloride transport. Of the two, LPE probably had the capability for achieving higher purity though VPE was better at controlling doping levels and layer thicknesses. It also had the edge in terms of surface smoothness—important for the large area photolithography necessary to a production process. An interesting problem which had to be overcome was a tendency for the free carrier density n_0 to be non-uniform, particularly close to the interface between layer and substrate (i.e. at the anode), where there was often a dip. This could be catastrophic because it produced a high resistance (and therefore high field) region at the anode and led to domain formation there, rather than near the cathode, as actually required. The famous dip could conveniently be monitored by Schottky barrier C–V profiling of the carrier concentration (see Section 5.4) as it turned out that the maximum depletion depth obtainable with a 'good' barrier was slightly greater than the active layer thickness and this held for a wide range of devices because of the n_0L condition on Gunn mode operation (equation (B5.4)).

Smooth, flat doping profiles were the aim during the preliminary establishment of the Gunn diode in the commercial environment but growing understanding of the various possible modes of operation led to experiments with graded profiles and, particularly with 'hot electron' injecting cathode contacts. As operating frequencies increased, it became clear that an extended region at the cathode end of the device was, in a sense, being wasted—it required a quite significant time for electrons to acquire sufficient energy to permit inter-valley transfer and they travelled a considerable portion of the active layer thickness in doing this. The answer lay in 'shooting' them in with appropriate energy from the cathode and this was achieved with the use of graded barrier contacts, made from AlGaAs with carefully selected band gap. Electrons then entered the active layer at a preselected energy above the bottom of the GaAs Γ minimum—they therefore hit the GaAs lattice running (!) and needed little further stimulus to transfer into the upper minima. Such was the ingenuity needed to obtain acceptable efficiency at high frequencies but, perhaps of greater interest was the question: what is the highest frequency that can be generated? The limit is set by the time taken for inter-valley scattering (transfer) and this depends on the strength of the interaction between free electrons and phonons. It is a complex topic and we shall make no attempt to discuss it—all that matters is the final outcome which is a cut-off frequency of about 100 GHz for GaAs. But even this was not allowed to represent an ultimate limit—by clever circuit design, it was found possible to extract surprising amounts of power at the second harmonic of the natural frequency making it possible to generate a few milliwatts at frequencies as high as 150 GHz. In fundamental mode typical modern GaAs devices produce up to 100 mW CW at 90 GHz, 500 mW at 35 GHz, and 1 W at 10 GHz, figures which reflect as much on the care devoted to heat-sinking as that given to electrical design. It nevertheless represents a splendid achievement.

Before closing this account we should make two final comments. First, these figures should be set against those appropriate to a quite different device, the IMPATT (impact ionization avalanche transit time) diode which typically produces power levels about 10 times greater, though because it depends on an avalanche process, with much higher noise level. Second, as we shall see later (Section 5.7), InP Gunn diodes have even greater high frequency potential than those made with GaAs. However, in practice, each device finds its own application niche and the GaAs Gunn diode still commands a significant fraction of the overall market.

We have dwelt at some length on the transferred electron device not only because it represented an important step forward for III-V semiconductors but also because it epitomizes the somewhat chaotic style of so much technological advance. Nevertheless, we must now move on to discuss an even more important development, that of the GaAs FET which has made a major impact in microwave integrated circuits and in the area of low noise high frequency amplifiers.

The story of the field effect transistor began, as we have already seen in earlier chapters, as early as 1930, though it was largely to the credit of William Shockley that it came to prominence. In 1952 Shockley published a paper in which he developed the so-called 'gradual channel' model of the FET which is still used to describe its operation today. What was more, the team at Bell Labs also appreciated the need to find a way round the inhibiting effect of surface states which so frustrated their early attempts to fabricate a practical device and this led to the realization of the junction FET, or JFET, in which the gate electrode took the form of a *p-n* junction, effectively burying the gate below the semiconductor surface, remote from the screening effect of surface states. Application of a reverse bias to the gate electrode caused the junction depletion layer to expand, progressively cutting off the channel, thereby controlling its conductance. Several JFET structures were reported in the 1950s, based first on Ge, then on Si but their performance lagged considerably behind that of the bipolar transistors which were already carving a niche in the growing electronics market and, in consequence, they were rapidly abandoned. It was only the happy discovery of the thermal oxide on silicon at the end of 1950s which revived interest in the FET, leading, as it did, to the now ubiquitous MOS transistor and all that this has entailed in the dramatic development of silicon integrated circuits. However, there were many technologists who recognized the potential of GaAs to provide improved high frequency performance as a consequence of its significantly greater electron mobility and the 1960s saw the vindication of their hopes. The first practical GaAs Metal Semi Conductor Field Effect Transistors (MESFETs) became a reality in 1969, with a high frequency cut-off of 10 GHz, some four times higher than comparable silicon devices, and by the mid-1970s GaAs was well established as the premier material for microwave FETs. In addition, their application to high speed logic circuits had also been demonstrated. Much hard work was to follow but there can be no denying the validity of their present position as the key devices for today's broadband microwave systems, mobile phones, satellite communications, and as the electronic components of fibre-optic communication systems. Operating frequencies have reached 100 GHz and the annual market for microwave FETs now runs at a level of well over $1 billion and is growing rapidly—but this is to anticipate!

The GaAs story really begins in the mid-1960s when the first FETs were reported with large gate lengths $L \sim 50\ \mu m$ and correspondingly low cut-off frequencies of order 100–200 MHz. (The speed of an FET depends on the time it takes free electrons to be injected into, or be swept out of the channel region in response to changes in gate voltage and a large gate length directly increases this transit time.) Not surprisingly, the first devices were based on existing silicon devices which meant JFET and MOSFET structures, though, in the latter case, employing a deposited film of SiO_2—native oxides on GaAs being notoriously unaccommodating! These were not, of course, microwave transistors but served to demonstrate principles. Indeed, in 1968 Jim Turner and Bryan Wilson at the Plessey lab in Caswell were already able to reveal one particularly important principle, namely that, in short gate length devices ($L < 10\ \mu m$), performance was limited not by the electron mobility but by the saturation drift velocity v_s. Reference to Figure 5.20 shows that, though the peak velocity in GaAs is greater than $2 \times 10^5\ m\ s^{-1}$, the saturation value is no more than $1 \times 10^5\ m\ s^{-1}$ and, in this respect, it differs rather little from silicon—in one fell swoop the major advantage of GaAs appeared to have been swept unceremoniously away! However, as we shall see, this is not the whole story, the little suspected phenomenon of 'velocity overshoot' (Ruch 1972—see van de Roer 1994: 78) coming nobly to the rescue in short channel transistors, effectively maintaining the status quo!

An important breakthrough came in 1967 when Hooper and Leherer at Fairchild reported the first GaAs MESFET in which the gate consisted of an evaporated metal film functioning as a Schottky barrier (see Box 5.6). Such a structure had been suggested only the previous year by Mead, thus pointing out the urgency with which the practical realization occurred. The principal reason for this development lay in the difficulties inherent in forming the gate junction in a JFET. It was necessary to diffuse zinc into the *n*-type channel to form the junction

Box 5.6. The GaAs MESFET

In this box we offer a brief outline of MESFET operation and design for those wishing to understand in greater detail the comments made in the text. However, it should not be treated as a substitute for the more complete accounts given in textbooks such as van de Roer (1994) or Shur (1990).

A schematic diagram of the MESFET structure is shown in Figure 5.22. For the most part we refer to the simple structure of Figure 5.22(a). As can be seen, the depletion region under the gate tends to cut off the channel and reduce its conductance. Application of a negative bias to the gate causes the depletion region to expand until the channel becomes completely pinched-off, at which point the source–drain current reaches saturation. The peculiar shape of the depletion region arises because the positive S–D voltage effectively adds to the gate–channel voltage which therefore increases towards the drain end. The need for the depletion layer to punch completely through the channel layer implies a relation between the layer thickness and its doping level N_D. This follows from the expression for depletion width:

$$W = \{2\varepsilon\varepsilon_0 V / eN_D\}^{1/2}. \tag{B5.29}$$

If we take the pinch-off voltage V_0 to be 5 V and $N_D = 10^{23}$ m^{-3}, we find $W \approx 0.25$ μm, which is readily achieved by epitaxy. However, it points up the need for precise control of the buffer–channel interface—the doping must rise from essentially 0 to $N_D = 10^{23}$ m^{-3} in a distance of order 0.01 μm (i.e. 10 nm) and this gradually led to the replacement of LPE and VPE by the better controlled methods of MBE or MOVPE.

The second important design criterion concerns operating frequency. The cut-off frequency f_{co} is determined by the transit time for free carriers to traverse the gate length L and here we must discriminate between long and short gate devices. If $L = 50$ μm, for example, we can write the drift velocity v_D in terms of the low field mobility μ and the electric field in the channel $E \sim V_0/L$. Thus:

$$\tau = L/v_D = L/\mu E = L^2/\mu V_0 \tag{B5.30}$$

from which:

$$f_{co} = 1/2\pi\tau = \mu V_0/2\pi L^2. \tag{B5.31}$$

Substituting $L = 50$ μm, $V_0 = 5$ V and $\mu = 0.5$ m^{2}V^{-1}s^{-1}, we find $f_{co} =$ 160 MHz which agrees well with early experimental data on such long gate devices. However, when gate lengths are reduced below 10 μm, this 'classical' theory fails—the drift velocity saturates because of the high electric field in the channel and we must use the alternative relation $\tau = L/v_s$, which leads to:

$$f_{co} = v_s/2\pi L. \tag{B5.32}$$

Substituting $v_s = 10^5$ m s^{-1} and $L = 1$ μm leads to $f_{co} = 15$ GHz, again in good agreement with measured values. This also suggests cut-off frequencies

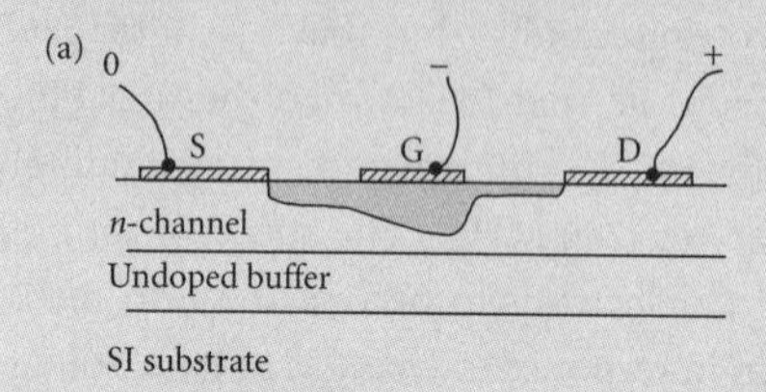

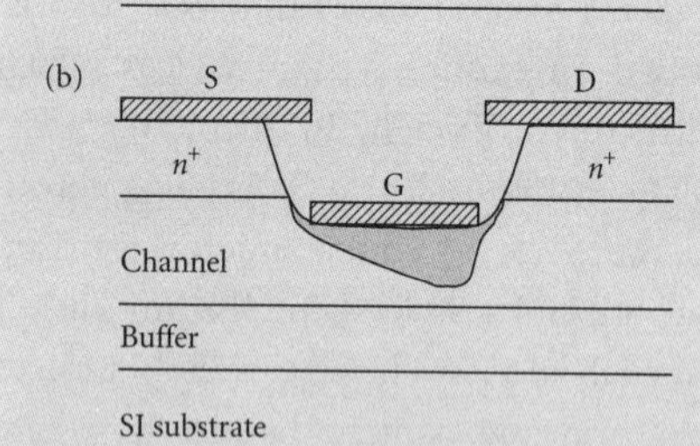

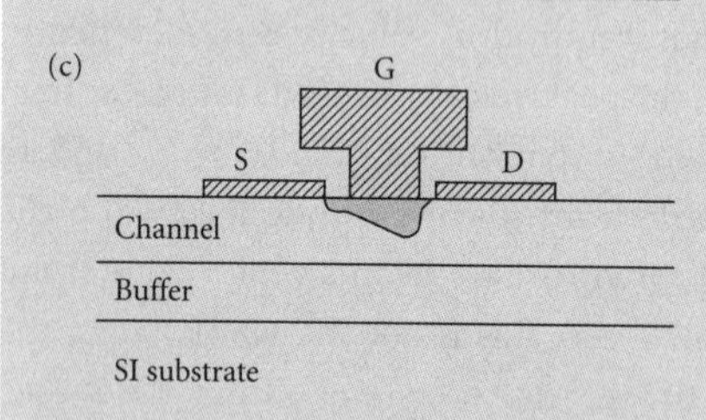

Figure 5.22. Three typical structures of GaAs MESFETs. In (a) is shown a basic structure in which the gate is significantly shorter than the source–drain spacing, while in (b) the use of a self-aligned gate minimizes the 'dead space' between the gate and the other two electrodes. The shaded area below the gate represents the depletion region whose depth is controlled by the applied gate voltage. (c) Illustrates the use of a T-shaped gate metallization which reduces the resistance in the gate circuit.

Box 5.6. Continued

of order 150 GHz for $L = 0.1$ μm devices but the reality is, if anything, even better as a result of 'velocity overshoot'. In a word, when gate lengths are below about 1 μm and the operating frequency is of the order 100 GHz, the average velocity during a half-cycle may be significantly greater than the saturation value, lying somewhere between the final saturation value of 1.1×10^5 m s^{-1} and the peak velocity of 2.2×10^5 m s^{-1} (see Figure 5.20). In other words, the electrons in the channel are in a non-equilibrium state, the time for them to reach their maximum velocity being of order half an oscillation period. In this respect GaAs has a further advantage over silicon—there being no transferred electron effect in silicon, the maximum possible velocity in silicon is the saturation velocity.

Two further design features of high frequency FETs need clarification. These involve parasitic circuit elements associated with the resistances in series with, on the one hand, the channel and, on the other, the gate. A glance at Figure 5.22(a) shows that the portion of the channel immediately beneath the gate is in series with two regions of partially depleted channel material which represent series resistances in the source–gate (i.e. input) and drain–gate (output) circuits. These parasitic resistances serve to reduce the overall performance in terms of gain, noise figure, and cut-off frequency so must be minimized as far as possible and Figure 5.22(b) illustrates one approach to meeting this requirement. It employs a recessed gate technology in which the gate is placed within an etched region of the channel, implying that the parasitic contacts are formed by thicker channel regions, thus reducing their resistances. In addition, they are formed from heavily doped n^+ material which reduces contact resistance and, finally, the gate is defined by a self-aligning process which minimizes the length of the parasitic channel elements. It makes use of the fact that etches nearly always undercut the etch mask used to define their area of application, so the source and drain metals are used as masks and the gate metal is evaporated through the gap between them. The problem of gate resistance is one resulting from the extremely small length of the gate in high frequency devices. The resistance of the gate metallization itself is significant when using the standard thickness of evaporated metal, an effect which is countered by building up the thickness and forming a Tee-shaped cross-section to further increase the cross-sectional area of the metallization, as illustrated in Figure 5.22(c).

As the principal application of the MESFET is that of a first stage amplifier in a microwave receiver, it is important that it provides useful gain while introducing the minimum amount of noise. Its capability in this respect is usually measured by its 'noise figure' F which is defined as:

$$F = (P_S/P_N)_{In}/(P_S/P_N)_{Out}, \tag{B5.33}$$

where P_S and P_N are the signal and noise powers, respectively and the subscripts 'Out' and 'In' refer to the FET output and input. F is usually measured as so many decibels (dB)—thus if we define the noise figure in dB as F', we can write:

$$F' = 10 \log_{10} F. \tag{B5.34}$$

A perfect amplifier has a noise figure of $F' = 0$ dB but, since any real amplifier must introduce some noise, F is inevitably greater than unity and, in a low noise FET, F' is typically of order 1 dB which implies $F = 1.26$. In other words, the S/N ratio at its output is roughly 25% worse than at its input.

and this was a high temperature process ($T \sim 800°$C) which caused unwanted changes in other parts of the structure, whereas Schottky barrier formation required temperatures no greater than 150°C, well below any 'damage' threshold. It was of particular relevance for GaAs on account of the latter's large band gap. In Si or Ge, barrier heights were much smaller (they tend to be limited to a fraction of about half the band gap) so that significant gate leakage current will flow under gate bias. In GaAs eV_B is typically about 0.8 eV for several metals which reduces gate leakage to insignificance. Together with the ready availability of SI GaAs substrates, which isolate the active part of the device from other parts of the circuit, the structure of the GaAs FET was complete. It was only necessary to reduce gate lengths to sub-micron dimensions and develop defect-free channel material to obtain perfect microwave transistors! If the material Gods had smiled benignly on silicon in granting it such a cooperative native oxide, they seemed inclined to redress the balance in allowing GaAs its isolating substrate and well behaved Schottky gates.

In 1970, the Plessey group reported MESFETs with 1 μm gate lengths courtesy of electron beam lithography and achieved gains of 8 dB at 6 GHz and cut-off frequencies as high as 20 GHz, while Doerbeck (Texas Instruments) reported $f_{co} = 30$ GHz using a self-aligned gate technology. Four years later Nozaki et al. from the Japanese telecom. firm NEC introduced another important innovation in the form of a 10 μm undoped buffer layer grown by VPE which removed the channel layer from contact with the substrate and considerably reduced parasitic effects due to the latter's indifferent electrical quality, a practice which rapidly became standard for all GaAs microwave transistors. They achieved values of f_{max} (maximum oscillation frequency) of 60 GHz from 1 μm gate devices. The latter half of the 1970s revealed a worldwide interest in all aspects of MESFET development. Gate lengths crept below the 1 μm barrier, f_{max} crept up to nearly 100 GHz, noise figures reached values of 1.4 dB at 10 GHz—thus revealing the long-hoped-for low noise amplifier for microwave receiver front ends—and power transistors made their bow on the world stage with outputs as large as several watts at frequencies approaching 10 GHz. The microwave world was clearly the GaAs oyster! But it required one further development to prize open the shell—that of the integrated circuit. The first practical advance came in 1974 when Charles Liechti at Hewlett-Packard reported on a digital IC using GaAs MESFETs but perhaps of even greater significance was the announcement in 1976 by Ray Pengelly and Jim Turner of Plessey of a monolithic X-band MESFET amplifier, the first monolithic microwave integrated circuit (MMIC). Once again, the GaAs SI substrate proved its value, this time as the insulating layer in microstrip circuitry which made the whole thing viable. Yet another innovation then made its appearance, the use of ion implanted channel regions, resulting in much improved uniformity over the wafer. Though of marginal interest for microwave circuits, this was clearly of importance for integrated logic circuits where it is vital to keep individual devices within a tight specification. From this point, therefore, the two technologies have tended to diverge, though in both cases with conspicuous success.

One final 1970s initiative merits a mention. In 1975 Barrera and Archer reported on an InP MESFET. More of this anon, but already it was clear that InP offered prospects of even higher speeds than GaAs. This was later confirmed when its saturation drift velocity was measured as being some 30% greater.

And now, to complete this rather lengthy section, we should at least note one other important application of GaAs in microwave systems. This is to the development of detector and mixer diodes. Chapter 2 describes the early work on microwave diodes, then in the form of point contact devices, which played an important part in Second World War radar systems. These were based on silicon or germanium as the only (reasonably!) well controlled semiconductor materials then available and, though they clearly represented a major success for the new-fangled semiconductors, they were still of doubtful reliability and reproducibility, and, relying, as they did on cat's whisker technology, they were rather noisy. Two features of GaAs were appealing to those concerned to improve the performance of microwave systems—the high electron mobility and the facility with which good Schottky barriers could be made. The former resulted in low values for series resistance and the second to much greater reproducibility and stability in the diode characteristics. Noise figures were also very much improved. The key technology consisted in the low temperature (typically 200°C) formation of Schottky barriers with barrier heights of order 0.7–0.9 eV by evaporation of suitable metals such as gold, platinum, or nickel onto a chemically cleaned GaAs surface, followed by photolithographic definition of diodes with diameters of order 10 μm. These were then contacted with fine gold wires and mounted in standard microwave packages such as used for the old point contact devices, this approach being well established by the middle of the 1960s. The advent of microwave stripline technology in the 1970s was quite compatible with the use of hybrid mixer diodes, combined with Gunn diode local oscillators as was the later development of MMICs. It was obviously convenient to make such diodes from the same material as used for the even more important MESFET.

5.7 Indium-phosphide

The development of semiconductor technology and its application to an ever increasing range of electronic and opto-electronic system requirements is inevitably littered with false starts, unjustifiably wild hopes, difficult choices, and conflict between the established moderate performer and the obviously talented but immature newcomer. That GaAs finally established its credentials as an *electronic* material (it had less difficulty in opto-electronics) was, in many ways, something of a triumph for the combination of faith and good fortune, allied to sheer dogged tenacity. That it possessed obvious 'talent' in the form of its high electron mobility was undeniable but its advantages came with the concomitant disadvantages of a more difficult technology and a significantly higher price. While silicon is one of the most widely available elements on earth, neither Ga nor As can lay similar claim—a 6 in. diameter silicon wafer costs roughly £5, a similar GaAs substrate some ten times more. Successful GaAs devices rely almost entirely on epitaxy—many silicon applications can be satisfied without. Add to this the inevitable 'sailing ship' effect (silicon just got better) and one can appreciate the nature of the struggle for ascendancy. It is hardly surprising, therefore, that similar considerations applied to InP when it came on the microwave scene. Yes, it could demonstrate some, perhaps, marginal advantages over GaAs but at what price?—the price, basically, of developing a whole new technology. Only, here, the battle is still raging. It remains far from clear just how great an inroad the new boy InP can make into the realm of the now well-established incumbent GaAs, though one must suppose that InP can hope for no more than a modest fraction of the GaAs market, just as GaAs takes no more than a few percent of that commanded by silicon. Had InP been able to make its claim first, of course, things might have been very different—such is the fascination of the market place! However, this is to put the commercial cart before the scientific horse—what are the relevant technical considerations?

InP has a superficial similarity to its rival GaAs in so far as it possesses a direct band gap of similar magnitude ($E_g = 1.35$ eV) and its band structure is also broadly similar. The electron effective mass (a measure of the conduction band curvature near its minimum) is fairly close to that of GaAs—$m_e/m = 0.077$ (cf 0.065 for GaAs) which implies a comparable, though somewhat smaller, electron mobility—0.5 $m^2 V^{-1} s^{-1}$ in pure material, compared with 0.9 $m^2 V^{-1} s^{-1}$ for GaAs. It is also important that InP can be obtained in semi-insulating form—by doping it with iron impurities, in this case—though the slightly smaller band gap implies a larger intrinsic carrier concentration which results in maximum resistivity about one order of magnitude less than SI GaAs. However, on the face of it, InP appears to offer little of advantage—indeed, the fact of its being a phosphide actually represents a negative factor. The vapour pressure of phosphorus being significantly greater than that of arsenic, means that InP is much harder to grow as a bulk crystal, even though its melting point is slightly lower than that of GaAs. Why, then, should anyone be prepared to invest in yet another new material whose vital statistics appear less exciting than those already provided by a well-established rival? In attempting to provide an answer, we shall see that the development of InP followed a rather tortuous path and, incidentally, demonstrated just how 'non-linear' scientific and technological progress can often be.

The story centres on the application to the transferred electron effect. It turned out, in practice, that, as far as transistors were concerned, the straightforward InP MESFET was never a serious contender—it was only in the 1980s and 1990s, with the advent of LDS that InP FETs began to challenge their GaAs forebears and, as these form the subject of Chapter 6, we shall say no more about them here. InP was, of course, one of the materials used by Gunn in his pioneering studies of the TEE in 1963 (even though he was unaware at the time of this explanation of his observations!) but we should be clear that his samples were fairly primitive—the development of practical Gunn diodes still awaited the advent of LEC bulk crystals and the application of epitaxy, which only became widely available towards the end of the 1960s and it was hardly surprising that the initial effort went into GaAs. Any serious development of InP clearly required some special stimulus and this came in 1970 with the publication of a theoretical paper by Cyril Hilsum and David Rees of RRE, Malvern. In this they made a bold prediction to the effect that InP Gunn oscillators functioned in a fundamentally different manner from the established GaAs devices on account of a different alignment of the three conduction band minima. In GaAs, it was generally believed that the order of the minima was (in ascending order) Γ–X–L and that the L minima played no part in the transferred electron effect—hot electrons transferred from the central Γ minimum directly into the X minima, where their drift velocities were strongly reduced, giving rise to the required negative differential resistance. The coupling between Γ and X was thought to be strong—expressed in terms of a 'deformation potential' D of about 1×10^{11} eV m^{-1}—and, using this model, it was possible to calculate a velocity–field characteristic for GaAs in good agreement with the experimental curve shown in Figure 5.20. Hilsum and Rees not only claimed that the ordering of the minima in InP was different, the L minima lying below the X, but that the coupling between Γ and L was much weaker ($D = 1 \times 10^{10}$ eV m^{-1}) than those between Γ and X and between L and X. In their model of InP, hot electrons transferred strongly from Γ to X, then from X to L where they remained, being unable to transfer back to X because of the energy difference ($(E_X - E_L) \gg kT$) and unable to transfer into Γ on account of the weak coupling. (Note that the transition rate between two minima is proportional to the square of the appropriate deformation potential). This 'three-level' mechanism, they suggested, would give rise to more efficient transfer into the low mobility minima, resulting in a larger negative resistance and a larger 'peak-to-valley ratio' (maximum-to-minimum average drift velocity). Figure 5.23 hopefully makes all this clear and includes an example of the predicted velocity–field characteristic taken from Hilsum and Rees' paper.

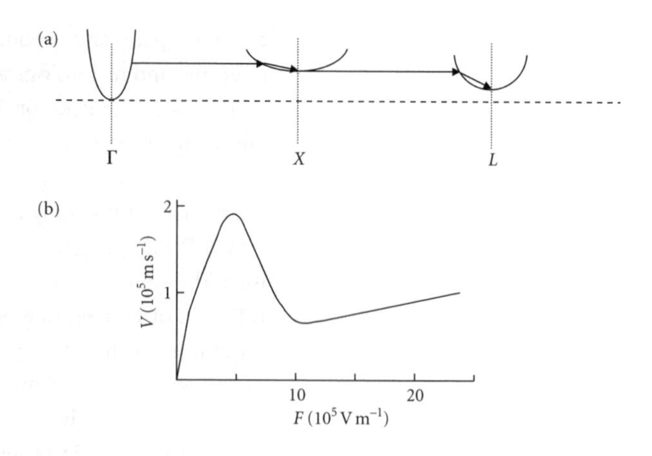

Figure 5.23. In (a) is shown the transferred electron mechanism suggested by Hilsum and Rees for the case of InP Gunn diodes. The coupling between Γ and X minima and between X and L minima is supposed to be strong, while that between Γ and L is weak. This results in a very efficient transfer of electrons into the low mobility L minima. (b) Shows an example of the velocity–field characteristic calculated by them using this model. From this they predicted that InP Gunn diodes would show significantly higher efficiencies than their GaAs counterparts.

Assuming their model to be correct, one could anticipate Gunn oscillators with higher efficiency which tended to operate as bulk negative resistance devices, without the travelling dipole domains described in Section 5.6 and this certainly served to stimulate research within the United Kingdom, if not elsewhere. The development of the LEC process for the growth of bulk InP crystals was reported in 1970 by Brian Mullin and collaborators at RRE and both LPE and VPE techniques were being explored at approximately the same time. Within a matter of 12 months material of suitable quality to test these predictions was fairly widely available. Results were somewhat equivocal—though there was some evidence of improved efficiency, there still appeared to be a correlation between device length and operating frequency which suggested some form of travelling domain but what was more puzzling, perhaps, were the results of velocity–field measurements. These were reported from a number of different laboratories, on material obtained from various sources and, while demonstrating acceptable consistency, showed no hint of agreement with the Hilsum–Rees curve—a typical example is shown in Figure 5.24, from which we can adduce the following discrepancies: the threshold field for the onset of negative resistance is much larger than predicted ($F_{\mathrm{thr}} = 1.2 \times 10^6\ \mathrm{V\,m^{-1}}$, compared with the $5 \times 10^5\ \mathrm{V\,m^{-1}}$ predicted), there is no suggestion of the minimum velocity predicted above threshold (in fact, the InP curve tends to gradual saturation in much the same manner as does the GaAs curve) and the measured peak velocity appears to be close to $3 \times 10^5\ \mathrm{m\,s^{-1}}$, against the prediction of slightly less than $2 \times 10^5\ \mathrm{m\,s^{-1}}$. Not all this was bad, of course. While throwing doubt on the Hilsum–Rees model, these results did show a marked increase in peak velocity and threshold field, compared with the GaAs values, which certainly encouraged hope of higher power generation. Also, though not apparent from Figure 5.24, there was a clear suggestion that relaxation effects were less serious in InP, thereby implying a potential for higher ultimate operating frequencies.

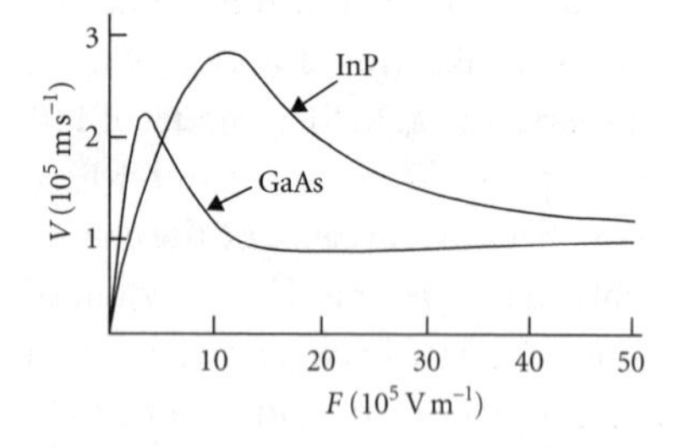

Figure 5.24. Comparison of the measured velocity–field characteristics of GaAs and InP. The peak velocity in InP is seen to be significantly greater, though the threshold field for electron transfer is also considerably larger. It is significant that the minimum in the InP (v–F) curve predicted by Hilsum and Rees is not found experimentally.

Here the matter rested for some years but enough had been done to encourage the belief that InP might yet become the preferred material for Gunn devices, at least at millimetre wavelengths, thus leading to fairly widespread development of both material and circuit applications. Indeed, there the matter may have continued to rest, had it not been for an unexpected development on the GaAs front. It was 1976 which revealed the ultimate irony in the form of a series of papers, which proved beyond reasonable doubt, that the accepted ordering of the conduction band minima in GaAs was incorrect—the L minima were found to lie *below* the X minima by approximately 0.2 eV. GaAs, itself, was therefore transformed overnight into a three-level system such as had been envisaged for InP and the two materials were seen to be qualitatively identical. Small differences in the energy separations of the minima and in intervalley relaxation times, together with the slightly different low-field mobilities could account for the observed differences in device characteristics—everything else was purely imaginary! Perhaps the key to reconciling the apparent discrepancies came with the recognition that the Γ–L deformation potential was, in neither material, as small as originally anticipated, the currently accepted values being close to 1×10^{11} eV m^{-1}, rather than the much smaller value proposed by Hilsum and Rees for InP. This implies, of course, that the concept of the three-level oscillator simply does not apply—neither material represents a true three-level system! But nor can it alter the fact that InP is, today, the preferred material for Gunn oscillators operating in the 100–300 GHz spectral region.

Was it all a rather expensive red herring? Not in the least! This kind of occurrence is surprisingly common in any scientific or technological sphere and, though the Hilsum–Rees theory may now lie in tatters, when compared with reality, it was nontheless an excellent idea which stimulated further work on the development of InP which might, at very least, have had to wait for some considerable time. When InP was really needed, as we shall see that it certainly was, the material was available in acceptable quality—it might well not have been so.

We have already noted the importance of InP in the area of LDS but, perhaps the largest market for InP nowadays is in the field of optoelectronics where it is desired to integrate electronic and optical devices on the same substrate but this story must wait until Chapter 9. Our present chapter ends here.

Bibliography

Agrawal, G. P. and Dutta, N. K. (1993) *Semiconductor Lasers*, Van Nostrand Reinhold, New York, USA.

Astles, M. G. (1990) *Liquid-Phase Epitaxial Growth of III-V Compound Semiconductor Materials and their Device Applications*, Adam Hilger, Bristol.

Blood, P. and Orton, J. W. (1992) *The Electrical Characterisation of Semiconductors: Majority Carriers and Electron States*, Academic Press, London.

Casey, H. C. and Pannish, M. B. (1978) *Heterostructure Lasers* (Part A Fundamental Principles, Part B Materials and Operating Characteristics), Academic Press, New York, USA.

Dupuis, R. D. (1987) *IEEE J. Quantum Elect. QE-23*, 651.

Gallium Arsenide and Related Compounds', Institute of Physics Conference Series, IOP Publishing Ltd, Bristol, England.

Gooch, C. H. (1973) *Injection Electroluminescent Devices*, John Wiley & Sons, London.

Gunn, J. B. (1976) *IEEE Trans. Electron Devices ED-23*, 705.

Hilsum, C. (1995) 'The Use and Abuse of III-V Compounds', in *Advances in Imaging and Electron Physics*, Vol. 91, p. 171, (ed. P. W. Hawkes) Academic Press, New York.

Hilsum, C. and Rose-Innes, A. C. (1961) *Semiconducting III-V Compounds*, Pergamon Press, New York.

Hollan, L., Hallais, J. P,. and Brice, J. C. (1980) 'The preparation of gallium arsenide', in *Current Topics in Materials Science*, 5, 1, North Holland, Amsterdam. (ed. E. Kaldis)

Hurle, D. T. J. (1994) *Handbook of Crystal growth, Vol. 2: Bulk Crystal growth, Vol. 3: Thin Films and Epitaxy*, North Holland, Amsterdam.

Look, D. C. (1989) *Electrical Characterisation of GaAs Materials and Devices*, Wiley, Chichester.

Madelung, O. (1964) *Physics of III-V Compounds*, John Wiley, New York.

Orton, J. W. (1971) *Material for the Gunn Effect*, Mills and Boon Ltd, London.

Orton, J. W. and Blood, P. (1990) *The Electrical Characterisation of Semiconductors: Measurement of Minority Carrier Properties*, Academic Press, London.

Perkowitz, S. (1993) *Optical Characterisation of Semiconductors*, Academic Press, London.

van de Roer, T. G. (1994) *Microwave Electronic Devices*, Chapman and Hall, London.

Runyan, W. R. and Shaffner, T. J. (1998) *Semiconductor Measurements and Instrumentation*, 2nd edn, McGraw-Hill, New York.

Shroder, D. K. (1998) *Semiconductor Material and Device Characterisation*, 2nd edn, John Wiley, New York.

Shur, M. (1990) *Physics of Semiconductor Devices*, Prentice-Hall, NJ.

Stradling, R. A. and Klipstein, P. C. (eds.) (1990) *Growth and Characterisation of Semiconductors*, Adam Hilger, Bristol, UK.

Townes, C. H. (1999) *How the Laser Happened*, Oxford University Press, Oxford, UK.

Welker, H. J. (1976) *IEEE Trans. Electron Devices ED-23*, 664.

Willardson, R. K. and Goering, H. L. (eds.) (1962) *Compound Semiconductors, Vol. 1, Preparation of III-V Compounds*, Reinhold Publishing Corporation, New York.

CHAPTER 6

Low dimensional structures

6.1 Small really is beautiful

As any semiconductor scientist can tell you, low-dimensional structures (LDS) is a drug. It came to the notice of the scientific community in the early 1970s and, since then, the number of addicts has grown in a quite alarming fashion. The great developments of the 1960s were concerned with epitaxy, which led to the first well-controlled heterostructures and, in 1970, the ultimate success of the DH laser. The following decade 1970–80 saw the extension of these skills to encompass the growth of a very special type of heterostructure, characterized by its hitherto undreamed-of *smallness*. The double heterostructure (DH) laser was based on a structure with an active GaAs layer as thin as 0.1 μm (100 nm)—during the next decade this was to be dwarfed (on an inverse scale!) by structures with dimensions in the range 1–10 nm. Indeed, it was not long before dimensions were being quoted in 'monolayers', rather than nanometres, a monolayer of GaAs being a single molecular layer, one plane of Ga atoms plus one of As atoms, having an overall thickness (in the (0 0 1) direction) of 2.83 Å (0.283 nm). But what was the point? In a word, the point was that such miniscule amounts of material demonstrate physical properties, which are distinctly different from those of bulk material and these new features can be put to a surprisingly wide range of uses, both scientific and technological. It is a moot point whether the introduction of these so-called 'Low Dimensional Structures' has had greater impact in the sphere of new semiconductor physics or of new semiconductor devices but it is not a question which will long detain us—we simply welcome both aspects, while recognizing once again the vitality of the interaction between them.

In attempting to understand these new structures, the two-questions confront us: why is size relevant? and what order of size is important? There are, in fact, two complementary ways in which we can approach these questions, both depending on the quantum mechanical description of electron behaviour. The first of these concerns the property of electron 'tunnelling' through a potential barrier and the second depends on the relationship between the size of the space allowed to an electron and its energy. Tunnelling came first so we will deal with it first. In classical mechanics, applied to macroscopic particles, the presence of a potential

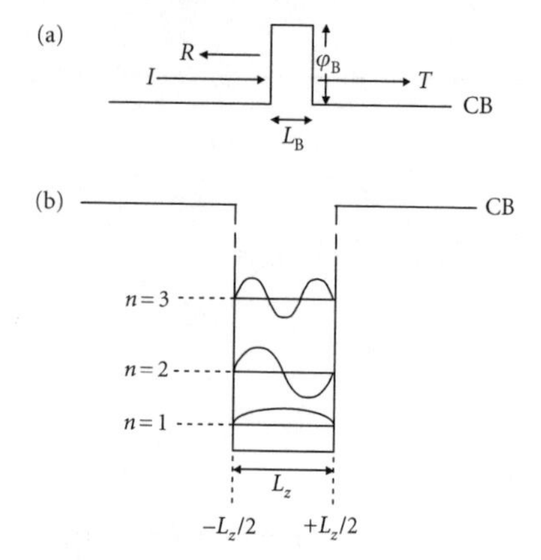

Figure 6.1. Energy diagrams to illustrate (a) electron tunnelling through a square conduction band barrier of height $\varphi_B \sim 1$ eV and thickness $L_B \sim 1$ nm, (b) electron confinement in an infinitely deep square quantum well of width L_z. In (a) we assume that an electron wave impinges on the barrier from the left and that there are probabilities for reflection R and transmission T, which depend on barrier height, and thickness. In (b) we illustrate the amplitudes of the wavefunctions, describing the probability that the electron is to be found at any point z within the well, for the first three confined energy states E_n (E_n is measured upwards from the bottom of the well).

barrier (Figure 6.1(a)) implies that the particle can only pass through that region provided it has sufficient energy to go over the barrier—there is no concept of it leaking through! However, in quantum mechanics, there is a finite probability of this leakage actually occurring. This, we express in terms of the probability that the particle, making impact with one side of a barrier, may be found on the far side of the barrier even though its energy is significantly less than the barrier height. This probability depends on a number of factors, such as the mass of the particle, the barrier height and thickness. In semiconductor structures we are concerned with electrons of effective mass m_e and with barrier heights, which are typically of order 1 eV or less so we can use the standard formulas of quantum tunnelling theory to estimate how thin a barrier has to be in order for there to be a significant probability for an electron to tunnel through. The answer turns out to be about 1 nm and, because the tunnelling probability varies exponentially with thickness, we can effectively ignore the effect for thicknesses much greater than this. Already we can appreciate the need for extremely thin semiconductor films.

The complementary approach concerns electron confinement by two potential barriers such as shown in Figure 6.1(b). If, for simplicity, we suppose the barriers to be infinite in height, the electron is confined entirely to the region of the semiconductor between the barriers (there is zero probability for it to be in the barrier regions) and we can confine it to a smaller and smaller space by reducing the thickness of the intermediate layer. What, then, happens? To see what we should expect, it is helpful to consider the case of an electron bound in an atom, for simplicity, the hydrogen atom. There is a clear correlation between the electron energy and the size of its orbit, that is, the space within which it is confined—the closer the electron is to the nucleus, the larger is its binding energy but not all energy values are possible, of course; only certain allowed energies are allowed by quantum mechanics. The ground state corresponds to the case where the electron wavefunction has just one wavelength round the orbit, the first excited state has two wavelengths, the second, three wavelengths, and so on (see Box 6.1). In the case of an electron confined between barriers, a similar correlation holds, the electron energy increases as the confined space becomes smaller, there being (in this case) a series of energy states corresponding to an integral number of *half*-wavelengths between the barriers. As we show in Box 6.1, the energies of these states are given by:

$$E_n = n^2h^2/8m_eL_z^2, \tag{6.1}$$

where L_z is the thickness of the confining layer (note that the subscript z implies that the well thickness is taken as the z coordinate of a system of cartesian coordinates—this somewhat unusual terminology has been sanctified by use so we adopt it in line with general practice), m_e is

Box 6.1. Particle in a box

The confinement of an electron in a potential well forms a parallel with a classic elementary problem in wave mechanics which finds a niche in most quantum mechanics textbooks—that of a particle in a finite box. The quantum well problem is actually somewhat simpler on account of its one-dimensional nature. It consists of solving the Schroedinger equation for a free electron which is constrained within coordinates $z = -L_z/2$ to $z = +L_z/2$ (see Figure 6.1)—in fact, the physics is really contained in the boundary conditions. If we assume that the well is infinitely deep, there is zero probability of finding the electron inside the barrier so, in order to match the wavefunctions in the well and in the barrier, it is necessary to specify that $\psi(z) = 0$ at $z = \pm L_z/2$. To give the problem maximum simplicity, we define the zero of energy to lie at the bottom of the well, so that, within the well, $V(z) = 0$ everywhere. The Schroedinger equation then takes the form:

$$-(h^2/8\pi^2 m_e)\, d^2\psi_n/dz^2 = E_n\psi_n, \tag{B6.1}$$

where we treat the electron as essentially free but take account of the fact that it resides in a semiconductor crystal by giving it a mass m_e, rather than the free electron value m. It is easy to see that this has solutions of the form:

$$\psi = A \cos kz \text{ or } A \sin kz, \tag{B6.2}$$

where $|\psi(z)|^2$ represents the probability of finding the electron at the point z.

Consider the cos kz solution. In order to satisfy the boundary conditions we require that $\cos(kL_z/2) = 0$, or that $kL_z/2 = \pm\pi/2, \pm 3\pi/2, \pm 5\pi/2$, etc. In other words, $k = n\pi/L_z$ where $n = 1, 3, 5$, etc. These are the solutions with even symmetry:

$$\psi_n = A \cos(n\pi z/L_z) \quad \text{with } n = 1, 3, 5, \text{ etc.} \tag{B6.3}$$

Similarly, the odd solutions are:

$$\psi_n = A \sin(n\pi z/L_z) \quad \text{with } n = 2, 4, 6, \text{ etc.} \tag{B6.4}$$

The constant A is not completely arbitrary because the probability for the electron to be somewhere within the well must be unity so we can write:

$$\int \psi^2 dz = 1 \tag{B6.5}$$

from which it follows that $A = \{2/L_z\}^{1/2}$, though this is not a result we shall make much use of.

Much more importantly, we can obtain the allowed energies by substituting these wave functions back into equation (B6.1) and find:

$$E_n = h^2k^2/8\pi^2 m_e = n^2h^2/8m_eL_z^2 \tag{B6.6}$$

Note that these energies are measured upwards from the bottom of the well—they are referred to as 'confinement energies'—and form a sequence with increasing spacing as the energy increases (because of the proportionality to n^2). Because the well is infinitely deep, there is an infinite number of confined states.

Box 6.1. Continued

This account shows some resemblance to the solution of the simple harmonic oscillator, which corresponds to the case of a parabolic well, rather than the square well assumed here. In that case, of course (see any text on quantum mechanics), the energies are proportional to n (rather than to n^2) and are therefore equally spaced but again there is an infinite number of allowed states.

It may be helpful to think of these solutions as representing an electron bouncing backwards and forwards between the impermeable barriers at $z = \pm L_z/2$. By virtue of the motion, it possesses momentum p which varies with z and reverses at $z = \pm L_z/2$. The average value of p must be zero because the electron spends as much time travelling towards positive z as towards negative z. However, the average of p^2 is clearly non-zero, in fact (see Box 3.1):

$$<p_n^2>_{av} = 2m_e E_n = n^2h^2/4L_z^2. \tag{B6.7}$$

This result is interesting in that it enables us to find a value for the de Broglie wavelength of the electron λ_n (Box 3.1):

$$\lambda_n = h/p_n = 2L_z/n. \tag{B6.8}$$

In other words, $L_z = n\lambda_n/2$, showing that the de Broglie wave of the electron corresponds precisely to the wavefunction found from the Schroedinger equation.

A similar concept of wavefunction matching emerges from the old quantum theory of the hydrogen atom. The quantum condition is introduced into this model by way of the Planck condition on the angular momentum p_φ:

$$\int p_\varphi d\varphi = nh \tag{B6.9}$$

and, as angular momentum is a constant of the motion, $\int p_\varphi d\varphi = p_\varphi \int d\varphi = 2\pi p_\varphi$. Therefore:

$$p_\varphi = nh/2\pi. \tag{B6.10}$$

Using the relation $p_\varphi = ap$, where a is the radius of the orbit, allows us to write the expression for the de Broglie wavelength as:

$$\lambda_n = h/p = ha/p_\varphi = 2\pi a/n. \tag{B6.11}$$

Thus, the Planck condition implies that the circumference of the orbit must be an integer number of wavelengths, $2\pi a = n\lambda_n$.

the electron effective mass and $n = 1, 2, 3, \ldots,$.etc. (Note that E_n is measured upwards from the bottom of the bulk semiconductor conduction band.) For a GaAs layer 10 nm thick, where $m_e/m - 0.067$, we find $E_1 = 56$ meV, $E_2 = 224$ meV, etc., energies which are significantly greater than kT, even at room temperature, so nearly all the electrons within the 'well' will be in the ground-state level E_1 (under thermal equilibrium conditions). In practice, we must remember that any real quantum well will be characterized by finite well depth so this elementary theory is no longer exact and, in general, confinement energies are smaller than that estimated above. Nevertheless, we may still draw the conclusion that these confinement energies become important when the thickness of the well material is reduced below about 10 nm—simply by controlling the layer thickness we can control the electron energy, thus giving us a very flexible method of adjusting material properties. This, then, explains the second demand for ultrathin layers.

The history of LDS can be traced to a paper by Esaki and Tsu, published in the *IBM Journal of Research and Development* in 1970, which was concerned with the properties of what have come to be called semiconductor 'superlattices', a series of two-dimensional layers in which the band gap changes up and down in a regular sequence. Electron transport through such a structure involves tunnelling through barrier layers, into adjacent wells, which implies they be of the same order of thickness as discussed above, though the theoretical description of their properties must also take account of their regular periodicity. In many ways such structures look like a semiconductor with ordinary crystalline properties in the plane of the layers but with a different 'crystal structure' normal to the plane, and Esaki and Tsu predicted some rather unusual properties for them. In particular, it was anticipated that they might show a negative resistance as a result of their unusual 'band structure'. There was, however, a major problem in so far as it was impossible to grow such a structure at that time—epitaxy had come a long way by 1970 but the challenge of producing well-controlled smooth layers some few hundred monolayers in thickness had to await the development of molecular beam epitaxy (MBE) and metal-organic vapour phase epitaxy (MOVPE). Leo Esaki was well aware of this and was already setting up an MBE facility at IBM when Cho and Arthur grew their first successful MBE films at Bell Labs, and he was later able to demonstrate the basic correctness of his original prediction using a superlattice grown by MBE, in 1974. Indeed, it was the pioneering work on MBE at Bell Labs and IBM, which was crucial to the dominance of these two laboratories in the early development of LDS. The year 1974 saw another major turning point when Ray Dingle and co-workers, at Bell, reported measurements of the optical properties of quantum well structures, which showed the presence of the series of confined energy states predicted by the theory outlined in Box 6.1 and this work rapidly led to the development of the quantum well semiconductor laser diode (first reported in 1975) which has become a standard over a wide range of applications. An important example of the Bell Labs MBE capability was revealed in 1976 when Art Gossard and colleagues reported the growth of monolayer superlattices based on AlAs/GaAs multilayers, while Gossard and Dingle were both involved in the first reports (1978–9) of two-dimensional conduction in a so-called two-dimensional electron gas (2-DEG) which led the way to an exciting new form of field effect transistor (FET) structure, the high electron mobility transistor (HEMT) first demonstrated in 1980 in Japan by Mimura and colleagues at the Fujitsu laboratories in Kawasaki. From 1975 onwards other groups joined what was obviously a highly stimulating and potentially lucrative field of activity— first came Joyce and Foxon at Philips (Redhill), then Klaus Ploog (Stuttgart), and Takahashi and Gonda in Japan, followed by many more worldwide MBE-based activities. Later MOVPE was also to demonstrate a capability for growing these ultrathin structures, and the rivalry between the two methods became a potent force in semiconductor physics and technology throughout the 1980s and 1990s. LDS expanded and flourished as few branches of solid-state physics have ever done and, even today, there is little sign of diminution.

In terms of dimensionality, the quantum well structure we have just described can be seen to be two-dimensional because electrons are free to move in either of two directions in the plane of the well and are constrained only in the third dimension. Thus, their behaviour in the plane of the well is similar to that which would be appropriate to a bulk sample of the well material—the energy assigned to the confinement effect actually represents the bottom of the two-dimensional conduction band in the well material. The confined energy states should not, therefore, be construed as representing sharply defined single levels. But why stop at two-dimensionality? It was very quickly recognized that yet more interest (and commercial gain) might accrue from reducing the electron's freedom even further. 'Quantum wires' (one degree of freedom) and 'quantum dots' or 'quantum boxes' (zero degrees of freedom) promised yet more scientific riches which physicists were all too eager to possess (and, in some cases, sell!). The 1980s and 1990s were to see their fair share of excitement, too. To drop a few names for future reference, we might mention 'the fractional quantum Hall effect', 'ballistic transport', 'mesoscopic devices', 'electron interference', 'strained layer superlattices', 'single electron effects', 'coulomb blockade', 'resonant tunnelling', 'strained layer QW lasers', 'vertical cavity lasers', 'heterojunction bipolar transistors (HBTs)', and, of course, HEMTs—a veritable feast for the funding agencies to sink their dollars into (not to mention their Yen, Pounds, and Euros!). But it should not be supposed that any of these treasures were to be acquired without a struggle—MBE may have been tailor-made to grow monolayer-thick films but the lithography required to produce confinement in the second and third dimension was another matter—in this case one was faced with a technology which dealt in microns, rather than nanometres and a good deal of ingenuity was to go into obtaining genuine one-dimensional structures, never mind zero-dittos. We shall examine some, at least, of these developments in the following sections. In the meantime, it may be well to draw attention to the numerous books which cover the field with considerable eclat. It is impossible to list them all but the following can be highly recommended: Willardson and Beer (1987), Bastard (1988), Jaros (1990), Weisbuch and Vinter (1991), Kelly (1995), and Davies (1998).

6.2 The two-dimensional electron gas

The 2-DEG started life in the silicon metal oxide silicon field effect transistor (MOSFET), at the interface between silicon and silicon oxide (see Section 4.6) but really came to prominence in 1978 at Bell Labs when Ray Dingle and co-workers developed the concept of 'modulation doping' in a $GaAs/Al_{0.3}Ga_{0.7}As$ multi-layer structure. This ingenious idea was proposed as a method of increasing the electron mobility above its value in doped *n*-type GaAs where, particularly at low temperatures, the mobility tends to be limited by ionized impurity scattering. For every mobile electron in the conduction band there must exist an ionized donor atom (from which that electron has been removed) and it is the electric field associated with these positively charged ions which dominates scattering of the conduction electrons in moderately-to-heavily doped material. For a very long time it had been assumed that there was no way of avoiding such scattering processes—the doping required to generate significant densities of conduction electrons inevitably led to reduced electron mobility and corresponding reduction in electrical conductivity. But, in a flash of inspiration, the Bell Scientists demonstrated a way round this apparent impasse—they placed the Si dopant atoms in the AlGaAs layers, while the resulting electrons resided in the GaAs layers, which formed quantum wells between the wider band gap alloy layers (see Figure 6.2(a)). Conduction

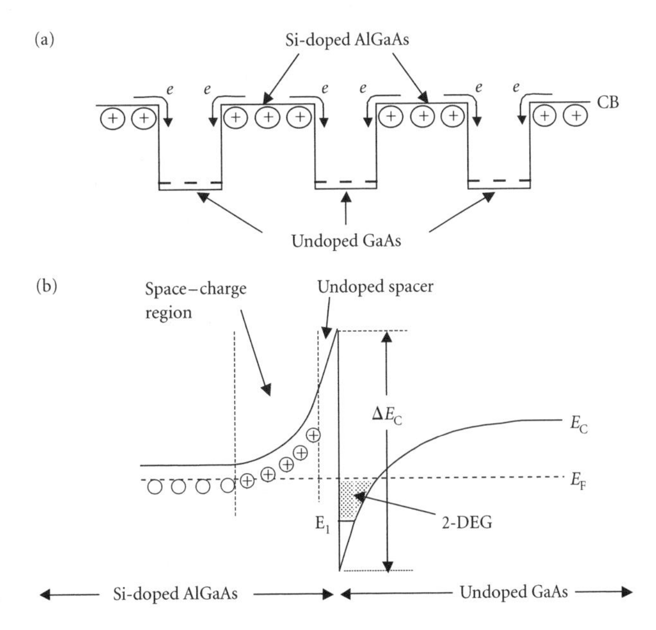

Figure 6.2. Schematic energy band diagrams for (a) a modulation-doped AlGaAs/GaAs multi-layer structure and (b) a single heterostructure, in greater detail. Doping is confined to the AlGaAs barrier regions, while the resulting conduction electrons reside in the undoped GaAs, due to the potential difference between the two regions. ΔE_c is the 'conduction band offset' between the two layers, resulting from the difference in their band gaps. In a typical case, ΔE_c would be about 0.3 eV.

normal to the plane of the layers was inhibited, of course, by the AlGaAs barriers but conduction in the plane (i.e. two-dimensional conduction) remained unencumbered. Electrons generated within the barriers, transferred readily into the wells on account of the potential difference between the layers (as will be clear from a glance at Figure 6.2(a)), thus neatly separating them from the ionized donors whence they originated. At a stroke, ionized impurity scattering was reduced to the role of impotent bystander! Well, nearly so! Because the electrostatic interaction between electrons and donors is essentially a long range effect, their limited spatial separation could do no more than effect a reduction—the oppositely charged species could still interact, even though to a much reduced extent.

Dingle et al. were, of course, well aware of this and introduced a further subtlety into their structures in the form of undoped spacer regions within the AlGaAs barriers, the dopant atoms being placed (as only MBE could place them!) in the central region of each barrier layer. In fact the barriers had an undoped region some 6 nm-wide adjacent to each interface, thus further separating the charged species by a carefully controlled amount. Of course, the proof of their multilayer pudding had to be in the eating—they measured a low temperature electron mobility $\mu_e = 2\ \mathrm{m^2 V^{-1} s^{-1}}$ which was to be compared with a typical value of $\mu_e = 0.2\ \mathrm{m^2 V^{-1} s^{-1}}$ for a comparably doped GaAs layer. Convincing proof if one were needed! The advantage was far smaller at room temperature, where lattice scattering tends to dominate, but, even there, some significant improvement was apparent. Once again, Bell was leading the field (as we shall see) into unimagined new realms.

Once the principle of modulation doping had been established, it soon became common practice to apply it to a single heterojunction, rather than in a multilayer. Once again, it was the Dingle group which led the way (having been joined by a visitor from Grenoble, Horst Stormer who was later to share a Nobel prize for his work on such structures)—they reported the first single heterojunction 2-DEG in 1979. Figure 6.2(b) shows a band diagram for such a structure, where Si donor atoms are located in the AlGaAs barrier layer, though again spaced some convenient distance from the interface with the undoped GaAs layer. This band diagram perhaps requires some explanation. As already pointed out, conduction electrons near the AlGaAs/GaAs interface tend to drop into the GaAs conduction band, leaving positively charged donors in the AlGaAs which form a space–charge region near the interface. This, in turn, results in band bending in the AlGaAs which forms a barrier to further electron transfer (another way to look at this is to say that the separation of electrons and donors sets up an electric field opposing further transfer), while the electron cloud on the other side of the interface causes band bending in the GaAs in the opposite sense. This forms an approximately triangular quantum well, which confines the electron cloud close to the interface (typically within about 10 nm). Equilibrium is reached when the electron gas fills all the available states within the well, up to the electron Fermi level, which, in turn, must be continuous across the interface and, finally, must lie close to the donor levels in the bulk AlGaAs on the left of the diagram. Note that the GaAs quantum well has a confinement energy E_1 associated with it in much the same way as the square well discussed above, so the electrons occupy levels only above this minimum value. All this can be calculated in a self-consistent manner but the calculations are somewhat complicated and we shall not attempt any more detailed account here. In practical terms, the doping level in the AlGaAs was close to 10^{24} m^{-3}, resulting in a sheet carrier density in the 2-DEG of about 1.1×10^{16} m^{-2}.

Once the appropriate technology had been demonstrated by the Bell group, it was not long before other MBE growers realized their own 2-DEG structures with similar electrical properties—Hadis Morkoc at the University of Illinois led the way but by 1984 there were activities at Rockwell, IBM, Fujitsu, Philips, University of Tokyo, the German Post Office, and the Max Planck Institute in Stuttgart (and by 1987 in France and Greece) and the race was then on to obtain both greater low temperature mobilities and greater understanding of the scattering mechanisms involved. There was, needless to say, considerable interest, too, in the capabilities of the new high mobility transistor, based on the 2-DEG but we shall leave that discussion until Section 6.5, concentrating here on the basic physics.

It seems something of a coincidence, perhaps, that at this very time (1980), when the AlGaAs/GaAs structure was beginning to stretch its two-dimensional wings, that von Klitzing et al. should discover the quantum Hall effect in the original two-dimensional conductor—that associated with the silicon MOSFET. But, given such a stimulus, it was scarcely surprising that someone should seek a similar phenomenon in the new two-dimensional system and even less surprising that those 'somebodies' should work at Bell Labs. Dan Tsui and Art Gossard met the challenge and published their results only a year later, pointing out that the GaAs-based system had the advantage of a much smaller effective mass (0.067 m, compared with 0.19 m in silicon), which made it much easier to observe the Hall plateaux. Whereas the MOSFET studies required temperatures below 2K and magnetic fields greater than 15 T, the corresponding figures for GaAs were 4.2K and 4.2 T. Properly to appreciate this, we need to understand the origin of the effect, itself, which demands an element of concentrated thought. Here goes!

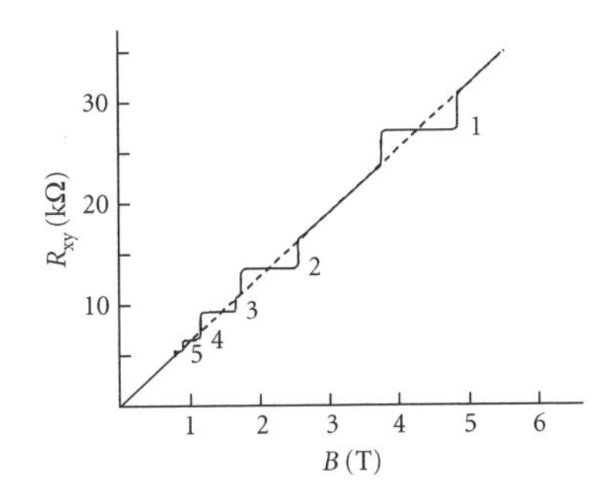

Figure 6.3. Typical experimental data illustrating the formation of quantum Hall plateaux at filling factors $i = 1, 2, 3, 4, 5$ for an AlGaAs/GaAs 2-DEG at 4.2K. The sheet carrier concentration is $n_s = 1 \times 10^{15}$ m^{-3}. R_{xy} is the Hall resistance ($R_{xy} = V_H/I_x$), measured in kilohms. B is the magnetic induction in Tesla.

Let us first be clear about the experimental observations. An example of quantum Hall plateaux measured on an AlGaAs/GaAs 2-DEG, with $n_s = 1 \times 10^{15}$ m^{-2}, is shown in Figure 6.3 where it can be seen that at low magnetic fields the Hall resistance R_{xy} $(=V_H/I_x)$ increases linearly with magnetic field B as described for the classical Hall effect outlined in Chapter 2 (Box 2.2) whereas, at higher fields, well resolved plateaux appear at regular intervals. In fact, these are characterized by values of $R_{xy} = h/ie^2$, where h is Planck's constant, e is the electron charge and i is an integer ($i = 1, 2, 3$, etc.). (Note that the fundamental constant $h/e^2 = 25812.807\ \Omega$ and is, of course, independent of the sample used for the measurement.) The plateaux in Figure 6.3 are identified by the appropriate values of i. Notice that, as i increases, the plateaux become less well resolved and merge into the classical straight line, though it is found experimentally that more plateaux can be resolved when the sample is cooled to lower temperatures. The one parameter which *does* appear to be sample-dependent is the *width* of the plateaux.

A full explanation of the effect would take us well beyond the ambition of this book but we can certainly gain some understanding in a fairly straightforward manner. The essential feature concerns the form of the electron states in the structure in question. In the direction normal to the 2-DEG plane carriers are constrained by the electrostatic potential well, with which we are now familiar. In the plane of the well they are free to move within the conduction band of the GaAs. However, when a high magnetic field is applied at right angles to the plane, electrons are also constrained in the plane because of the cyclotron orbits we referred to in Section 3.3. As explained there, the magnetic field splits the conduction band states into a series of Landau levels whose energies are given by

$$E_n = (n + 1/2)(h/2\pi)\omega_c, \tag{6.2}$$

the corresponding radii of the orbits being

$$a_n = \{(2n + 1)h/2\pi eB\}^{1/2}, \tag{6.3}$$

where the cyclotron frequency ω_c is given by

$$\omega_c = eB/m_e \tag{6.4}$$

and $n = 0, 1, 2, 3$, etc. In the two-dimensional case with which we are concerned here, because electrons are confined in all dimensions, these Landau levels are sharp, precisely defined levels, just like energy states in an atom. We show a plot of E_n against magnetic field in Figure 6.4 for the case of the GaAs effective mass $m_e = 0.067$ m.

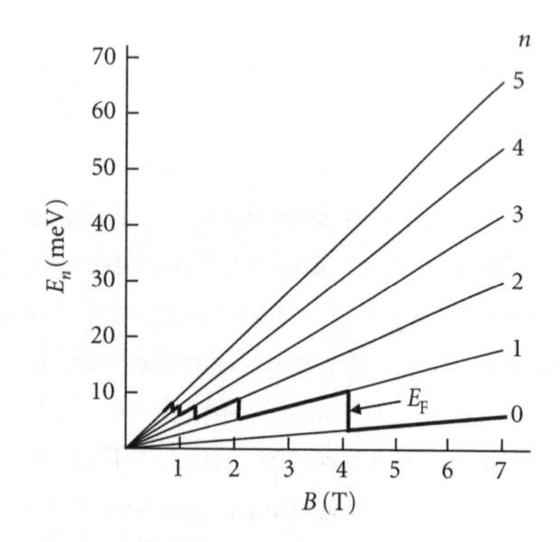

Figure 6.4. Plot of Landau level energy E_n as a function of magnetic induction B for the case of an AlGaAs/GaAs 2-DEG. The heavy line shows the position of the electron Fermi level, which follows each level in turn as the field increases until that level becomes empty, when it drops to the next lower level, finally settling in the lowest level $n = 0$. Note that, on the scale of the diagram, the thermal energy kT at $T = 4.2$K is about the thickness of one of the lines, representing the energy levels.

The second important property of the Landau levels is the fact that their degeneracy N_s (i.e. the number of allowed electron states per square metre, associated with each one) increases linearly with increasing magnetic field. Thus

$$N_s = eB/h, \tag{6.5}$$

which means that we can write $N_s = 2.42 \times 10^{14}\ \text{m}^{-2}$ for $B = 1$ T. This implies that the n_s conduction electrons in the 2-DEG will be distributed over the various Landau levels in a manner which changes as the magnetic field changes. In particular, at low temperatures and at values of B greater than about 4.13 T, all the $n_s = 1 \times 10^{15}\ \text{m}^{-2}$ electrons will be accommodated in the lowest level ($n = 0$). In this case, the electron Fermi level which defines the level occupation (see Box 5.3) lies close to this Landau level. If we now imagine B to be reduced, N_s will decrease accordingly and it is no longer possible for all the electrons to be accommodated in the $n = 0$ level. Some of them must therefore transfer to the $n = 1$ level and the Fermi level will then lie close to this second level. As B decreases further, some fraction of n_s will transfer to level three ($n = 2$), then level four, and so on. As can be seen from Figure 6.4, the Fermi level traces a zigzag path between the levels, jumping abruptly upwards at specific values of magnetic induction B, and these values of B are given by the condition that $n_s = (n + 1) \times N_s$, from which we find:

$$B_i = hn_s/ie, \tag{6.6}$$

where the so-called 'filling factor' $i = (n + 1)$, from which $i = 1, 2, 3$, etc. When $i = 1$, all the electrons are in the lowest level, when $i = 2$, they are shared between the lowest two levels, and so on.

We can then obtain an expression for the corresponding values of R_{xy} by noting that the straight line in Figure 6.3 is given (for a two-dimensional system) by:

$$R_{xy} = V_H/I_x = B/en_s \tag{6.7}$$

Thus, the value of R_i which corresponds to B_i, is:

$$R_i = B_i/en_s = h/ie^2. \tag{6.8}$$

This is an exciting result because it shows that the occurrence of the Hall plateaux corresponds to those values of R_{xy} at which the Fermi level jumps between Landau levels. At the same time, it is a somewhat frustrating result because it does *not* explain why the plateaux occur! According to this argument, the plateaux should have zero width—because the Fermi level jumps are strictly vertical in Figure 6.4—and should therefore be unobservable! We clearly need to introduce some important modification to the model if we are to understand why the plateaux are finite in width.

The dilemma results from our assumption that the Landau levels are infinitely sharp in energy which is a result of over-idealizing the experimental situation. Real samples are imperfect in various ways and, in practice, it appears that each Landau level is 'inhomogeneously broadened'.

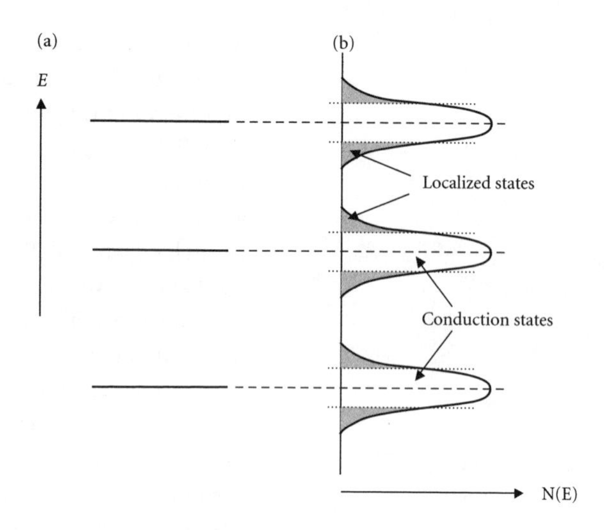

Figure 6.5. Illustration of the distribution of Landau level energies which results from random imperfections in the plane of the AlGaAs/GaAs interface. In (a) are shown the sharp Landau levels which formed the basis of our elementary model, in (b) The modified model including inhomogeneous broadening. Not only are the conducting states broadened, but in the wings of the distributions there exist localized states which make no contribution to conduction. Note that the diagram refers to a fixed value of magnetic field.

Suppose, for example, that the interface between the AlGaAs and the GaAs layers is not perfectly flat but that the two layers interpenetrate to a small random extent. Effectively, then, the band gap of the GaAs and the height of the interface barrier will show random variations over the area of the interface and the energy of each Landau level will show variation in consequence. It is no longer valid to represent each level as being sharp in energy and we must allow for them to be broadened, as shown in Figure 6.5. What is more, it seems likely that there will exist locations in the interface plane where these random imperfections can result in carrier localization—the potential wells resulting from this 'disorder' being deep enough (greater than kT) to trap electrons and prevent them from taking part in the conduction process (and, of course, in the generation of a Hall voltage). This is indicated in Figure 6.5 by the shaded energies which correspond to these localized, non-conducting states in the wings of the Landau energy distributions. The good thing about this modified model is that it is not only more realistic but it has within it the seeds of a complete explanation for the experimentally observed plateaux. In particular, as the Fermi level moves between Landau levels, the transition is no longer abrupt but occurs over a finite range of magnetic fields. In other words, it takes a finite change of magnetic field to cause the Fermi level to move from one set of conducting states to the next and it is this range of fields which represents the Hall plateau.

There is more to the argument than this but we must leave it there and return to the comment we made above about the greater ease of observing the quantum Hall effect in AlGaAs/GaAs structures, because we are now in a position to understand it! Two criteria exist for the observation of the effect—first, that the electrons complete several cyclotron orbits before they are scattered (this is the condition for being able to define Landau levels) and, second, that the energy separation of the levels is very much greater than kT (this prevents thermal transfer of electrons between levels, which would compete with the quantum effects). The first condition can be expressed as $\omega_c \tau \gg 1$, where τ is the scattering time, and, as ω_c increases with increasing magnetic field, there is a minimum value of B below which we could not hope to see any quantum phenomena. Writing the scattering time in terms of the electron mobility μ_e, this condition can be written as:

$$\mu_e B \gg 1, \tag{6.9}$$

showing that the larger mobility in GaAs (consistent with its much smaller effective mass), compared with silicon gives it an initial advantage. The second criterion is influenced by the rate at which the Landau levels diverge as a function of magnetic field. The energy separation between adjacent levels is simply $(h/2\pi)\omega_c = (h/2\pi)eB/m_e$ so, once again, the much smaller effective mass in GaAs implies that the second criterion can be satisfied at higher temperatures than is the case in silicon (or, for the same temperature, at much lower magnetic fields). In any event, it is the AlGaAs/GaAs structure which has been employed in the great majority of quantum Hall studies. Two final comments can be added: first, it is important to note that the magnetic field required in order to reach the $i = 1$ plateau (i.e. in order to have all the electrons in the lowest level) depends on the sheet carrier density n_s so this must be carefully controlled, though it is relatively easy to obtain values in the vicinity of $n_s = 1 \times 10^{15}$ m^{-2}; second, it is important that the material quality is not *too* good—as we saw above, the existence of well defined plateaux requires a certain element of disorder, which tends to have a retrograde effect on the mobility. Life is full of compromises!

At magnetic fields large enough that all the 2-DEG electrons are contained in the lowest Landau level, we should expect no additional features in the R_{xy} vs B curve because the Fermi level can no longer move between levels. It was therefore a major surprise when (in 1982) the Bell group, in the form of Dan Tsui, Horst Stormer, and Art Gossard reported the existence of several new plateaux at fields significantly higher than those required for the $i = 1$ plateau. These new features showed very similar behaviour to the, by now, familiar 'integer' effects but were characterized by filling factors which were fractional, that is $i = 1/q$ and $i = (1 - 1/q)$, where q is an odd integer). Obviously such features could not be explained on the basis of transitions between Landau levels but surely needed some form of energy gap within the lowest Landau level whose nature was totally unknown. Thus was born the 'fractional' quantum Hall effect (FQHE) which opened a whole new field of physics, and still yields unexpected new results even today. Gradually more and more new fractional filling factors were discovered, some with even, rather than odd, denominators. It also became apparent that the mysterious quasi-particles, which were associated with these strange features, carried an electric charge which was one-third of that on an electron, in complete conflict with scientists' long-held belief that the charge 'e' was the basic unit of electric charge. Condensed matter theoretical physics had been thrown such a juicy bone that it was to provide many years of luscious pickings and, let it be said, many successes. The understanding now acquired is based on the theory of many-body interactions—the quasi-particles effectively consist of groups of strongly interacting electrons which do, indeed, behave as though they possessed fractional electron charge. The associated energy gaps have actually been measured. The fractional filling factors can be predicted. And the work still goes on. We shall attempt no deeper penetration into the dense thickets of theoretical afforestation—we can do no more than emphasize the importance of this amazing new realm of physics which was beyond anyone's wildest dreams before 1982. Tsui, Stormer, and Laughlin (whose many-body wavefunction first allowed daylight into the depths of the theoretical forest) were awarded the Nobel Prize for Physics in 2000—poor Art Gossard, whose crystal growing skills made it all possible, was not! Crystal growers the world over should be up in arms in defence of their profession.

It rapidly became clear that the 2-DEG samples required to observe the FQHE had to be of higher quality than those which showed well-defined integer plateaux and this led to strenuous efforts to improve low-temperature electron mobilities—the 'million mobility' chase was on! (Mobilities in those days were always measured in units of $cm^2\ V^{-1}\ s^{-1}$, so 'million' refers to $10^6\ cm^2\ V^{-1}s^{-1}$.) For example, the first observation of the integer effect in AlGaAs/GaAs employed a sample with $\mu_e = 2\ m^2\ V^{-1}\ s^{-1}$, whereas the fractional effect was discovered in a sample with $\mu_e = 10\ m^2\ V^{-1}\ s^{-1}$ and there was an obvious correlation between sample mobility and the discovery of greater and greater detail in the fractional Hall 'spectrum'. Not that it was an easy challenge to the MBE growers—as can be seen in Figure 6.6, the chase begun in 1979 did not come to a complete stop even in 1987 when the initial target had been satisfactorily met. Further fine-tuning of sample purity went on almost to the end of the millennium, when low temperature ($T \sim 1$ K) mobilities had passed the '10-million' mark. Another striking feature of Figure 6.6 is the truly worldwide nature of the endeavour—the number of MBE facilities involved ran well into double figures, involving not less than six different countries; USA, Japan, England, Germany, France, and Israel. Of particular interest is the dominance of industrial laboratories over universities in what was very largely an academic exercise. Although it was apparent that there may be some long-term possibilities for ultrasmall electronic devices using very few (even single) electrons, the principal driving force towards high mobilities was certainly the exciting new physics to be gleaned from studying the fractional quantum hall effect (FQHE). Two factors probably came into play—first, the industrial laboratories were

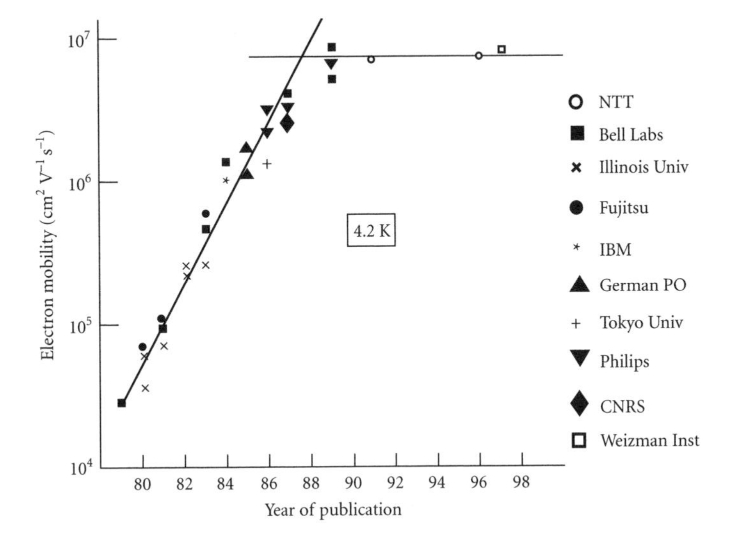

Figure 6.6. The best 2-DEG electron mobilities measured at 4.2K as a function of year of publication, showing the gradual improvement during the 1980s and effective saturation thereafter. In the best material mobility at 4.2K is limited to about 5–7 $\times$ 10^2 $m^2 V^{-1} s^{-1}$ by acoustic phonon scattering. At temperatures below 1K, mobilities tend to become independent of temperature with values as high as 1.5 $\times$ 10^3 $m^2 V^{-1} s^{-1}$.

more likely to be equipped with the expensive MBE machines, which seemed essential to the production of such esoteric epitaxial samples, and second, the 1980s probably marked the hey-day of 'blue skies' research in industry. The euphoria over the virtue of pure research in industrial laboratories which had been triggered by the fairy-tale success of transistor development was still riding high, though the 1990s were to see an element of backlash in the face of tougher commercial competitiveness and it now seems highly unlikely that a similar academic thrust would be led from an industrial base.

The two-and-a-half orders of magnitude improvement in electron mobility presented in Figure 6.6 represents a dramatic enhancement over what was far from a primitive starting point. How was it obtained? To appreciate the nature of the achievement we must examine the various scattering mechanisms involved in limiting electron mobility and learn something of their characteristics. These may be listed as follows: acoustic phonon scattering, remote ionized impurity scattering, local background impurity scattering, interface roughness scattering, and alloy disorder scattering. Let us begin by making a few general comments on each, noting first that in two dimensions, they show different characteristics (e.g. temperature dependence) than for the more common three-dimensional case. Acoustic phonon scattering is beyond the control of the crystal grower and must be accepted as inevitable. However, it shows a temperature dependence of the form $\mu \propto T^{-1}$ so can be minimized by reducing the temperature—in the best samples this may mean below 1K. As we saw earlier, the ionized donors in the AlGaAs barrier layer can still interact with electrons in the 2-DEG but samples may be designed with large undoped spacers which push these positively charged donors further away from the interface and minimize this scattering process. There is, however, a cost. Thicker spacer regions lead to a reduction in the density of free electrons at the interface which leads to a corresponding decrease in mobility. This is the result of lowering the kinetic energy of electrons at the Fermi level (and therefore their thermal velocity), while ionized impurity scattering increases as the electron velocity is reduced. The net effect is that, as the spacer layer thickness is increased, the electron mobility goes through a maximum for a value of thickness in the region of 50–100 nm so all the really high mobility samples have been designed with spacers of this order. Ionized impurities within the GaAs come into close contact with two-dimensional electrons so are particularly undesirable when high mobilities are sought after. The answer is, of course, to remove them by growing under conditions of utmost purity, ultralow residual gas pressures in the MBE chamber, high purity sources of Ga, Al, As, and optimized growth temperatures. The key, as realized eventually by all growers, is to keep the MBE machine under vacuum continuously for a large number of growth runs, there being a steady clean-up of the system with the number of layers grown—the very best samples appear after upwards of 50 layers. Interface roughness has been demonstrated to influence mobility in some cases but can be improved by optimizing growth conditions. One effective method consists of growing an AlGaAs/GaAs superlattice before growing the undoped GaAs layer. This has been shown to lead to monolayer smoothness in subsequent interfaces and has the further benefit of gettering background impurities by trapping them at the multiple interfaces of the superlattice. Alloy scattering due to the random nature of the AlGaAs alloy seems inevitable but there is no clear evidence that it plays a significant role in limiting mobility.

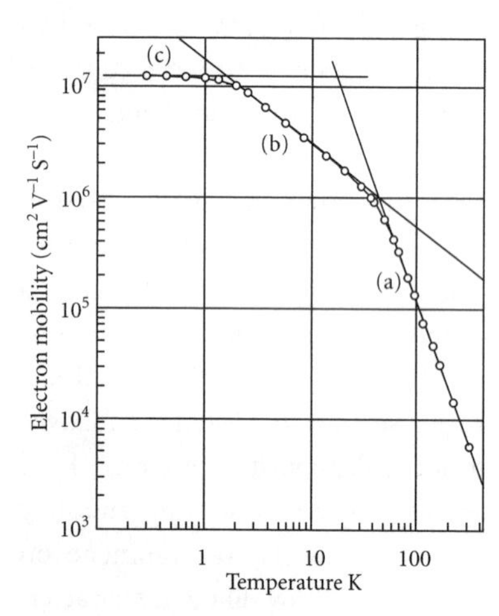

Figure 6.7. Electron mobility vs temperature for an AlGaAs/GaAs 2-DEG sample grown at Bell Telephone Laboratories in 1989. This represents one of the best 2-DEG samples ever grown, with an estimated background impurity level in the GaAs below 2×10^{19} m^{-3}. Note the three distinct regions of temperature, corresponding to dominant (a) optic phonon, (b) acoustic phonon, and (c) ionized impurity scattering. (From L. Pfeiffer, K. W. West, H. L. Stormer, and K. W. Baldwin (1989) *Appl. Phys. Lett. 55*, (1888), figure 1.) Reprinted with permission from the American Physical Society.

In Figure 6.7 we show some data from a 1989 paper by Loren Pfeifer and colleagues from Bell Labs which represents one of the best ever 2-DEG samples. It is straightforward to distinguish three temperature regimes—above about 50K, mobility is limited by optical phonon scattering, between 2 and 50K, by acoustic phonon scattering (note the approximate T^{-1} dependence of mobility), while below 2K, mobility becomes independent of temperature at a value slightly greater than 10^3 $m^2 V^{-1} s^{-1}$ (10^7 $cm^2 V^{-1} s^{-1}$). The authors attributed this limiting value to ionized impurity scattering and suggest their background impurity level must, therefore, be less than 2×10^{19} m^{-3}. Somewhat similar data have also been reported by the groups in England (Philips), Japan (NTT), and Israel (Weizmann Institute). One final comment is necessary—the ideal sample for FQHE studies may not *always* be the one with maximum mobility. Remember that the aim of getting all the electrons into the lowest Landau level is more easily satisfied in a sample with low electron density and this may be of overriding importance, even though it means a corresponding reduction in mobility. The ultimate challenge to the MBE grower was to produce samples with high mobility, combined with low electron density.

Initially, as we have already made clear, the motivation for improving 2-DEG mobility was the search for more detail in the FQHE spectrum and this was very successfully achieved, but it soon became apparent that it was not the end of the story—ballistic transport and mesoscopic systems were other important outcomes (see Section 6.3) which depended on the surprisingly long 'mean free path' for electrons in 2-DEG structures at low temperatures. The mean free path l is related to the mean scattering time τ (which we have already met on several occasions) in the obvious way:

$$l = v_{\mathrm{T}}\tau, \tag{6.10}$$

where v_{T} is the thermal velocity of an electron. Bearing in mind that τ is also related to electron mobility as $\mu_{\mathrm{e}} = e\tau/m_e$, we can write the mean free path as:

$$l = \mu_{\mathrm{e}} m_{\mathrm{e}} v_{\mathrm{T}}/e = (\mu_{\mathrm{e}}/e)\{2m_{\mathrm{e}}E_{\mathrm{K}}\}^{1/2}, \tag{6. 11}$$

where E_{K} is the kinetic energy of the electrons. In Section 3.3 we used this approach to calculate a mean free path for electrons in a normal three-dimensional sample, where $E_{\mathrm{K}} = kT$ but, in the case of the 2-DEG where the electrons are in a degenerate electron gas, E_{K} must be identified with the Fermi energy E_{F} which takes a typical value of $E_{\mathrm{F}} = 30$ meV (see Figure 6.2(b)). Using $\mu_{\mathrm{e}} = 10^3\ \mathrm{m}^2\ \mathrm{V}^{-1}\mathrm{s}^{-1}$, we find $l = 150$ μm. Contrast this with the value of 0.1 μm estimated in Section 3.3. Even for slightly less esoteric samples, the high mobilities lead to mean free paths of order 10 μm. The true significance of these long mean free paths is that (a) physical dimensions much less than this can be defined using lithographic techniques and (b) electron trajectories can be resolved experimentally to an accuracy much less than this, but detailed discussion forms the subject of our next section, Section 6.3.

6.3 Mesoscopic systems

The word 'mesoscopic', from the Greek word meaning 'middle-sized' was coined to cover those physical phenomena which relate to structures having sizes between the microscopic (i.e. atomic) and macroscopic (i.e. familiar, everyday). This hardly provides a precise definition but (conveniently, perhaps) permits the inclusion of a rather wide range of effects. In this section we look at just four of these; ballistic transport, quantum interference effects, resonant tunnelling, and the so-called Coulomb blockade which pertains to single electron effects. These are all related to electron transport (we leave optical effects until Section 6.4.)

Having mastered the art of making high mobility 2-DEG structures, the next cast of the LDS dice was aimed at controlling electron motion in the plane of the structure. How might it be possible to confine electrons so as to allow them to flow in only one direction? How might it be possible to make a 'quantum wire'? One way of looking at this was to think in terms of electrons being confined within an electron waveguide, the electron version of a microwave or optical guide. To answer this question we need to know something of the electron wavelength—the de Broglie wavelength which we covered briefly in Section 4.3, and again in Section 6.2 (Box 6.1). We can write $\lambda_n = h/p$, where p is the electron's momentum, $p = \{2m_e E_K\}^{1/2}$, so:

$$\lambda_n = h/\{2m_e E_K\}^{1/2}. \tag{6. 12}$$

E_K is the kinetic energy of the electron which, in the case of the 2-DEG, is to be identified with the Fermi energy E_F and, as we pointed out above, this typically has a value of about 30 meV, leading to a wavelength $\lambda_n = 28$ nm. Could one fashion a waveguide with lateral dimension as small as this? Present-day lithography allows feature sizes of less than 100 nm so, in principle, one could expect to get fairly close—but how? The obvious method was to etch away the material above a 2-DEG in order to leave a narrow stripe, as indicated in Figure 6.8(a) and this was tried in various laboratories (e.g. Cambridge University, Glasgow University, IBM, Bell, and others) during the period 1986–9 but with very limited success. The problem encountered was associated with the generation of damage by the ion beams used as etchants. This surface damage and its associated high surface-state density produced strong band bending which tended to deplete the wires of free carriers—any wire of width less than 100 nm was essentially completely empty!

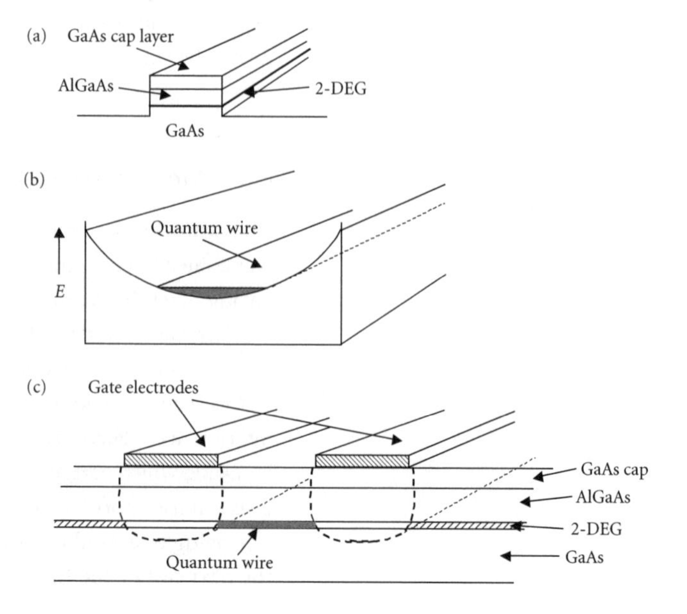

Figure 6.8. Illustration of methods for obtaining a quantum wire using a 2-DEG structure. In (a) the wire is formed by etching away the 2-DEG structure to leave a narrow mesa, (b) shows an energy diagram which represents the effect of band bending from the etched sides of the mesa as a result of the surface damage produced by the etch. The electron gas is confined laterally by the potential well formed by the band bending, and (c) illustrates an alternative method which uses two Schottky barrier gate electrodes separated by a small amount. Negative bias on these two electrodes depletes electrons from below them, leaving a narrow quantum wire between the edges of the depletion regions.

This, however, was not the end of the matter—uniformity (or, rather, the lack of it) made the situation much worse. Had the width and surface-state density been perfectly uniform, it would have been possible to use wider stripes which contained an undepleted central region (see Figure 6.8(b)), the boundaries of the wire being quadratic potential barriers, rather than physical edges. This, incidentally, isolates carriers from the surface states and surface roughness but the inevitable non-uniformities resulted in variable depletion widths and concomitant undulations in potential (of order 0.1 eV) along the length of the wire—not at all desirable for a conductor designed for the propagation of ballistic electrons at low temperatures (where kT was of order 0.1 meV).

A better approach, which avoided the unwanted effects of surface damage, was found when controlled carrier depletion resulted from the deposition of split Schottky barrier gate electrodes, as suggested in Figure 6.8(c). The application of negative gate potentials expanded the depletion regions under each electrode and facilitated control of the wire width in the 2-DEG between them. Once again, non-uniformity in separation between the gates led to some variation in wire width but to a much smaller extent than in the case of an etched stripe—sufficiently smaller to allow a wide range of experimental studies of one-dimensional conductivity. However, before going on to describe them, we should first note an essential feature of electron propagation along such a waveguide—reflection from the sidewalls. The random nature of thermal velocities implies that electrons in the guide will be travelling in all directions and must, in general, make frequent contact with the walls. It is important to recognize that these collisions are 'elastic', that is that they conserve the energy of the electron. The momentum changes, of course, because the direction of motion changes, but not the magnitude of the velocity and, therefore, not the energy. The significance of this observation is that the phase of the electron wave is conserved—or, put more graphically, the electron *remembers* its phase through the collision. This is the condition for ballistic transport in a waveguide. The mean free path which is relevant here is the distance between collisions which destroy the phase memory—collisions in which the electron energy changes.

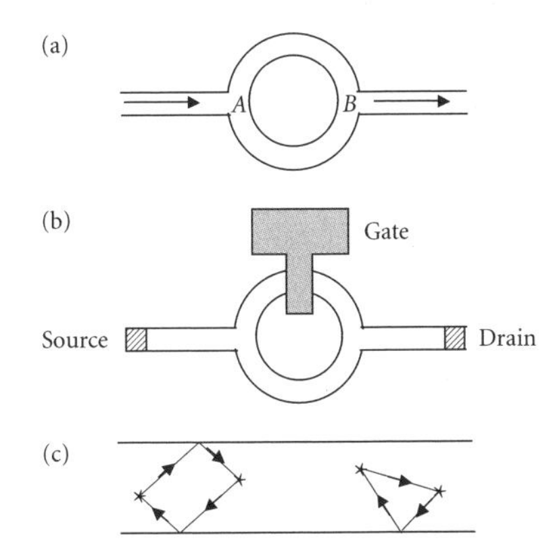

Figure 6.9. Idealized 'Aharonov–Bohm' ring structure for demonstrating electron self-interference. An electron current is supposed to be flowing from left to right. Interference at *B* is constructive when the two arms of the ring are equal in length, in which case the current is enhanced. In (b) a reverse-biased Schottky gate is used to change the electron wavelength in one half of the ring, changing the interference condition and thereby controlling the drain current, and (c) illustrates the formation of microscopic Aharonov–Bohm rings within a single wire, which result from elastic scattering at impurity sites or potential barriers within the wire.

Having established the ground rules, we may now look at an interesting experiment in which the wave nature of the electron makes itself apparent—the so-called Aharonov–Bohm effect. In Figure 6.9 we show an idealized quantum wire circuit in which an electron propagating from left to right meets, at point A, a circular roundabout on which there is no preferred direction of travel—the electron may take either the clockwise (British) or the anti-clockwise (American) route with equal probability. In a macroscopic world, this would mean it would proceed *either* one way *or* the other but in the quantum world it goes *both* ways and interferes with itself at point B. Because the alternative paths are equal in length, this interference is constructive—the probability of the electron finding itself at B is therefore enhanced. On the other hand, if we could in some way change the effective path in one direction by half an electron wavelength ($\sim$14 nm), the interference would be destructive and the probability of transmission from A to B would be reduced. In other words, the electron current flowing through the device could be controlled. In fact, two different methods of changing the path length have been demonstrated, one electrical and one magnetic.

In the first case, a Schottky gate is arranged to cover part of one side of the ring (see Figure 6.9(b)) so that, applying a reverse bias changes the density of free carriers in that region. This, in turn, affects the electron wavelength, thus changing the number of wavelengths in the upper arm of the ring and shifting the phase of the wave at point B with respect to that for the lower half. In principle, it should be possible to modulate the source-drain current by 100% but, in practice, the electrons in typical quantum wires are distributed over several one-dimensional sub-bands, which implies that they have different wavelengths, so it is impossible to achieve a condition for which the required π radians of phase shift occur for all the electrons involved. There are other problems with these 'quantum interference transistors' (not least the fact that they require temperatures below 10K for their operation) which make it unlikely they will ever achieve practical application, though they certainly provide a vehicle for interesting new scientific studies. The other method of changing the path length uses a magnetic field normal to the plane of the ring, when it is found that a change of magnetic flux through the ring of $\Delta\varphi = h/e$ (the fundamental flux quantum—see Box 6.2) changes the relative phases of the two electron waves by 2π radians—that is from constructive interference to destructive and back again to constructive. This is the Aharonov–Bohm effect which was first demonstrated in different circumstances (i.e. in vacuum) in 1959. For a typical ring diameter of 1 μm, it is easy to show that the change in magnetic flux density ΔB required to change the phase by 2π is approximately 5.3 mT so the effect is not difficult to demonstrate in a quantum wire structure. In 1989 the Cambridge group were able to obtain many oscillations of conductance with a period similar to this, using an AlGaAs/GaAs 2-DEG suitably gated to form the necessary quantum wire ring.

Yet another fundamental phenomenon explored in quantum wires, during the 1980s was that of 'universal conductance fluctuations'. Early attempts to demonstrate the mesoscopic *A–B* effect were confused by the observation of apparently aperiodic fluctuations of conductance of order e^2/h (the fundamental unit of conductance—see the discussion of the integer quantum Hall effect, above) and it gradually became clear that these resulted from the existence within a quantum wire of microscopic *A–B* loops as illustrated in Figure 6.9(c). Such loops were the result of *elastic* scattering events in which an electron bounced off impurity atoms within the wire and off the walls of the wire so as to complete

Box 6.2. The quantum of magnetic flux

In the discussion of the Aharonov–Bohm effect we introduce the concept of magnetic flux quantization—the total magnetic flux threading through the conducting ring changes by a fixed amount $\Delta\phi = h/e$ when the phase of the electron wave changes by 2π. This small increment of flux represents the fundamental unit of flux—the flux quantum—and in this box we examine its origin.

The motion of an electron in a large magnetic field B consists of circular orbits in a plane at right angles to the field, and the wave-mechanical description of this motion shows that the electron energy does not vary continuously but is quantised, with allowed values:

$$E_n = (n + 1/2)(h/2\pi)\omega_c \tag{B6.12}$$

where ω_c is the cyclotron frequency eB/m_e and n = 0, 1, 2, 3, etc. These energies are the Landau level energies which we have alluded to several times already. Each energy corresponds to a different sized orbit where the areas of the orbits are given by:

$$A_n = (n + 1/2)h/eB. \tag{B6.13}$$

As the magnetic flux density B is assumed to remain constant, it follows that the total flux threading each orbit is given by:

$$\varphi_n = A_n B = (\mathrm{n} + 1/2)h/e, \tag{B6.14}$$

so we see that h/e is the flux change between adjacent orbits, known as the flux quantum. Putting in values for h and e, shows that $\Delta\varphi = 4.136 \times 10^{-15}$ T and that, in any macroscopic circuit, the value of n is very large indeed. However, in the small cyclotron orbits associated with free electron motion, $n = 0, 1, 2, 3$, etc and, if $B = 1$ T, the radius of the ground state ($n = 0$) is $a_0 = \{A_0/\pi\}^{1/2} = \{h/2\pi eB\}^{1/2} = 2.566 \times 10^{-8}$ m $= 25.66$ nm.

It is, perhaps, far from obvious that the ratio h/e has dimensions of magnetic flux but we can see the correctness of this as follows:

The units of h/e are joule-seconds/coulombs = Joules/amps = volts-amps-seconds/amps = volt-seconds. If we now refer to Faraday's law of electromagnetic induction: emf $= -d\varphi/dt$, we see that the units (volt-seconds) are equivalent to those of magnetic flux QED!

a microscopic loop while maintaining phase coherence. These random interference effects combined to give random (i.e. aperiodic) fluctuations in the conductance of the wire in the absence of a magnetic field and are therefore of fundamental significance in experiments involving any form of ballistic transport but they can best be demonstrated by measuring the conductance as a function of magnetic flux density. The *A–B* effect in each loop modulates the conductance at each point in the wire but, because of the random variation in loop size, with random period. This was widely observed in experiments using patterned AlGaAs/GaAs 2-DEG structures towards the end of the 1980s.

The range of novel physics based on ballistic transport which was explored during the latter half of the 1980s is impressive, to say the least, and any reader wishing to make a serious attack on the subject might consult the review article by Cees Beenakker and Henk van Houten in the '*Solid State Physics*' series (Beenakker and van Houten 1991). For our purposes, it is probably sufficient to describe just two further aspects, those of the 'quantum point contact' and of electron focusing. The Philips group in Eindhoven invented this method of injecting electrons into a wire or 2-DEG in the mid-1980s when high mobility 2-DEG samples first became available from the Redhill laboratory. As illustrated in Figure 6.10, it consists of a pair of electrodes spaced by a distance small compared with the electron inelastic mean free path. Application of a reverse bias to both electrodes expands the depletion regions beneath them and tends to cut off the channel between them but, as this bias is reduced, the channel opens and electrons are allowed to flow through. If we assume that the density of electrons on one side of the contact is slightly greater than that on the other, electrons will flow down the concentration gradient and it is found that the conductance G of the channel is quantized in units of $\Delta G = 2e^2/h$, a result of the theoretical requirement that the component of electron momentum in the y-direction (Figure 6.10) is constrained to 'fit' the width W of the channel.

To examine the nature of the injected electron beam it was convenient to use a pair of point contacts such as shown in Figure 6.10(b), one acting as an emitter and the other as collector. A magnetic field applied to this system normal to the plane of the 2-DEG produces circular electron trajectories for which the detector current depends critically on the relationship between the field and the distance between the contacts. The first maximum corresponds to the condition that the cyclotron *diameter* $2mv/eB$ equals the contact separation L, or

$$B = 2mv/eL. \tag{6.13}$$

Assuming a Fermi velocity of $v = 4 \times 10^5\ \mathrm{m\,s^{-1}}$ and a contact separation of $L = 3\ \mu\mathrm{m}$, requires a value of $B = 0.1$ T. A second maximum will

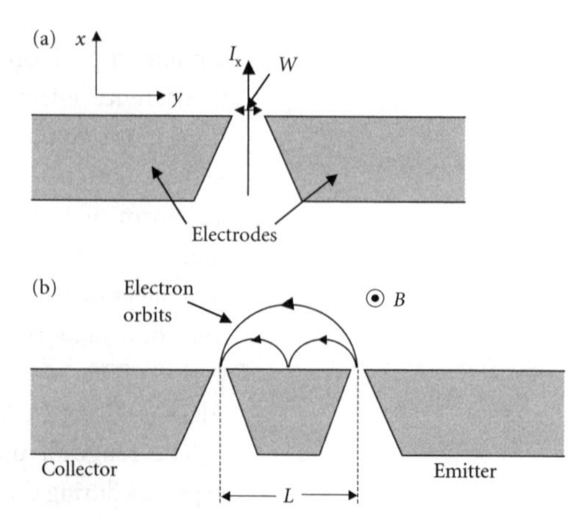

Figure 6.10. (a) Schematic outline of a quantum point contact electron emitter. The electrodes represent Schottky contacts which may be biased to control the effective separation of the points through which the emitted electrons must pass. (b) An example of how a pair of point contacts can be used as emitter and collector. The electron orbits are controlled by varying a magnetic field normal to the plane of the diagram. The first two orbits which result in current maxima at the detector are shown.

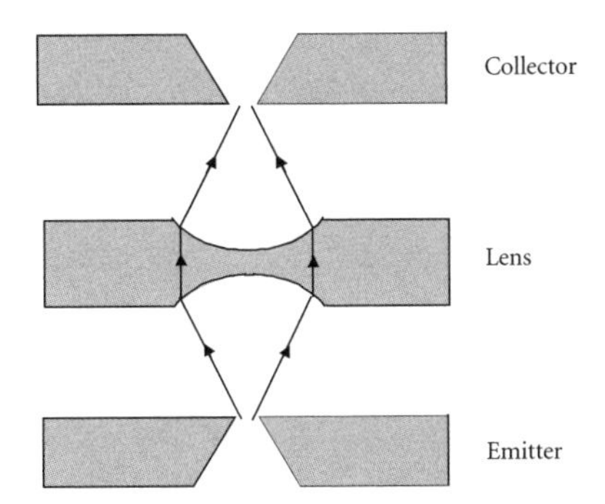

Figure 6.11. Schematic representation of electron focussing using a bi-convex lens in the form of a Shottky contact. The depletion under the lens electrode reduces the free carrier density which, in turn, reduces the electron velocity and causes the beam to bend away from the normal to the electrode edge. As can be seen, this results in focussing of the beam onto a second point contact, acting as collector.

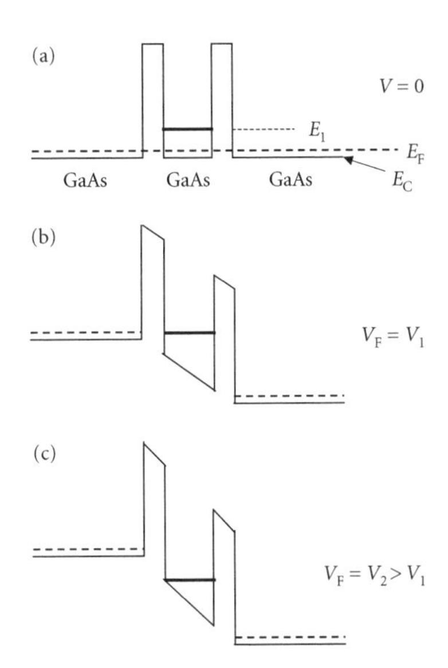

Figure 6.12. Conduction band energy diagram to illustrate resonant tunnelling through a double barrier structure. In case (a), at zero bias, electrons at the Fermi level in the highly doped GaAs are unable to tunnel into the quantum well because there is an energy mismatch with the confined energy level E_1. In case (b), where the applied bias $V = V_1$, these electrons can tunnel into states at the bottom of the two-dimensional band in the well, while in case (c), $V = V_2$, tunnelling is no longer allowed because it would involve transitions into well states with $k > 0$. This implies that, between $V = V_1$ and $V = V_2$, the current–voltage characteristic shows a negative differential resistance.

occur when L corresponds to two cyclotron diameters, a third to three diameters, and so on, so we expect current peaks at $B = 0.1$ T, 0.2 T, 0.3 T, etc. Experimentally, a series of current peaks was detected as the field was increased, precisely as predicted. Clearly, this represented a solid-state version of a well-known experiment formerly performed in vacuum. The ability to generate such a beam of electrons also facilitated the demonstration of electron focussing. Because the electron velocity depends on the Fermi energy (and therefore on the electron density), it was found possible to change the velocity by altering the local electrostatic potential. Then, because the direction of the beam depends on the beam velocity, in just the same way as that of a light beam traversing two media of different refractive indices, it was found possible to bend the electron beam by passing it across an interface between regions of the 2-DEG having different free carrier densities. Figure 6.11 shows a schematic diagram of an electrode shaped so as to function as a converging lens in such an experiment, yet another example of solid-state sleight of hand being used to mimic typical vacuum techniques.

All the above examples were concerned with electrical conduction in the plane of a quantum confined electron gas. We now move on to consider the orthogonal case of electron conduction normal to the plane of quantum well systems, conduction which depends on the possibility of electron tunnelling through narrow barriers. We begin by looking at the example of resonant tunnelling through a quantum well formed between a pair of barriers such as shown in Figure 6.12. For simplicity, we consider only the lowest confined energy state within the well E_1. At zero bias, electrons at the Fermi level in the n^+ GaAs to the left of the structure are unable to tunnel into the well because there are no electron states available to them. As the forward bias V_F is increased to the point $V_F = V_1$, the Fermi level in the GaAs is aligned with the bottom of the two-dimensional conduction band in the well and tunnelling is allowed. Maximum tunnelling probability occurs under conditions of energy and momentum conservation, which are both satisfied in Figure 6.12(b)—electrons at $k = 0$ in the GaAs conduction band tunnel into states with $k = 0$ in the well. For further increase in V_F, electron states in the well are available but they are characterized by $k > 0$ (because they have energies $E > E_1$ and the relation between E and k in the two-dimensional band follow the relationship $E = E_1 + h^2k^2/4\pi^2m_e$). Thus, as V_F increases above V_1, the tunnel current through the structure decreases, so the I–V characteristic shows a region of negative differential resistance. In a manner similar to the operation of a Gunn diode, this can be used as the basis of a high frequency oscillator and many examples have been reported. It is hoped that resonant tunnelling devices may eventually demonstrate significantly higher speeds than can be achieved with the transferred electron effect and the demonstration by a group at MIT of NDR effects at frequencies in the Terrahertz region supports this. In practical terms, the highest frequencies generated

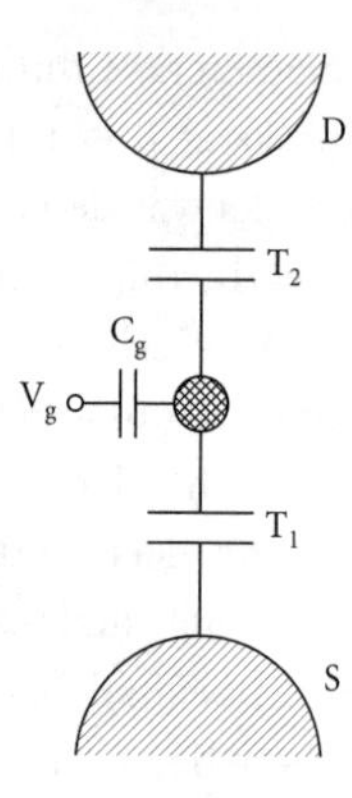

Figure 6.13. Schematic diagram of a single electron transistor (SET). The source S and drain D represent large reservoirs of electrons which are connected to a small island via tunnel junctions T_1 and T_2. The island is also connected to a gate voltage V_g through a gate capacitor C_g. Charge induced on the island by appropriate choice of gate voltage allows electrons to tunnel from S to D, just one at a time.

(up to about 1995) lie in the region of 400–700 GHz—420 GHz for a GaAs/AlAs structure and 712 GHz for the InAs/AlSb system, both by the MIT group, but power levels are relatively low, of order 0.2 μW. Harmonic generation from Gunn oscillators operating at 100–200 GHz may prove a viable alternative for such power levels so it is not clear at the moment how this field will develop.

The inherent speed of tunnelling devices may find a much more important application, however, in futuristic transistors designed to take over from conventional devices when dimensions become so small as to make present principles impracticable. A major difficulty with the steady reduction in dimensions concerns power dissipation, and leads to the conclusion that the ideal device should operate with single electrons, rather than the large densities implicit in present-day designs. In other words, the ultimate digital bit should be represented by the presence (or absence) of a single electron charge. In fact, this is not at all fanciful—the single electron transistor has existed since the early 1990s, though operating only at very low temperatures. A schematic circuit of such a transistor is shown in Figure 6.13. It consists of a small island which can charge and discharge through a pair of tunnel junctions which are connected to large reservoirs of electrons. In addition, the island can also receive charge from a capacitive gate electrode. Note that there is an important difference between this gate and the tunnel connections—the latter depend on tunnelling and the amount of charge is therefore quantized in units of the electron charge e, whereas the gate does not permit electrons to flow to the island—its operation depends on induced polarization which implies that the charge can be varied continuously and may be less than e. Furthermore, the island is made so small that its capacitance C is absolutely minute, something of the order 10^{-17} F. This is to ensure that the amount of energy required to charge it with a single electron is large compared with thermal energy, kT. Thus

$$e^2/2C \gg kT. \qquad (6.14)$$

For $C = 10^{-17}$ F, $e^2/2C = 2.56 \times 10^{-21}$ J so the above inequality is satisfied when $T \ll 186$K. In fact, much of the experimental work has been performed at temperatures below 1K.

What, though, is the point of this improbable requirement? It means that we must think carefully about where the charging energy can come from when a single electron tunnels onto the island. Clearly, it cannot be thermal energy—we have ensured that this is far too little. In point of fact, the answer is that it cannot come from anywhere—the electron is effectively prevented from making the transition at all! This unusual condition is known as the 'Coulomb blockade'—the coulomb energy associated with charging the island capacitor is so great that the tunnelling transition is blocked. No current flows between source and drain—the device is in the 'off' state. What use, now, can be made of the gate electrode to change the situation? Suppose we apply a positive gate voltage V_g, which induces an effective charge on the island of $q = C_g V_g$ and suppose that we make $q = e$. The result is effectively to cancel the negative charge due to an electron tunnelling from the source reservoir—the charging energy is effectively zero and *one* electron can make the transition— current flows and the device is now in the 'on' state. There is no reason for the electron not to tunnel to the drain reservoir, so, once it has done so, another electron can tunnel on from the source, and so on. But note that only one electron charge moves at a time, in marked contrast to the situation in a conventional FET where (even for a very short gate device) literally thousands of electrons are involved. Thus, we have invented the single electron transistor and we shall, of course, refer to it as SET.

The only question remaining is how to make an island small enough that its capacitance is of the order 10^{-17} F! It can actually be done by arranging a set of gates on the surface of a 2-DEG structure so designed that the application of negative gate voltages squeezes the area containing charge down to a minute size usually referred to as a 'quantum dot'. (We shall discuss more of these in Section 6.4.) This raises certain difficulties as to the distinction between tunnel junctions, control gate, and quantum dot-making gates but technology can achieve remarkable successes when it is really pushed to the limit and SETs have actually been made to work at milli-kelvin temperatures—whether they can be persuaded to function at room temperature seems, to say the least, a rather tall order, as the island would have to be made at least two orders of magnitude smaller. Perhaps a compromise temperature of 77K might be acceptable but it is too early to say at present. What is certainly true, though, is that wonderful progress has already been made, and it would be a rash forecaster who said 'never'.

6.4 Optical properties of quantum wells

Section 6.1 served to introduce the concept of confinement energy in a quantum well but it is now time to examine in greater detail how these ideas were used in understanding the optical properties of quantum well samples and developing new applications for those properties. In terms of timescale, we move rapidly back to the publication of Dingle's seminal paper of 1974 which set out the basic theory of optical absorption by quantum wells and demonstrated its essential correctness. In order to describe optical absorption we need to consider transitions between electron states in the valence band and those in the conduction band—in the case of a QW sample these states take the form of confined states in the two wells, one associated with the conduction band and the other with the valence band. We need to extend the account given in Section 6.1 in two respects—first to include both conduction and valence bands, and second, to allow for the fact that real quantum wells are not infinitely deep (as we assumed there) but have depths typically of a few tenths of an electron volt.

An important difference introduced by the finite well depth is that electron (or hole) wavefunctions are no longer confined entirely to the well material but also penetrate (tunnel) a little way into the barriers. Within the well, the appropriate functions look very much like those shown in Figure 6.1(b), while in the barriers they fall off exponentially to zero. The narrower the well, the greater is the penetration into the barriers, which implies that carriers spend an increasing proportion of the time in the barriers. In fact, when the well width actually goes to zero, it must be that the wavefunction becomes that of a free carrier in the barrier material. Perhaps of more obvious importance, is the fact that the number of confined states is now finite, rather than infinite and, in practice, certainly for narrow wells, is a fairly modest number. Figure 6.14 illustrates this point for a set of GaAs quantum wells within $Al_{0.3}Ga_{0.7}As$ barriers. As can be seen, the (conduction band) well depth in this case is 250 meV and a 15 nm well contains just four confined states, while for narrower wells the number is reduced still further. Another feature brought out by Figure 6.14 is the manner in which each state is pushed closer to the top of the well as the width L_z is reduced. Clearly, the energy of the confined state in very thin wells approaches that of the barrier, as we must expect. There is, however, a general theorem which indicates that there will always be at least one state in any quantum well.

Similar considerations apply to hole states but with two provisos: first, the valence band well depth is roughly half that in the conduction band, and second, the valence band is considerably more complex. As shown in Figure 5.1, the GaAs valence band contains three branches, light and heavy holes which are degenerate at $k = 0$, together with the spin-orbit split-off band some 350 meV lower in energy. Because its separation from the other two bands is larger than the well depth, we can conveniently forget about the split-off band, while, in the reduced symmetry of the quantum well, the light and heavy holes are no longer

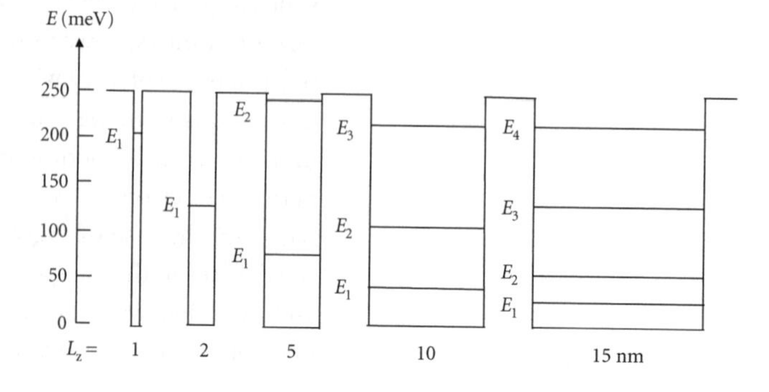

Figure 6.14. Calculated energy levels in GaAs quantum wells with $Al_{0.3}Ga_{0.7}As$ barriers. The well depth is 250 meV. As the well width L_z becomes smaller, energy states are pushed towards the top of the well and spill out so as to reduce the number of confined states in narrower wells. There is, however, always at least one state in even the narrowest well.

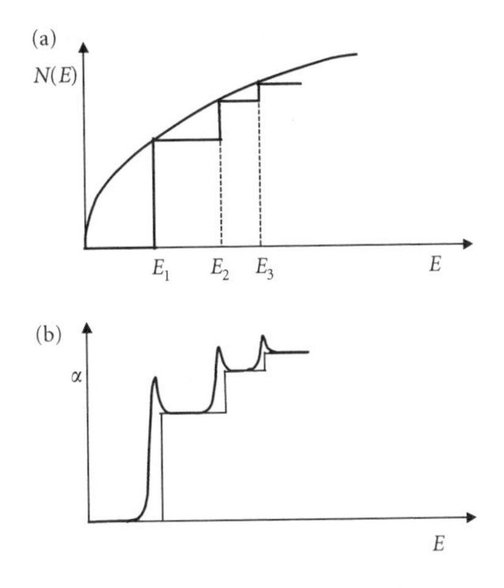

Figure 6.15. (a) shows an example of the stepped density of states for a typical quantum well sample. The steps occur at energies E_n which are the confinement energies (see, for example, Figure 6.14), (b) illustrates the way in which exciton absorption modifies the absorption spectrum, adding a sharp exciton peak below each step.

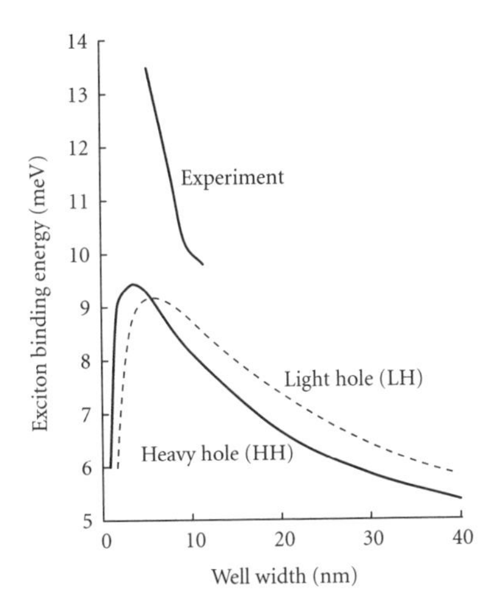

Figure 6.16. Calculated exciton binding energy E_x vs well width L_z for an $Al_{0.3}Ga_{0.7}As$/GaAs quantum well. The solid line is for the $(HH_1 - E_1)$ exciton, the dashed line for the $(LH_1 - E_1)$ exciton. Note that the maximum binding energy occurs in the region of $L_z = 3$–6 nm. (From R. L. Greene, K. K. Bajaj, and D. Phelps (1984) *Phys. Rev. B* 29, 1807.) The experimental curve represents data on the $(HH_1 - E_1)$ exciton from P. Dawson, K. J. Moore, G. Duggan, H. I. Ralph, and C. T. B. Foxon (1986), *Phys. Rev. B* 34, 6007, for $Al_{0.35}Ga_{0.65}As$/GaAs MQW samples (see text).

degenerate. Also, because their effective masses differ, they give rise to two quite separate sets of confined states (note that confinement energies are proportional to the reciprocal of the effective mass in equation (B6.6)). In the absorption spectrum of a GaAs quantum well we might therefore expect to measure optical transitions between both sets of hole states and each electron state. However, there is a 'selection rule', which allows only transitions for which the quantum number n does not change—for example, the transition $(HH_1 - E_1)$ (heavy hole 1 to electron 1) is strong, whereas $(HH_1 - E_2)$ (heavy hole 1 to electron 2) will be weak (often so weak as to be undetectable). In addition, we must also take account of the density of states in the various two-dimensional bands, an example being given in Figure 6.15(a). Note that the density consists of a series of steps, each one corresponding to a confined energy state, while the envelope of the step heights coincides with the three-dimensional density of states for the well material (curved line in Figure 6.15(a)). This implies that the absorption spectrum will contain corresponding steps in intensity at the onset of each new confined state, rather than the smooth rise in absorption shown in Figure 3.11.

This is not, however, the end of the story—to complete the picture, we must include exciton effects and it turns out that excitons are even more important in quantum wells than they are in bulk material. The reason for this is that the confining potential of the barriers squeezes the electrons and holes closer together than they would be in bulk material, thereby increasing the electron–hole binding energy. Thus, for GaAs, the exciton radius in bulk material is about 10 nm so any well narrower than 20 nm will clearly cause some perturbation of the binding energy. In fact, it can be shown that, for an ideal two-dimensional exciton the binding energy E_X is just four times that of a bulk exciton—for GaAs this means an increase from about 5 V to 20 meV—which implies that the exciton is much more stable against thermal dissociation in a quantum well. Even at room temperature, we can expect to see exciton effects. In practice, however, wells are of only moderate depth and may have significant widths (significant in comparison with the exciton dimension) so the enhancement of exciton binding is somewhat reduced. Figure 6.16 shows a typical example of the calculated exciton binding energy as a function of well width—note that E_X decreases again for very narrow wells, as must be the case because at zero well width the exciton is a bulk exciton again, though now associated with the barrier material, rather than the well.

Finally, we are in a position to understand the nature of the absorption in a real quantum well. It consists of *two* series of strong exciton peaks superimposed on the stepped density of states curve, as shown in Figure 6.15(b). However, we should also be prepared to observe additional 'forbidden' transitions where the quantum number n changes by 1 or even 2, though these are likely to be much weaker than the $\Delta n = 0$ transitions. Experimentally, it is the exciton peaks which are used to characterize a QW sample—the stepped density of states produces a rising background but the steps are never sharp enough to provide adequate resolution in the absorption spectrum. Remember that the exciton appears at an energy E_x below the step edge and this must be allowed for when making comparison with the calculated confinement energies which, of course, refer to the steps. Note, too, that the presence of forbidden transitions can be extremely helpful in obtaining a reliable interpretation of a particular measurement.

It would take us far beyond the scope of a brief overview to attempt a detailed analysis of such spectra but it is probably worthwhile trying to understand something of the problems faced by those who were concerned with such analysis in the 1970s and 1980s. The essential difficulty lay in the number of a priori unknown parameters which determine the photon energies involved. We have spoken so far in terms which imply that the well width L_z is reliably known, that the well depths can be inferred from the fraction x of Al in the Al_xGa_{1-x} As barriers (and, indeed, that x itself is accurately known), that the exciton binding energy is well established, that well width is constant throughout the sample, that the interface between well and barrier is sharp on the atomic scale, and so on. In fact, none of these supposed 'facts' could be taken on trust. For example, crystal growers may have firmly believed that they could control the growth conditions so well as to be able to predict layer thicknesses with sufficient accuracy, but experimental checks were sometimes found *not* to confirm this! Careful studies of X-ray diffraction and transmission electron microscopy also suggested that even the best interfaces were *not* atomically flat but that interpenetration of the two materials usually occurred over one or two atomic planes either side of the notional interface (not very important in a 20-nm well, perhaps, but crucial in one having $L_z = 1$ nm!). While the Al fraction x might be approximately known from growth parameters, how was it actually to be measured? Again, what were the relative fractions of the band gap difference appearing in the conduction and valence bands (i.e. what was the 'band offset ratio')? From the theoreticians' point of view there were other uncertainties, such as the appropriate values to use for hole effective masses—bulk values certainly did not apply nor was it clear how to obtain reliable values to plug into the many calculations being made around the world. All these things just had to be winkled out—which probably accounts for the time it took.

Let us briefly examine two of these points. Dingle et al., in 1974, reported the band offset ratio for AlGaAs/GaAs as $Q_C : Q_V = 85 : 15$ with considerable confidence, yet in 1985 Geoff Duggan from the Philips Redhill laboratory appraised the situation in a paper published in *J. Vac. Sci. Technol., B* with the following comment: 'An offset ratio of ~60 : 40 is currently more consistent with the majority of recent determinations.' Not only was the nominal ratio considerably different, but, even after 11 years of endeavour, its value was still hedged with uncertainty! Geoff, himself, had pioneered a method of fitting measured absorption spectra with theoretical parameters but had never found it possible to obtain the offset ratio with anything like the confidence he (or any of us) would have hoped for. It was only in 1987 that his colleague Phil Dawson, while a visitor at Bell Labs, Murray Hill, could provide a direct and reasonably unequivocal method of measuring the valence band offset in AlGaAs / GaAs using the spectroscopy of so-called 'type II' quantum wells. Initially, these were AlAs / AlGaAs structures containing thin QWs in which the single confined electron state in the AlGaAs lies above the X minimum in the AlAs barrier, leading to a luminescence emission line corresponding to the transition X_C(AlAs)–Γ_v(AlGaAs) but these experiments were later extended at Redhill to include AlAs / GaAs structures. We shall not attempt a detailed analysis here but simply quote the result of these latter studies to the effect that Q_v was found to lie in the range 0.33–0.34 (and therefore $Q_c = 0.66$–0.67). At last (1987) the uncertainties were down to 3%! It has been suggested that science is 1% inspiration and 99% perspiration—here we appear to have experimental support for such a hypothesis.

The other question we shall examine concerns the exciton binding energy (Box 6.3). As already outlined, theoretical estimates of E_x suggest values in the range 5–10 meV which represent significant corrections to be applied to calculated confinement energies. But can we not *measure* E_x experimentally? Again, it took a long time but the answer was finally 'yes' when, in 1986, the Philips group reported measurements of well-resolved emission and absorption lines from *both* the 1s *and* the 2s exciton states. Being rather like a hydrogen atom, a light electron orbiting a heavy hole, the exciton can exist in a whole series of energy states, the lowest (or ground state) being designated 1s, the first excited state 2s, the next 2p, 3s, 3p, etc., and it is possible to calculate their relative energies in terms of fractions of the ionization energy for the 1s state. In particular the Philips group calculated the 2s binding energy for the two-dimensional exciton which was a relatively small 1.5 meV (so, even if the calculation was not very accurate, it made rather little difference to the final result). By adding this to their *measured* 1s–2s splitting, they arrived at the first reliable values for the exciton 1s binding energy in four multiple quantum well (MQW) samples with well widths in the range 5.5–11.2 nm. Their results for the HH exciton are plotted as the 'experimental' curve in Figure 6.16 and demonstrate just how difficult it can be to make sound calculations in these two-dimensional systems! The peak value of binding energy appears from the experiments to be at least 14 meV, some 50% larger than the calculated value (and, incidentally, much closer to the simplistic two-dimensional value for GaAs of about 20 meV). Why did it take so long to arrive at this point? Once again, it was a question of waiting for sufficiently good quality samples to be grown and, ironically, perhaps, it was the effort devoted by Tom Foxon to obtain high mobility 2-DEG samples which opened the way. These MQW samples were grown after the MBE machine had produced over 100 μ of GaAs in that particular run and the background impurity level was down to about 10^{20} m^{-3}. Only then were optical linewidths small enough to obtain the required resolution in the appropriate spectra. Science moves, as we have seen before, in mysterious ways! However, even more interesting is the fact that these important experimental results appear to have escaped the notice of all the book writers who have otherwise so thoroughly described this exciting field. One wonders how many workers still believe that Greene et al.'s calculations represent the best estimate of exciton binding energy available? But, unless this smacks a little of special pleading, let me make the point that it can only be one example of many where important information fails to obtain the attention that it surely deserves—the scientific literature in a field such as this is so vast that probably no one can hope to provide a complete account, and there must be very many cases where, with the best will in the world (and the results *were* published in the world's most prestigious physics journal, *Physical Review*), information simply gets lost.

Of the many aspects of the optical properties of low-dimensional structures, which could appropriately be discussed in this section, we shall concentrate on just one other, the subject of quantum dots. From the early days of LDS, it was appreciated that quantum wells were only one of a set of structures which promised exciting possibilities for new physics and for new, or improved devices. It was inevitable that quantum wells would take precedence simply on the grounds that they were straightforward structures to grow by MBE (and, later, by MOVPE). We saw something earlier, in Section 6.3, of the additional complexity inherent in forming one-dimensional structures, or quantum wires—how much more difficult again might it be to make zero-dimensional structures, or dots? Nevertheless, it was apparent from a theoretical point of view that dots would be of special interest because their density of states would consist of sharp, atomic-like levels, quite unlike those associated with wells (Figure 6.15). In fact, we have already come across one example of such levels, Landau levels in a 2-DEG with a high magnetic field normal to the plane of the structure. In this case, the field results in localization of electrons in the plane—a field of $B = 1$ T confines electrons to a ground-state orbit of diameter approximately 50 nm (see Box 6.2). In a quantum dot it is an electrostatic confining potential (due to surface band-bending) which is responsible, but the problem here is that of defining a dot small enough to show quantum confinement in all three dimensions. In fact, a surprising variety of approaches has been followed (Banyai and Koch 1993)—these include clusters of impurity atoms dissolved in glass, chemical precipitation, sol–gel methods, incorporation of semiconductor material in small cavities in host materials such as zeeolites and the growth of very small crystallites. In the spirit of this section, we shall concentrate on two methods of forming dots which depend on conventional semiconductor processing techniques. The first is based on the application to a two-dimensional structure of Schottky gates in the form of a grid (this being an extension of the technique used to form wires), the second makes use of certain unusual aspects of crystal growth which result in 'self-organized' dots. This latter method is very attractive in so far as it requires no lithography and can be employed to produce three-dimensional arrays of dots, resulting in a pseudo-three-dimensional material with one-dimensional properties.

Our story begins in 1986 when groups at Bell and at Texas Instruments reported dot arrays based on AlGaAs/GaAs QW structures where the dots were formed using gates defined by electron beam lithography. These had lateral dimensions of order 100 nm but were small enough to show significant shifts in the energy of their photo- or cathodo-luminescence peaks, compared with the emission from the unmodified quantum well, shifts of the order 10–50 meV towards higher energy. The shift increased as the dots were made smaller, consistent with the idea of quantum confinement. A year later, another Bell group reported similar results from a lattice-matched InGaAs/InP structure. This began a worldwide effort to control and understand the properties of these zero-dimensional pseudo-atoms. It seems that theory moved somewhat faster than experiment—by about 1990 there existed a comprehensive body of theoretical understanding, comprehensively summarized in the book by Banyai and Koch (1993). Experiment was hindered by the complexity of processing required to achieve really small-scale lateral dimensions but began to take off in a serious manner with the realization in the early 1990s that dots could be formed during MBE crystal growth, without the need for lithographic definition. Ironically, it depended on reversing the well-established philosophy of MBE which aims always to achieve two-dimensional growth, that is, layer-by-layer growth, one monolayer being completed before a second one is deposited. RHEED patterns (Box 5.2) could be used to differentiate this mode from the undesirable three-dimensional, or 'island' growth which usually produced poor crystal quality.

The first of these self-organized dots resulted from the growth of InAs layers on a GaAs substrate. It was found possible to deposit approximately 1.5–2 monolayers of InAs in a strictly two-dimensional form but any further growth produced islands which grew larger in both thickness and lateral size until they coalesced to form continuous, though defective, InAs films. The secret of dot growth lay in stopping the deposition of InAs at a point when the islands were still small enough to produce quantum confinement, say at a lateral dimension of about 30 nm and then cover the dot material with a barrier layer of much wider band gap GaAs. In this way, the dots were completely immersed in GaAs and would (at least, in principle) remain unaffected by any subsequent growth. Perhaps the really remarkable feature of the process was its uniformity and repeatability. One might have anticipated a wide range of random shapes and sizes of dot but it turned out, on careful examination by scanning tunnelling microscopy and electron microscopy that the dimensional tolerance was no more than $\pm 10\%$. It was clear that some built-in control mechanism was operative and it was soon recognized that this must depend on the 7% lattice mismatch between the InAs and its GaAs substrate (InAs being larger). What appeared to be happening was an initial growth of InAs with an in-plane lattice parameter corresponding to that of GaAs, resulting in a highly strained layer. At roughly 1.75 monolayers of coverage, the strain energy became so large that it drove the growth process to produce islands of InAs with the normal InAs lattice constant, rather than a continuation of the strained two-dimensional film. In other words, it was energetically preferable for the subsequent InAs molecules to be deposited on top of one another, rather than in the plane. Again, remarkably, these islands turned out to be dislocation-free—the change of growth mode served to relieve strain without the usual dislocation generation. In fact, this was less a random process, much more a naturally controlled one, and this accounted for the relatively tight control over dot dimensions. Nature seemed actually to like the idea of dots! So much so, in fact, that dots were gradually found to grow in other mismatched material combinations such as SiGe/Si and InN/GaN—it was obviously a very general phenomenon.

Remarkable though the uniformity of these structures might be, however, the distribution of sizes was still wide enough to broaden any optical spectra to an undesirable degree. This was well illustrated by work performed at the laboratories of France Telecom. in 1994. In two letters, one in *Applied Physics Letters*, the other in *Physical Review Letters*, Moisson, Marzin, and co-workers described the growth of InAs dots on GaAs substrates. After 1.75 ML of two-dimensional growth their layers separated into faceted dots of typically 3 nm height by 24 nm diameter, spaced approximately 60 nm apart, which showed strong luminescence at a wavelength of about 1.1 μm (photon energy $h\nu = 1.12$ eV) with a line width of $\Delta h\nu = 50$ meV. (Compare this with low temperature line widths of order 0.1 meV obtained from high quality GaAs films.) To demonstrate the inhomogeneous nature of this linewidth, they used lithography to define mesas with diameters ranging from 5000 nm down to 100 nm and compared the resulting luminescence spectra. The smaller mesas contained only a few dots so it was possible to resolve the emission from each one, the resulting spectrum consisting of a number of sharp lines with widths less than 0.1 meV (the resolution of their apparatus), the measured distribution of photon energies being entirely consistent with the linewidth of 50 meV observed from large area mesas. This one experiment, therefore, proved the point that optical linewidths in quantum dots are extremely narrow (as expected from the atomic-like nature of their energy levels) and demonstrated the essential practical problem of using dots in optical devices. In order to obtain useful light intensities it is necessary to employ large numbers of dots but their inevitable (?) distribution in size results in undesirably wide emission (or absorption) lines.

These observations sparked wide-ranging interest in the detailed nature of the dots and recent work at Imperial College, London and in the USA at IBM and the University of Texas has shown that the mechanism of their formation is (inevitably?) more complex than originally thought. In particular, the dots (a) are partly buried within the InAs 'wetting layer', (b) contain significant quantities of Ga—that is, up to 30%, and (c) are non-uniform in composition, having a high In concentration in a pyramidal core region. All these imply that the confining potential for electrons and holes must differ very significantly from the idealized models often used, making it extremely difficult to predict the effective band gap of the dot material and the positions of the confined energy levels and, therefore, what the emission energy should be. This is an ongoing activity but we may note that ad hoc attempts to make InGaAs dots emitting at wavelengths close to 1.3 μm (of interest for fibre-optic communication lasers) have been quite successful. There is no sign yet of emission linewidths coming down below about 30 meV, though it does appear that linewidth is related to dot spacing—surface densities of dots may vary between about 1×10^{14} m^{-2} and 1×10^{15} m^{-2} and to obtain a linewidth of 30 meV demands a density near the lower end of this range.

The question of density may well be of major importance for practical applications, so it is important that it has been found possible to grow multiple layers of dots, thus forming a three-dimensional array. It is interesting, in this context, that these arrays show a tendency towards self-organization, the dots in one plane lying immediately above those in the next lower plane. This, again, appears to result from the presence of strain fields, the degree of perfection of the array increasing (a) as the layers are grown closer together and (b) as more layers are deposited. Typical volume densities of dots lie in the range

10^{22}–10^{21} m^{-3}, rather smaller than typical impurity doping densities. We shall come back to dots in our discussion of the quantum well laser in Section 6.6.

6.5 Electronic devices

I well remember during the 1980s, when the much vaunted UK LDS programme had been running perhaps 3 years, a university acquaintance of mine asking in a worried voice 'where are all these new devices which we were promised?' He may have been concerned for the future of his government-based funding if 'Industry' did not immediately come up with the goods to justify the investment but he was certainly being naive about the appropriate timescales. LDS offered exciting new technology but new devices required not only that but new applications which could benefit from those technologies. (The considerable investment required to turn research into production demanded clear commercial justification—UK industry was not much in the habit of wild speculation!). Time would tell, of course, but no one at that juncture envisaged the massive consumer demand for microwave systems in the guise of personal telephones which was to engulf us all in the 1990s. The high electron mobility transistor was first demonstrated in 1980 and found a niche as the signal amplifier at the front end of every satellite TV receiver, a market which has continued to grow steadily, if not explosively, ever since. (Though, from a chauvinistic British viewpoint, it was unfortunate, perhaps, that most of the profits went to Fujitsu, rather than to GEC!) Nevertheless, it was the development of the 'mobile' which really brought home the low-dimensional bacon and sparked a fascinating competition to realize the ultimate in noise performance (to reduce the crackling to a minimum, perhaps?). Given the understandable concern over microwave biological damage, it was natural to look for systems which could operate at the lowest possible levels of signal power, a concern which not only demanded low-noise amplifiers but also allowed semiconductor devices to replace the bulky and inconvenient travelling wave tube as power source at the transmitter. The HEMT turned out to be the ideal pre-amplifier but other functions might be performed by the newly re-emerging bipolar transistor, in the shape of the HBT. In this section we shall have something to say about both and, as an antidote to the cult of the popular, a device which lies at the very opposite end of the popularity spectrum, the quantum Hall effect resistance standard.

Not surprisingly, the first HEMTs were based on the well tried material system AlGaAs/GaAs and, in Figure 6.17 we show a typical structure for such a device. In this version, the gate is recessed so as to be located on the doped AlGaAs layer, while the source and drain contacts

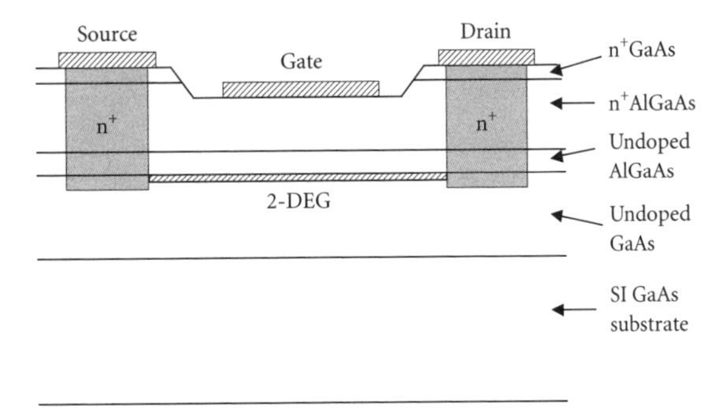

Figure 6.17. Schematic diagram of a recessed gate AlGaAs/GaAs HEMT. The source and drain contacts are placed on heavily doped GaAs regions to ensure low-resistance contacts and the connection to the ends of the 2-DEG are also made via heavily doped material to minimize access resistance at source and drain. An undoped spacer layer is included in the AlGaAs barrier to distance the 2-DEG from the ionized donors, which are the source of the free electrons.

are made to heavily doped GaAs contact regions, the connection to the two-dimensional electron gas being made via heavily doped ion-implanted pillars. An undoped AlGaAs layer is used to separate the 2-DEG from the ionized dopant atoms and the whole structure is grown on a SI GaAs substrate which serves to isolate the device electrically. This particular contact technology emerged as the result of various attempts to improve early device performance. Originally the metal contacts were deposited directly onto the AlGaAs layer which avoided the need to recess the gate but led to severe problems with contact resistance (i.e. 'acess resistance') to the 2-DEG channel. In particular, it reduced the device transconductance $g_m = dI_{SD}/dV_G$ and had a deleterious effect on high frequency performance (readily understood in terms of a simple equivalent circuit). In the original paper from Fujitsu, no radio frequency (RF) measurements were reported but following papers from Thomson CSF (Laviron et al. 1981) and Cornell University (Judaprawira et al. 1981) indicated that g_m was reduced by as much as 50% by the measured access resistances. Clearly something had to be done!—but it could only be done at the price of much more intricate processing. Perhaps something should also have been done about the rapid proliferation of names—the French group immediately renamed the device a two-dimensional electron gas field effect transistor (TEGFET), while the Americans, not to be outdone, adopted the alternative of modulation-doped field effect transistor (MODFET). More recently, a compromise terminology, heterojunction field effect transistor (HFET) has been introduced by some authors but this well-meaning attempt to find a middle way appears only to have compounded the problem! For our part, we shall stick to the original Japanese—HEMT! (even though its room temperature mobility is not all that high, its main advantage in modern narrow gate devices being an improved saturation drift velocity).

Call it what you would, the device had clearly caught the imagination of engineers around the world and progress was remarkably rapid, both in improving the performance of the original AlGaAs/GaAs structure and in the introduction of new materials. As early as 1983, two separate groups at Bell Labs had reported the first application of InGaAs to

HEMT technology, based on the reasonable argument that InGaAs has a smaller electron effective mass than GaAs and therefore a higher mobility (and higher saturation velocity). They made use of the already established fact that $Ga_{0.47}In_{0.53}As$ is lattice-matched to InP so they used Fe-doped SI InP as substrate but went one better by adding a barrier layer of doped $Al_{0.48}In_{0.52}As$ (similarly lattice-matched) to provide the free electrons for the 2-DEG. Then, in 1984, the Fujitsu group introduced the alternative of doped InP as the supply layer, InP also having a wider band gap (1.35 eV) than $Ga_{0.47}In_{0.53}As$ (0.65 eV). Finally, in 1985, researchers at IBM demonstrated the first pseudomorphic HEMT (or PHEMT) in which an InGaAs channel layer was combined with a conventional AlGaAs supply layer. The word 'pseudomorphic' refers to the fact that InGaAs is not lattice-matched to GaAs (or AlGaAs) and was grown as a thin (20 nm) strained layer, the InGaAs taking up the in-plane lattice constant of the GaAs on which it was deposited. To illustrate the level of interest, we note that, by the end of the 1980s it was easy to list (in no particular order), the following laboratories actively engaged in HEMT development: Bell Labs, IBM, Honeywell, Sandia, Fujitsu, Toshiba, Siemens, Thomson CSF, Hewlet Packard, Rockwell, General Electric, GEC, Philips, NEC, Cornell University, University of Illinois, Brown University, and, no doubt, many others. At the time of writing, some of the names have changed but the interest is still as intensive though with a much keener commercial bias. However, all this is to ignore the logic which defined the driving force—let us take a step back and examine some of the reasons why things happened as they did.

Progress in the development of high speed HEMT devices is well covered, to the end of the 1980s, in two review articles in the '*Semiconductors and Semimetals*' series, Morkoc and Unlu (1987), and Schaff et al. (1991). We shall attempt a brief survey here—readers seeking greater detail are recommended to these sources. Following the initial demonstration of DC characteristics at Fujitsu, it was clearly important to measure performance in microwave and high speed logic circuits, a task immediately undertaken at Fujitsu, itself but also at Bell Labs, Thomson CSF, and Honeywell. By 1984 AlGaAs/GaAs devices with gate lengths of 1 μm were showing switching speeds of about 10 ps and had been incorporated into frequency dividers operating at frequencies up to 5 GHz and, with gate lengths of 0.7 μm, up to 6.3 GHz. Low-noise microwave HEMTs with 0.35 μm gates were operating at Q-band (35 GHz) with noise figures as low as 2.7 dB and 2 years later General Electric reported noise figures of 0.8 dB at 18 GHz from 1-μm devices. Progress was certainly impressive but not everyone was happy with the overall picture. Device characteristics showed anomalies, some of which, at least, could be traced to the effect of what have come to be known as 'DX centres'. It transpires that Si dopant atoms in AlGaAs give rise to a species of deep impurity levels which cause undesirable high frequency effects by capturing and releasing electrons at a rate which does not follow the applied microwave signal. This meant that the fraction of Al which could be used in the barrier layer was limited to about 0.2–0.25 and this, in turn, limited the band offset to about 0.25 eV. This had the effect of limiting the maximum density of free electrons in the 2-DEG to $1 \times 10^{16}\,\mathrm{m}^{-2}$ and that, in its turn, limited the maximum source-drain current.

What was to be done? Various approaches were tried. As stated above, one method of removing the DX centre problem was to use a different material as barrier. AlInAs does not suffer from DX centres and has the added advantage of providing a larger band offset (~0.5 eV) at the AlInAs/GaInAs interface. Its use allowed larger values of sheet carrier density in the 2-DEG ($n_{\mathrm{max}} = 2.5 \times 10^{16}\,\mathrm{m}^{-2}$) and, together with the larger value of v_s in InGaAs, led to considerable improvement in both transconductance and saturation current. The downside of this approach was the increased difficulty in growing the required films and the need for an InP substrate, which was of poorer quality and, from a commercial viewpoint, more expensive than GaAs. Similar remarks apply to the use of an InP barrier layer and this approach seems never to have been followed up. The alternative to these radical solutions was to modify the AlGaAs itself, either by growing it in the form of an AlAs/GaAs superlattice and doping only the GaAs (thus avoiding DX centres) or by using a so-called 'delta-doping' scheme for incorporating the Si atoms. This involved stopping the MBE growth and depositing a single monolayer of Si at an appropriate point in the AlGaAs. Finally, the solution which appears to have stood the test of time (and commercial acceptance!) made use of a lower Al fraction (of the order 15–20%) in the barrier layer, together with an active layer of InGaAs (with 15–20% In). In this manner, it was possible to minimize the DX centre problem, obtain a larger band offset, a higher drift velocity, and to use a GaAs substrate. (As already mentioned, the InGaAs had to be deposited as a thin layer to avoid strain relaxation by the incorporation of dislocations but this was perfectly straightforward.) Though the growth of InGaAs-strained layers with even larger In concentration initially proved difficult, it was eventually (1988) found possible to incorporate 25% In and reach a value of $n_{\mathrm{max}} = 2.5 \times 10^{16}\,\mathrm{m}^{-2}$, equal to that available from the AlInAs/GaInAs combination. Consistently with all this, it became clear (Weisbuch and Vinter 1991: 151) that measured values of f_T (note that $f_T \propto v_s$) increased monotonically in the order: GaAs MESFET, AlGaAs/GaAs HEMT, AlGaAs/InGaAs PHEMT, AlInAs/InGaAs/InP HEMT, the overall improvement being slightly less than a factor of 3. (Box 6.3 attempts to make some of this a trifle clearer.) Gate lengths as short as 0.1 μm were reported to give values of f_T up to about 200 GHz which was progress indeed, and important from a practical point of view because certain military requirements asked for

Box 6.3. Theory of the microwave HEMT

To assist with the understanding of HEMT performance, it may be useful to present, here, a simplified model of its operation. Needless to say, there are numerous detailed modifications which have to be included in a proper engineering model but the account we give here at least makes clear the general relationship between the various parameters used to describe it.

We concentrate attention on the region of the device under the gate. This has length (in the direction from source to drain) L and width W. It contains a sheet electron density n_s (m^{-2}), which is controlled by the gate voltage V_G. We treat the AlGaAs barrier layer as an insulator of thickness d, in which case the gate electrode and the channel charge $q_s = n_s eLW$ act as a parallel plate capacitor C_G, given by:

$$C_G = \varepsilon LW/d, \tag{B6.15}$$

where ε is the dielectric constant of the AlGaAs layer ($\varepsilon \approx 10\varepsilon_0$). This gives us a simple relationship between n_s and V_G:

$$q_s = C_G V_G$$

or

$$n_s = (\varepsilon LW/d)V_G/eLW$$
$$= \varepsilon V_G/ed. \tag{B6.16}$$

For a short gate length device which operates under saturated drift velocity conditions, we can write the source drain current as

$$I_{SD} = en_s v_s W$$
$$= \varepsilon V_G v_s W/d. \tag{B6.17}$$

This gives us the important relationship for the transconductance of the device g_m:

$$g_m = dI_{SD}/dV_G$$
$$= \varepsilon v_s W/d. \tag{B6.18}$$

Finally, we need an expression for the limiting speed of the device, which depends on the transit time τ for electrons to pass under the gate:

$$\tau = L/v_s. \tag{B6.19}$$

The effective gain of the transistor will drop to unity at a frequency f_T given by $2\pi f_T \tau = 1$. Thus:

$$f_T = 1/2\pi\tau$$
$$= v_s/2\pi L, \tag{B6.20}$$

Box 6.3. Continued

which, using (B6.15) and (B6.18), can also be written in the form:

$$f_T = g_m/2\pi C_G. \tag{B6.21}$$

Equation (B6.20) shows that, for high operating frequencies one requires to maximize the saturation drift velocity v_s and to minimize the gate length L. The former is, of course, a pure material parameter which is larger in InGaAs than in pure GaAs, the latter is a matter for the technologist who has obliged with values of L down to 0.1 μm, no mean achievement.

It may be useful at this point to insert appropriate parameter values into some of these equations to see how they relate to practical device results. Taking $v_s = 2 \times 10^5$ m s^{-1} and $L = 0.1$ μm gives us the value $f_T = 300$ GHz (to be compared with the experimental value of about 200 GHz) or a switching time $\tau = 0.5$ ps. Typical transconductance values can be obtained from (B6.18), using an AlGaAs thickness $d = 20$ nm, as $(g_m/W) = 1000$ S m^{-1} (usually written as 1000 mS mm^{-1}), in other words, 1 S mm^{-1} of gate width (such values have actually been achieved). The saturation value of source-drain current is found from (B6.17), using values of $n_s = 2 \times 10^{16}$ m^{-2} and $v_s = 2 \times 10^5$ m s^{-1}, as $I_{sat} = 600$ Wamps—in other words, a gate width of 100 μm would give $I_{sat} = 60$ mA, an output power level of roughly 100 mW (100 W mm^{-1}). Note that the gate capacitance C_G takes values of about 1 pF mm^{-1} which implies actual capacitance of order 0.1 pF or less, demanding extreme care in minimizing stray capacitances.

transistors operating at 94 GHz. Let us never forget the part played by the Military in stimulating new (and often expensive) device development. Finally, we should emphasize the huge improvements in noise figure, values as small as 0.8 dB at 10 GHz, 1.5 dB at 60 GHz, and 2.5 dB at 94 GHz being available in the early 1990s. The HEMT was clearly a wonderful invention for application to low-noise amplifier circuits.

By 1990, it was apparent that the HEMT had come to stay and the subsequent growth in the mobile phone market not only emphasized its importance but brought out in sharp relief the need for optimization, there being nothing like a serious consumer market for emphasizing subtle differences of approach. An apparently minor change can save millions of dollars when multiplied by sales of some 500 million handsets per annum! and this has meant that, rather than resting on its obvious laurels, the HEMT has been pushed to achieve even better performance. In particular, the competition between the PHEMT on GaAs and the InP-based AlInAs/GaInAs has continued to run, with the various protagonists making ever more exaggerated claims. What is more, though, the technology has also moved on. InP-based devices now employ InGaAs layers with up to 80% In and have, in the process, become pseudomorphic structures themselves, while simple GaAs-based devices still incorporate no more than 30% In. The argument also rages around the question of cost—InP substrates are still considerably more expensive than GaAs, a difference further emphasized by the availability of 6 in diameter GaAs, compared with only 3 in for InP. But the story certainly does not end there—GaAs technology fought back again in the shape of a so-called 'Metamorphic HEMT', which employs a graded AlGaAsSb buffer layer between the substrate and the InGaAs layer, allowing up to 80% In to be used in the active layer and providing performance quite comparable with that of the InP-based device, *together with* a cost advantage! What will InP do now, one wonders? No doubt someone will think of something and this account may well be out of date before publication—I make no attempt to predict the ultimate outcome but continue to watch with considerable interest. For the record, cut-off frequencies are now routinely greater than 300 GHz and amplifiers have been reported at frequencies above 200 GHz. Noise figures are of the order 2 dB at 94 GHz.

Low-noise amplifiers may be the essence of microwave receivers but transmitters require power and the competition is no less fierce in this department. Once again, the InP PHEMT could outperform the straight GaAs-based PHEMT in terms of power added efficiency (PAE) by something like 50% but this particular argument looks like becoming irrelevant as a result of recent developments in SiC and in the Group III nitrides (InN, GaN, and AlN). Both SiC and GaN have a history in the field of light emitting devices which we shall explore in Chapter 7 but it became clear during the 1990s that they also had something of significance to offer in the way of high power microwave transistors, and discussion of these developments obviously belongs here. Both SiC and GaN are relatively wide-gap semiconductors which gives them an advantage with respect to breakdown field, which means that transistors made from these materials can be operated at significantly higher voltages than is the case for GaAs or InP-based devices. It also offers potential for high temperature operation which is of considerable interest in the automobile and aerospace industries. SiC also has a high thermal conductivity which is important for any kind of power device—the inevitably large amounts of heat generated must be conducted away to a convenient heat sink, otherwise the device will be destroyed. As we shall discuss in greater detail in Chapter 7, bulk crystals of GaN are not readily available for use as substrates so GaN structures are frequently grown on SiC substrates, giving them a similar advantage in thermal design. The chief difference between the two materials is that SiC devices are MESFETs or MIS (metal–insulator–semiconductor) devices, whereas the nitrides take the form of AlGaN/GaN HEMTs. At the time of writing, much yet remains to be clarified but, broadly speaking, it seems that SiC is likely to dominate the lower frequency range, up to about 10 GHz, while the nitrides, with their high saturation drift velocity have a clear advantage at higher frequencies. Typical power densities for current devices are 1.5 W mm^{-1} for GaAs at 10 GHz, 3 W mm^{-1} for SiC at 1 GHz, 7 W mm^{-1} for AlGaN/GaN at 10 GHz and 3 W mm^{-1} at 18 GHz. However, there appears to be considerable scope for further development of nitride devices, and it is still too early to say what their ultimate performance might be.

A fascinating feature of the AlGaN/GaN structure, which has excited considerable interest among physicists concerns the piezo-electric induction of electrons in the channel region. The nitrides crystallize in the wurtzite structure which has axial symmetry (rather than the cubic symmetry of the other Group III–V compounds) and they exhibit strong piezo-electric effects when strained. It is particularly significant, therefore that a relatively large lattice mismatch exists between AlN and GaN (approximately 2.7% in the in-plane lattice parameter) which produces strain in the AlGaN layer which, in turn, generates a piezo-electric field. The *change* in electric field at the AlGaN/GaN interface then implies that there must be a layer of charge at the interface—that is, a 2-DEG even in the absence of doping in the AlGaN layer. It is partly this and partly the fact that this material combination provides a very large conduction band offset that leads to extremely large values for the electron sheet density, values as high as $1.5 \times 10^{17}\ m^{-2}$ having been realized fairly routinely (roughly 10 times that available in the earlier devices). This, combined with the high electron drift velocity in GaN, is obviously favourable for obtaining large source-drain currents in power devices which, allied to high voltage operation, explains the potential for the generation of large power densities. Finally, we note that the lattice mismatch between AlN and InN (11%) is considerably larger than that with GaN which gives scope for even greater piezo-electric effects in this material system.

The development of GaN-based optical devices was, as we shall see, an essentially Japanese initiative so the dominance of the Americans in high power nitride device work is particularly striking. Rumour has it that, in the mid-1990s, America felt to be so far behind in the development of blue light emitting diode (LEDs) and laser diodes (LDs) that their funding agencies made a policy decision to concentrate on microwave devices instead—something not being seriously addressed by their rivals! Be this as it may, there can be no doubt in the leading role played by US universities and companies in demonstrating the potential of GaN HEMTs as microwave power sources. Lester Eastman's group at Cornell University (with a long history of microwave semiconductor device expertise, based on MBE growth) has, perhaps, led the way but others followed strongly, Hadis Morkoc, at the University of Illinois, and Steve den Barrs at the University of California, Santa Barbara, in particular. US companies involved include Nitronex, Cree, Sandia, RF Nitro, Sanders, GE, Epitronics, and HRL, names (not surprisingly) almost entirely unheard of during the early development of the semiconductor industry.

The MESFET and the HEMT clearly made the microwave revolution possible, leaving the bipolar transistor trailing in the MHz region as a mere adjunct to the metal oxide silicon (MOS) transistor in logic circuits!

One reason for this was, of course, the early realization that GaAs bipolar devices failed disastrously on account of their extremely short minority carrier lifetimes. Another reason was that, because high frequency transistors required narrow bases, they tended to display undesirably large base resistances. It was therefore necessary to dope the base region to a much higher level than required in low frequency devices and this, in turn, led to poor emitter efficiency. The emitter–base junction in an *n-p-n* transistor must be designed as an n^+-p junction to ensure that, under forward bias, current is carried almost entirely by electrons which are injected from the emitter into the base. Any holes flowing in the reverse direction must represent a tiny fraction of the injected electron flow. Doping the base inevitably increased this hole injection and degraded emitter efficiency, a dilemma which became more acute as the operating frequency increased. Surprisingly, perhaps, the answer was available many years ago. No less a luminary than William Shockley, himself, in a patent issued in 1951, proposed the use of a heterojunction emitter to counter the hole injection, while still allowing the base to be heavily doped. Nothing was done about it then for the simple reason that the technology was not available but Herb Kroemer took up the idea again in 1957 and later (in the 1980s when the technology was available) made a number of specific proposals for its implementation. In particular he proposed using a combination of InGaP/GaAs as emitter–base materials and grading the junction in the interest of improved injection efficiency. (We shall see why this is so). It is particularly interesting that, according to Hilsum and Rose-Innes (1961): 203, the first GaAs bipolar transistors to give gain were actually HBTs in which the emitter was formed from GaP but the first practical HBTs were probably those reported in 1981 by Peter Asbeck and co-workers at North American Rockwell, and recently, following rapid progress in the development of their high frequency capability, there has been a further surge of interest in their use in practical microwave circuits. Several material systems have been used, including one important non-III-V system—Si/SiGe. This latter has the obvious advantage of being compatible with conventional Si technology and therefore demands our attention, even in a chapter devoted largely to III-V compound semiconductors!

Strictly, the HBT is not a low-dimensional device, though it depends for its success on the same technology of hetero-epitaxial growth by MBE or MOVPE so it is convenient to discuss it here, all the more so because it serves as a rival to the HEMT, on which we have concentrated for much of this section. To help focus our ideas, Figure 6.18 illustrates the physical design of a typical transistor, together with a band diagram for the case of an AlGaAs/InGaAs emitter–base junction device. In addition to the important valence band step which counters hole current, two other features of the band diagram are significant: first the presence of a spike in the conduction band, and second the

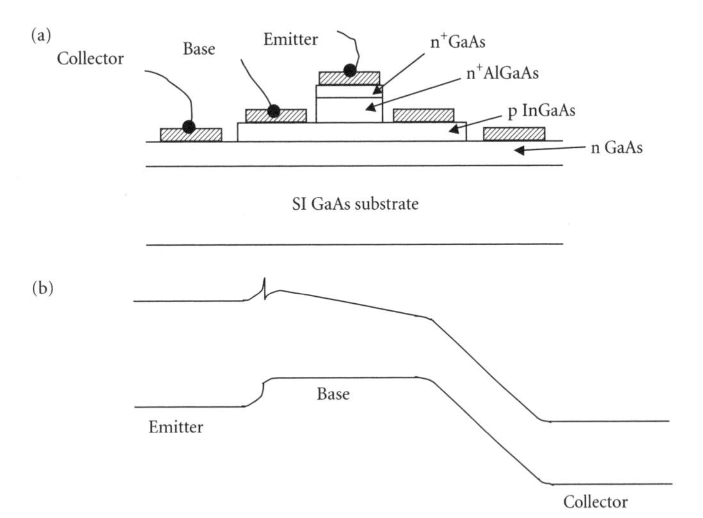

Figure 6.18. Structure and band diagram of a typical AlGaAs/InGaAs HEMT. The emitter/base junction is shown as an abrupt junction which incorporates a conduction band spike. The InGaAs base is graded to provide a built-in electric field to reduce the base transit time. Thicknesses of the various layers are as follows: n^+-GaAs 100, n^+-AlGaAs 250, p-InGaAs 50, and n-GaAs 500 nm.

grading of the base region (i.e. increasing *'x'* in $In_xGa_{1-x}As$ from emitter junction to collector junction). The spike is not altogether desirable because it reduces electron injection into the base but careful compositional grading of the junction region itself can be used to reduce, or even eliminate it. The grading of the base is valuable because it introduces an electric field which accelerates electrons through the base and offers a considerable speed advantage over the normal diffusive transport. (This innovation represents yet another of Herb Kroemer's contributions—it dates from 1954 and was eventually used in 1981!). In practice, it is used to allow the base thickness to be increased without sacrificing transit-time, thus reducing base resistance. The fact that values of f_T as high as 300 GHz have been achieved for graded base devices illustrates the effectiveness of such procedures. (Note that InGaAs is not lattice-matched to GaAs but, because the base is thin and the amount of In required is small, it grows as a pseudomorphic strained layer.)

As with HEMT technology, a number of different material systems have been tried, including the use of $In_{0.5}Ga_{0.5}P$ (lattice-matched to GaAs) as emitter, instead of AlGaAs. This results in a somewhat greater valence band offset and smaller conduction band spike and there are now moves to replace AlGaAs with InGaP in many HEMT designs—though the principal reason appears to be greater reliability and reproducibility. Once again Kroemer is vindicated—he made just such a recommendation in 1983! But, let us be fair, his main argument was based on the assumption of a conduction band–valence band split in the AlAs/GaAs system of 0.85/0.15. Now we know that the correct value is close to 0.67/0.33, this argument carries far less weight (if any?). Another important lattice-matched material system is $InP/In_{0.53}Ga_{0.47}As$ which, as we saw above, has been used with considerable success for HEMTs. The high electron mobility in InGaAs is valuable for HBT operation too, and some of the most impressive performance figures relate to it. However, as with HEMTs, there are commercial arguments against any technology based on InP and it remains to be seen how the balance between financial and technical pressures is resolved. It may well be that most applications can be satisfied with GaAs-based devices apart from a few very high frequency requirements. (But remember here we are writing history, not speculating on futures!). What *is* abundantly clear is that HBTs are finding a niche in many microwave applications, particularly where power is important (the HEMT still has better low-noise performance) and, in this context, it may be well to keep an eye on AlGaN/GaN HBTs, particularly if high operating temperatures are envisioned. There is a problem with base contact resistance at present but if this can be solved —? (Again we are not in the business of forecasting!).

Last, but very far from least, is the Si/SiGe system which has been of interest almost since the early days of Si transistor development. For a long time it proved too difficult to grow satisfactory alloy material but the appearance of MBE and low-pressure chemical vapour deposition (LPCVD) has changed all that and heralded rapid development of HBT devices. Apparently, the first commercial SiGe devices to be produced came from a German firm called Temic (part of Daimler-Benz) early in 1998 but, as a result of Temic's being sold, their market lead was eclipsed by IBM's announcement towards the end of 1998 of SiGe RF circuits operating in the 1–2 GHz range. Though it is still too early to form a clear picture, there now seems every likelihood of SiGe making a strong bid for the 1–10 GHz device market, with the participation of many other manufacturers.

The essential characteristic of SiGe alloys is that they possess band gaps smaller than that of pure Si with the energy step at the Si/SiGe interface being almost entirely in the valence band, making them ideally suitable for the base region in an *n-p-n* Si/SiGe structure. By grading the Ge fraction in the alloy, it is also possible to form a graded base in the manner already described for AlGaAs/InGaAs and, finally, it turns out that SiGe can be doped *p*-type to very high levels in the interest of obtaining low base contact resistance. Nature clearly loves silicon (!) but there *is* one problem—the lattice parameter of Ge is some 4% larger than that of Si so it is necessary to use modest Ge fractions in order to achieve pseudomorphic growth of the alloy film—too much Ge and strain is relieved by dislocation generation with consequent degradation of device properties. However, there is yet more subtlety—for a base width of 50 nm with a fully stable strained base layer, the Ge fraction should be kept below about 10% but layers with as much as 30% Ge remain *metastable* as long as the structure is not subjected to high temperatures. This has led to diverging strategies depending on whether or not it is desired to incorporate SiGe devices on the same chip with conventional Si circuitry. If so, silicon processing temperatures demand the low percentage option, if not, it is possible to use much higher Ge fractions and, incidentally, to employ higher *p*-type doping. The available device performance therefore depends quite markedly on the answer to the above question and it is important to be aware which regime is relevant when making judgement. Both types of device are now commercially available and SiGe will certainly make a strong challenge for some parts of the high frequency market previously thought to belong to the III-Vs. Such is the way of electronics—such has always been its way!

Happily, not all of life's many facets are as viciously competitive as high speed circuitry. We now turn attention to a more gentlemanly and much slower moving aspect of semiconductor development, the application of the quantum Hall effect as a resistance standard. Almost exactly 10 years elapsed between the discovery of the Quantum Hall Effect in 1980 and its formal acceptance as a resistance standard on 1 January 1990. Nor should it be supposed that this represents the end of the quest—work is still very much in progress to perfect appropriate techniques. Some of the early work is described in a series of papers published in volume 34 of the *IEEE Transactions on Instrumentation and Measurement* in 1985, a useful introductory account is available in Hartland 1988, while recent developments can be found in the paper by Witt 1998.

Prior to the acceptance of the quantum Hall resistance (QHR), resistance standards took the form of wire-wound resistors immersed in temperature-controlled containers, the wire being either manganin or nichrome alloys in the interest of achieving a low temperature coefficient of resistivity. These temperature coefficients were in the region of $10^{-6}\ \mathrm{K}^{-1}$ and the temperature was stabilized at 'room temperature' to an accuracy of 1 mK (though 'room temperature' meant slightly different things to different laboratories—anything between 20°C and 28°C). Usual values were $R = 1\ \Omega$ and $R = 10\ \mathrm{k}\Omega$. Much study had been made on the detailed properties of these resistors. They were found, for instance, to drift monotonically with time at a rate of order one part in 10^7 per annum! There was evidence of a hysteresis behaviour with temperature cycling. There were clear differences in measured values as a function of frequency when the measurements were made with alternating currents, rather than DC—the equivalent circuit contained, not surprisingly, a series inductance and a shunting capacitor. All these features had to be allowed for, when comparing any secondary standards with the primary versions held in standards laboratories round the world. Indeed, the question of accurate comparison was (and is) central to the whole business, a challenge which has stimulated the development of extremely accurate bridges, capable of comparing two resistors to within a few parts in 10^9. Clearly, it was essential for at least some of this equipment to be transportable between different laboratories in the interest of maintaining worldwide standards. The net result of this earnest endeavour was agreed standards with an accuracy of order 1 part in 10^7.

It was scarcely surprising that the discovery of the quantum Hall effect (QHE) by von Klitzing et al. in 1980 sparked an immediate interest in standards laboratories across the globe. Their obvious question was: can this new effect provide a standard which is independent of individual Hall samples and give an accuracy at least as good as that currently available. Bearing in mind, too, that the measurement involved some quite esoteric and expensive equipment, it was hoped that it would actually offer real improvement in accuracy and reliability. An international comparison of standards, calibrated against QHR measurements was undertaken in 1990 when it was found that the mean of the various standards agreed with QHR measurements within a few parts in 10^8. Comparison between various different QHR measurements in different laboratories (which involved the use of a transportable QHR equipment) showed that, provided certain guidelines were followed, the different values of R_H agreed within a few parts in 10^9 which was a very satisfactory two orders of magnitude better than the original accuracy available from wire-wound resistors. Care was necessary to use only moderate measurement current because the QHE breaks down at high current, it was necessary to measure both the $i = 2$ and $i = 4$ plateaux to check that leakage currents were minimal, measurement temperatures were to be kept low (preferably 1.3K) and it was important to check the value of the longitudinal resistance R_{xx} (within a Hall plateau, this should be as close to zero as possible). Of course, it was still necessary to use the QHR to calibrate standard resistors and this involved several steps—the QHR was first compared with a resistor of value close to $R_H(i)$ ($i = 2$ or 4), then this resistor was compared with a 100-Ω resistor which, finally, was compared with the standard 1-Ω resistor. An overall accuracy of two parts in 10^9 was obtained in one particular experiment.

Prior to 1992, all measurements were made at DC (or, in one case, at $f = 4$ Hz) but, in view of the trend to using a calibrated 'calculable capacitor' as an impedance standard, it has become desirable to perform measurements at kilohertz frequencies, so there is now considerable interest in making QHR measurements with AC, rather than DC. At the time of writing, it appears the AC (1.6 kHz) and DC measurements agree within one part in 10^8, though there is little theoretical help towards understanding how closely these measurements can be expected to agree. Further progress, though measured, can be anticipated with confidence.

6.6 Optical devices

It is a moot point whether the introduction of LDS systems had a greater influence on electronic or optical devices (and it probably does not matter very much!). In any event, we shall close this chapter by discussing two quite different optical devices which *have* clearly benefited.

Much the more significant of these is the semiconductor laser which has been considerably improved by the introduction of LDS versions and has seen a rapid growth in market, while the optical modulator is, perhaps, slightly more of a scientific curiosity.

In Chapter 5 we took the laser story to the point in 1970 where CW operation at room temperature had been achieved, with threshold current densities of about 2 kA cm^{-2}, and in 1973 where the separate confinement DH laser had further reduced threshold currents to 500 A cm^{-2} (compared with the figure of 50 kA cm^{-2} measured on the original homojunction devices in 1962). The virtues of the good lattice match between AlAs and GaAs were clearly demonstrated by these figures but this was by no means the end of the development story—the introduction of quantum wells was eventually to lead to $J_{th} \sim 50$ A cm^{-2}, three orders of magnitude smaller than those found for homojunction lasers in 1962, a truly remarkable improvement over a period of less than 20 years. This, together with development of the vertical cavity laser in the 1990s has revolutionized the commercial position of the semiconductor laser almost beyond belief and led to rapidly expanding markets, currently about $4 billion but growing rapidly. Broadly speaking, we can divide the market among high power lasers (mainly GaAs), optical communication lasers (mainly InGaAsP), and optical disc (read and write) lasers (GaAs, InGaP, and GaN). We shall deal with the GaAs QW laser in this section, leaving other materials until the appropriate chapters—InGaP and GaN in Chapter 7 and InGaAsP communications lasers in Chapter 8.

The first reports of laser emission from AlGaAs/GaAs QW structures appear to be those from Bell Labs in the period 1975–6. Miller et al. and van der Ziel et al. observed laser action in optically pumped structures which did not contain a *p-n* junction, it only being necessary to add a pair of reflecting mirror facets to a standard QW structure, originally used for photoluminescence studies. Though it is hardly surprising that these reports should emanate from Bell, where the first luminescence studies on quantum wells were made in 1974, the speed with which it took place was certainly to be admired. The samples were, as we have seen, grown by MBE which was, by that time, a well developed capability at Murray Hill, as illustrated by the fact that Cho and Casey had already used MBE to grow DH injection lasers showing performance close to that obtained from liquid phase epitaxy material. Other labs which might have wished to study QW structures were prevented by their lack of suitable growth facilities, though 3 years later Kolbas et al. from the University of Illinois reported similar optical pumping experiments on samples grown by the MOVPE method. This was particularly interesting because it heralded a long-fought battle for supremacy between the two growth methods which lingered on well into the 1990s, when it finally became clear that there really was nothing to

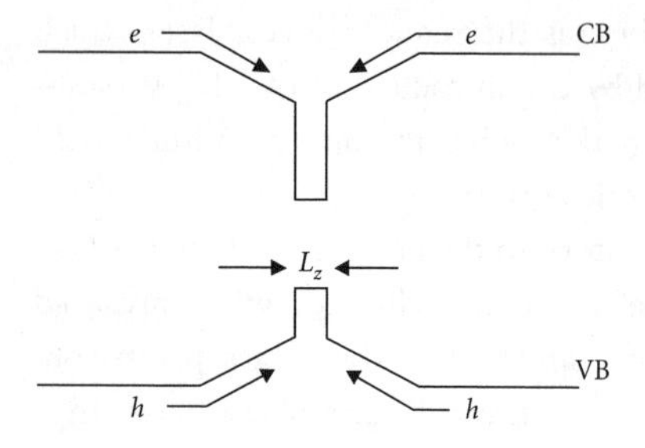

Figure 6.19. Schematic band diagram to illustrate the use of a GRINSCH structure to improve carrier capture into a quantum well. The graded AlGaAs layer acts rather like a funnel to accelerate both electrons and holes into the well, where they recombine.

choose between them—both techniques were capable of growing good lasers for very much the same financial investment!

Yet another MBE activity was established at Murray Hill by Won Tsang who, in 1979, reported threshold current densities of 2 kA cm^{-2} from AlGaAs/GaAs *p-n* junction MQW structures, comparable to the values obtained on conventional DH lasers at that time. It was quickly appreciated that one problem with the use of quantum wells lay in the poor capture probability for minority carriers into the wells and Tsang (1981–2) seems to have been first to find a solution, in the form of the GRINSCH structure, which horrible-sounding acronym stands for 'graded index separate confinement heterostructure'. A representative band structure is shown in Figure 6.19, where it will be seen that the region of graded band gap material serves like an electronic funnel to draw carriers of both persuasions into the central well. It clearly worked—Tsang reported threshold current densities as low as 250 A cm^{-2} from such devices and really put the quantum well laser on the opto-electronic map.

But all this ignores the important question: why should anyone wish to explore the use of quantum wells in lasers in the first place? The answer was clear from the very early days: quantum well lasers were expected to have properties differing significantly from those of the conventional DH laser. In a short review of QW properties in 1975, Ray Dingle pointed out that, because the lasing transition was associated with the $n = 1$ quantum confined states, QW lasers could be tuned by the simple means of altering the well width L_z, while, as we show in Box 6.5, the gain is expected to remain unchanged. He also suggested that they would show important differences in current-gain characteristic as a direct result of the step-like density of states (see Figure 6.15) in both valence and conduction bands. In particular, it turns out that this feature leads to lower threshold current density on account of its giving higher optical gain (see Box 6.4 for an explanation) and, furthermore, because the two-dimensional density of states varies linearly with temperature, compared with a $T^{3/2}$ dependence in three-dimensions, the temperature-dependence of J_{th} is reduced, an important advantage in many applications. Nor should we overlook the rather obvious factor that use of quantum wells leads to reduced threshold current density, simply on the grounds that the *volume* of material being pumped is greatly reduced. This also implies, of course, that threshold current is correspondingly lower, which may be an important advantage, but we should recognize, on the debit side, that the output power is also correspondingly less. Clearly, the way in which one regarded these features depended on the requirements for any particular application but it was clear that QW lasers promised several possible advantages compared with their three-dimensional counterparts. What, then, of their realization?

Box 6.4. Luminescence line widths

The width of luminescence emission lines from semiconductors depends on the two parameters, temperature and conduction (valence) band density of states. As an illustrative example, we shall assume that the optical transition involved is one in which an electron in the conduction band recombines with a hole trapped on an acceptor level (see Figure 6.20). There is then no dispersion in the hole states (i.e. all the hole states are at the same energy). For the case of a bulk semiconductor, the conduction band density of states $N(E)$ varies as $(E - E_C)^{1/2}$, where E_C represents the bottom of the conduction band. In a luminescence experiment, electrons are injected into the conduction band and then rapidly *thermalize* towards the bottom of the band so that they occupy states up to energies of order kT from the bottom. In fact, we can write the thermal distribution of electron energies as:

$$\begin{aligned} n(E) &= N(E)\exp\{-(E-E_C)/kT\}, \\ &= A(E-E_C)^{1/2}\exp\{-(E-E_C)/kT\}, \\ &= A(kT)^{1/2}x^{1/2}e^{-x}, \end{aligned} \tag{B6.22}$$

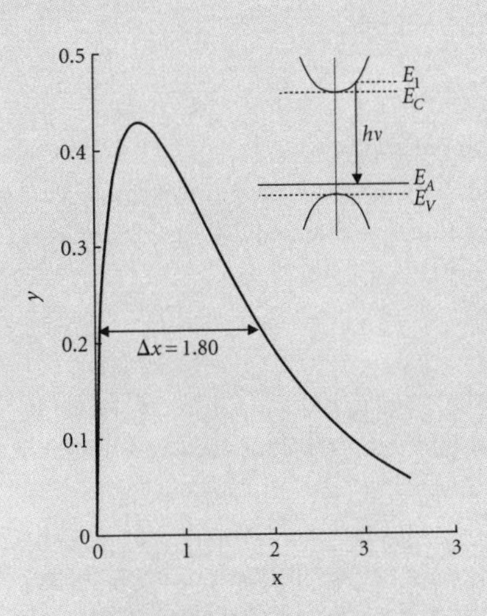

Figure 6.20. The line shape function for a conduction band-to-acceptor emission line in bulk GaAs (or other direct gap material). The parameter $x = (E - E_C)/kT$ and the emission intensity as a function of energy is proportional to $y = x^{1/2}e^{-x}$. As can be seen, the peak of the emission occurs at $x = 1/2$ and the half-height full width is $\Delta x = 1.80$. Thus, the peak of the emission line energy is $hv_{pk} = (E_C - E_A) + kT/2$. These results are based on the assumption of a conduction band density of states varying as $N(E) = A(E - E_C)^{1/2}$, which is appropriate to a pure, defect-free sem.iconductor (and is a good approximation to many real situations).

where $x = (E - E_C)/kT$. The shape of the emission line will then follow this function which is plotted in Figure 6.20. As can be seen, it is an asymmetric function with a peak value at $x = 1/2$ and a half-height width of $\Delta x = 1.80$, i.e. $\Delta E = 1.80$ kT. This dependence on T is a reflection of the fact that electrons occupy states up to energies of the order kT above E_C. At room temperature it implies a luminescence linewidth of about 47 meV, at 4K of about 0.62 meV. There may, of course, be other contributions to the linewidth which, particularly at low temperatures, mask the thermal contribution. What we should say, is that 1.80 kT represents the minimum possible width for this type of transition. Note that a similar, though slightly more involved, calculation shows that a conduction band-to-valence band transition has a shape function of xe^{-x}, rather than $x^{1/2}e^{-x}$, which peaks at $x = 1$ and has a half-height width of $\Delta x = 2.45$.

In the case of emission from a quantum well, the density of states function differs in that it is constant (i.e. independent of energy) above each confined energy step. In this case, the line shape function has a sharp cut-off on the low-energy side and an exponential tail to higher energies. The half width is easily seen to be $\Delta x = 0.693$, giving $\Delta E = 18$ meV at room temperature, about 3.5 times smaller than a band-to-band transition in bulk material (though, in practice, QW emission is usually dominated by exciton recombination). We should emphasize here that a very similar argument applies to the gain spectrum of a QW laser which is correspondingly narrower than that of a bulk laser. This implies that the value of injected carrier density required to reach a particular *peak* gain is significantly lower and this implies a lower threshold current density, a major reason why QW lasers have attracted so much attention.

Box 6.5. The QW laser

To understand the behaviour of QW lasers we must look carefully at a simple model of the interaction between the injected free carrier densities and the light wave within the optical cavity. Figure 6.21 is designed to help. We make the assumption that the optical gain occurs within a single quantum well of width L_z which is situated at the centre of an optical waveguide of width W and, for simplicity, we assume the waveguide to support a plane wave which propagates between the ends of the optical cavity of length L. A fraction R of the light intensity is reflected back from each cavity mirror, while $(1 - R)$ is transmitted to the outside world. In Box 5.4 we showed that the threshold condition for lasing to begin is given by:

$$G = \alpha + (1/L)\ln(1/R), \tag{B6.23}$$

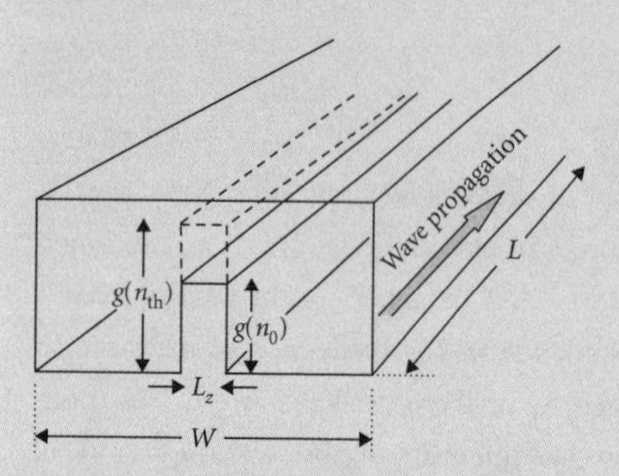

Figure 6.21. Illustration of the relative positioning of the gain region within the optical waveguide in a single QW laser. Assuming the light wave propagates as a plane wave, the filling factor is readily seen to be $\Gamma = L_z/W$, where W is the width of the waveguide. L is the cavity length, light being partially reflected from the mirror facets at each end.

where G is known as the 'modal gain', α represents the internal cavity losses and the final term represents the loss by way of emission from the cavity end facets. G is an effective average gain across the whole of the waveguide cross-section. Because the actual gain (or 'local' gain) g occurs only over the quantum well, we define a 'filling factor' Γ such that G is given by

$$G = g\Gamma. \tag{B6.24}$$

It is clear from Figure 6.21 that, on our simple model, $\Gamma = L_z W^{-1}$ and is a rather small number. In practice $L_z \sim 10$ nm, while $W \sim 1$ μm so $\Gamma \sim 0.01$.

In elementary models of laser action, it is assumed that the local gain g is a linear function of the injected carrier density n. Thus

$$g = a(n_{th} - n_0), \tag{B6.25}$$

where n_{th} is the value of n at threshold and n_0 is the transparency value (discussed in Box 5.4). The proportionality factor 'a' is constant across the width of the QW. Using this expression, we can obtain an equation for the relationship between the modal gain G and the threshold current density J_{th}. Let us examine the flow of current into and out of a unit area of the well just below threshold. The input is simply the current density J, while the outgoing flux is due to electron–hole recombination. Various recombination processes may occur but it is usually the case that the recombination lifetime is largely determined by non-radiative recombination, for which we can write the rate as An (where A is a constant). There is also a significant amount of radiative recombination (spontaneous emission) but having a rate much smaller than An. (In other words, the radiative efficiency is low.) On this model, the recombination current density is given by $J_R = edn_s/dt = eAnL_z$ so we can write, for any value of J:

$$J = eAnL_z. \tag{B6.26}$$

Finally, using (B6.25) and (B6.26), we arrive at the relation between G and J:

$$G = g\Gamma = (aL_z/WeAL_z)(J_{th} - J_0)$$
$$= (a/WeA)(J_{th} - J_0). \tag{B6.27}$$

We make no pretence of calculating a numerical value for G from equation (B6.27) but simply draw the important conclusion that G is predicted to be independent of well width. This implies that we should be able to tune the laser wavelength by varying L_z, without affecting the gain (and, hence, the threshold current). Notice that this conclusion depends on the assumption that A can be taken as constant, independent of well width—in other words, that we can treat recombination within the well as though it were bulk recombination. As we comment in the text, this may not necessarily be true.

The most straightforward aspect to be demonstrated was the reduction of lasing wavelength with decreasing QW width. This followed approximately the behaviour predicted from the increase in confinement energy as L_z was reduced, a result dramatically illustrated in 1984 by work at the Philips Labs in Redhill—Karl Woodbridge and co-workers observed visible (707 nm) laser emission from a 1.3 nm GaAs QW, the effective band gap of the GaAs having been increased by as much as 320 meV simply by making it thin enough! The same effect could, in principle, have been achieved by using $Al_{0.26}Ga_{0.74}As$ as active material in a conventional DH laser but this overlooks the fact that it is difficult to grow AlGaAs of sufficient quality—Al tends always to getter oxygen, and oxygen in AlGaAs forms a deep level which acts as a recombination centre and strongly reduces the radiative efficiency, whereas a GaAs QW suffers this degradation to only a minor degree (because the confined carrier wavefunctions penetrate some little way into the AlGaAs barrier). A compromise lay in the use of a modest Al content in a QW active layer, and wavelengths as short as 650 nm were obtained in this manner. However, in either case, there was clear evidence that, as the emission wavelength decreased, the measured threshold current increased, in contradiction to the prediction of simple models (Box 6.5). Two factors come into play: first, the assumption of a stepped density of states is probably an oversimplification—we have already indicated that real interfaces are not infinitely sharp and it turns out that well widths are not quite constant over the plane of the well, a variation of one to two monolayers being quite usual, and this has the effect of *inhomogeneously* broadening the confinement energy (the energy takes a slightly different value at different 'geographical' positions). In other words, the step edges are blurred and the calculated emission linewidth calculated in Box 6.4 is an underestimate of the actual linewidth. This, in turn leads to the actual threshold current being greater than the value calculated for the ideal model. What is more, because the variation in L_z of 1–2 ML is more significant for *narrow* wells, the effect becomes greater as L_z (and λ) become smaller. Second, because the carrier wavefunctions penetrate (relatively) further into the barriers as L_z is reduced, the probability of non-radiative recombination in the barrier material increases and our assumption in Box 6.4 of a constant value for 'A' breaks down. In fact, if recombination were to be *dominated* by barrier recombination, 'A' would be proportional to L_z^{-1}, leading to a threshold current density J_{th} which also varies as L_z^{-1}. It is hardly surprising, therefore, that J_{th} increases with decreasing wavelength in real QW lasers.

The predicted advantage of improved temperature-dependence of J_{th} is actually observed for lasers with wide wells ($L_z \sim 10$ nm) but, again, when L_z is reduced to achieve shorter wavelengths, the improvement tends to disappear. The effect may be quite complex (like most aspects of

QW laser performance!) but is probably dominated by thermal emission of carriers from the wells, which accentuates recombination in the barrier region. As L_z is reduced, the confined electron states move upwards, towards the top of the well, making it so much easier for thermal emission to occur.

Another feature of QW lasers which appears to be supported by experiment is the large differential gain of the device above threshold (i.e. a large value of the parameter 'a' in equation B6.25). This is a function of the large density of states at the bottom of the two-dimensional conduction band (and top of the two-dimensional valence band) and, though the effect is somewhat reduced by the broadening effect referred to above, it is still significant, at least for wide quantum wells. Its main consequences are two: first, the high gain above threshold leads to improved efficiency of the laser at *high* currents (and therefore *high* power output). QW lasers have shown excellent performance at high power levels and have influenced the development of high power devices which now challenge CO_2 gas lasers in many applications. Second, the high slope efficiency has led to high frequencies (~30 GHz) for the 'relaxation oscillations' which are an intrinsic feature of all semiconductor lasers. This is important because it allows QW lasers to be modulated at significantly higher frequencies than their bulk rivals, an important consideration in light of the ever-increasing bit rates used in today's optoelectronic systems.

While the QW laser has already proved its worth, laser development has not stopped there. It was recognized quite early that, if quantum wells could offer improved performance over bulk material, quantum wires and dots might offer even more. This conclusion follows directly from the way in which the density of states (DOS) changes with further confinement of carriers. We are already familiar with the stepped DOS appropriate to two-dimensional confinement and (see Box 6.4) with the resulting reduction in linewidth of the optical transitions involved in laser emission—in principle, we can expect even greater advantages to accrue from further localization. The DOS function associated with a quantum wire consists of a sharp step at $E = E_i$ on the low-energy side, followed by a fairly rapid fall-off on the high energy side ($N(E) \propto (E - E_i)^{-1/2}$) ($E_i$ being the energy corresponding to the bottom of the appropriate one-dimensional energy band) and this fairly narrow range of energies significantly sharpens the associated optical transitions, leading to enhanced local gain and correspondingly lower threshold current. This is even more the case for a quantum dot, where the DOS is a sharp, monoenergetic level. Indeed, in calculations published by Asada et al. in 1986, it was predicted that J_{th} for bulk, well, wire, and dot structures should scale roughly as 1000 A cm^{-2}, 400 A cm^{-2}, 150 A cm^{-2}, and 50 A cm^{-2}, illustrating the considerable advantages inherent in the use of increased confinement. This could also be expected to ally with reducing temperature-dependence in so far as DOS $\propto T^{3/2}$, T, $T^{1/2}$, and 'constant', respectively. In particular, for this simple model, there should be no temperature-dependence for the QD laser threshold current. These results apply, of course, to an ideal situation in which it is possible to grow wires or dots with uniform physical dimensions. We have already seen that this is not quite true even for wells, so it is no more likely to be true for the smaller structures—indeed it is even less likely for these, on account of their greater complexity. So it proved. The inhomogeneous broadening associated with the almost inevitable spread in size of these fine structures has made life extremely difficult for anyone attempting to use them in practical devices. As we saw earlier, the highest degree of uniformity obtained with dots was about 10%, corresponding to luminescence linewidths of the order 30–50 meV, slightly greater than kT at room temperature and comparable to widths in bulk material. Similar remarks can therefore be made concerning the width of the gain spectrum in a laser. Clearly, the technological challenge of growing uniform wire or dot structures was a major one and progress was correspondingly less rapid than was the case with QW lasers.

Having said this, progress has certainly not been totally lacking. The first luminescence measurements on quantum dots were made in 1986 but it took another 8 years before a QD laser was made to operate CW at room temperature, the satisfaction falling to what, at first sight, appears to be the somewhat surprising combination of groups at the Ioffe Institute in St Petersburg, in Berlin and Weinberg, Germany. In fact, it reflected the strongly felt need for western scientists to come to the aid of their Russian colleagues, left in dire financial straits following the end of communist rule in the late 1980s. Several collaborative schemes were set up at this time and the first QD laser was one of the early fruits of such collaborative ventures. It was a happy feature, too, that 1 of the 13 authors was none other than Zhores Alferov who was intimately concerned with the invention of the DH laser back in 1970 and who was honoured with the Nobel Prize in 2000. The laser in question was based on self-organized InAs quantum dots in an AlGaAs/GaAs structure, and emitted radiation at a wavelength close to 970 nm and registered a threshold current density of just under 1 kA cm^{-2}, not very competitive with available QW lasers but undoubtedly an exciting first shot. Better was to come—the same combination reported in 1999 an InAs QD laser operating at a wavelength of 1.3 μm with a threshold current as low as 65 A cm^{-2}, now very competitive with QW lasers, and workers at the University of New Mexico in Alburquerque measured J_{th} = 24 A cm^{-2} on a similar structure at temperatures up to 100°C. This device also showed an admirably low temperature-dependence, as predicted by theory. After many years of effort, there was now real hope that QD lasers would make a serious impact on ultimate performance. It is a constantly changing scene and we can only wait in hope for further improvements.

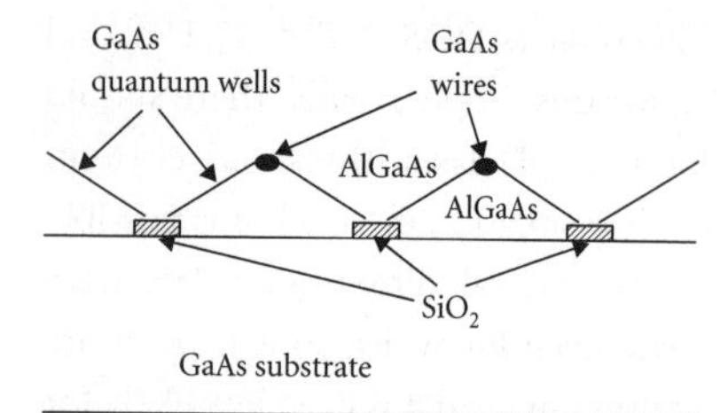

Figure 6.22. Outline of the technique used by Arakawa to grow arrays of GaAs quantum wires. The GaAs substrate was first patterned with stripes of silicon oxide, then a layer of AlGaAs grown in the shape of a pitched roof (the result of the SiO_2 modifying the growth conditions). A thin GaAs quantum well was then grown on top of this, followed by a final cladding layer of AlGaAs. At the apex of the "roof" the well is somewhat thicker than elsewhere and electrons within the well become localized at these thicker regions which therefore behave as quantum wires. (Note that the confinement energy within the well is smaller in these thicker regions so the electron energy is correspondingly lower.)

In the meantime, quantum wire lasers had also made progress. The same problems of obtaining adequate uniformity applied here too and it needed a major breakthrough in material growth before results of practical significance could be achieved. The early work on QWR lasers was based on the formation of wires at the apex of V-grooves on patterned substrates, a typical form of the technology (described by Arakawa in 1994) being illustrated in Figure 6.22. Silicon oxide stripes were deposited on the GaAs substrate to modify the subsequent growth of an AlGaAs film, the result being a series of triangular section stripes. Growth of a GaAs quantum well on top of this produced a local region of large well width at the apex of the stripes, where the free carriers were localized. It was clear that the luminescence observed originated from these wires but uniformity was disappointing, linewidths being of the order 100 meV. Another difficulty was concerned with the rather small number of wires (typically 100) which could be incorporated into a laser structure. Greater success attended an alternative technique, which depended on the growth of quantum wells on 'high index' planes. An example of this was reported by Higasiwaki and co-workers at Osaka University in Japan who grew GaAs QWs by MBE on (775)B-orientated GaAs substrates. They reported the growth of QWRs in 1997, followed by laser diodes in 1999. Conventionally, MBE films are grown on (1 0 0) crystal planes which provide a uniform growth rate over the whole surface but these high index surfaces are characterized by different growth rates in different crystal directions, resulting in corrugated surfaces, as shown in Figure 6.22. The crucial feature seems to be that growth of AlGaAs results in more-or-less flat surfaces, while GaAs grows with a corrugated surface (this is related to the different surface mobilities of Al and Ga). Higasiwaki et al. were able to grow AlGaAs/GaAs quantum wells with a periodic variation in well thickness, the period being 12 nm and the maximum and minimum thicknesses being 2.7 nm and 1.5 nm. Carriers were confined in the region of the well where it was thickest, forming wires normal to the plane of the diagram. It was also found possible to grow self-organized stacks of these wires, allowing upwards of 10^4 wires to be contained in a single laser structure. Luminescence emission was at 670 nm (a very short wavelength for a GaAs well), with a linewidth (at low temperature) of 15 meV, illustrating the high degree of uniformity achieved. The best laser thresholds were about 1.5 kA cm^{-2} (which is quite a respectable figure for such a short wavelength) and the temperature-dependence shows promise, being less than observed on QW lasers made at the same time on (1 0 0) substrates.

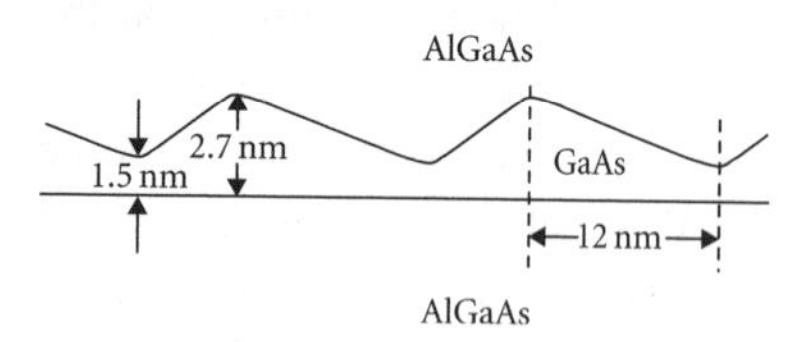

Figure 6.23. 'Corrugated' MBE growth of a GaAs quantum well on the high index (775)B plane of a GaAs substrate. AlGaAs tends to grow with more or less smooth surfaces, whereas the GaAs thickness varies according to the position of steps on the high index surface. Electrons are localized in the thicker parts of the well, which results in wire-like behaviour.

Low-dimensional semiconductor laser devices have certainly made a tremendous contribution to improved laser performance, from low-power devices with threshold currents as low as 1 mA (suitable for incorporation as signal links in high speed integrated circuits) to high power devices capable of cutting and machining steel. Single GaAs lasers are now able to deliver output powers of order 10–15 W, while phased arrays of lasers have been reported to produce as much as 150 W. A specific application for high power devices is that of pumping the erbium fibre amplifiers being used in increasing numbers in fibre-optic communication systems. The incorporation of Er as a dopant in glass fibre allows the fibre to function as a laser which may be used as a low-noise amplifier in repeater stages in long haul communication systems, and is absolutely vital to submarine cables. However, the erbium laser has to be pumped optically at a wavelength of either 980 nm or 1480 nm, with some preference for the former. To generate 980 nm emission demands InGaAs, rather than GaAs, but usually in the form of a strained layer quantum well. The trick is to deliver as much power as possible (of the order of 1 W) into the end of an optical fibre—more pump power means higher inversion and improved noise performance—improved noise performance means repeaters can be spaced further apart, which is, needless to say, a considerable commercial virtue! The market for these pump lasers runs at about $0.5 billion and is growing rapidly. Another important market for shorter wavelength (e.g. 780 nm) lasers is in laser printers, where a mechanical scanning system scans the laser beam across a page to induce an electric charge 'image' of the required information which is then turned into visible form by electrostatic attraction of toner particles. This market, too, is growing, though it remains an order of magnitude less than that for pump lasers and pales into insignificance when compared with the demand for read and write lasers for optical disc storage which is already well past the $1 billion mark.

All the lasers we have discussed so far have been stripe lasers—that is, they are made in the form of long, thin bricks, with the light being emitted through the ends of the brick which form the mirrors of the optical cavity. Typical dimensions are $500 \times 10 \times 2$ μm, though the whole structure is grown on a conducting substrate (e.g. Si-doped GaAs), which may be 200 μm thick. This structure is then mounted on a copper heat sink which, in turn, is mounted on some form of support which allows electrical connection to top and bottom of the brick (e.g. a transistor header). Lasers are produced in large numbers on a single slice of GaAs (or other suitable substrate) and must then be separated into individual devices by scribing and cleaving. Such technology is convenient for the manufacture of laser diodes to be used individually (for instance, in a CD player) but is less suitable for applications which call for arrays of devices or even for individual devices which are required to emit light normal to the plane of the semiconductor slice—for this, we require a 'vertical cavity surface emitting laser' or VCSEL, which is made using a planar process, compatible with the fabrication of large arrays. Applications of such devices include parallel signal processing where the signals are represented as parallel beams of light, rather than as electric currents, or high speed optical coupling of signals between integrated circuits (using a laser on the first circuit and a photo-detector on the second). A further advantage of the VCSEL is its relatively small cross-sectional area, which minimizes the operating current, and, yet again, its circular emission pattern which is immediately compatible with coupling into optical fibres (the conventional laser produces a strongly elliptical output pattern which must be corrected with a suitable lens). Interest in, and demand for such lasers grew rapidly during the 1990s and involved considerable development effort. Current markets for VCSELs amount to about $1 billion and are projected to exceed 10 times this level by the year 2010.

The VCSEL is rather more than a conventional laser turned on its end! We first need to recognize that any vertical structure grown by some form of epitaxy is limited in thickness to only a few microns, thus severely limiting cavity length. It follows, therefore, that, if the gain is similar to that available in a conventional laser, the mirrors which form the optical cavity must have significantly greater reflectivity. We can easily quantify this by referring to equation (B6.23). It is easier if we rewrite it in the form:

$$R = \exp\{-L(G - \alpha)\}, \tag{6.15}$$

where R is the mirror reflectivity, L the cavity length, G the modal gain, and α the loss. In a VCSEL the filling factor Γ (see Box 6.5) is close to unity, so the modal gain is equal to the local gain g, which may typically be of the order $10^5\ \mathrm{m}^{-1}$. Assuming $G \gg \alpha$, and putting $L = 1\ \mu\mathrm{m}$, we find that $L(G - \alpha) = 10^{-1}$, so we may write equation (6.15) as:

$$R \approx 1 - L(G - \alpha). \tag{6.16}$$

Thus, we obtain the result that $R = 1 - 0.1 = 0.9$. This is clearly much larger than the value of $R = 0.3$ which is typical of a cleaved laser facet used as the partial mirror in a conventional design. To obtain such a reflectivity suggests the use of metal reflectors and this is exactly what Soda et al. did in 1979 when they demonstrated the very first VCSEL. (The semiconductor laser, as such, may have been predominantly an American invention but the VCSEL was equally a Japanese invention, the idea having been mooted as early as 1977 by Iga et al. of the Tokyo Institute of Technology.) The active region in this case was a 1.8-μm thick layer of GaInAsP (sandwiched between InP contact layers) which showed pulsed laser action at 77K with the rather formidable threshold current density of 11 kA cm^{-2}, resulting from the use of such a large active volume—$V = 4 \times 10^{-15}\ \mathrm{m}^3$, compared with about $6 \times 10^{-16}\ \mathrm{m}^3$ for a typical DH stripe laser ($300 \times 10 \times 0.2\ \mu\mathrm{m}$) or $3 \times 10^{-17}\ \mathrm{m}^3$ for a 10 nm single quantum well device. (But remember the first homojunction lasers operated at about 50 kA cm^{-2}.) The way forward lay in two directions,

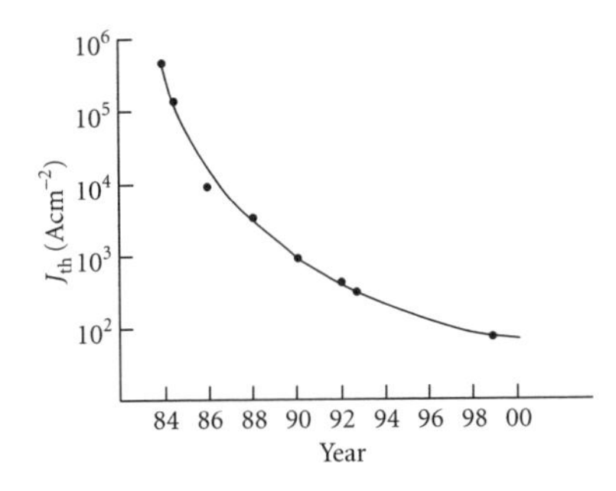

Figure 6.24. Plot of measured threshold current densities reported by selected VCSEL groups during the period 1984–99. As can be seen, J_{th} came down by slightly more than three orders of magnitude over the 15 years, a rate of progress even more dramatic than that experienced by stripe lasers over the period 1965–80 when J_{th} improved from about 50 kA cm^{-2} to 500 A cm^{-2}.

first, to change the active material to GaAs (and the contacts to AlGaAs), and second, to reduce the active volume, the logical conclusion being, again, a single 10-nm well. However, there would be a price to pay for this. Returning to equation (6.16) makes clear that, if $L = 10$ nm, the reflectivity required becomes $R = 0.999$—that is 99.9%! and even gold mirrors, having maximum reflectivity of about 98%, balk at such a challenge. Life was not going to be that simple!

Nevertheless, progress *was* made, as can be seen by looking at the reduction of threshold current density over the period 1984–99 (see Figure 6.24). In 1984 the Tokyo group demonstrated pulsed room temperature operation of an AlGaAs/GaAs/AlGaAs structure with gold mirrors but again with high threshold current ($\sim 10^5$ A cm^{-2}) due to the large thickness (2.5 μm) of active material. The emission wavelength was 874 nm and the far-field pattern was circular, with an angle of divergence of less than 10°. They also showed that the separation between adjacent longitudinal modes of the laser was $\Delta\lambda = 11$ μm, consistent with the 7 μm length of the optical cavity and implying that it would be relatively easy to obtain single-mode operation. Things were coming together but it was still essential to reduce the threshold current before the device could be taken seriously by anyone looking for practical applications. Both threshold current and current density could be reduced by reducing the active layer thickness, while reduction of the device area would also reduce the threshold current but not, of course, the current density. Needless to say, reducing the device current inevitably reduces the light power, though many applications require only modest powers and so this may not be serious.

As pointed out above, thinner active layers lead to a reduction in total gain which must be balanced by higher mirror reflectivity. In the limit of a single quantum well, this demanded a totally new approach to mirror design and it very soon became common practice to use multilayer Bragg reflectors, rather than metal mirrors. Figure 6.25(a) illustrates the principle, using the example of an AlAs/GaAs stack, which may conveniently be grown by MBE or MOVPE. At each interface there is a change in refractive index which gives rise to a small amount of reflection and, as indicated in the figure, if the thickness of each layer is a quarter wavelength of the light being emitted from the active layer, the reflected waves from adjacent interfaces will interfere constructively. In a multilayer stack this will be true for every pair of interfaces so, even though only about 0.6% is reflected at each interface, the overall reflectivity rises quite strongly as the number of pairs of layers increases (see Figure 6.25(b)). To achieve 99.9% reflection requires roughly 30 AlAs/GaAs pairs, each layer being approximately 80 nm thick (a total thickness of about 5 μm).

Having reached this point, it is high time that we looked at a typical VCSEL structure employing Bragg mirrors, such as used by Geels et al. from Larry Coldren's group at the University of California, Santa Barbara

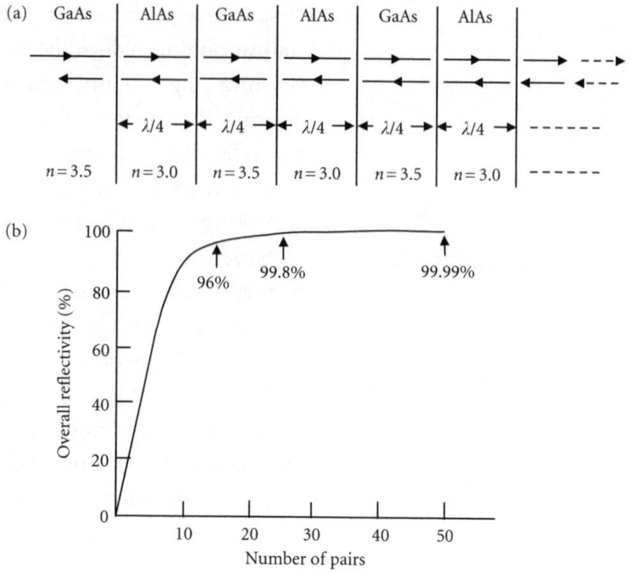

Figure 6.25. Schematic diagram of an AlAs/GaAs multilayer Bragg mirror. Light incident from the left is partially reflected at each interface, with a phase change of π at each AlAs/GaAs step (where the refractive index step is positive), while without phase change at each GaAs/AlAs step (where it is negative). When the layers are each a quarter wavelength thick, light waves reflected from each interface interfere constructively and the overall reflectivity is a strong function of the number of AlAs/GaAs pairs, as can be seen in Figure 6.25(b).

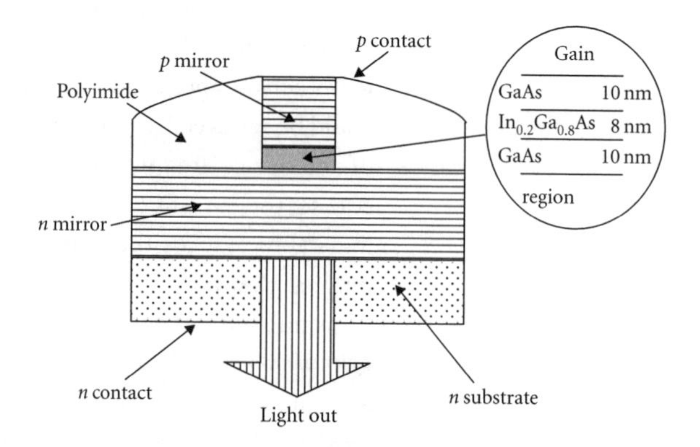

Figure 6.26. Schematic diagram of the VCSEL structure used by Geels et al. to demonstrate the first room temperature CW operation. The active region consists of a single InGaAs QW located between a pair of AlAs/GaAs Bragg mirrors. The cross-sectional area is defined by reactive ion beam etching, the exposed surfaces being passivated by a film of polyimide. Because the photon energy is less than the band gap of GaAs, the laser output beam may be transmitted through the substrate.

when they demonstrated the first VCSEL to operate CW at room temperature in 1990. Just as the laser world had hailed the first CW-RT stripe laser in 1970, here was the corresponding breakthrough for the VCSEL, and it depended on an equivalent degree of sophistication. Figure 6.26 illustrates the basic features. The whole structure was grown by MBE, the active region of the device consisting of an 8-nm strained $In_{0.2}Ga_{0.8}As$ QW sandwiched between two 10-nm GaAs barrier layers, resulting in an emission wavelength $\lambda = 963$ nm. Note that the photon energy is significantly lower than the band gap of the Bragg mirror materials which minimizes absorption loss in the mirrors, important for the achievement of high reflectivity. The n-type mirror stack consisted of 28.5 AlAs/GaAs pairs, doped at a level of $n = 4 \times 10^{24}$ m^{-3}, while the p-type mirror contained 23 pairs, doped at a similar level. (As is clear from Figure 6.26, the drive current flows through the mirror stacks so they must be doped as highly as possible to minimize series resistance.) The *p*-side of the device was etched by reactive ion etching to define a square cross-section in the range 12×12 to 1×1 μm^2 and the resulting pillar passivated by depositing a film of polyimide (a convenient insulating material). Threshold currents were close to 1.1 mA for a number of 12×12 μm^2 devices, the corresponding current density being 800 A cm^{-2}, and the threshold voltage was 4.0 V, somewhat greater than the band gap of the active layer due to the remaining series resistance in the structure. This points up a fundamental problem associated with the use of Bragg mirrors, namely that they introduce a large number of steps in both conduction and valence bands, each of which represents a small voltage drop. Geels et al. minimized their effect by grading the Al content at each step and by including a graded short period AlAs/GaAs superlattice—one reason why I referred above to the sophistication of their design. Doing a rough sum, based on the information given in their paper, we can estimate the total number of distinct layers to be grown as about 935! Needless to say, the MBE machine was computer-controlled, the complete structure being pre-programmed, prior to growth.

These developments certainly established the VCSEL as a serious practical device and set the tone for future progress. Threshold currents continued to improve as a result of improved optical design (making the active region into a resonant optical cavity) and improving the technology for defining the electrical cross-sectional area. This latter process now depends on the introduction of a single thick layer of AlAs which is oxidized from the edge of the device to form an insulating 'collar' (or 'current aperture') close to the active region, an approach which avoids the effect of surface damage introduced during the reactive ion etching process by keeping such damage well away from the active region. As a result, the lowest threshold current densities are now in the region of 100 A cm^{-2} and threshold currents have been reduced to about 20 μA in 2×2 μm^2 devices which is of importance for the application of VCSELs to signal transfer between high speed circuits. Ideally, one wishes to use the digital signals themselves to drive the laser, rather than to use a separate bias. But, in case this gives the impression that the VCSEL is useful *only* at ultralow light levels, let it be said that small arrays (19 lasers) have recently been developed at the University of Ulm, Germany which give output powers of over a Watt at 980 nm wavelength, suitable for pumping Er fibre lasers. Other major developments have been associated with the application of VCSELs in the field of optical communications, i.e. at wavelengths of 1.3 and 1.55 μm, but we shall leave this discussion until Chapter 8.

That the VCSEL story represents a remarkable success is beyond doubt. The optoelectronic community now has available to it a range of light emitters, covering a wide range of wavelengths and output powers

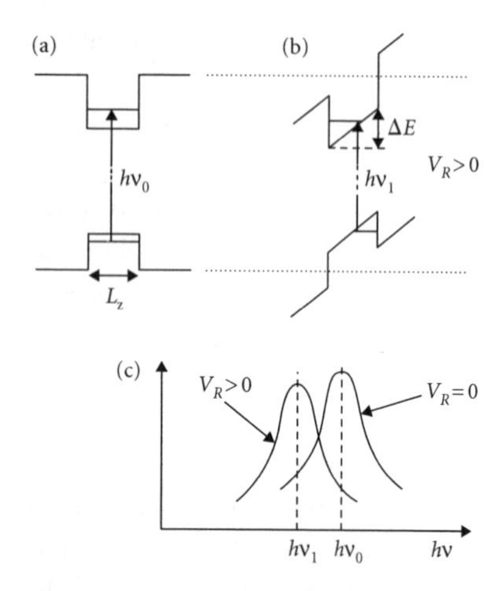

Figure 6.27. Illustration of the physical principles behind the QCSE optical modulator. In (a) we show the optical transition between the $n = 1$ heavy hole state and the $n = 1$ electron state in a quantum well. Exciton absorption occurs at an energy below this corresponding to the exciton binding energy. In (b) we show the effect of applying a large electric field normal to the plane of the well. The resulting tilt of the band edges causes electrons and holes to move towards opposite sides of the well, with corresponding changes in their confinement energies. The net effect is to reduce the photon energy appropriate to exciton absorption, compared to its zero field position, and (c) illustrates the use of this exciton shift in realizing an optical modulator, based on the optical frequency ν_1. At zero bias absorption is low, while, with an appropriate reverse bias V_R applied across the *p-i-n* diode, the exciton absorption shifts downwards in energy, causing a marked increase in absorption loss at the operating frequency ν_1.

and having symmetrical output beam shape and relatively narrow beam divergence, ideal for coupling into optical fibres. VCSELs have a number of other advantages over conventional stripe lasers in so far as their processing is simpler, they naturally emit in a single longitudinal mode and they can readily be prepared in the form of large, planar arrays, suitable for parallel optical signal processing. In addition, their planar technology makes them compatible with other planar electronic devices which allows ready combining of different electronic and optical functions on a single semiconductor slice. Nor should we lose sight of the fact that their design strategy depends on the use of quantum wells. The VCSEL therefore provides an excellent argument with which to reassure my doubting university friend—it may have been a long time in development but really good things are surely worth waiting for.

Our final topic in this chapter on LDS concerns the discovery and preliminary exploitation of an interesting physical effect which is unique to these structures, namely the 'quantum confined Stark effect'. The Stark effect is well known in several fields, in particular, the spectroscopy of atoms in the gas phase. It refers to the effect of large electric fields in modifying atomic energy levels, which may be detected via their emission or absorption spectra. The case which concerns us here differs only in so far as it involves the energy states of quantum wells. To understand the effect we should look again at the nature of light absorption in a quantum well. As we saw earlier, the absorption edge corresponds to an optical transition between the $n = 1$ heavy hole state in the valence band and the $n = 1$ electron state in the conduction band (for convenience, we illustrate this again in Figure 6.27(a)) but we should also remember that exciton effects are important in two-dimensional systems. The exciton photon energy is given by the sum of the band gap of the well material plus the two confinement energies, minus the exciton binding energy. What happens when we apply an electric field F normal to the plane of the well? The answer is given in Figure 6.27(b)—the band edges are tilted as shown and, in particular, the bottom of the well is no longer flat, but slopes by an overall amount (in energy) of:

$$\Delta E = eFL_z \tag{6.17}$$

Two things happen: first, the confinement energies are increased (with respect to the lowest conduction band/highest valence band energy), and second, the electrons and holes are pulled apart so that they collect near opposite sides of the well. It is important to recognize that they are still held close together by the well potential so excitons still exist (this is quite different from the case of a bulk semiconductor, where excitons are dissociated by the field), and therefore exciton absorption still takes place. However, the photon energy is reduced (largely due to the slope of the band edges) and the spatial overlap of the electron and hole wavefunctions is also reduced. In other words, the exciton peak moves to lower energy as the electric field is increased, while the intensity of the absorption decreases. It is possible to use the downward shift of the exciton peak to make an optical modulator. If, as shown in Figure 6.27(c), we shine a light beam through the structure, at an energy $h\nu_1$ just below the exciton absorption there will be rather a small absorption loss but application of an appropriate electric field will shift the exciton absorption downwards until the peak lies at our chosen energy and the absorption loss will be much increased. By swinging the field amplitude up and down in sympathy with an appropriate signal, the signal will be amplitude modulated onto the light beam. Potentially, this can be a very fast process because it is limited only by the time it takes electrons and holes to move the rather small distance across the well.

In order to design such a modulator, we need to know what size of electric field is needed. Clearly, it should be large enough to shift the exciton peak by an amount roughly equal to the linewidth (typically about 10 meV in a 10-nm well) and, if we use equation (6.17) to give us a (very!) rough estimate of the shift, we obtain $F \sim 10^{-2}/10^{-8} = 10^6\ \mathrm{V\,m^{-1}}$. This is inevitably an underestimate because it neglects the increase in confinement energy we mentioned above, which works in the opposite sense and reduces the shift, so, in practice, somewhat larger fields, of the order $5 \times 10^6\ \mathrm{V\,m^{-1}}$ are required. The question then arises: how can we conveniently generate such fields? and the answer is that we should use the field existing in a reverse-biased *p-n* junction. Even better, we should use a p^+-i-n^+ junction because the field in the undoped *i*-region is uniform. In order to increase the maximum absorption to a useful level, it is desirable to employ many QWs (say 50) so we need a field which is uniform over a thickness of at least $50 \times 20\ \mathrm{nm} = 10^3\ \mathrm{nm} = 1\ \mu\mathrm{m}$. Wood et al. from Bell Labs first demonstrated a modulator based on these principles in 1984. They obtained a factor two change in absorption by applying 8 V reverse bias to a GaAs *p-i-n* diode containing a 50-period AlGaAs/GaAs MQW and measured a switching time of 2.8 ns (limited by the RC time constant of the diode) and this has been the model for numerous later experiments.

The obvious weakness of the simple transmission modulator was the difficulty in obtaining a large on/off ratio but this was later much improved by the development of reflection modulators based on the use of Bragg mirrors. The incident beam traverses the modulator MQW twice, being reflected back by the Bragg stack. Detailed analysis of such a structure shows that the net reflectivity is very sensitive to loss within the MQW and overall on/off ratios greater than 100 : 1 were reported by Whitehead et al. of University College, London in 1989. However, the single-pass modulator still has an important role to play in the form of an integrated laser–modulator unit where part of the MQW structure is forward biased to produce laser action, while an adjacent region is reverse biased to act as a modulator. This arrangement has the considerable advantage that the light beam travels in the plane of the wells, thus increasing the absorption length very considerably. There is much current interest in such structures in the business of fibre-optical communications.

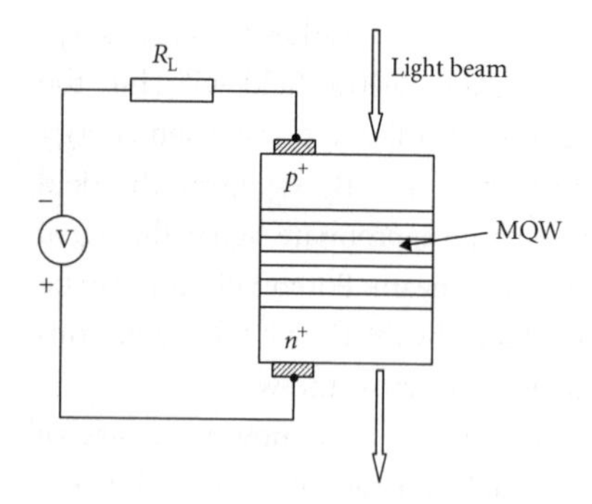

Figure 6.28. Circuit diagram to illustrate the operation of a SEED optical switch. The device switches between a high-impedance–high-transmission state and a low-impedance–low-transmission state as a result of two effects: (i) the photocurrent generated by optical absorption in the MQW region of the diode and (ii) the energy shift of the MQW exciton absorption with diode reverse bias.

The same basic principle of the MQW modulator has been applied in an interesting way to an optical switch. It was first described by David Miller and co-workers from Bell Labs in 1984 who christened it the 'Self-Electrooptic Effect Device' or SEED. It has been improved and modified in various ways since, though it is still not clear whether optical signal processing and optical computing, to which it is relevant, have a significant commercial future. The new feature of the device depends on the fact that a MQW structure within a *p-i-n* diode works as an efficient light detector—when light is absorbed in the MQW layers, a photocurrent flows through the diode and through any external circuit. Consider the circuit shown in Figure 6.28. In the absence of light, the diode presents a high impedance and nearly all the applied voltage is dropped across the diode. Suppose now that light of photon energy corresponding to the zero-bias absorption peak impinges on the diode—there is some small amount of absorption and the photocurrent generated effectively reduces the diode impedance and more voltage is dropped across the load resistor, less across the diode. This results in an upward shift of the absorption peak which increases the optical absorption (as explained above), more photocurrent is produced and even less voltage is dropped across the diode. Clearly, there is positive feedback in the system and the diode will switch abruptly from a high-impedance/ low-absorption state to a low-impedance/high-absorption state, which is stable as long as the light continues to impinge. Reducing the light intensity causes the SEED to switch back to its original state. Switching times as low as 30 ps have been measured for some SEED circuits, though the fact that switching depends on electronic circuitry is seen by some as a serious drawback—all-optical switching would be much preferable in many applications.

It is here that we leave the subject of LDS but not without emphasizing that LDS ideas and technology have made an impact on a great many other areas of semiconductor studies. There are many examples but we can illustrate the point well enough by instancing the recent surge of interest in light-emitting devices based on the Group II-VI compounds and the Group III nitrides (see Chapter 7). It was clear from the word 'go' that LEDs and LDs based on these materials would employ quantum wells (or even quantum dots) in their active regions. Once the basic behaviour of LDS had been established in the original material systems such as AlAs/GaAs, it was second nature to apply them to other materials and to other problems, and with equal success. Without doubt, LDS is here to stay. Perhaps one final comment is in order before we move on. The perspicacious reader would have noticed that there is no section in this chapter devoted to 'physics'. Why not? Because physics is apparent in every aspect of LDS. It seemed quite inappropriate to try to separate any particular aspect and call it 'physics'—we have been living with physics all the time.

Bibliography

Banyai, L. and Koch, S. W. (1993) *Semiconductor Quantum Dots*, World Scientific Publishing Company, Singapore.

Barnham, K. and Vvedenski, D. (eds.) (2001) *Low Dimensional Semiconductor Structures*, Cambridge University Press, Cambridge.

Bastard, G. (1988) *Wave Mechanics Applied to Semiconductor Heterostructures*, Edition de Physique Les Ulis France.

Beenakker, C. W. J. and van Houten, H. (1991) *Solid State Physics*, (eds. Ehrenreich and Turnbull) Vol. 44, p. 1.

Davies, J. H. (1998) *The Physics of Low Dimensional Structures*, Cambridge University Press.

Hartland, A. (1988) *Contemp. Phys.*, 29, 477.

Hilsum, C. and Rose-Innes, A. C. (1961) *Semiconducting III-V Compounds*, Pergamon Press, Oxford.

Jaros, M. (1990) *Physics and Applications of Semiconductor Microstructures*, Clarendon Press, Oxford.

Kelly, M. J. (1995) *Low Dimensional Semiconductors*, Clarendon Press, Oxford.

Morkoc, H. and Unlu, H. (1987) *Semiconductors and Semimetals*, (eds. R. K. Willardson and A. C. Beer). Vol. 24, p. 135.

Schaff, W. J., Tasker P. J., Foisy, M. C., and Eastman, L. F. (1991) *Semiconductors and Semimetals*, (eds. R. K. Willardson and A. C. Beer). Vol. 33, p. 73.

Weisbuch, C. and Vinter, B. (1991) *Quantum Semiconductor Structures*, Academic Press, San Diego, CA.

Willardson, R. K. and Beer, A. C. (eds.) (1987) *Semiconductors and Semimetals*, Vol. 24, Academic Press, San Diego, CA.

Witt, T. J. (1998) *Rev. Sci. Instrum*, 69, 2823.

CHAPTER 7

Let there be light

7.1 Basic principles

One of the more exciting and appealing aspects of semiconductor technology concerns the possibility of converting electrical power directly into light power. We have already met the GaAs light emitting diode (LED) in Chapter 5. A forward biased *p-n* junction emitting 'light' in the infrared part of the spectrum ($\lambda = 880$ nm), it made its debut in 1955, though this was simply a natural development of an already well-established tradition of light emitting devices. The first report of electroluminescence appears to have occurred as early as 1907 when an Englishman, H. J. Round (who had been at one time an assistant to Marconi) reported the results of experiments on SiC, observations which were later given a more scientific basis by the Russian, Oleg Losev of the Nizhegorodskaya Radio Laboratory who recognized, during the 1920s, the distinction between forward and reverse bias. However, it was not until 1951 that Kurt Lehovec and colleagues (Fort Monmouth Signal Corps Engineering Laboratory) discussed the effect in terms of minority carrier injection across a *p-n* junction (using insight gained from their work on the still new transistor). This was followed soon afterwards by the observation of forward bias electroluminescence in Ge and Si by Haynes and Briggs at Bell Laboratories and, by 1955, in a number of III-V materials such as GaAs, GaSb, and InP.

It was very quickly realized that, if similar devices could be made using semiconductor materials with wider band gaps, they would emit radiation within the visible waveband ($\lambda = 700$–400 nm, corresponding to $h\nu = 1.77$–3.10 eV). Early examples were SiC, ZnS, and GaP, though, because ZnS could not be doped *p*-type, the emission, in this case, took the form of AC electroluminescence, widely developed during the 1950s. The bright yellow emission was pleasant to the eye but technical difficulties and lack of proper understanding of its mechanism militated against its wider acceptance. SiC *p-n* diodes were capable of generating blue light, though with very low efficiency, while it was not until the 1960s that GaP became sufficiently well developed as a material to allow its red and green luminescence to be commercialized. Both SiC and GaP are indirect gap materials which make it difficult to generate light with high efficiency—the first direct gap visible emitter was GaAsP which could be persuaded to emit in the red part of the spectrum and which was used (as we saw in Chapter 5) by Holonyak in his pioneering laser work. The advent of liquid phase epitaxy (LPE) growth of AlGaAs for heterojunction lasers led to the development of more efficient red LEDs in the early 1970s, and this was followed shortly afterwards by its principal competitor InGaAlP, also a direct gap material but having a slightly larger band gap and able to generate red, orange, and yellow light.

This situation lasted for some time—red LEDs with efficiencies of several per cent were readily available, rather inefficient green devices based on GaP and very inefficient blue devices based on SiC made up the spectrum of colours which could be generated using forward biases of a few volts and currents of a few tens of milliamperes. The red emitters were very quickly adopted as indicator lamps and, in the shape of a seven segment format, in a wide range of digital displays. More recently, both red and yellow emitters have also begun to take over the role of brake and signalling lights in automobiles and trucks (cars and lorries). Because of the eye's much greater sensitivity in the green part of the spectrum (see Box 7.1), GaP devices appeared comparably bright and were also quite widely used, but the lack of an efficient blue emitter precluded the possibility of developing practical full-colour displays or, for that matter, white light emitters. Liquid crystal displays provided major competition in many applications on account of their much lower power consumption and the development of flat panel liquid crystal TV displays in the 1980s gave liquid crystals a major advantage over LEDs in the display field. Various forms of plasma display also competed strongly, helping to keep a tight rein on the commercial development of LEDs. Nevertheless, the relatively efficient, self-luminant properties of LEDs gave them an appeal which ensured their continuing presence in the market place—even without an efficient blue device the LED market had climbed above $1 billion by the beginning of the 1990s and is currently well above $3 billion.

The desired capability for full-colour LED display finally emerged in the middle of the 1990s with the development of the Group III nitride materials. The research group under Shuji Nakamura, at the Nichia Chemical Company, near Tokushima in Japan demonstrated both blue and green LEDs with efficiencies in the few per cent range in 1994–5 and full-colour, outdoor displays appeared in many Japanese shopping streets within a matter of months (or so it seemed!). Experiments with blue/green traffic lights were also undertaken with comparable expedition and LED traffic signals are now appearing across most of the developed world, their improved efficiency and greater reliability making significant financial savings for local authorities with appropriate initiative. The even more desirable blue laser followed shortly afterwards. It was nothing less than a fairy tale breakthrough, after a 25 year wait! Of possibly even greater significance, the availability of an efficient blue emitter led to the development of white light LED sources which began appearing

Box 7.1. LED efficiency

The efficiency of an LED is clearly an important parameter, determining the perceived brightness in terms of the electrical power used in driving it. Not surprisingly, the literature is well laced with references to 'efficiency' but we need to be careful concerning the precise meaning to attach, there being several different definitions. We must first differentiate between 'quantum efficiency' and 'power efficiency', then bear in mind the, possibly large, difference between 'internal efficiency' and 'external efficiency' (see Box 7.2). Finally, we have to recognize that light output may be expressed in terms of watts (i.e. power) or lumens (i.e. luminous flux) and the use of the latter can easily lead to confusion! Whole books are written on the subject of light units (see, for example, Keitz 1971) so we can hope to do no more than scratch the surface of the subject here—but hopefully even a scratch may help to avoid total confusion.

To begin with, let us consider *internal* quantum efficiency. The operation of an LED involves a current of electrons flowing from one side of the junction, meeting with a hole current flowing from the other side, electrons and holes recombining near the junction to generate both light (photons) and heat (phonons). The quantum efficiency for light generation is defined in terms of the ratio, (number of photons)/(number of photons plus number of phonons). (Note here that it is implicit that the total diode current involves electron–hole recombination.) We can also define an internal power efficiency which is the ratio of the light power generated (in watts) to the electrical power input to the diode. Needless to say, these two efficiencies are closely related. Let us suppose that N_{eh} electron–hole pairs recombine per second while N_{ph} photons are generated per second from the junction area. It is clear that the internal quantum efficiency η_i^Q is given by:

$$\eta_i^Q = N_{ph}/N_{eh}. \tag{B7.1}$$

In the same terms, we can write the diode current I as $I = N_{eh}e$ and, therefore, the electrical power as:

$$P = IV = eN_{eh}V, \tag{B7.2}$$

where V is the forward bias on the diode. The light power L generated is:

$$L = N_{ph}h\nu, \tag{B7.3}$$

from which the power efficiency is:

$$\eta_i^P = L/P = N_{ph}h\nu/N_{eh}eV = (h\nu/eV)\eta_i^Q, \tag{B7.4}$$

where $h\nu$ is the photon energy of the light. Under practical operating conditions, both eV and $h\nu$ will be fairly close to the band gap energy E_g so the power and quantum efficiencies differ by a relatively small factor (rarely greater than 2) but we should nevertheless be aware of the distinction.

From the application viewpoint, it is usual to consider the *external* efficiency which is related (in both cases) via the 'light extraction factor', F, an example of which we calculated in Box 7.2. Thus, we can write:

$$\eta_{ext} = F\eta_i, \tag{B7.5}$$

and it will be apparent that equation (B7.4) holds equally well for external efficiencies. We should emphasize that it is the *total* light power emitted from the diode which concerns us here—we are not interested in its angular distribution (though this may well be of importance in its own right). A frequently used expression applied to practical LEDs is the

Box 7.1. Continued

'wall plug efficiency', which is simply the external power efficiency (i.e. light power in watts divided by electrical power in watts) but an important variant on this is the external efficiency measured in lumens per watt. To appreciate its significance demands some further understanding and leads us into the realms of photometry.

Perhaps the first point to make is that photometry introduces units which measure the effect of light on the eye—thus, radiation is measured according to the visual sensation produced (a somewhat woolly concept, perhaps (!) but entirely appropriate to the output from a LED because it is obviously designed to be seen by the eye). To a first approximation, we may treat the LED as a point source but one which *does not* radiate uniformly in all directions (see, for example, Box 7.2). If, therefore, we examine the light flux Φ, we shall find that it depends on the direction from which we make the observation (referring to Figure 7.2, Φ is a function of the angle θ). More precisely we should define the flux within a fixed solid angle Ω, in a specific direction θ, in which case, the *luminous* intensity I from the LED is given by:

$$I(\theta) = \mathrm{d}\Phi / \mathrm{d}\Omega. \tag{B7.6}$$

The unit of intensity is 'candella', which is defined in terms of the radiation from a black body at the temperature of melting platinum. The unit of flux is 'lumen' which is the flux per unit solid angle from a source of strength 1 cd. Note that the use of the word 'luminous' implies that the unit is to be interpreted as representing a 'visual sensation' rather than one expressed in watts. We obviously need to know how to relate 'lumens' to 'watts' and we shall see how this is done further, but first it is useful to explain the relationship of intensity and 'brightness', brightness, or 'luminance', being the term commonly used to express the apparent strength of radiation from a source. In fact, brightness is the intensity per unit area of an extended source which, again, is a function of the direction of viewing. Brightness is variously measured in units of lumens per square metre ('apostilb'), lumens per square centimetre ('Lamberts'), or lumens per square foot ('foot Lamberts'), it being understood that each of these refer to unit solid angle. While the SI unit is the apostilb, the literature abounds with Lamberts and foot Lamberts so it is better to be aware of their definitions. However, so, as long as we are thinking of our LED as a point source, it is preferable to retain the concept of luminous intensity—measured in lumens.

In relating these parameters to the overall efficiency of a device, we first note that the total light flux is given by:

$$\Phi_{\mathrm{T}} = \int \mathrm{d}\Phi, \tag{B7.7}$$

where the integration is to be taken over all solid angles (in theory 4π steradians—in practice, for many LEDs, 2π steradians). This is to be compared with the total light power L which we introduced earlier. We can now express the relation between light power and flux in terms of the following equation:

$$\Phi_{\mathrm{T}} = KL, \tag{B7.8}$$

where K is the required conversion constant with units 'lumens per watt'. However, this apparently simple relation is complicated by the fact that the eye sensitivity is a strong function of wavelength, the relative sensitivity being given graphically in Figure 7.1. Clearly, the conversion constant must also be a strong function of wavelength, having a maximum value $K_{\mathrm{m}} = 683\ \mathrm{lm\,W^{-1}}$ at $\lambda = 555$ nm in the green part of the spectrum. In practice, we need to write K as a function of λ, which we do as follows:

$$K(\lambda) = K_{\mathrm{m}} V(\lambda), \tag{B7.9}$$

Box 7.1. Continued

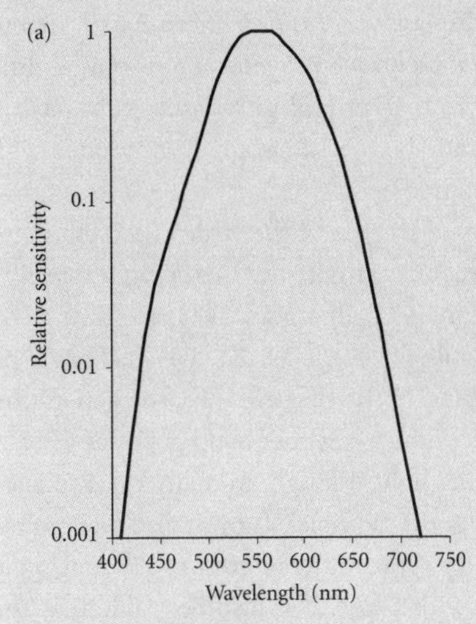

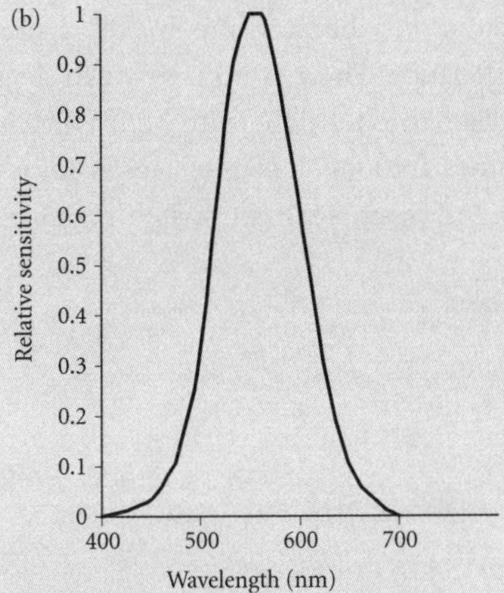

Figure 7.1. (a) The standard curve representing the relative sensitivity of the human eye to light of different wavelengths. (b) The peak in the curve occurs at $\lambda = 555$ nm at which point the luminous efficiency has a value of 683 lm W^{-1}. The two curves represent plots of the same data on logarithmic and linear scales, respectively.

where $V(\lambda)$ is the eye sensitivity function plotted in Figure 7.2. Thus for red light at $\lambda = 650$ nm, $K = 73.1$ lm W^{-1}, while for green light at 555 nm, $K = K_m = 683$ lm W^{-1}, and for blue light at 440 nm, $K = 15.7$ lm W^{-1}.

The question then arises, how do we deal with white light, containing a wide spread of wavelengths? Clearly we must express the relationship in terms of an integral over all wavelengths:

$$\Phi_T = K_m \int_0^{\infty} (dL(\lambda)/d\lambda) V(\lambda)\, d\lambda, \qquad \text{(B7.10)}$$

where $dL(\lambda)$ is the light power (in watts) over an increment of wavelength $d\lambda$ at λ and $V(\lambda)$ is the eye sensitivity curve given in Figure 7.2. In order to perform the integration, we need to know the detailed make-up of the light—that is, the precise form of $L(\lambda)$. The simplest case concerns equal light power at all visible wavelengths, in which case we need to simply integrate $V(\lambda)$. This can easily be done numerically or graphically and gives an effective value of $K_{av} = K_m V_{av} \approx 300$ lm W^{-1}. It suggests, therefore, that the 'conversion factor' for white light takes roughly this value, that is, 300 lm W^{-1}, though it should be borne in mind that 'white light' may be made up in a variety of ways and the above result should be regarded as no more than a useful guide.

Finally, we take the opportunity of defining yet another efficiency, the luminous efficiency η^L of a particular emitter. Again its units are lumens per watt but we note that it takes account of both the external quantum efficiency of the emitter *and* the eye sensitivity curve. Mathematically, we write:

$$\eta^L(\lambda) = \eta^P_{ext} K_m V(\lambda). \qquad \text{(B7.11)}$$

It is clear that η^L is measured externally to the device so we can safely omit the suffix 'ext'. More importantly, we must remember that η^L is a strong function of wavelength through the eye sensitivity curve $V(\lambda)$.

during 1996–7. White light can be produced either by combining red, green, and blue LEDs or by using an LED to pump a suitable phosphor, the key issue here being one of efficiency—if the overall efficiency can be raised to about 50%, white LEDs will become seriously competitive with conventional light sources. Excitement runs high. Though we should not lose sight of the fact that the LED market is still dominated by the sale of the cheap, modestly efficient GaAsP and GaP devices developed in the 1960s (so long as they provide enough light to act as effective indicators and digital displays there is little incentive for manufacturers to invest in their more expensive progeny) there can be little doubt that the super-bright LED market will grow and grow—it can surely be little more than a decade before LEDs are synonymous with lighting in nearly all its manifestations.

That, in a nutshell, is the story of the rise of the visible LED but there is much detail to be added (see, for example, the following books and review articles—Bergh and Dean 1972; Bhargava 1975; Craford 1977; Willardson and Weber 1997–9: vols 44, 48, 50, 52, 57). We shall begin by examining some general principles. In the first instance, the generation of visible light demands a semiconductor whose band gap is at least as large as the photon energy of the light, though, in many cases, it may be significantly larger (GaP is a good example). It must also be possible to dope the material both *n*- and *p*-type, in the interest of forming a *p-n* junction. Based on experience with GaAs, it is also desirable that the active material should have a direct gap because this makes it much easier to obtain high internal efficiency. There is no law which denies the possibility for indirect materials to emit light with high efficiency but, because their radiative lifetimes are long, it is much easier for non-radiative recombination processes to compete. If we write the radiative recombination rate as:

$$-(\mathrm{d}n/\mathrm{d}t)_{\mathrm{R}} = n/\tau_{\mathrm{R}} \tag{7.1}$$

and the non-radiative rate as:

$$-(\mathrm{d}n/\mathrm{d}t)_{\mathrm{NR}} = n/\tau_{\mathrm{NR}}, \tag{7.2}$$

then the internal quantum efficiency of the radiative process is:

$$\begin{aligned}\eta_{\mathrm{i}}^{\mathrm{Q}} &= (n/\tau_{\mathrm{R}})/\{n/\tau_{\mathrm{R}} + n/\tau_{\mathrm{NR}}\}\\ &= 1/\{1 + \tau_{\mathrm{R}}/\tau_{\mathrm{NR}}\},\end{aligned} \tag{7.3}$$

which makes clear that high efficiency requires the ratio $(\tau_{\mathrm{R}}/\tau_{\mathrm{NR}})$ to be very much less than unity. As non-radiative processes tend to be characterized by short lifetimes, it is obviously necessary for the radiative lifetime to be even shorter—thereby favouring direct gap materials. However, because most non-radiative processes involve recombination via deep levels within the band gap, if the material can be prepared with very high crystal quality and with an absolute minimum of unwanted impurity atoms, it may be possible to increase τ_{NR} to the point where it exceeds τ_{R}, even for an indirect gap semiconductor. Such was the case for GaP. However, it should be appreciated that these requirements make the task of maintaining material quality under production conditions very much harder than would be the case for a direct gap semiconductor where there is a much greater margin of acceptability.

It is also convenient if the active region can be sandwiched between two layers with larger band gap; it then being possible to employ heavily doped regions which inject both types of carrier into an undoped active layer where they recombine with high efficiency. We saw that, in the case of GaAs, these injecting layers consist of AlGaAs and this principle can be extended to cover the case where AlGaAs forms the recombination layer—it merely requires the injecting layers to have higher Al content. However, this may not always be possible and some LEDs, such as GaP and SiC employ simple homostructures which involve a compromise between the need for heavy doping to encourage radiative recombination and the likelihood that too heavy doping will lead to poor material quality, thus encouraging non-radiative recombination.

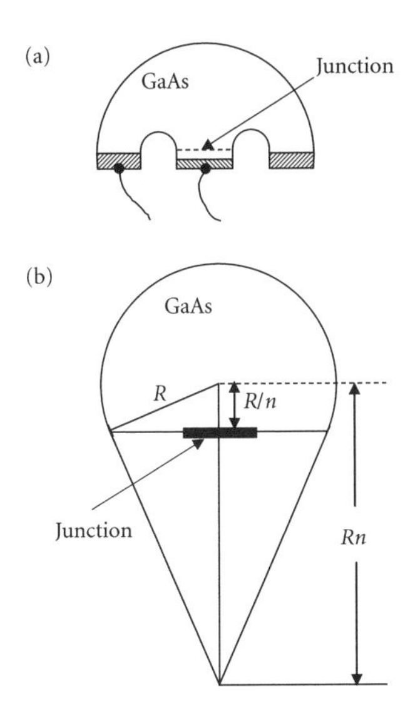

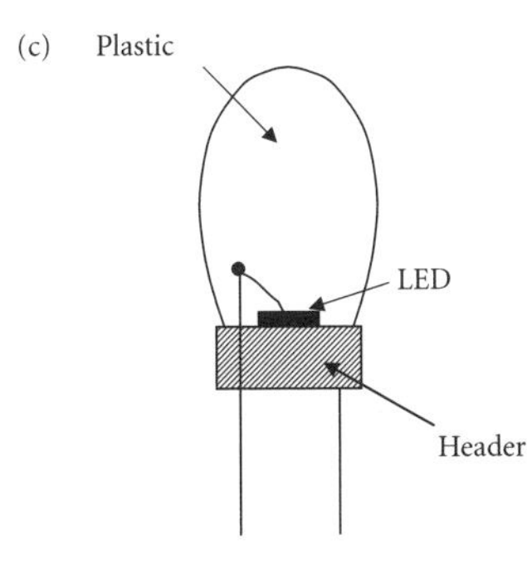

Figure 7.3. Some examples of LED structures designed to increase the amount of light emitted into the air. (a) This illustrates the use of a semiconductor hemisphere with the junction located near the centre of curvature. All the upward emission strikes the semiconductor–air interface close to normal incidence. (b) The Wierstrass sphere is shown here. It serves to focus the emitted light so as to give increased brightness when the lamp is viewed normally. In practice, many commercial LEDs are encapsulated within a plastic dome such as that illustrated in (c). This is less effective in increasing the light extraction factor but is considerably cheaper to manufacture.

So much for questions of internal efficiency. It will be apparent that the ultimate objective is to collect as much light as possible *external* to the device structure and this brings with it its own design rules. Two factors are important: first and foremost, the ratio of external to internal light power and, second, the spatial distribution of the light emitted. In Box 7.2 we present a simple calculation of the fraction of light which is transmitted through a plane semiconductor/air interface and, because most semiconductors have refractive indices greater than 3, this results in a maximum possible external efficiency for simple plane structures of $\eta_{\text{ext}} \sim 2\%$. As can be seen from Figure 7.2, most of the light generated within the body of the semiconductor is lost by total internal reflection at the top surface of the device. It also follows that the light intensity has an angular distribution of the form $I = I_0 \cos^2 \theta$, where θ is the angle between the direction of viewing and the normal to the surface. If this were to be the end of the story, it would take much of the excitement out of the LED romance—something clearly had to be done to improve the light extraction factor F and many things have, indeed, been done.

The simplest solution in principle is to arrange for the junction to lie approximately at the centre of a hemispherical semiconductor dome, as demonstrated on a GaAs diode by Carr and Pittman of Texas Instruments in 1963 (see Figure 7.3(a)). This means that all the light emitted within the upper hemisphere will strike the interface close to normal incidence and at least 70% of it will be transmitted. The introduction of a reflecting metal film immediately below the junction improves matters still further by returning the downward emission so it, too, impinges on the domed surface—external efficiencies of over 50% appear possible with this geometry. There are two drawbacks: first, the emitted light has to traverse a considerable thickness of semiconducting material which

Box 7.2. Light emission from a plane LED

The fact that semiconductors have relatively large refractive indices gives rise to a significant problem in extracting light from an LED. Figure 7.2 illustrates the effect of refraction of light emitted from a point source P below the surface of the material. Using Figure 7.2(a), we can apply Snell's law of refraction to relate the exit angle of the light θ to the incident angle φ:

$$\sin\theta = n\sin\varphi \tag{B7.12}$$

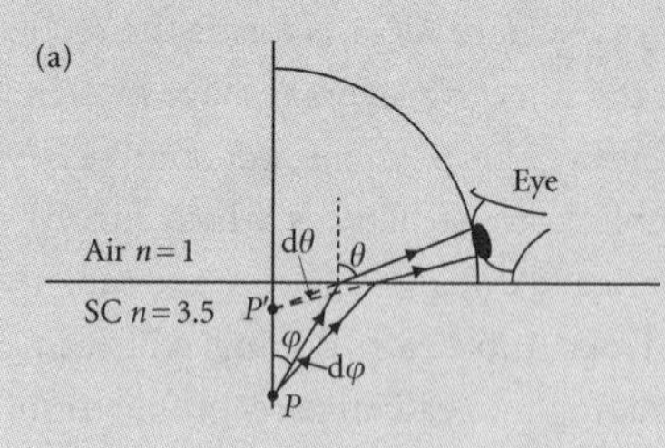

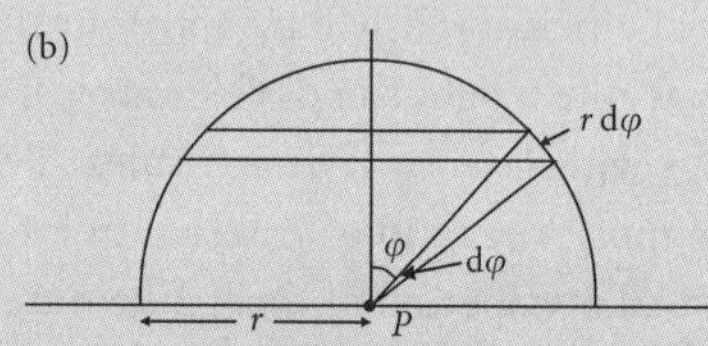

Figure 7.2. Schematic diagram to illustrate the manner in which light is emitted from the plane surface of an LED. In (a) light rays are shown taking a path from the LED at P to an observer's eye, the pupil subtending an angle $d\theta$ in the air, with corresponding angle $d\varphi$ within the semiconductor. (b) This illustrates the calculation of the total light flux emitted between angles 0 and φ, within the semiconductor. The maximum value of $\varphi(\varphi_{max})$ is determined by the onset of total internal reflection when $\theta = 90°$.

and then use this to calculate the light power emerging from the top surface as a fraction of the total light power emitted from the point P. We assume that P emits uniformly in all directions, so that (in Figure 7.1(b)) we can write the fraction of total light from P making an angle φ with the normal as:

$$\begin{aligned} dF &= 2\pi r\sin\varphi \cdot r\,d\varphi/4\pi r^2 \\ &= \sin\varphi\,d\varphi/2. \end{aligned} \tag{B7.13}$$

The fraction of light power emerging from the top surface is given by integrating this expression from $\varphi = 0$ to $\varphi = \varphi_{max}$ where φ_{max} is the maximum value of φ, corresponding to $\theta = \pi/2$. Thus:

$$\sin\varphi_{max} = n^{-1}\sin\pi/2 = n^{-1}. \tag{B7.14}$$

For the case of a semiconductor with $n = 3.5$, $\varphi_{max} = 16.6°$, so it is at once clear that only a small fraction of the light will escape from the surface, the rest being totally internally reflected. By integrating equation (B7.13), we find:

$$F = (1 - \cos\varphi_{max})/2. \tag{B7.15}$$

And, inserting the above value of φ_{max}, we obtain $F = 0.021$, that is, approximately 2% of the light escapes. The remainder is reflected back into the bulk of the device, where it may be absorbed, scattered, or reflected again from the back contact. Some small proportion may reach the top surface again and make a second attempt to escape, so the above estimate of the escaping fraction may be a *slight* under-estimate. Finally, we note that, though this result refers specifically to a point source of light, we may treat the actual emission from a LED as made up of an assembly of point sources, thus allowing us to extend it to cover the LED too.

It will be clear from Figure 7.2(a) that the light which does escape is emitted over the full 180° arc above the surface and we can use the same ideas to work out the appropriate angular distribution. Imagine an observer viewing the emission from above, at a constant radial distance from the image point P' (Figure 7.2(a)). The observer's eye pupil will accept light from a constant solid angle defined by the small angle $d\theta$, the solid angle being proportional to $(d\theta)^2$—we need to calculate the relative amount of light within this solid angle as the angle θ changes. What we know is that the light intensity *inside* the semiconductor is proportional to $(d\varphi)^2$ and is independent of φ. We must therefore calculate an expression for

Box 7.2. Continued

the value of $d\varphi$ which corresponds to a fixed value of $d\theta$, as a function of θ. This is obtained straightforwardly enough by differentiation of equation (B7.12). Thus:

$$\cos\theta\cdot d\theta = n\cos\varphi\cdot d\varphi \tag{B7.16}$$

from which we obtain:

$$\begin{aligned} d\varphi &= \cos\theta\cdot d\theta / n\cos\varphi \\ &= \cos\theta\cdot d\theta / n(1-\sin^2\theta/n^2)^{1/2} \\ &= \cos\theta\cdot d\theta / (n^2-\sin^2\theta)^{1/2}. \end{aligned} \tag{B7.17}$$

Given the large values of n for typical semiconductors, the denominator in equation (B7.17) varies rather little with θ (typically between 9 and 10) so we can make the approximation that $d\varphi \approx \cos\theta\cdot d\theta$ and, finally, that the light intensity detected by the observer's eye varies as:

$$I = I_0\cos^2\theta, \tag{B7.18}$$

where I_0 is the intensity observed along the normal to the semiconductor surface.

exaggerates any absorption loss and, second, the necessary thickness of material is incompatible with the use of epitaxial techniques. While the former difficulty can be overcome by using a double heterostructure so that the photon energy of the emitted light is well below the band gap of the cladding layers, the latter is unavoidable and more or less damning! A typical junction dimension of 200 μm implies a radius of say 300 μm for the hemisphere but, what is more, it implies that each diode must be individually shaped, a production manager's nightmare! Similar objections apply to a number of other geometries such as the Weierstrass sphere (Figure 7.3(b)) which has the property of focusing the emitted beam to some extent towards the normal direction, thus improving the brightness when viewed directly from above. In practice, most commercial devices employ the compromise solution illustrated in Figure 7.3(c) where the diode is encapsulated in some form of plastic dome. Typical refractive indices for such materials are 1.5–1.6 which result in improvement over the plane device geometry of about a factor 3. Efficient reflection of the downward-going emission then allows external efficiencies of order 10–15% which, for a long time, represented the best performance commercially available. In principle, light extraction factors up to a factor 2 better still are possible using special glasses, such as those developed at RCA in 1969 by Fischer and Nuese, with refractive indices in the range of 2.4–2.9, but these appear not to have been widely used because of large thermal expansion mismatch with the semiconductor material. Finally we should note that, when the semiconductor material does not absorb the emitted light (as in the case of GaP, for example) it may be possible to collect light which is emitted sideways and downwards using an external curved reflector, thus restoring a significant part of the light to the emitted beam. In this way, efficiencies of the order of 10% were achieved for GaP red-emitting diodes as early as 1970.

Having made clear that LEDs function by the radiative recombination of carriers injected across a *p-n* junction, we should recognize that the precise nature of the recombination process is, in itself, a matter of considerable interest. In fact, a great deal of work to establish radiative processes was performed on II-VI compound materials such as CdS, ZnS, and ZnSe during the period 1955–65 and extended to the III-Vs during the 1960s. GaP proved particularly fruitful in providing results of scientific interest and we shall consider these in the appropriate section, but anyone seriously involved in the field must acknowledge a lasting debt to the work on II-VI compounds which elucidated so much of the detail concerning the nature of radiative recombination. Unfortunately, from the point of view of practical device development, these compounds suffered from the disadvantage that they could only be doped *n*-type so, though efficient light emission was readily obtainable, its application to LEDs remained, for a long time, a tantalizing chimera. It was not until the early 1990s that the application of molecular beam epitaxy (MBE) growth allowed ZnSe to be doped *p*-type with N acceptors, making it a competitor to the Group III nitrides as a wide band gap light source. This story will unfold in Section 7.4 along with that of the nitrides but first we must go back to 1962 and the struggle to develop GaAsP as the first practical LED material.

7.2 Red-emitting alloys

Once reasonably efficient GaAs LEDs had been demonstrated in the early 1960s, the goal of extending this performance into the visible region suddenly became more urgent. The question was: how to obtain a direct gap material able to generate visible light? and, bearing in mind the eye sensitivity curve shown in Figure 7.1, how to make the wavelength as short as possible? A device emitting at 670 nm, for example, can be seen to have nearly 10 times the luminous effect of a similar one (with equal quantum efficiency) operating at 700 nm. Thus, a small (4.5%) increment of band gap ($h\nu$ increasing from 1.77 to 1.85 eV) could provide the same visual enhancement as months of work to improve internal efficiency and light extraction factor. A glance at the 'available' III-V compounds quickly revealed only three candidates with wider band gaps, AlP ($E_g = 2.45$ eV), GaP ($E_g = 2.26$ eV), and AlAs ($E_g = 2.17$ eV) but unfortunately all three were indirect. An alternative approach involved

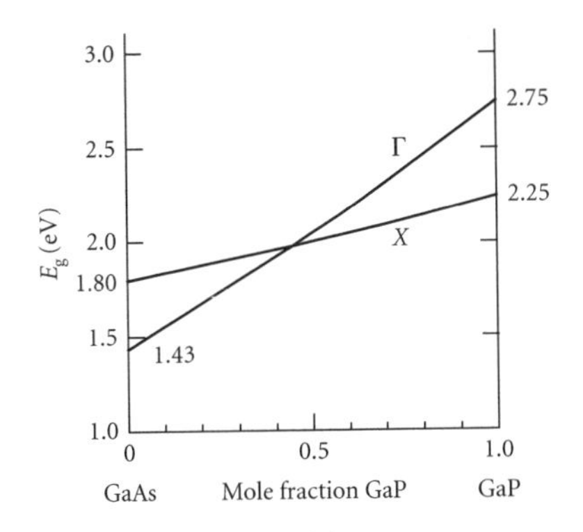

Figure 7.4. Energy band diagram to illustrate the variation of the energy gap E_g of the alloy semiconductor GaAsP with changing mole fraction 'x' of GaP. The alloy has a direct gap for values of x between 0 and 0.45. For larger values of x, the indirect X conduction band minimum lies below the central Γ minimum. The cross-over occurs at an energy gap $E_g = 1.95$ eV.

searching for suitable alloys and one, in particular, GaAsP was already available. (Two others, AlGaAs and InGaP were to emerge in due course.) The substitution of P for some of the As in GaAs increased the direct band gap approximately linearly with P concentration, but only up to a point—as can be seen from Figure 7.4, at a concentration close to 45% the gap changed from direct to indirect, the X conduction band minimum lying below the central Γ minimum for all higher P fractions. The maximum direct gap was 1.95 eV, corresponding to a wavelength ($\lambda = 640$ nm) well into the red part of the spectrum. (All this was clearly established by Holonyak's group as early as 1963.) Herein lay hope for an efficient visible emitter and, as we have already seen, for the first ever (semiconductor) red laser.

It is interesting to discover just how rapidly the technical environment changed at this time. In their book surveying the III-V compound scene, published in 1961, Hilsum and Rose-Innes make no mention of the LED (indeed, the word 'electroluminescence' is not even indexed!)—they include sections on varicap diodes, microwave diodes, switching diodes, tunnel diodes, but *not* LEDs. Yet the semiconductor laser was demonstrated just 1 year later! It is interesting, too, to learn that when Holonyak took up with GaAsP it was not with red light in view but rather in hope that the enhanced band gap might enable him to make better tunnel diodes! So, indeed, it did, but one can only admire the speed with which he changed course to enter the race for the first semiconductor laser, being in at the kill in September 1962. Such is the manner of technological advance—empiricism and opportunism so often taking precedence over long-term planning! Interestingly, apart from a few demonstration devices, he appears to have given remarkably little attention to the red LED, remaining faithful to his laser diode (LD) programme while others took up the LED cudgels.

As with any device, the key to success in LEDs turned in the lock of growth technique. Holonyak made his lasers from polycrystalline GaAsP grown by vapour transport of GaAs and GaP together in a closed ampoule. The *p-n* junction was formed by subsequently diffusing Zn into *n*-type material to a depth of about 25 μm. Ainslie et al. at RCA obtained similar quality boat-grown material some 2 years later (1964) but the basis for a serious commercial process, epitaxial deposition of controlled thin films from the vapour phase, had to wait until 1966. Tietjen and Amick (also at RCA) demonstrated the growth of GaAsP on a GaAs substrate using a current of HCl gas over a Ga boat to transport Ga and the gases arsine and phosphine to transport the Group V elements. Three years later Manasevit et al. of North American Rockwell reported the successful growth of GaAsP by metal-organic vapour phase epitaxy (MOVPE) (tri-methyl gallium, arsine, and phosphine) and commercial developments have been based on one or other of these processes ever since. The first commercial red-emitting devices (based on bulk crystal growth) were offered by General Electric as early as 1962, though the activity was on a rather small scale and it was not until 1968 that Hewlett-Packard and Monsanto commercialized the halide process to generate a serious high-volume business. Today there are companies worldwide (dominated by Japan, Taiwan, America, and Germany) competing strongly for a still rapidly growing market. Though most workers preferred to form the *p-n* junction by diffusing zinc into *n*-type GaAsP epilayers, Neuse and colleagues at IBM demonstrated the equally effective use of epitaxially grown junctions. In the former method, individual devices were defined by diffusion through a suitable diffusion mask—in the latter, by mesa etching through the grown junction, there being little to choose between the two technologies.

The use of GaAs substrates was obviously convenient on account of their ready availability but brought with it a problem that had been automatically avoided by the use of bulk crystals, that of lattice-mismatch. If GaAsP is grown directly on GaAs, the resulting layers contain high densities of dislocations which seriously degrade LED performance (there is a 4% lattice-mismatch between GaP and GaAs). The solution (perhaps 'partial' solution would be more accurate) lay in growing first a graded buffer layer, some 10 μm thick in which the P content was smoothly increased from 0% to 40% (or whatever value was appropriate to the device in question) before growing the 20 μm thick uniform GaAsP layer, into which the Zn diffusion was made. The second design problem concerned the maximum amount of P which could be introduced before the influence of the X minimum became significant. Because the effective mass of electrons in the X minimum is much larger than that associated with the Γ minimum, the X minimum density of states is also much larger so, if the two minima coincided in energy, nearly all the electrons would reside in the X minimum. It was therefore found necessary to adjust the P content so as to keep the Γ minimum at least 0.05 eV (i.e. 2 kT) below X. In fact, the quantum efficiency started to degrade for $[P] > 0.3$ ($h\nu = 1.77$ eV) but, because the eye sensitivity curve compensates for this, the measured brightness peaked at about $[P] = 0.4$ ($h\nu = 1.91$ eV) which lay comfortably within the red spectral region ($\lambda = 650$ nm). The best quantum efficiencies of these red diodes were about 0.05%, resulting in overall luminous efficiencies of 0.035 lm W^{-1} (brightness ~700 ft L at $J = 10\ \text{A cm}^{-2}$)—commercial devices were slightly less good, giving $\eta^L = 0.02\ \text{lm W}^{-1}$—though perfectly adequate for a wide range of applications.

This was the situation at the beginning of the 1970s at which point competition from GaP red diodes was beginning to look serious—there was definitely a need for improvement and, as it turned out, not one but two innovations came to the rescue. First, in 1973 George Craford and his group at Monsanto reported the use of N as a dopant for their GaAsP which had the effect of improving the efficiency of diodes emitting at shorter wavelengths and allowed them to produce both orange and yellow lamps with luminous efficiencies of nearly 1 lm W^{-1}. This was an idea borrowed from GaP where N-doping was used to obtain green emission, as we discuss in Section 7.3—N forms a shallow impurity level which traps electrons from the conduction band and holes from the valence band to form bound excitons which recombine radiatively with high efficiency, even in material with an indirect band gap. Second, Sorenson and his group at Hewlett-Packard reported the use of GaP substrates, instead of the usual GaAs and achieved an order of magnitude improvement in external efficiency in consequence, the transparency of GaP allowing them to collect a much larger fraction of light. This raised the luminous efficiency of red diodes to about 0.4 lm W^{-1} and, though this was still considerably lower than the best GaP red emitters, it was sufficient to save the day for GaAsP on account of the latter's simpler technology and greater reliability. However, it is also noteworthy that the use of GaP substrates resulted in diodes being roughly 10 times more expensive than those grown on GaAs, so the original technology continued to be used for applications where ultimate performance was not essential—the main advantage of GaP substrates was to provide flexibility of choice, customer preference always being of paramount importance. According to Bhargarva, in 1975, large scale production of LEDs was still based on GaAsP, when he reported production as $\sim 10^6$ square inches per year (approaching 10^3 m^2pa), though competition was certainly building. GaP was capable of higher brightness in the red and, more importantly, acceptable green emission, while AlGaAs and InGaP were potential rivals for the mass market in red devices. Perhaps they would offer higher efficiencies? Certainly there was plenty of scope—a few tens of percent should surely be possible.

The material development associated with the use of AlGaAs in DH GaAs lasers in the late 1960s provided the opportunity for people interested in improving red LED performance (or cost) to explore this alloy system along the same lines as previously undertaken for GaAsP. The AlGaAs system was found to mirror that of GaAsP in so far as there was a direct–indirect cross-over in the conduction band structure at an aluminium fraction near 40% and it would be necessary to optimize the alloy composition in precisely the same way. The initial breakthrough in growth technology was achieved at IBM by Ruprecht and colleagues in 1967. They modified the 'dipping' liquid phase epitaxy LPE process originated by Nelson for GaAs growth and developed a method of growing both *n*- and *p*-type AlGaAs in one growth run. Initially, the gallium melt was doped with Te donors so that, as it cooled slowly, the first layer to grow was *n*-type, then, at the mid-point of the run, the melt was counter-doped with Zn to form a more heavily doped *p*-region. This allowed them to grow the desired *p-n* junction in a single deposition. They demonstrated the ability to make diodes emitting at 1.70 eV ($\lambda = 730$ nm) with external efficiencies as high as 1.2% and, when encapsulated in suitable epoxy resin domes, up to 3.3%. A glance at the eye sensitivity curve in Figure 7.2 shows that these diodes were not very bright to the eye (0.02 lm W^{-1}) but the efficiency was nonetheless very creditable for a first attempt. It was clear that the LPE process produced relatively defect-free films which favoured high radiative efficiency—the drawback lay in a tendency to yield less than ideally smooth surfaces which made subsequent photolithographic processing somewhat unpredictable. It also proved less easy to industrialize than the alternative vapour phase growth processes.

On the physics front, the question remained, of course, of what was the precise position of the direct–indirect cross-over and of what the corresponding maximum direct gap was? Interestingly, this was a matter for quite lengthy discussion, new formulations of the (E_g vs x) plots appearing well into the 1980s, but the source of the difficulty lay rather in the measurement of x (the mole fraction of Al in the alloy) than in that of band gap. By the early 1970s, there was a reasonable consensus concerning the optimum photon energy for red LEDs. The room temperature cross-over was recognized to occur at a band gap of $E_g = 2.0$ eV, suggesting a maximum possible direct luminescence transition at $h\nu = 1.93$ eV ($\lambda = 640$ nm). At this wavelength an external efficiency of 1% would correspond to a luminous efficiency of 1.4 lm W^{-1}. The group at IBM perfected their techniques and reported (in 1969) 6% external efficiency at a photon energy of 1.65 eV ($\lambda = 752$ nm) and 0.8% at 1.83 eV ($\lambda = 678$ nm). The corresponding luminous efficiencies were 0.033 and 0.11 lm W^{-1}, illustrating the importance of obtaining as short a wavelength as possible for optimum visual effect. In 1971, Diersche et al. at Texas Instruments, using LPE growth and Zn diffusion, reported $\eta_{ext} = 4\%$ at 695 nm (0.14 lm W^{-1}) and 0.4% at 675 nm (0.07 lm W^{-1}). At the same time, Blum and Shih from IBM reported somewhat similar performance. In spite of all these efforts, the results were still not competitive with the improved GaAsP diodes and it took several more years before the ultimate performance was approached. The development of high quality DH lasers during the 1970s led to similar structures being used as LEDs and by the early 1980s efficiencies of a few per cent at optimum wavelengths of 650–660 nm were being obtained reproducibly in Europe, Japan, and in the United States. For example, in 1983, Ishiguro et al. of Matsuchita grew a thick AlGaAs substrate by LPE and etched away the original GaAs substrate to obtain a DH diode with fully transparent cladding layers. They thereby achieved an external efficiency of 8% at $\lambda = 660$ nm and luminous efficiency of 4.4 lm W^{-1}. Further steady progress led to efficiencies of 10–15% ($\eta^L = 7$–10 lm W^{-1}) becoming routinely available in the 1990s and improved growth techniques produced material smooth enough for lithography, enhancing the possibility of commercial exploitation. By this time GaAsP had been left well behind but, in the meantime,

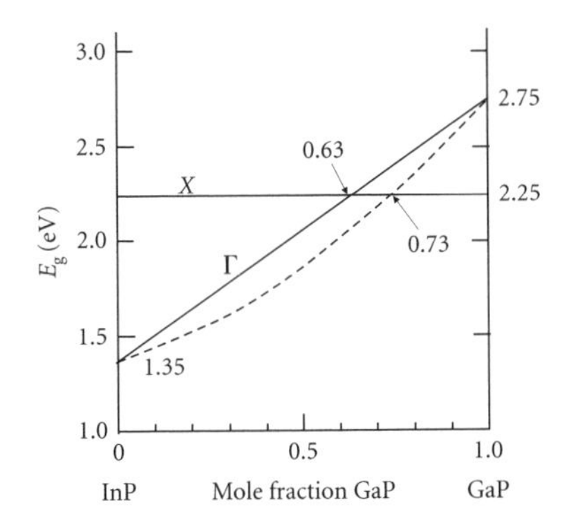

Figure 7.5. Energy band diagram for the ternary alloy $Ga_xIn_{1-x}P$, showing the conflicting experimental results obtained by the European and American 'camps' for the variation of E_g with Ga fraction x. In particular, the direct/indirect cross-over points differed significantly—0.63 and 0.73, respectively.

another competitor was demanding attention—AlGaInP was promising to outperform even AlGaAs.

Once again, the story of AlGaInP began in the 1960s and it began with a controversy. First to be explored was the ternary alloy $Ga_xIn_{1-x}P$ and, based on the fact that InP is a direct gap material, while GaP is indirect, we might expect the plot of band gap vs Ga fraction x to look rather like that of GaAsP as shown in Figure 7.4. Indeed, the first such plot reported by the 'European' combination of Hilsum and Porteus at the 1968 'Physics of Semiconductors' conference in Moscow did so—the direct Γ minimum varied approximately linearly as indicated by the solid line in Figure 7.5, crossing the (almost flat) X minimum at the point $x_c = 0.63$, $E_g = 2.25$ eV. However, later that same year a group at the IBM Watson Research Centre in the United States published an *Applied Physics Letter* with a surprisingly different set of data. Their results followed a significantly bowed curve (the dashed line in Figure 7.5) which gave a cross-over point of $x_c = 0.73$, $E_g = 2.26$ eV. At first it appeared possible that the discrepancy resulted from the difference in measurement technique—the former group using optical absorption, the latter electroluminescence—but time and a wide range of further measurements from both Europe and America gave the lie to this simple explanation. It became clear that luminescence and absorption data yielded closely similar values of band gap energy but the discrepancy remained—one group of publications favoured $x_c = 0.63$, another $x_c = 0.73$ and the argument was still in full flow as late as 1978 when Casey and Pannish published their seminal book on *Heterostructure Lasers*. The likely explanation appeared to lie with the quality of the material—various methods of crystal growth were employed, bulk growth, epitaxy on mismatched substrates (e.g. GaAs) and epitaxy on matched substrates (GaAsP)—but what seemed strange about this was the fact that the 'best', lattice-matched material gave lower band gap energies, counter to one's intuitive feeling about such things. Of course, it could be argued that there was no practical point of concern because the maximum direct energy gap involved was essentially the same in both cases $E_g = 2.25$ eV, suggesting, for efficient luminescence, a maximum photon energy of about 2.2 eV ($\lambda = 560$ nm) in the green part of the spectrum and very close to the peak in the eye sensitivity curve. This was clearly exciting, whatever the explanation of the above discrepancy and the IBM group measured LED efficiencies as high as 3×10^{-4} which corresponded to luminance values of 0.2 lm W^{-1}, remarkably bright compared with earlier diodes based on GaAsP. The problem, as ever, came down to that of growing high-quality material with this energy gap. Bulk growth (a version of the Bridgman method) was possible but tended to produce polycrystalline material, epitaxy on matched GaAsP substrates (actually graded GaAsP epitaxy on GaAs) resulted in moderately high dislocation counts, epitaxy on mismatched substrates (GaAs or GaP) provided even worse quality. The goal of high luminescence efficiency seemed frustratingly out of reach. What was to be done?

The virtues of lattice-matched AlGaAs/GaAs pointed the way. It was possible to grow lattice-matched $Ga_xIn_{1-x}P$ on GaAs substrates when the value of x was close to 0.5 (a paper by Stringfellow in 1972 made clear that it was vital to get the composition just right) and a glance at Figure 7.5 shows that this alloy could be expected to show a band gap of either 1.92 or 2.08 eV (photon energies of about 1.85 or 2.00 eV), according to one's allegiance! However, subsequent data obtained from a very wide range of experimentalists concerned with developing both LEDs and LDs strongly favoured the lower value, corresponding to a wavelength of about 670 nm. The origin of the alternative value predicted from the 'European' data is still something of a mystery. It may have been good to reach a consensus but it was disappointing that it should restrict the shortest wavelength to the red spectral region. The promise of yellow and green suggested by Figure 7.5 was tantalizing, to say the least, and the thought that all this honest endeavour should end there, was scarcely acceptable. Fortunately, neither was it necessary.

The remedy was already at hand in the shape of data (also obtained by the IBM group in 1970) on the similar ternary alloy $Al_xIn_{1-x}P$. In this case the E_g vs x plot followed the anticipated (approximately) linear behaviour and showed a cross-over at $x_c = 0.44$, $E_g = 2.33$ eV. (Later measurements by Bour and Shealey at Cornell (1987) essentially confirmed their data.) This, on the face of it, was also frustrating because it showed that the alloy composition lattice-matched to GaAs was indirect and was unlikely to offer high luminescence efficiencies. However, it suggested the possibility of exploring the quaternary system AlGaInP—by substituting some of the Ga in $Ga_{0.5}In_{0.5}P$ by Al, one had the prospect of an increased direct band gap while maintaining the lattice-match condition. All that was necessary was to establish just how much Al could be substituted before the band gap became indirect. This was finally accomplished in 1982 by Asahi et al. of NTT, the Japanese national telecommunications company. They grew the quaternary by MBE and demonstrated that it remained direct for Al : Ga ratios up to 0.35 when the band gap had increased to 2.3 eV, offering the hope of high luminescence efficiency at wavelengths as short as 550 nm, in the green spectral region. The flood gates were open for a concerted attack on the red, yellow, and green targets for both LEDs and LDs—we shall discuss lasers in the final section of this chapter but the story of the LED continues here.

Let it be said that AlGaInP is a singularly difficult alloy to grow and progress depended on a considerable effort to control composition and maintain an adequate lattice-match to the GaAs substrate. In contrast to the case of AlGaAs, it was not practical to grow films by LPE, on account of the large ratio of Al in the melt to that in the grown crystal, so the field was open to MBE and MOVPE. Both approaches were adopted and both have been successful—both, indeed, are in commercial use today, though MOVPE appears to dominate the LED activity—after overcoming an interesting surprise set-back. MOVPE was first introduced for growth of AlGaInP towards the end of the 1980s when it was discovered that the band gap varied with the crystallographic orientation of the GaAs substrate! When the usual (1 0 0) orientation was used the measured band gap was found to be some 100 meV smaller than measured on a similar sample grown by LPE. It was then discovered that growing on significantly disoriented substrates led to an increase in gap—in particular, if the (1 1 1)B or (5 1 1)A orientations were used, the gap was restored to its accepted value. This was eventually traced to the phenomenon of segregation—in the case of GaInP, the Ga and In atoms tended to form an ordered structure, rather than be randomly distributed throughout the crystal. In other words, the crystal took the form of a short period superlattice which (as expected) had a slightly smaller effective band gap than the equivalent random alloy. It had been established in 1977 by Bruhl et al. that segregation occurred in GaInP and some possible evidence for its occurrence in MBE grown samples was noted in the early 1980s, but the full appreciation of its importance only emerged towards the end of the 1980s when MOVPE was introduced by a group of Japanese workers. This might have provided an explanation for the controversial results of E_g vs x measurements during the 1970s except that ordering resulted in a band gap even *less* than the smaller of those values! It was nevertheless important to find a means of eliminating it in the interest of obtaining high brightness LEDs and lasers. It was fortunate, then, that the use of disorientated substrates inhibited the ordering process and allowed the sample to show its 'natural' band gap.

Once the ability to master crystal growth problems became widely established, it was possible to push ahead with developing high brightness diodes and the 1990s saw tremendous strides taken. The work was based on the application of double heterostructures similar to those used for laser diodes, allowing the use of an undoped active layer of high crystal quality together with cladding layers which were transparent to the emitted light. Red light ($\lambda = 660$–70 nm) could be generated by structures employing $Ga_{0.5}In_{0.5}P$ active layers, shorter wavelength orange, yellow, and green emitters required variable amounts of Ga to be replaced by Al. According to the group at Hewlett-Packard led now by the ubiquitous George Craford, it required as much as 70% Al to achieve a wavelength of 560 nm in the green (perhaps the original work by Asahi et al. was in error here?) but the efficiency dropped severely, compared with values obtained at longer wavelengths (640–590 nm). This points to yet another controversy over a Γ–X cross-over. The original work by Asahi et al. suggested that it occurred at $x = 0.35$, $E_g = 2.3$ eV, whereas Craford seems to be putting it nearer $x = 0.7$! There was obviously a need for careful measurements to

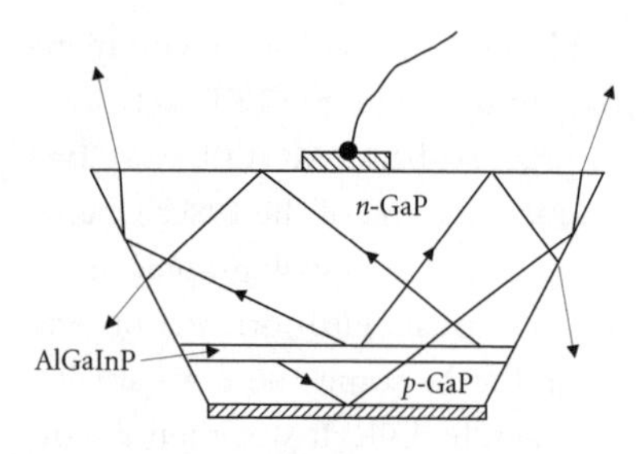

Figure 7.6. One example of the many different structures employed in efforts to improve the light extraction factor from light emitting diodes. This is the so-called 'truncated inverted pyramid' structure, used by researchers at Hewlett-Packard to enhance the brightness of orange AlGaInP devices. The maximum dimension is about 1 mm. Some typical light rays are shown to illustrate how they are reflected before being emitted. It is important that the material of the pyramid is transparent to the emitted light. External efficiencies in excess of 50% have been achieved with this technology.

resolve the discrepancy and results from un-ordered material at the University of Sheffield in England in 1994 provided reliable values of $x = 0.5$, $E_g = 2.25$ eV which are consistent with measurements on AlGaInP lasers at the University of Wales in Cardiff. This suggests a maximum photon energy for direct material as being about 2.20 eV ($\lambda = 564$ nm). On this basis, the Hewlett-Packard material would obviously be indirect which accounts for the reduced efficiency.

Once the material was under control, further advances depended on structural innovation and quite remarkable improvements in external efficiency were obtained during the 1990s. For example, a thick layer of GaP grown on top of the diode served as a window, allowing η_{ext} to exceed 6%, while removal of the absorbing GaAs substrate and its replacement with a wafer-bonded GaP substrate effected a further factor of 2 improvement. Diodes with over 30% efficiency were achieved by using an active region composed of multiple thin layers ($L_z = 50$ nm), rather than a single thick layer and values of $\eta_{ext} > 50\%$ by employing a 'truncated inverted pyramid' structure (see Figure 7.6) to optimize the light extraction factor. The net result of all these modifications has been to produce LEDs with luminous efficiencies of 100 lm W^{-1} at a wavelength of 610 nm (orange), though falling fairly steeply below 600 nm. Clearly there are still problems associated with using large amounts of Al but, even in the green spectral region, luminous efficiencies of nearly 10 lm W^{-1} have been demonstrated, some 10 times greater than available from the standard GaP : N green diodes. We leave further discussion of these developments until we have completed our survey of all the LED technologies currently available—suffice it to say that this dramatic progress is putting pressure on several well-established lighting technologies.

7.3 Gallium phosphide

Gallium phosphide was one of the first light emitting materials to be investigated. Electroluminescence from rather primitive structures was reported as early as 1955 and by 1961 'serious' LEDs had been made at Bell Labs, emitting in both the red and green spectral regions, albeit with rather low efficiencies ($10^{-8}-10^{-4}$). Somewhat similar results appeared in Europe in 1963. In this early work, GaP films were deposited by vapour phase epitaxy (VPE) methods on GaAs substrates and it was only with the availability of Czochralski-grown GaP substrate material in 1968 that significant progress could be made towards commercially practical devices. The other important development was the use of LPE which produced purer films with reduced densities of non-radiative recombination centres. The improvement in efficiency was then dramatic—in 1969, Saul et al. at Bell reported efficiencies for red lamps as high as 7% and in 1972

Solomon and Defevre at Fairchild achieved 15%. Corresponding results for green diodes were less impressive, the best from Bell and from RCA being about 0.7%, though Bergh and Dean, in their 1972 review article, were confident that GaP offered the best red and green devices available at that time. Allowing for the strongly wavelength dependent eye sensitivity, the green diodes appeared somewhat brighter than the red, in spite of the difference in their efficiencies.

A point of particular interest in the development of GaP as a light emitting material concerned the rapid improvement of our understanding of recombination mechanisms. As we have made clear, the fact of its having an indirect energy gap may have been a disadvantage for the development of commercial LEDs, but it turned out to be advantageous when unravelling the complex physics involved in radiative recombination. We shall therefore take this opportunity of filling in some necessary background which we have so far neglected. Whether the recombining carriers are injected by forward biasing a *p-n* junction, as in an LED, by excitation with a high energy electron beam in cathodoluminescence, or with a light beam having a photon energy greater than the band gap, as in photoluminescence, we are concerned with a wide range of possible recombination mechanisms. In practice, the rate at which electrons and holes are scattered by interaction with lattice vibrations (phonons) is much faster than typical recombination rates, so we can reasonably assume the process to start from a situation where we have distributions of electrons thermalized near the bottom of the conduction band and of holes near the top of the valence band. One possible recombination process is clearly that, often referred to as 'intrinsic', in which a conduction band electron recombines directly with a valence band hole. Particularly in an indirect gap material, this process is less probable than certain others which involve free carrier capture by inter-gap recombination centres. It is appropriate here to examine some of these in greater detail.

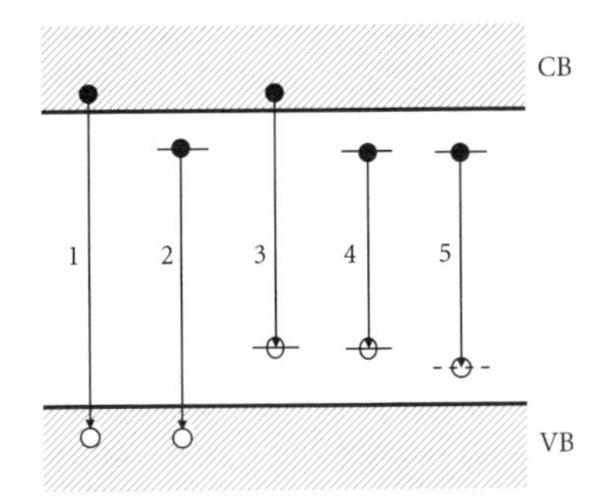

Figure 7.7. Energy diagram to illustrate possible radiative recombination mechanisms. (1) Intrinsic, band-to-band recombination; (2) Donor-to-valence band transition; (3) conduction band-to-acceptor transition; (4) donor-to-acceptor transition; (5) Donor-bound exciton recombination.

Figure 7.7 illustrates the possibilities—and let me emphasize that we are concerned with *radiative* processes—there are, of course, various non-radiative processes which compete with the light-generation mechanisms, but these are generally less well understood and we shall not attempt any detailed description here. Having already disposed of the band-to-band process, we may move on to the second process illustrated in Figure 7.7, recombination of an electron captured by a shallow donor with a free hole in the valence band. Note that this represents the recombination of a mobile particle (the hole) with a *localized* particle (the electron). The electron localization is important in the sense that, once bound to the donor, the electron is no longer free to find one of the many non-radiative centres which compete with radiative processes to reduce the radiative efficiency. Assuming the electron remains bound, it must eventually recombine radiatively so, in terms of the competition between radiative and non-radiative processes, the important parameter is not the recombination time but the capture time which is likely to be considerably shorter. But this begs the question as to whether the electron stays bound to the donor—in general, there is always a finite probability for it to gain sufficient energy from lattice vibrations to escape back into the conduction band. It is then free to move, making it susceptible to capture by another donor or, alternatively, by a non-radiative centre. Because donor binding energies in GaP are typically of order, 100 meV (i.e. $E_D \sim 4kT$ at room temperature), and the escape probability contains a factor $\exp\{-E_D/kT\}$, there is a fair likelihood that the electron will remain bound long enough for radiative recombination to occur, even in an indirect gap semiconductor. We can, at least, say with confidence that radiative recombination is a good deal more likely via this process than by way of the band-to-band process. It is also clear from the figure that the photon energy of the emitted light will be somewhat less than the band gap—in fact, the peak of the emission energy occurs at $h\nu = E_g - E_D + kT/2$ (see Box 6.4).

The third recombination process is a mirror image of the second—it involves a bound hole recombining with a free electron and much of the above comment applies to it equally. However, many of the common acceptors in GaP have binding energies E_A close to 50 meV which makes the escape probability significantly larger—we can therefore be less confident that the radiative process will dominate at room temperature. (Non-radiative centres have an advantage here because they usually lie much deeper in the gap—once they capture a free carrier, they just do not let go!) Clearly, a reduction in temperature can have a major influence and there is plenty of evidence for low temperature luminescence resulting from such processes, though, as the measurement temperature is raised towards room temperature, the intensity tends to fall sharply as holes become thermally delocalized. A final comment—if the material is doped *n*-type (for example), there will be an abundance of free electrons available, so it is the fate of the minority holes which determines the radiative efficiency and this third process is therefore particularly relevant to such a situation. In *p*-type material, the second mechanism is important.

Moving again to the right in Figure 7.7, we meet a fourth radiative process which has given hours of fun to semiconductor scientists faced with unravelling its intricacies! Here we have a situation where both types of carrier are localized, electrons on donors, holes on acceptors, the recombination process being referred to as donor–acceptor (D–A) pair recombination. It was first recognized in the mid-1950s by people working on II-VI compounds, more specifically by Prenner and Williams (1956) in ZnS but its understanding came to glorious fulfilment in the work on GaP undertaken during the 1960s and early 1970s. The very first observation of D–A pair lines in the emission spectrum of a

GaP sample was reported in 1963 by Hopfield, Thomas, and Gershenzon of Bell Labs and the flood gates were thereby opened to an outpouring of literally hundreds of subsequent papers exploring, explaining, confirming, and refining the basic concepts to an unbelievably high degree of sophistication. Why was this apparently simple phenomenon treated to such hero-worship? There were two reasons: first GaP was an important material in the quest for efficient semiconductor light emitters and therefore claimed a large share of attention in a number of industrial laboratories round the world (it meant, for instance, that crystal growth was well developed, making available a wide range of carefully controlled samples) and, second, the scientific culture of the time was such that basic research to understand commercially relevant materials and phenomena was strongly supported by R&D managers on both sides of the Atlantic—in this particular case, Bell Labs, RCA, and Philips were strongly involved. The afterglow of euphoria following the tremendous surge in semiconductor science responsible for the invention of the transistor and its younger sibling, the integrated circuit was still influencing R&D funding. So long as a promising commercial outcome could be recognized by management, eager young scientists were given their head. That explains the commercial interest—where, though, lay the scientific?

The unique feature of the D–A pair recombination process lies in its dependence on a *pair* of impurity centres which may take up a great many different relative positions in the crystal lattice, from nearest neighbour to very distant pairing. In each particular sample, therefore, there are very many distinct pairs, each of which makes a contribution to the luminescent spectrum and this implies a wealth of detail in such a spectrum, detail which is peculiar to the particular donor and acceptor species involved. To understand this, we need to examine the expression which represents the photon energy of the radiation emitted by each D–A pair:

$$h\nu = E_g - (E_D + E_A) + e^2/4\pi\varepsilon\varepsilon_0 r - C/r^6 \tag{7.4}$$

where r is the separation of the donor and acceptor and ε is the relative dielectric constant of the semiconductor. The first two terms on the right of this expression will be self-evident from a glance at (4) in Figure 7.7, the remaining two require some explanation. The term in r^{-1} represents the Coulomb energy of a pair of electronic charges separated by a distance r and makes its appearance because, before the recombination transition, the two centres are electrically neutral, whereas, after the transition, they carry equal, but opposite electron charges. Before the transition there is no Coulomb energy associated with the pair—after the transition there obviously is. Work has been done by the introduction of these two charges and this contributes to the overall energy of the emitted photon.

The final term represents the effect of polarization—it is small except for very close pairs and we shall neglect it.

It follows from equation (7.4) that, for a specific pair, having particular values of E_D and E_A, there will be a large number of emission lines associated with all the possible values of r which are determined by the crystal structure and lattice parameter of the host crystal. Thus, a typical photoluminescence spectrum looked like the example shown in Figure 7.8, each of the sharp lines corresponding to one specific value of the pair separation r, while at large separations, the individual lines could no longer be resolved and merged into a broad band. It is not difficult to imagine the concentration of effort which went into their detailed interpretation (Box 7.3 offers a little further explanation). What is more, there were six distinct donor and seven acceptor species, all with different ionization energies (see Table 7.1), which were available for study, making a total of 42 different pairs, each with its own characteristic set of lines. And, hidden in this statement lurks yet another subtlety—there are four distinct *kinds* of pairs. If we look at the donors, for example, O, S, Se, and Te act as donors when substituted on the

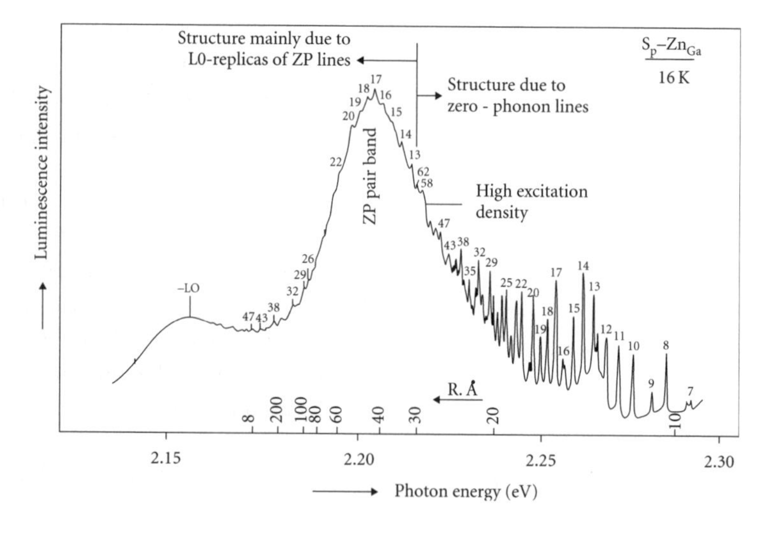

Figure 7.8. Example of a (D–A) pair photoluminescence spectrum measured at low temperature. The donors are S atoms on P sites, the acceptors Zn atoms on Ga sites (a Type I spectrum). The spectrum shows sharp lines due to relatively close pairs, a broad line due to distant pairs plus weaker sharp lines due to 'phonon replicas' (transitions involving the simultaneous emission of both photons and phonons). The numbers against each sharp line are the 'shell numbers' (see Box 7.3). (From Vink, A. T. (1974) Ph.D. Thesis, Technical University of Eindhoven, p. 12). Courtesy of Bibliotheek Technische Universiteit Eindhoven.

Table 7.1. Donor and acceptor binding energies in GaP

Donor energies (meV)		Acceptor energies (meV)	
O	895.5	Be	50
S	104.1	Mg	53.5
Se	102.7	Zn	64
Te	89.8	Cd	96.5
Si	82.1	C	48
Sn	65	Si	203
		Ge	300

Box 7.3. Donor–acceptor pair luminescence

The key to interpreting the details of the fine structure in (D–A) pair spectra lies in the evaluation of the Coulomb term in equation (7.4). This, in turn, involves finding an expression for the radii of the various 'shells' in the zinc blend lattice, corresponding to (1) the nearest neighbour separation, (2) the next nearest neighbour separation, (3) the next but one nearest neighbour separation, etc. The numbers 1, 2, 3, etc. are referred to as the shell numbers 'm' and we need a relationship between m and the corresponding radii r, bearing in mind that there are two kinds of pairs—those in which the donor and acceptor are on different lattice sites (Type I) and those in which they occupy the same sites (Type II). This problem was dealt with in the early literature—see, for example, Thomas et al. 1964 who derived the following expressions:

$$\text{Type I} \qquad r(m) = (m/2)^{1/2} a_0, \tag{B7.19}$$

$$\text{Type II} \qquad r(m) = (m/2 - 5/16)^{1/2} a_0, \tag{B7.20}$$

where a_0 is the lattice parameter ($a_0 = 0.545$ nm for GaP).

Let us consider the Type I spectra. We can evaluate the Coulomb term for nearest neighbour pairs using the value $r_{nn} = 0.236$ nm and $\varepsilon = 11.0$ for the low frequency dielectric constant of GaP as follows: $e^2/4\pi\varepsilon\varepsilon_0 r_{nn} = 0.555$ eV. If we then take a typical value of $(E_D + E_A) = 0.15$ eV for GaP we obtain a value for the photon energy of about 0.4 eV *above* the band gap! This serves to tell us that we cannot expect to see luminescence from *nn* pairs. In fact, to reduce $h\nu$ to below the band gap, we need to increase r to a value greater than about 0.9 nm, nearly four times the *nn* distance (or about $1.7a_0$). This corresponds to the sixth shell for Type I pairs. Thus, in practice, we can expect to see sharp lines from about the seventh shell upwards until the lines become so close together as to be impossible to resolve, when they merge into a broad 'distant pair' line.

We can estimate a typical energy separation of adjacent lines as follows:

$$\Delta E_{8,9} = (E_8 - E_9) = (e^2/4\pi\varepsilon\varepsilon_0 a_0)\,[(2/8)^{1/2} - (2/9)^{1/2}]$$
$$= 6.86 \text{ meV}, \tag{B7.21}$$

or approximately 7 meV between adjacent lines at the high energy side of the spectrum. The separations will gradually decrease as we move to larger shell numbers. For shell numbers greater than 50 the separation will have dropped to about 0.5 meV (the term in square brackets in equation (B7.21) becomes $[(2/50)^{1/2} - (2/51)^{1/2}]$) and individual lines will not be resolved. At the low energy limit, the emission will tail-off because the recombination lifetimes become extremely long and the emission rate tends towards zero (see main text). Thus, at least in general terms, we can explain all the features of Figure 7.8.

Group V site (i.e. when substituted for P), while Si and Sn, being in Group IV, are Group III site donors. Similarly, Be, Mg, Zn, and Cd are Group III site acceptors, while Si and Ge are Group V site acceptors, so there was a choice of Group V donor with Group III acceptor, Group V donor with Group V acceptor, Group III donor with, . . . , all of which added spice to the analytical cake. 'Eager young scientists' had much, indeed, to be grateful for!

Nor was this the end of the excitement. When measurements of recombination lifetime were made, yet further subtleties emerged. Because each emission line corresponded to a different pair separation, it was characterized by a different degree of overlap between the trapped electron and hole wavefunctions. As we saw in our very first chapter (Box 1.4), an electron trapped on a donor can be regarded as a kind of hydrogen atom, each bound state having a hydrogen-like wavefunction which represents the probability of the electron being found at a specific distance from the 'nucleus' (donor atom). In particular, the average distance for the ground state is characterized by a Bohr radius which depends on the electron effective mass and, therefore, on the donor binding energy but there is an exponentially diminishing probability of the electron being found at significantly greater distances from the donor. Similar comments apply to the hole bound on the acceptor, so it will be apparent that each (D–A) pair will be characterized by a wavefunction overlap which depends on (1) the nature of the donor and acceptor involved, and (2) on each specific value of separation. This is important because the recombination lifetime for each pair is inversely proportional to the square of the wavefunction overlap and this means that we can expect a considerable variation in lifetime as a function of the pair separation. Indeed, this was precisely what was found in practice—typically, time varied from 10 μs for fairly close pairs ($r = 1$ nm) to 1 s for more distant pairs ($r = 5$ nm) and was measured on a wide range of (D–A) spectra. Note that these times are relatively long for GaP because of its indirect gap—similar measurements on a direct gap material such as GaAs resulted in times some three orders of magnitude shorter (though it should also be said that the fine structure seen so clearly in GaP spectra is not resolved in GaAs).

One final comment is necessary. All these measurements were performed at low temperatures because, as the temperature was raised towards room temperature, the shallower of the two centres (donor or acceptor) tended to ionize and luminescence intensity was transferred to the appropriate free-to-bound transition. To be specific, if we consider the case of (S–C) pairs in GaP, where the donor energy is much larger than the acceptor energy, holes were thermally excited from the C acceptors, leaving the electrons still trapped on the donors, so the dominant luminescence transition became that labelled (2) in Figure 7.7. What is more, it will be clear, for this same reason, that (D–A) recombination does not play any serious role in the operation of LEDs at room temperature. Its main virtue has been the considerable light it has thrown on our understanding of the properties of GaP, in particular, and on the nature of radiative recombination, in general.

What, then, *is* the nature of the radiative processes responsible for GaP LED emission? Let us first consider green emission. The energy gap of GaP at room temperature is $E_g = 2.26$ eV, while the green emission peaks at about 2.21 eV ($\lambda = 560$ nm), suggesting the involvement of a relatively shallow recombination centre. Given that the ionization energies of several acceptors in GaP lie in the range 50–60 meV, one might have guessed that the appropriate transition was of the type 'free electron-to-bound hole' (e-A), an eminently reasonable guess, were it not for the low probability for such a transition to occur (a result of the indirect band gap in GaP). The surprising feature of the green emission was its relatively high efficiency, in spite of the indirect nature of the GaP band structure, seeming, therefore, to imply some novel process—and novel, indeed, it was. It depended on the incorporation of a small amount of nitrogen in the GaP crystals, N forming what came to be known as an 'isoelectronic centre', 'isoelectronic' because N comes from the same column in the periodic table as P and therefore has a similar configuration of outer electrons. In this, it differs from the donors and acceptors with which we are now familiar—they come from an *adjacent* column and have either one electron more or less than the host atom which they replace—so N is clearly neither a donor nor an acceptor. However, surprisingly enough, it *is* able to bind an electron from the conduction band with a binding energy of about 10 meV. A similar centre had been recognized in 1964 when Aten and Haanstra (of Philips Research Labs) reported on Te centres in CdS so there was already a precedent when, just 1 year later, Thomas et al. (at Bell) discovered similar properties for N in GaP.

The theory of how these centres can bind electrons is complex—suffice it to say that it depends on the relative size and electronegativity of the impurity, compared with the host atom—but one feature which we do need to emphasize is the strong degree of localization involved. The electron is confined very closely to the N centre and, in this respect, it is significantly different from an electron bound to a donor atom by the Coulombic attractive force appropriate to that case. This is important because it implies something about the corresponding momentum of the bound electron. If we recall one method of stating the Heisenberg uncertainty principle:

$$\Delta x \Delta p = h/2\pi, \tag{7.5}$$

to the effect that precise knowledge of the electron's position x implies large uncertainty in its momentum p, we immediately see that strong localization of the bound electron implies a large spread in the possible values of its momentum. In other words, the electron wavefunction contains values of the momentum vector drawn from right across the Brillouin zone—the free electron may be characterized by only those momentum values close to the edge of the zone at the X point but the bound electron wavefunction contains all values of k, including those from the zone centre at the Γ point. This is crucial because it implies a very much stronger probability for the optical transition between the bound state and the top of the valence band, at the Γ point—in other words, the probability of radiative recombination is very much greater than for a normal band-to-band process (or for transitions involving shallow donors or acceptors). This is just what is needed for efficient light emission and goes much of the way towards overcoming our (justified!) prejudice against indirect gap semiconductors. Nature loves to surprise us but the principal surprise in this case lies in the fact of her making life *less* difficult than we anticipated!

There is just one more subtlety involved in the GaP green emission. The complex formed by an electron bound to a N isoelectronic centre is negatively charged (another difference from a donor-bound electron), so it can rather easily attract a free hole to form a bound exciton which recombines to give green light. It was clear, then, as early as 1966 that this was the process responsible for the emission from green LEDs. Note that it benefits equally from the strong localization of the electron, important because the total binding energy is still only a few tens of milli-electron volts and the complex can readily be thermally dissociated at room temperature. It therefore behoves the centre to recombine radiatively as rapidly as possible before such dissociation blights its chances in favour of some lurking non-radiative centre. (The experimental observation that the best efficiencies rarely exceed 1% is indicative of the fact that, in spite of the best efforts of materials scientists, this does often happen!) Note that the presence of something like 10^{25} m^{-3} N atoms in the GaP lattice has no effect on the free carrier density—it is necessary to dope the material *n*- and *p*-type to form the necessary *p-n* junction, recombination occurring on either side of the junction depending on the ratio of *n*- and *p*-type doping used. However, the density of N atoms incorporated did produce an important effect. As early as 1966, Thomas et al. at Bell labs had observed and explained the presence of many sharp lines in the absorption spectrum of heavily N-doped GaP crystals as due to the formation of NN pairs which also bound excitons and gave rise to emission lines at wavelengths in the yellow spectral region. Yellow LEDs were developed post haste, though they never showed very high efficiency and have now been replaced by the much more efficient AlGaInP devices.

One of the more interesting historical aspects of the story concerns the manner in which nitrogen was first incorporated. In 1964 Frosch, at Bell labs, reported a new VPE process for growing GaP films which used the reaction between GaP and water vapour at temperatures near 1000°C. Passing water vapour over a boat containing GaP chips produced Ga_2O plus P_2 which were transported to the GaP substrate where the reaction was reversed, depositing an epitaxial GaP film. When the open tube in which this took place was made from quartz, the films contained no nitrogen but if boron nitride tubing was used, water apparently attacked the tube to produce NH_3 which resulted in N being incorporated in the GaP film. In this rather fortuitous manner was the connection between the green luminescence and N established. In future methods (particularly when using LPE growth) N was introduced either by deliberately introducing NH_3 gas or by dissolving a small amount of GaN in the solution. It turned out that gaseous nitrogen, N_2 was far too chemically stable to be a practical source of N centres.

The reader who has survived this dose of explanatory physics will have little difficulty in understanding the origin of the GaP red emission. It was discovered at SERL Baldock, United Kingdom in 1962, quite early in the development of GaP, that strong red emission was produced when the material was doped with Zn (to make it *p*-type) in the presence of oxygen but it was some time before it was accepted that this resulted from a subtlely different form of bound exciton recombination. At typical (epitaxial) growth temperatures, both O and Zn are present in the crystal lattice as ions, O^+ and Zn^- which therefore enjoy a mutual attraction. It was not surprising, then, that they showed a tendency to form close pairs (i.e. to occupy nearest neighbour lattice sites), thereby forming electrically neutral centres. What was more, there was proof of this from a sophisticated form of infrared spectroscopy—the measurement of local mode vibrations. A Zn–O *nn* pair, though electrically neutral overall, actually represents an electric dipole which can be made to oscillate by the application of an oscillating electric field in the form of an electromagnetic wave. At resonance, when the frequency of the wave corresponds to the natural vibration frequency of the pair of atoms (a frequency lying in the infrared region) energy is transferred from the wave to the mechanical vibration, giving rise to an absorption spectrum. If the radiation is plane polarized, it is clear that the absorption will be a maximum when the electric vector is aligned with the axis of the dipole (a (1 1 1) crystal direction for a *nn* pair) and this confirms the orientation of the pair. A measurement of the resonant frequency can then be compared with the calculated vibrational frequency of a Zn—O pair to verify the nature of the centre.

Having thus confirmed their nature, it required only a minor leap of the imagination to recognize them as being similar to the isoelectronic centres we have just discussed, and the application of a little high powered theoretical effort to demonstrate that they too could bind excitons. The essential difference was that the binding energy was considerably larger—something like 300 meV—sufficient to shift the wavelength to 690 nm. The increased binding energy reduced the likelihood of thermal dissociation, leading to the much improved efficiencies to which we have already referred. The story was complicated, however, by the discovery of two competing radiative processes. Though work at IBM by Morgan et al. in 1968 demonstrated that red emission did, indeed, result from this bound exciton process, related work at Bell by Henry et al. (also in 1968) showed that an alternative mechanism, recombination of an electron trapped on a (Zn–O) centre with a hole trapped on an isolated Zn acceptor, also generated red light. It was left to yet another

Bell group, under the Englishman Paul Dean, to prove that it was the bound exciton process which dominated at room temperature. So, finally, it was clear that recombination of excitons bound to isoelectronic centres was responsible for both green and red emission in GaP LEDs at room temperature. Also clear was the depth of understanding of radiative processes in GaP which had emerged from over a decade of concentrated scientific effort.

7.4 Wide band gap semiconductors

The task of generating light over the red, orange, yellow, and green parts of the spectrum was successfully accomplished by around 1975, at least to the point of achieving luminous efficiencies of about 1 $\mathrm{lm\,W^{-1}}$. Red diodes would be made from AlGaAs, green from GaP:N, and the intermediate colours from either GaP:NN or GaAsP:N. The major problem which remained was that of obtaining a balancing performance from blue and violet emitters. The best available blue LEDs were made from SiC but showed quantum efficiencies of order 0.02%, yielding luminous efficiencies of about 0.004 $\mathrm{lm\,W^{-1}}$, some 250 times below those available in the red–green region. This made it impracticable to use LEDs in full-colour displays, seriously limiting their range of application, a situation which would only be remedied by the advent of efficient GaN LEDs in the mid-1990s. The present section is chiefly devoted to telling the story of GaN but we shall begin with a brief account of those other wide gap materials which had potential for filling this particular role.

Silicon carbide (see Bergh and Dean 1972; Morkoc et al. 1994; Willardson and Weber 1998: vol. 52), with a band gap of order 3 eV, which makes it suitable for visible LEDs, probably has the distinction of being the first semiconductor material to demonstrate electroluminescence. This was in 1907 when H. J. Round observed yellow luminescence from a SiC radiowave detector. Subsequently Losev and others during the 1920s observed similar luminescence and developed some limited understanding of the relation between light emission and rectification. In this early work, the material was relatively uncontrolled, the first serious attempts to grow good quality single crystals being made in 1955 when Lely, in Germany, investigated the sublimation method. However, as with most compound semiconductors, effective progress with devices depended on the introduction of epitaxy. In 1969, Bob Brander and colleagues at the GEC Hurst Research Centre in London developed an LPE process, using substrates grown by sublimation, while Potter et al. at the American GE Company employed a form of VPE. Over the next 30 years a variety of alternative methods was explored, both for substrate growth and for epitaxy—CREE, who are probably the leading exponents of SiC growth today, employ chemical vapour deposition (CVD) on sublimation-grown substrates.

The problems with SiC as an optoelectronic material are many: first it exists in a number of different polytypes (differing in the stacking arrangement of adjacent layers), each having a different band gap; second, these all possess indirect band gaps (with no indication of helpful isoelectronic centres such as those which bailed out GaP); third, though its melting point is sometimes quoted as 2800°C, there seems to be some doubt whether a liquid phase exist at all; and fourth, its extremely strongly bonded structure requires almost any processing to be performed at temperatures of order 2000°C. For example, the crystals grown by GE in the United States, during the 1960s, in an attempt to provide green and blue emitters required temperatures of 2500°C, while diffusion of acceptors to form the necessary *p-n* junction was performed at 2200°C. Such high processing temperatures made it extremely difficult to produce material with the degree of structural and chemical perfection appropriate to efficient LED production. For example, any impurities in quartz or other containers would be highly likely to be transported into the growing SiC crystals—purity would always be at risk.

During the 1960s, quite serious attempts to tame the material were mounted at GE, at GEC, in the Uinited Kingdom and in Soviet Russia, and most of this work was concerned with the so-called '6H polytype' which has a band gap of 3.0 eV. The *p-n* junctions were made using nitrogen donors and aluminium acceptors, though N actually resulted in two donor levels and the Al acceptor level was rather deep (200 meV) which limited the free hole densities obtainable at room temperature. N-doped material gave blue luminescence, Al-doping resulted in green emission, while red LEDs were made by doping with both Al and B, though efficiencies remained consistently low, at typically 3×10^{-5} or worse. Clearly, these devices were unable to compete with GaP, GaAsP, and AlGaAs. The small size of available substrates provided yet another barrier to commercial exploitation. However, the possibility of obtaining blue diodes proved sufficient incentive for continuing effort and their efficiencies gradually improved—in 1969 Brander reported a figure of 10^{-5}, in 1978, Munch (Hanover) claimed 4×10^{-5}, in 1982, Hoffman (Siemens) claimed 1×10^{-4}, and in 1992 Edmond (CREE) realized a figure of 3×10^{-4}. Needless to say, commercially available diodes were rather less good than this but good enough, nonetheless, to maintain a reasonably steady (though modest) market level. However, the desired objective of bright, full-colour LED display was still a long way off and the chief advantages of SiC, its chemical and thermal stability, and large thermal conductivity, seemed more likely to find application in high power microwave devices which could be operated up to temperatures of perhaps 300–400°C (see Section 6.5).

Two other groups of materials possessed band gaps wide enough to yield blue luminescence—the Group III nitrides, GaN and AlN, having $E_g = 3.4$ and 6.2 eV, respectively and the II-VI compounds ZnSe (2.7 eV),

ZnS (3.7 eV), and CdS (2.5 eV). There were several similarities between the two groups, such as the fact that they crystallized in both cubic (zinc blende) form or hexagonal (wurzite) form and the fact that it proved extremely difficult to dope any of the wide gap materials both *n*- and *p*-type, though, historically, the II-VI compounds were developed much earlier than the nitrides. For example, (D–A) luminescence in ZnS was recognized as early as 1956, exciton recombination at isoelectronic centres in CdS in 1964 and, by 1967, the II-VI compounds were enjoying the luxury of having their own conference (Thomas 1967). By contrast, the Group III nitrides merit scarcely a mention in Madelung's 1964 summary of the properties of III-V compound semiconductors, nor can they be found in the proceedings of the early 'GaAs and Related Compounds' conferences (1966, 1968, 1970)—it was not until the 1970s that serious semiconductor physics was done on any of the nitrides. It is interesting, therefore, that the important technological breakthrough, leading to the development of visible LEDs and LDs, namely, the ability to dope both *n*- and *p*-type, occurred at almost exactly the same time for both groups of materials. Nevertheless, we shall follow historical precedence and discuss the II-VI compounds first. Reviews can be found in the articles by Morkoc et al. (1994) and Willardson et al. (1997: vol. 44).

It will evoke no degree of surprise to learn that the early work on II-VI compounds was performed on fairly crude bulk crystal samples grown, for the most part, by a vapour transport method at relatively high temperature. Uncontrolled impurity concentrations were high, typical samples being rather strongly *n*-type (e.g. CdS, ZnS, ZnSe) or *p*-type (ZnTe), and apparently nothing could be done about that. Indeed, nothing ever was done to change these propensities of bulk crystals—it was not until the development of MBE and MOVPE epitaxial growth in the late 1970s that hope of type conversion revived and not until 1990 that success was finally achieved. However, a great deal of useful physical understanding was elucidated by measurements on bulk samples, including the observation of bright visible luminescence, characteristic of the direct energy gaps confirmed in all these compounds, making it even more frustrating that *p-n* junctions were impossible of attainment. Two possibilities existed to explain this: on the one hand, it could be the result of unwanted donor (or, in the case of ZnTe, acceptor) impurities being present in too large quantity or, (and for many years this was the accepted wisdom) the problem lay with some form of 'auto-compensation' due to lattice defects which countered any attempt to introduce *p*-type doping. The former difficulty could, in principle, be overcome by improved crystal growth techniques, whereas the latter had the appearance of complete intractability. Overall, it took something like 40 years to prove the correctness of the first explanation! Such is the tenacity of scientists when faced with a near-impossible task.

The first step required the selection of a suitable acceptor (we shall concentrate from hereon on the wide gap materials CdS, ZnS, and ZnSe). For some time most people thought this should be Li, which acts as an acceptor when incorporated onto the Group II site, but little success in *p*-type doping was actually achieved. In 1983, Paul Dean (having returned from the United States to RRE Malvern) demonstrated the acceptor properties of N in ZnSe. Though it still proved impossible to achieve type-conversion, he was able to measure the acceptor binding energy (111 meV) by analysing (D–A) pair spectra and to conclude, importantly, that failure to obtain *p*-type material was the result of excessive uncontrolled donor impurities, rather than to any in-built malevolence of the host crystal. Efforts to reduce this background were therefore redoubled and it fell to the lot of MBE growth to find a viable solution. MBE has the great virtue of allowing good control over the stoichiometry of the growing film and it was this which led to the desired reduction in donor levels. However, there was still a problem of how to incorporate the N acceptors—the use of N_2 gas proved totally ineffectual and it was only with the development of so-called 'plasma nitrogen sources' that success in growing *p*-type ZnSe was finally achieved. The principle depended on establishing a nitrogen plasma by means of a radio frequency discharge, thus creating a beam of *atomic* (rather than molecular) nitrogen. Park et al. in America demonstrated the technique in 1990, obtaining free hole densities up to 3.4×10^{23} m^{-3} and making a prototype LED which emitted blue light ($\lambda = 465$ nm). This was followed almost immediately by a successful application of MOMBE (metal organic MBE) by Taike et al. at the Hitachi laboratories in Japan who obtained *p*-type ZnSe with $p = 5.6 \times 10^{23}$ m^{-3}, using NH_3 gas as the *p*-type dopant. A year later Qiu et al. achieved $p = 1 \times 10^{24}$ m^{-3} but it became clear that this required $[N] > 10^{25}$ m^{-3} and any attempt to increase the hole density further actually led to a *reduction* in p. While this was certainly not catastrophic from the point of view of making suitable junctions, it led to serious problems in making low resistance ohmic contacts to the *p*-side of the junction and later success had to rely on a variety of subtle tricks which make life less than comfortable for the device technologist.

Most of the above work was performed by growing ZnSe films on substrates of GaAs (to which it is very nearly lattice-matched), GaAs being available in much larger and cheaper wafers than the existing, somewhat less well developed bulk ZnSe crystals. However, there is a significant drawback to the use of GaAs in so far as there is a serious mismatch of thermal expansion coefficients. Even if the films are matched at the growth temperature, strain will surely exist in the film at room temperature. If high quality, large size ZnSe crystals can be produced commercially, they may well be preferred to GaAs in the long run—only time will tell.

Once the breakthrough had been made, it was rapidly followed by attempts to make LEDs and LDs based on ZnSe. In fact, quite complex structures have been employed to accomplish both carrier and light confinement but we shall leave their description until Section 7.5 when we discuss visible lasers. For the moment, we should simply comment that bright LEDs have been made by several groups in America and Japan. In 1995 Eason et al. (North Carolina State University and Eagle Pitcher) described both blue and green LEDs made by MBE growth on ZnSe substrates. Cl was used for *n*-doping and N for *p*-doping. Blue diodes employed an active region consisting of ZnCdSe quantum wells, while green diodes used a single layer of ZnSeTe. Quantum efficiencies were 1.3 and 5.3%, respectively, resulting in luminous efficiencies of 1.6 and 17.0 $\mathrm{lm\,W^{-1}}$. The ZnSe substrates were 50 mm in diameter which might be adequate for an introductory commercial process but the operating lifetime of these diodes was still not sufficient, being only some 500 h. This pointed out a serious problem with II-VI devices which appear to be very sensitive to the presence of line defects (dislocations). It was far from obvious that these problems could be overcome but there have been recent reports from Sumitomo Electric that they intend to commercialize a white LED based on ZnSe—blue light is emitted from the active layer, while a broader yellow emission is generated from the ZnSe substrate, the overall appearance being white. While this is certainly interesting, if one were cynical, one might feel tempted to believe that its principle virtue lies in its ability to bypass a host of patents filed to cover similar devices based on the nitrides (to which we shall soon refer in greater detail). Finally, as we shall discuss in Section 7.5, Sony are also pressing ahead with a ZnSe-based laser so there are clearly possibilities for II-VI visible and ultraviolet (UV)-light emitters. How these materialize remains yet to be seen.

All of which brings us to the romance of the nitrides! As we said earlier, the nitride story started seriously in the early 1970s when GaN crystals with reasonable quality first became available, though in this case these were not bulk crystals but epitaxial films grown by the halide vapour phase epitaxy (HVPE) process on non-lattice-matched substrates, principally sapphire. This immediately points up a major technological problem which has dominated the whole body of work concerned with both the physics and the commercial exploitation of the Group III nitrides—the acute shortage of GaN (or other nitride) substrate crystals. Why? In a word, because bulk crystals are extremely difficult to grow. GaN, for example, melts at approximately 2800°C, at which temperature the vapour pressure of N_2 over GaN is no less than 45 kbar (45,000 atm), making Czochralski growth virtually impossible. In practice, bulk crystal growth has been the sole preserve of one research group, led by Professor Porowski at the High Pressure Research Centre in Warsaw, whose work grew out of a long-term interest in the properties of solids under high pressures. GaN growth was developed during the 1980s, the method adopted being based on solution growth from a saturated solution of GaN in liquid gallium, contained in a BN crucible, at temperatures of order 1500°C under nitrogen overpressures of 10 kbar. Growth depends on maintaining a temperature gradient along the furnace tube and crystals in the form of platelets approximately 100 μm thick can be grown over a period of about a day. In their early work, lateral dimensions were a few millimetres—after several years of development, they now reach 1–2 cm. Until recently, crystals were strongly *n*-type ($n \sim 10^{25}$ m^{-3}) which precluded their use as device material, though recently the group has developed a method of growing semi-insulating material as a result of doping with Mg. These crystals have undoubtedly made an important contribution to the whole field of nitride research but one cannot see them providing a basis for any commercial developments on account of their high cost. Numerous attempts have been made to find alternative methods of making substrates but, so far, all serious commercial programmes are based on some form of heteroepitaxy.

Returning to 1970, we note that the VPE process grew naturally out of earlier work on other III-V compounds, particularly GaAs. Gallium was transported in the form of GaCl by passing HCl gas over a heated boat containing liquid Ga. In the growth of GaAs, arsenic was provided as $AsCl_3$ whereas, in the case of GaN, there being no nitrogen halides, nitrogen was transported as NH_3 and, because NH_3 is a much more stable molecule than $AsCl_3$, it required a significantly higher growth temperature, 1000–1100°C, rather than 800°C, to break it down. This, in turn, demanded a substrate which remained stable at these temperatures, ruling out the III-V compounds and leading to the choice of sapphire, even though there was a large lattice-mismatch with the grown film. Perhaps the main virtue of the HVPE process was its rapid growth rate >10 μm h^{-1} which encouraged the growth of relatively thick films (50–100 μm), film quality gradually improving towards the top surface. This meant that material quality was adequate for a wide range of measurements to be made and allowed the electronic and optical properties of GaN to be reliably catalogued during the 1970s. Three groups were active: Ray Dingle and Marc Ilegems at Bell, Jacques Pankove at RCA, and Bo Monemar at the Lund Institute in Sweden, all making use of HVPE material.

The (direct) band gap of GaN was established to vary between 3.503 eV at 4K and 3.44 eV at room temperature, and the structure of the valence band was also explored by means of photoluminescence spectroscopy and optical absorption. This revealed the anticipated three separate valence band maxima at the Γ point of the Brillouin Zone—'anticipated' because GaN crystallizes preferentially in the hexagonal (wurzite) structure and the axial component of the 'crystal field' raises the degeneracy of the pair of valence bands which are degenerate in zinc blende crystals (see Figure 5.1). It also showed that films were strained to a variable extent as a result of the mismatch in lattice parameter and thermal expansion coefficient between GaN and sapphire (strain alters the band gap and shifts photo luminescence lines in sympathy). Various studies were made on the effects of doping GaN with impurities such as Mg, Zn, and Cd which were expected to act as acceptors. Results were less than ideally clear-cut but suggested that acceptor binding energies were unexpectedly large, being of order 250, 350, and 550 meV, respectively (compared with an effective mass value which should be approximately 100 meV). These values appear to be related to the electronegativities of the individual atoms, but there is still no widely accepted explanation. More important, perhaps, was the failure of any of these acceptors to make the material *p*-type—not altogether surprising when we note that nominally undoped films were found to be *n*-type with free electron densities typically in the range of 10^{24}–$10^{25}\,m^{-3}$. This was generally attributed to the presence of N vacancies in the crystals, V_N acting as a shallow donor (though this interpretation has been disputed more recently, preference being given to an explanation in terms of uncontrolled impurities such as Si or O—shades of the dispute which bedevilled the II-VI scene for so many years!).

The fact that GaN had been shown to be a very efficient visible light emitter made this apparent aversion to *p*-type conductivity all the more frustrating. In fact, as early as 1971, Pankove did actually demonstrate a form of blue LED based on an MIS structure (metal–insulator–semiconductor) though the efficiencies of such devices are rarely competitive and this was no exception! The need to obtain type conversion became ever more desperate as the years slipped by and there was a noticeable decline in interest in the nitrides during the 1980s which placed them in very much the same 'political' position as the II-VI compounds. Why pour endless cash into a lost cause? However, not quite everyone took such a negative view—the nitride flag was kept resolutely flying in Japan by Professor Isamu Akasaki at Nagoya University whose group explored the application of MOVPE growth to GaN. MOVPE had been introduced surprisingly early (1971) to the nitride scene by its inventor Manasevit—it consisted basically of using gallium triethyl (or methyl) for the transport of gallium, rather than GaCl, while still using NH_3 for nitrogen and growing at 1000–1100°C as before. Its principal virtue, its ability to provide much better control over the growth process, was to prove of considerable value in the development of low-dimensional structures (LDS) in the AlGaAs/GaAs system during the 1970s and 1980s. However, its impact on nitride growth was not immediate. Early attempts led to films which cracked badly on cooling to room temperature and it was some time before a solution could be found. It was only in 1986 that Amano et al. (of Akasaki's group) discovered the advantage of introducing a special buffer layer between the sapphire substrate and the GaN film, which took the form of a thin film of AlN. This was grown at low temperature (~500°C) which resulted in its being amorphous but the temperature was then raised to 1000°C or more for the growth of the GaN film. Quite what happened to the buffer layer during this stage was not altogether clear but the net effect was dramatic—high quality GaN films could be grown with thicknesses of a few microns and, what was more, the background carrier density came down to the more acceptable level of 10^{23} m^{-3}, which certainly improved the chances of successful *p*-type doping.

What happened next was somehow typical of the capacity for GaN research to spring surprises. Akasaki decided to attempt *p*-type doping with magnesium, initially with no greater success than had been achieved during the previous decade, but then he noticed that, following their study by scanning electron microscopy, some samples did, finally show the desired result—*p*-type conduction could, at last, be measured. The electron beam which had been scanned over the film surface had magically 'activated' the Mg acceptors in sufficient numbers to counter the background donors and turn the film *p*-type. This was in 1989. In no time at all the process was christened LEEBI (low energy electron beam irradiation) and was being copied in numerous other laboratories—interest in GaN was reviving rapidly. The possibility that *p-n* junctions could be fabricated suddenly changed everything. However, the mechanism whereby LEEBI could activate the Mg was not understood and it soon became clear that it only operated in a thin surface region, not throughout the bulk of the film, an inconvenience, to say the least, so it came as a considerable relief when an important further discovery was made—activation could also be achieved by a simple thermal annealing process which 'worked' right through the Mg-doped region. This looked, and, indeed, was, a much more practical technology. From that point onwards progress seemed unstoppable—high brightness blue LEDs were demonstrated in 1993 and the first blue laser in late 1995 (pulsed operation at room temperature). By this time half the laboratories in Japan and the United States had established GaN programmes of one sort or another, though Europe was more sceptical (or simply more hide-bound?) and seemed surprisingly slow to follow suit. Nevertheless, it eventually got its act together (the first European GaN workshop took place in 1996) and today 'Nitride Semiconductors' represents one of the most firmly established programmes, worldwide, with a dedicated international conference attracting more than 500 participants. It is also characterized by a remarkable number of review articles and books, some of the more recent of which we list in the reference list—Mohamad et al. (1995), Mohamad and Morkoc (1995), Akasaki and Amano (1996), Neumayer and Ekerdt (1996), Ponce and Bour (1997), Orton and Foxon (1998), Gil (1998), Edgar et al. (1999), Willardson and Weber (1997: vol. 50; 1999: vol. 57), and Nakamura et al. (2000). No serious student need feel deprived of adequate study material!

But we are getting ahead of ourselves. The story of *how* the brightest of all LEDs came to pass has its own fascination. It goes back to the year 1979 when a young physicist, Shuji Nakamura, with a first degree from Tokushima University joined a small Japanese company called the Nichia Chemical Company, just a few miles from Tokushima. His early research into conventional III-V materials made relatively little impact except that it clearly impressed the Company Chairman who, in 1988, took an amazing gamble in financing Nakamura to the tune of $3 million to undertake a programme of work on what was then a difficult and far from promising material—GaN. Nakamura spent a year at the University of Florida to learn the technique of MOVPE, then returned to Nichia, set up his own version of the process and started to grow GaN films on sapphire. His first innovation, in 1991, consisted in using a GaN amorphous buffer layer in preference to Akasaki's AlN and, in 1992, it was he who discovered the thermal annealing cycle which led to effective *p*-type doping with Mg acceptors. He also speculated that the physics behind the process involved hydrogen compensation of the acceptors—(H–Mg) complexes were formed during epitaxial growth (hydrogen is generated by the breakdown of ammonia in the MOVPE process) which were dissociated during the heat treatment. Later work demonstrated the essential correctness of this hypothesis and it was confirmed by the observation in 1993 by groups at Boston University (Ted Moustakas) and the University of North Carolina (Bob Davis) that Mg doping in MBE growth (where no hydrogen is present) leads to *p*-type conduction in as-grown material, no anneal being required.

The stage was set for the race to produce the first efficient blue LED, the only serious competitors both being Japanese! Now that a material of much improved quality was available, the principal problem was to engineer diodes which emitted at the appropriate wavelength. The band gap of GaN is 3.4 eV at room temperature, whereas the desired photon energy ($\lambda = 450$ nm) is 2.75 eV, some 0.65 eV smaller. In the first instance (1991) both groups made simple homojunction diodes doped on one side with Si and on the other with Mg. The structure is shown in Figure 7.9 (note that, because the sapphire substrate was electrically insulating, it was necessary to etch down to the *n*-type layer in order to make electrical contact). Blue emission in these diodes came from the *p*-side of the junction, being associated (in some slightly obscure manner!) with the Mg centres. Measured efficiencies were about 0.2%, somewhat better than could be achieved with SiC but still not good enough to excite universal interest. Akasaki and colleagues tended to stay with this simple homojunction device and eventually obtained efficiencies of 1% or more, while Nakamura experimented with a variety of structures. His first variation employed InGaN in the active region, the incorporation of In reducing the band gap towards the desired value. However, it required something of order 30% In to obtain a gap of 2.75 eV and, because of

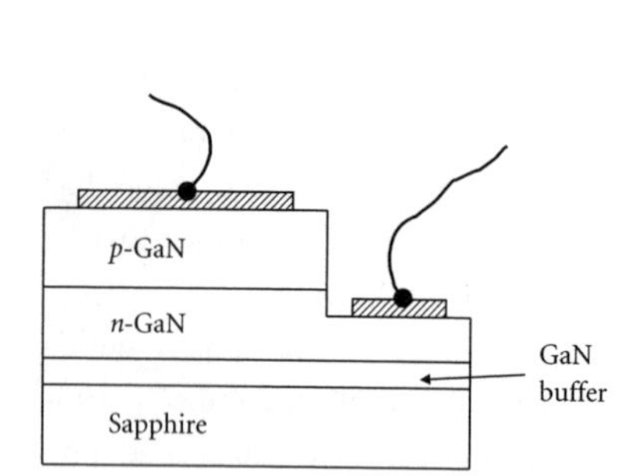

Figure 7.9. Schematic diagram to illustrate the structure of a simple homojunction GaN LED. A layer of Si-doped (*n*-type) GaN is grown on a sapphire substrate, using an amorphous GaN buffer layer to improve crystal quality. This is followed by a Mg-doped (*p*-type) layer which is partially etched away to allow an electrical contact to be made to the *n*-type side of the junction. The more usual back contact is impossible on account of the insulating properties of sapphire.

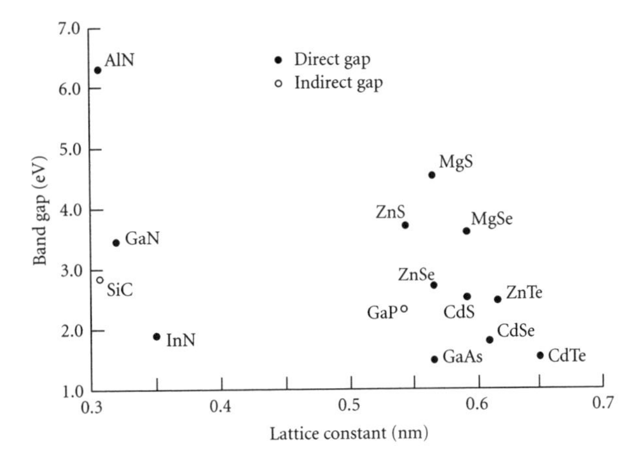

Figure 7.10. A plot of band gap energy vs lattice parameter for a range of semiconducting materials which are used in the fabrication of LEDs. The open circles represent indirect band gap materials. Note the large (11%) mismatch between GaN and InN and the smaller, but significant mismatch (2.5%) between AlN and GaN.

the 11% lattice-mismatch between GaN and InN (see Figure 7.10), this introduced too many defects, leading to relatively poor efficiency. His solution was to dope the InGaN with Zn which introduced a deep level, thus reducing the photon energy sufficiently that only a few per cent of In was then required to achieve the desired wavelength. Efficiencies of 3% were achieved with this structure but yet further improvements were made by the use of a narrow InGaN QW as the active layer, sandwiched between *n*- and *p*-type AlGaN cladding layers. This involved a high In fraction but, because of its extreme thinness, the layer was strained but contained relatively few dislocations. Nakamura achieved efficiencies of nearly 10% with this structure which compared favourably with those available from the best red LEDs of the time (1995). At about the same time, he also reported on green emitters (by increasing the amount of In in the well) and obtained 6% efficiency. (The details are spelled out at length in the book by Nakamura et al. 2000.) Clearly, material quality had improved enormously between 1991 and 1995 but one quite remarkable fact continued to challenge credulity—GaN films grown on sapphire substrates were universally found to contain something like 10^{14} dislocations per square metre, while nevertheless performing as efficient light emitters. Such densities result in complete darkness when located in most other semiconductors, whereas GaN not only shines brightly but continues to do so for extremely long operating lifetimes, degradation seeming to be a relatively minor problem. As we noted earlier, GaN still has the ability to surprise.

At all events, in a dramatic 4 year period, the outstanding LED problems had finally been solved and the way was now open for the development of full-colour, large area, outdoor displays. One need only walk through down-town Tokyo (or any other large Japanese city) to see the result—aesthetically horrific but technologically magnificient! In addition, by 1997, the Tokushima city council had mounted the first large scale experiment in the use of blue–green traffic lights, encouraged, no doubt, by the prospect of making considerable savings on their electricity bill, allied with much reduced maintenance costs. Similar experiments are now under way in many other large cities across the world. What is more, the existence of efficient red, green, and blue LEDs made it possible to fashion a bright white light source by mixing the colours in appropriate amounts. The result was quite startling and served to tempt those of an optimistic temperament to contemplate an attack on the (huge!) general lighting market. Before closing this section, we should briefly examine this development in greater detail, but, before doing so, it is important to emphasize that Nichia is no longer the only serious commercial practitioner in the field of blue and green LEDs. In Japan there is now Toyoda Gosei (a company with which Akasaki is associated), in the United States there are CREE and Hewlett-Packard and in Europe Osram, to mention only the bigger players. Competition for what promises to be a highly lucrative market is fierce and patent lawsuits frequent the technical journals almost as often as technical innovations! CREE have a special role here in that they also grow SiC which they use for substrates (rather than sapphire) giving them the advantage of a conducting back contact. It is no longer necessary to etch away the top layers of the diode to make contact to the *n*-side of the junction, making their devices compatible with standard packaging used for other LED types. The downside is that SiC is more expensive than sapphire which probably leaves the situation nicely balanced.

Figure 7.11 illustrates the quite amazing improvement in visual efficiency of LEDs over the period 1965–2000, a performance which has been neatly summed up in the form of 'Craford's Law' (formulated by George Craford in the late 1980s) which states that luminous efficiency increases by a factor of 10 in each decade. It has a clear symbiosis with that other famous generalization (Moore's Law of integrated circuits) which claims a doubling of the number of transistors per square inch

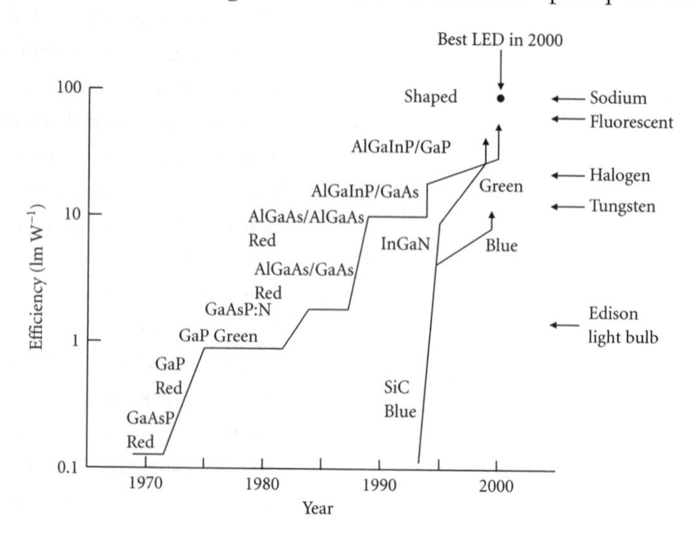

Figure 7.11. Illustration of 'Craford's Law' describing the evolution of visual efficiency of LEDs between 1965 and 2000—efficiency increases by a factor of 10 per decade. Note the dramatic rise in efficiency of nitride LEDs, on the right of the diagram, which has already left the tungsten filament lamp well behind. The maximum allowed value of visual efficiency is 683 lm W^{-1}, corresponding to an external quantum efficiency of 100%. (The diagram is copied from an article by M. Meyer appearing in *Compound Semiconductor*, March 2000, p. 26.) Courtesy of Lumileds.

each year. Moore's Law (in a slightly modified form of a factor of 1.6 per year) looks set to live for a few more years at least—Craford's Law reaches an inevitable buffer stop when diode efficiencies reach 100% and appears, therefore, to be close to its nemesis. In any case, from the viewpoint of general lighting, there are two factors to consider—luminous efficiency and cost, cost being a strong driving force in this context. As can be seen from Figure 7.11, the luminous efficiencies now being obtained from green and yellow LEDs have already outstripped that appropriate to the tungsten filament lamp and are currently (i.e. in the year 2000) challenging the performance of fluorescent tubes and sodium lamps. However, there is still some way to go before the performance of white light LEDs can hope to take a significant share of the lighting market. It is generally reckoned that an overall luminous efficiency of 200 lm W^{-1} constitutes a reasonable target to aim for and this will require individual efficiencies of 50% each for red, green, and blue LEDs (within a factor two of perfection!). As we saw earlier, this figure has already been met in one or two cases which suggests it to be a realistic aim for the relatively near future but whether it can be achieved at an acceptable cost is less easy to predict. Time will doubtless tell how this intriguing situation will develop, though white LEDs will soon find application in special situations where cost may be a secondary consideration—examples include personal reading lights in aircraft and lighting in hospital operating theatres.

No less than three different methods of producing white light have been demonstrated—the straightforward mixing of the three primary colours, the use of a suitable phosphor together with a bright blue LED and the use of phosphors with a UV LED. In the second approach, the phosphor absorbs some fraction of the blue light and emits a broad band of red and green light which, combined with the remaining blue light, produces white. In the case of the UV LED, three phosphors are used, one for each of the primary colours. In terms of *power* conversion, the process is inevitably less than 100% efficient, a blue photon having a larger energy than either red or green photons—in the case of red, the ratio is roughly 1.5 (i.e. a maximum efficiency of 67%) and this alone may be sufficient to rule out the phosphor approach in many applications. Similar remarks apply even more strongly to the use of a UV LED. It does, however, have the advantage of compactness—if an application calls for a small bright source, there is much to be said for it.

7.5 Short wavelength laser diodes

We left the laser diode story in Chapter 6 at the point where double heterostructure AlGaAs/GaAs devices, based on quantum well active regions were operating at wavelengths typically in the range 750–850 nm, the largest single application being the compact disk audio system, designed round a 780 nm laser. This was the situation in 1985. Quantum dot lasers came on the scene towards the end of the 1980s but were aimed specifically at longer wavelengths, appropriate to fibre-optic communication systems. The thrust towards *shorter* wavelengths began in the mid-1980s, still based on the AlGaAs material system. As we have seen, it was possible to reach wavelengths of about 700 nm with narrow GaAs quantum wells and 650 nm (well into the red spectral region) when the well material contained aluminium. However, it became clear that there was a penalty in the shape of much increased threshold current and there was little hope of any further wavelength reduction using this material. Further advance would obviously require a change in semiconductor. But we need to be clear about the motivation.

There was, of course the general interest in obtaining all three primary colours for application to display but this was overshadowed by the much more specific demands of optical disk recording systems. The CD audio system, introduced in 1979, had been such a major commercial success that it led inevitably to more advanced developments involving video storage, such as the CD interactive system and, more significantly, DVD video recorder/player equipment which was under development during the 1990s. (Whether one believes 'DVD' to stand for 'digital versatile disk' or 'digital video disk' is probably largely irrelevant—the system will without doubt be known as 'DVD' long after anyone can remember either interpretation!) At much the same time, the optical disk was introduced as a key element in data storage for personal computers and, in both applications, there was a need for much higher storage densities (i.e. more bits per square inch). This could, in part, be alleviated by clever data compression software but there was, nevertheless, considerable pressure to make the individual bits smaller, a demand which could only be met by the use of shorter laser wavelengths in both write and read functions. The audio system used a bit (burned into the master disk) which was 600 nm wide and read by a laser spot approximately 1500 nm in diameter (at the half-power level). These limits were defined by diffraction effects and implied that, to double the disk capacity would require a reduction in wavelength by a factor of $\sqrt{2}$, demanding a laser wavelength of 550 nm, well beyond the capability of the AlGaAs system. What were the alternatives?

The answer to this question is already clear from our discussion of LED materials, with the additional condition that laser action is only possible in direct gap materials. This rules out GaP and leaves us with the AlGaInP system, the II-VI materials or the nitrides. Prior to the 1990s, neither the II-VI materials nor the nitrides could be doped *p*-type so the only immediate possibility of developing an injection laser lay with AlGaInP whose maximum direct gap was close to 2.25 eV, with a corresponding minimum wavelength for electroluminescence of about 560 nm (though, when one allows for the requisite confining and cladding regions, *laser* emission is limited to wavelengths of 620 nm or longer). Indeed, work to develop laser diodes continued in parallel with the LED work during the 1980s and reports of the first room temperature CW laser were published by Kobayashi et al. of NEC and Ikeda et al. of Sony as early as 1985 (followed in 1986 by yet another Japanese group, this time from Toshiba). In common with much of the early work, these devices employed an active region of $Ga_{0.5}In_{0.5}P$, lattice-matched to a GaAs substrate and emitted at wavelengths in the region of 680 nm. AlGaInP was used only for the confining regions. Shorter wavelengths would require either the use of quantum wells (lattice-matched or strained) or the incorporation of aluminium in the active region.

We have come to recognize that first attempts to make any new type of laser are likely to result in devices with undesirably high threshold currents and these were no exception—values of $J_{th} = 4\ \text{kA cm}^{-2}$ were typical of these early results, compared with then current values of about $100\ \text{A cm}^{-2}$ for 800 nm AlGaAs/GaAs lasers. There was clearly scope (and need) for improvement. Two important steps had been made by the end of the 1980s: first, it was appreciated that the laser current contained a component which represented leakage of electrons through the structure, not involving recombination in the active region (thereby increasing the threshold current unnecessarily) and, second, quantum wells were incorporated into the active region. To appreciate the first point, we need to look at the energy level diagram of a typical laser under forward bias, as illustrated in Figure 7.12. This shows that electrons which are thermally excited out of the active region may recombine in the confining barrier region or be further excited over the barrier between the confining and cladding layers. In either case, this current does not contribute to laser action and, what is more, because it depends on thermal excitation, it increases significantly with increasing temperature. This accounts for the general observation that threshold current increased strongly with temperature above room temperature, an undesirable feature of lasers based on the AlGaInP system and one which became significantly worse as the operating wavelength was reduced towards 630 nm. Nevertheless, once this feature was understood (and the details were only clarified by Peter Blood's group at the University of Wales in Cardiff in 1995), it became possible to minimize its effect by maximizing the energy barrier. This, it was realized, required high *p*-type doping in the $Al_{0.5}In_{0.5}P$ cladding layer, a requirement not immediately straightforward of attainment. The commonly used acceptor, Zn, was difficult to incorporate in alloys containing large fractions of aluminium and, in any case it was characterized by a rather deep acceptor level. Fortunately, Be and Mg acceptors suffered less from these drawbacks and enabled acceptable doping levels of order $2 \times 10^{24}\ \text{m}^{-3}$ to be obtained. In particular, the fact that Be was (and is) the standard *p*-type dopant used

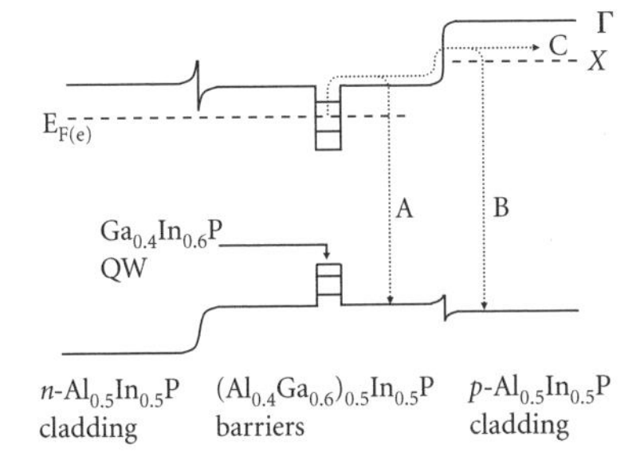

Figure 7.12. Band diagram of a typical AlGaInP quantum well laser under forward bias. The dotted lines represent electron loss processes which result from electrons thermally excited out of the well. 'A' corresponds to recombination in the *p*-type barrier region, 'B' to recombination in the *p*-cladding region, and 'C' to minority carrier current flowing to the *p*-type contact. Note that the processes in 'B' and 'C', involve electrons in the *X* minima, rather than those in the Γ minima, the *X*-minima being below the Γ in $Al_{0.5}In_{0.5}P$.

in MBE growth, led to many groups opting for this method of material preparation, in preference to MOVPE.

Quantum wells made an impressive contribution to the reduction of threshold current but in a rather subtle manner which requires yet another technical detour in pursuit of understanding. We have already noted in Chapter 6 that the use of quantum wells reduces threshold currents simply because there is a reduced volume of material to pump but another important factor depends on an idea put forward by Alf Adams of Surrey University, UK in 1986, based on the use of *strained* wells. (At the time, he was concerned with long wavelength InGaAsP lasers, which we shall be discussing in Chapter 8, but the principle is the same when applied to AlGaInP devices.) In other words, the well material was chosen not to be the lattice-matched composition $Ga_{0.5}In_{0.5}P$ but some slightly different composition such as $Ga_{0.4}In_{0.6}P$. Because InP has a larger lattice parameter than GaP (see Figure 7.10), this alloy would have an unstrained lattice constant greater than that of the GaAs substrate. However, provided the well thickness is small enough, the well material is constrained to having an in-plane lattice constant equal to that of GaAs and (related by the Poisson ratio) a perpendicular constant significantly greater—it is therefore under biaxial compression and this has two important effects. First, it increases the band gap of the well material and, second, it changes the relative positions of the light and heavy hole valence bands so that the uppermost band has light hole character. This means that it has a relatively low density of states and, in turn, this implies that it requires a smaller density of holes to separate the electron and hole quasi-Fermi levels sufficiently to allow laser action (see the discussion in Box 5.4, particularly equation (B5.21)). The overall effect is a reduction in threshold current density and a shorter operating wavelength, both of considerable interest for optical storage applications.

In practice, the reduction in threshold was dramatic, while the dominant effect on wavelength was actually in the opposite sense, resulting from the *reduction* in band gap due to the compositional change (less Ga, more In, reduces the gap towards that appropriate to InP—see Figure 7.5).

Japanese interest in short wavelength lasers was clear from the fact that the first three reports of room temperature CW lasers came from that country. The groups at NEC, Sony, and Toshiba continued to report improved performance, with threshold currents down to 1 kA cm^{-2} by 1988, but they were not allowed to keep the field to themselves. Hagen et al. from Philips reported $J_{th} = 1.2$ kA cm^{-2} in 1990, while a group from McDonell Douglas achieved $J_{th} = 425$ A cm^{-2} in 1991 and Bour et al. of Xerox achieved values below 200 A cm^{-2} in 1994, all for relatively long wavelengths, let it be said. Wavelengths shorter than 670 nm could be obtained at the expense of increased threshold current, Bour et al. reporting a device which operated at 623 nm with $J_{th} \sim 1.5$ kA cm^{-2}.

As it became clear that useful devices could be produced over the wavelength range of 620–690 nm, a variety of further applications came into view, one particular goal (if I am allowed to mix metaphors?) being achieved when the wavelength crept down to 633 nm. The He–Ne gas laser, which emits at 632.8 nm, had established a firm hold on many application areas over a long period of time (it was invented in 1960) and was used, for instance, in the first Philips video disk system, known as 'Laser Vision'. Needless to say, semiconductor laser enthusiasts were keen to offer a very much smaller and more convenient alternative—the advent of the 633 nm AlGaInP laser finally allowed them to do so, some 30 years after Ali Javan's first demonstration of the gas laser. The second important goal was that of high power output. To read the data from a video disk required no more than 10 mW of light power but writing the master disk demanded considerably more, there being a fairly obvious relationship between writing speed and laser power. This, though was only one spur towards higher powers—there were others, concerned with pumping solid state lasers and with laser printing. We came across the use of semiconductor lasers as pumps for Er-doped fibre-optic amplifiers in Chapter 6—other important examples (see Smith 1995: ch. 8) concern neodymium-doped YAG (yttrium aluminium garnet) and chromium doped Li[Ca,Sr]AlF_6 solid state lasers. In all cases, the key to efficient pumping lies in achieving a good match between the energy levels of the active ions in the host material and the pumping wavelength. While Nd:YAG lasers must be pumped at 809 nm (using AlGaAs lasers), chromium based lasers require wavelengths in the region of 670 nm which are only available from AlGaInP devices and this led in the early 1990s to the development of devices giving more than 1 W of output power. The quest for higher resolution in laser printing also relies on these same developments and, once again, the need for high speed demands relatively high laser power.

Even this is not the end of the AlGaInP story. Other applications are as light pens, as optical sources for short haul plastic-fibre communication systems, where fibre loss is at a minimum in the 600 nm band, and in the field of photodynamic therapy. (Photodynamic therapy is a method of treating cancers with drugs which are light sensitive, their activity being localized by shining light of appropriate wavelength onto the cancerous cells via an optical fibre. The relevant absorption bands of many such drugs are in the red spectral region.) It is remarkable how rapidly the number of applications for a new laser technology grows once that technology is properly established, in marked contrast to the situation following the invention of the first lasers in 1960. At that time, the common reaction to their advent followed the somewhat cynical line that the laser was 'a solution in search of a problem', whereas, today, we appear to have learnt the art of visualizing applications even before the necessary devices are off the drawing board. However, in spite of this wide range of uses, there can be little doubt that the DVD player constitutes by far the most important market for these laser diodes. At the time of writing (i.e. in 2002) the first generation of DVD players is attracting consumer interest in a serious way, each machine employing either a 650 or 635 nm AlGaInP laser (according to manufacturer) which allows each disk to hold as much as 4.7 GB of information (compared with 650 MB for CDs). The second generation will probably use a blue nitride laser giving each disk a capacity of 15 GB—but we shall come to that in a moment.

It was the DVD player which also stimulated interest in the development of ZnSe-based lasers. The band gap of ZnSe, at 2.7 eV, suggests possible laser emission at 460 nm, and even shorter wavelengths would be possible by alloying it with ZnS to increase the band gap. While photo-pumped laser action was reported by several groups during the 1980s, the possibility of an injection laser emerged only in 1990 with the achievement of *p*-type doping in MBE growth. This immediately led to a surge of interest in suitable II-VI combinations to provide QW, barrier, and cladding materials, together with experiments designed to select an optimum substrate. The choice lay between using bulk ZnSe crystals, which were of modest size and quality, and GaAs to which ZnSe is nearly lattice-matched (the mismatch is approximately 0.25%). In practice, neither solution was without difficulties—homoepitaxy proved problematic until suitable techniques for surface preparation of the substrate could be established and heteroepitaxy was complicated by the fact that the atoms involved in substrate and epilayer differed in valence. In addition, both approaches were bedevilled by the introduction of dislocations emanating from the substrate–epilayer interface, a problem of serious import in the II-VI materials which appear to be particularly sensitive to such defects.

Following our earlier discussion of strained quantum wells in AlGaInP lasers, we may anticipate the use of a similar technique here. In this case, the well material was chosen to be ZnCdSe which, as can be seen in Figure 7.10, had a band gap smaller than ZnSe and allowed the latter to be used as the barrier material, approximately lattice-matched to GaAs. Choice of precise composition and of well width provided scope for varying the operating wavelength in the range 490–530 nm (blue–green). However, two other problems soon became apparent in attempts to design a laser based on ZnSe, those of providing a suitable cladding layer and of making electrical contact to the *p*-side of the structure. The first report of a blue/green laser came in 1991 from Haase et al. of the 3M Company in St Paul, Minnesota, using a single ZnCdSe quantum well, ZnSe barriers, and ZnSSe cladding. They observed pulsed lasing action at 77K with an encouragingly low threshold current density of 320 A cm^{-2}. The incorporation of ZnS in the cladding layers certainly increased the band gap but fell short of the ideal in so far as the conduction band offset between ZnSe and ZnSSe was found to be very small (most of the band gap difference appearing at the valence band edge). This resulted in poor electron confinement, in the same way as found in AlGaInP lasers but to a significantly exaggerated extent. While it just sufficed in the case of 77K operation, there was little hope of its being successful at room temperature and any attempt to increase the electron barrier by increasing the ZnS fraction was likely to be frustrated by the increased lattice-mismatch (see Figure 7.10). Some considerable improvement was essential. Nor was it long in materializing—later in 1991 Okuyama et al. of the Sony Corporation proposed the use of an innovative quaternary alloy ZnMgSSe which would not only allow the band gap to be 'tuned' between 2.8 and 4.0 eV, while maintaining lattice-match to the GaAs substrate, but would provide a much improved electron barrier *and* a dielectric constant smaller than that of ZnSe, as required for effective optical waveguiding. All this, and it was also found possible to dope the alloy *p*-type, thus allowing holes to be injected into the active region. (The often misused colloquialism 'all singing, all dancing' appears more than justified here!) This led to the sophisticated laser structure shown in Figure 7.13 which was favoured by all the subsequent contenders. Note the use of $ZnS_{0.06}Se_{0.94}$ barriers, rather than pure ZnSe, to ensure a perfect lattice-match to the GaAs substrate. Sony were able to capitalize on this in developing the first room temperature CW laser which they reported in 1993 but they were not alone—groups at Philips, 3M, and Purdue and Brown Universities achieved room temperature pulsed operation during the period 1991–3 while the University collaborators also reported CW operation in 1993. However, none of these early CW devices lived very long. Typical lifetimes were of the order of 10 s, catastrophic degradation being related to the presence of 'dark line defects' (dislocations) introduced during the early stages of epitaxy ('continuous' operation might be seen as something of a misnomer, perhaps?).

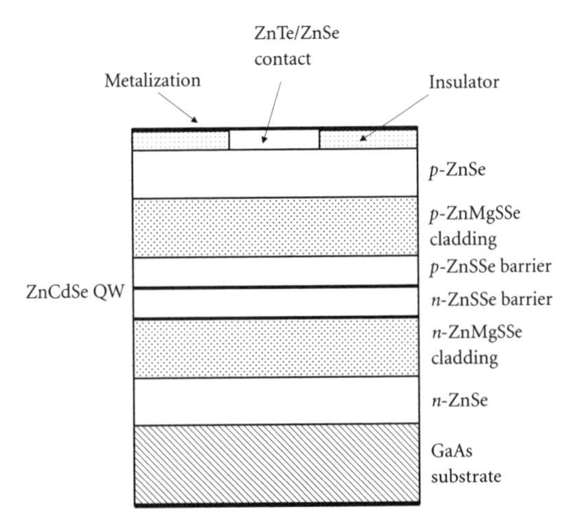

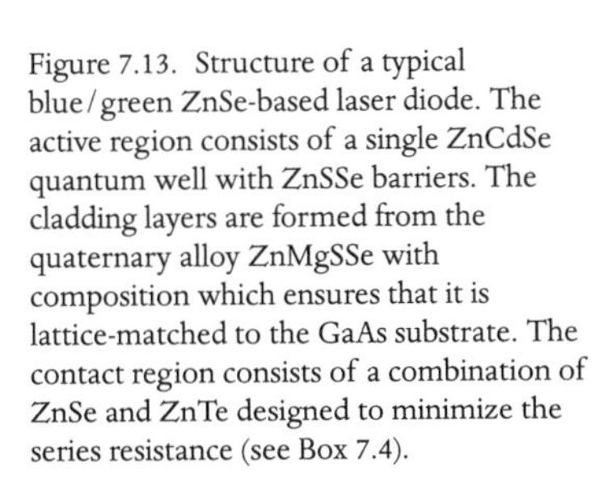
Figure 7.13. Structure of a typical blue/green ZnSe-based laser diode. The active region consists of a single ZnCdSe quantum well with ZnSSe barriers. The cladding layers are formed from the quaternary alloy ZnMgSSe with composition which ensures that it is lattice-matched to the GaAs substrate. The contact region consists of a combination of ZnSe and ZnTe designed to minimize the series resistance (see Box 7.4).

The other obvious problem associated with the early lasers concerned the operating voltage. In theory, this should take a value close to the band gap of the active material, typically of the order of 2.5–3.0 V, whereas, in practice, values as high as 20 V were measured experimentally. The difficulty lay with the high series resistance associated with the *p*-type contact which tended to drop most of the applied voltage. We discuss this in rather more detail in Box 7.4 but, briefly, it stems from the difficulty in achieving adequate hole densities in the *p*-ZnSe contact region (shades of the AlGaInP laser diode but for a completely different reason). The discovery of effective *p*-type doping with active (i.e. atomic) N represented a huge step forward in II-VI technology but the maximum hole density was still limited to about 5×10^{23} to $1 \times 10^{24}\,m^{-3}$ (compared, for example, with values of $3 \times 10^{25}\,m^{-3}$ obtainable in ZnTe—which only exists in *p*-type form). In fact, the solution made practical use of the high doping available in ZnTe, by inserting a layer of ZnTe between the metallization and the ZnSe contact layer. That alone would not have been enough because it would only have transferred the problem to the ZnTe/ZnSe interface—in practice a graded layer of ZnTe/ZnSe was necessary between the two which was optimized by making it in the form of a ZnTe/ZnSe superlattice with graded layer thickness. This, of course, made an already sophisticated structure even more so but it achieved the necessary reduction in voltage, down to values of about 4.5 V, there being a small voltage drop in the bulk of the structure.

The remaining problem was clearly one of increasing the operating lifetime—a value of 10,000 h continuous operation being required before lasers could be seriously considered for commercial applications. This amounted to a struggle to reduce defects of one kind and another in the active region of the device, the immediate goal being to improve the quality of the substrate/device interface from which propagated most of the lethal line defects. Two approaches were followed, one to replace the GaAs substrate with ZnSe, the other to incorporate an epitaxial GaAs buffer layer before growing the device structure. This latter involved setting up double chamber MBE equipment, one chamber being used for GaAs growth, the second for the II-VI layers. New labourers in the II-VI vineyard, from the Sumitomo laboratories, followed the first path, having developed the technology to grow their own ZnSe substrates. They demonstrated the ability to achieve CW laser operation in 1998 but still with lifetimes of little more than a minute. The old firm of Sony stayed with GaAs substrates and made steady improvements, reporting in 1998 lifetimes in excess of 100 h, still well short of the desired goal but certainly some considerable way towards it. It appeared at this point that they may have overcome the interface problem, the remaining degradation mechanism being associated with point defects, such as Se vacancies. Significantly, threshold current densities have come down considerably in

Box 7.4. Making electrical contact to II–VI lasers

The problem of contacting II-VI lasers which reveals itself in the form of high operating voltages stems from the difficulty of obtaining a high *p*-type doping in the ZnSe contact region (see the structure in Figure 7.13). In this box we look a little more carefully at the nature of the contact and how it was subsequently improved. We begin by considering a direct metal–ZnSe contact, then look at the way in which ZnTe can be incorporated, to reduce contact resistance.

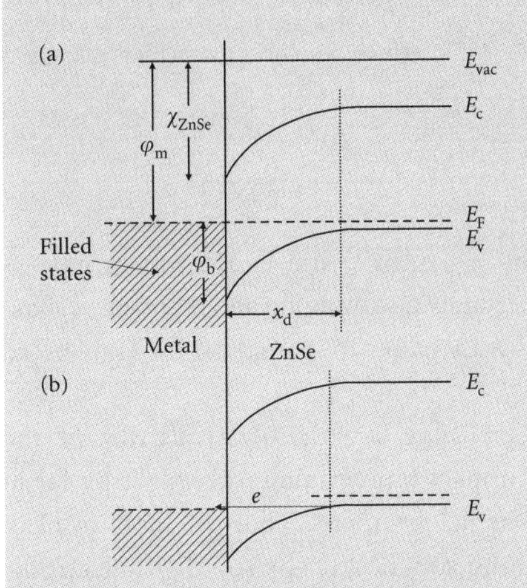

Figure 7.14. Band structure diagram to illustrate the nature of the contact between a metal such as Au or Pt and p-type ZnSe. The metal and the bulk of the ZnSe are separated by a barrier region of height φ_b and thickness x_d. (a) corresponds to zero applied bias while (b) illustrates electron tunnelling from the ZnSe valence band into an empty state at the metal Fermi level when a small forward bias exists across the contact.

Figure 7.14(a) shows the band-bending associated with the contact between a metal and *p*-type ZnSe under zero bias. In this case, there is a barrier φ_b to current flow at the valence band edge which is given by:

$$\varphi_b = E_g + \chi_{ZnSe} - \varphi_m, \tag{B7.22}$$

which gives values of $\varphi_b \sim 1$ eV for typical metals. Our earlier discussion of metal/semiconductor contacts in Section 3.3 suggested that the presence of interface states might render this equation invalid and such is, indeed, the case for Ge, Si, GaAs, and many other materials but, for II–VI compounds, equation (B7.22) does, in fact, represent a good approximation. Thus, the metal and the bulk of the semiconductor are isolated by a depletion region of thickness x_D, where x_D is proportional to $\varphi_b^{1/2}$ and to $N_A^{-1/2}$ (where N_A is the acceptor density in the ZnSe). If current is to flow between them, electrons must tunnel through the barrier as shown in Figure 7.14(b) which illustrates the situation when a small forward bias is applied. Note, first, that electrons from the valence band can only tunnel into *empty* states in the metal just above the metal Fermi level and, second, that an electron current from ZnSe to metal represents a hole current from metal to ZnSe which is in the correct sense to result in hole injection across the adjacent *p-n* junction (not shown in the figure).

The key fact in relation to contact technology is the dependence of x_D on N_A. In order to encourage tunnelling, it is necessary for the barrier to be as thin as possible, which implies that N_A must be as large as possible, a serious problem for ZnSe where it is difficult to incorporate more than about 10^{24} m^{-3} acceptors. On the other hand, the ease with which ZnTe can be doped with acceptors makes it relatively easy to obtain a low resistance contact between ZnTe and high workfunction metals such as Pt or Au (equation (B7.22) shows that φ_b is minimized by making φ_m as large as possible). In fact, a specific contact resistance as low as 5×10^{-6} Ω cm^2 can be obtained, which results in a voltage drop of only 5 mV for a current density of 1 kA cm^{-2}. The question remains, however, how to translate this excellent result into an acceptable metal/ZnSe contact resistance. The problem is that the valence band offset at the ZnTe/ZnSe interface is surprisingly large, that is, $\Delta E_v \sim 0.8$ eV which again results in a serious interface barrier, the problem having been shifted from the

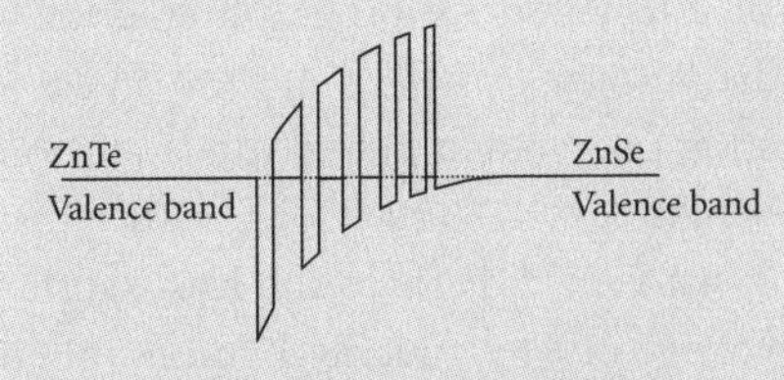

Figure 7.15. Band-bending diagram to demonstrate the principle of a graded superlattice used to maximize the tunnelling current between *p*-type ZnTe and *p*-type ZnSe. The superlattice well widths are carefully designed so that the confined energy states lie at a constant energy, corresponding to the ZnTe valence band edge. This allows resonant tunnelling of electrons from the ZnSe into the ZnTe at constant energy. This corresponds to an equivalent hole current into the ZnSe, as required by the laser diode.

Box 7.4. Continued

metal/semiconductor interface to the ZnTe/ZnSe interface! Fortunately, this can largely be overcome by grading the composition gradually from ZnTe to ZnSe but, unfortunately, the lattice-mismatch between the two materials produces structural defects which can be a source of degradation in the laser diode. The ultimate solution makes use of resonant tunnelling between adjacent wells in a graded ZnSe/ZnTe superlattice as illustrated in Figure 7.15. Because the ZnTe wells are thin, they are strained rather than introducing dislocations at each interface, a solution which smacks of sheer genius!

this later work, Sony reporting 430 A cm^{-2} and Sumitomo as little as 220 A cm^{-2}. These results are certainly encouraging and represent a significant improvement compared with values measured on the GaN-based lasers which we now describe.

The development of the GaN blue laser hardly rivals that of the point contact transistor in its impact but certainly exceeds it by far in the bizarre circumstances of its achievement. That this holy grail of optoelectronics should be so long sought after by top flight electronic companies, yet first be reached by an obscure Japanese chemical company with little or no research pedigree will remain forever a classic example of the well-known axiom that truth is stranger than fiction. We have already outlined the background to the story in Section 7.4 so there is no need to repeat it here. We take it up at the point in 1995 when Nakamura was able to report success in developing 'super-bright' green and blue LEDs. He was already well advanced in a programme to extend his group's remarkable material growth skills to the even more ambitious project of laser development, as is proved by the timing of their first laser report. This occurred at the end of 1995, perhaps a trifle unconventionally, in the form of a Japanese TV programme, prior to the publication of a paper in the *Japanese Journal of Applied Physics* in January 1996. The laser in question operated in pulsed mode at room temperature with output wavelength $\lambda = 417$ nm (violet, rather than blue) and threshold current $J_{th} = 4$ kA cm^{-2}. Perhaps the least satisfactory feature was the threshold voltage of 34 V, roughly 10 times that to be expected in an ideal device and yet again a reflection of the difficulty in making adequate electrical contact to *p*-type wide gap materials. The structure of the laser followed quite closely that developed for the later quantum well LEDs—in the active region it employed 26 periods of 2.5 nm $In_{0.2}Ga_{0.8}N$ quantum wells with 5 nm $In_{0.05}Ga_{0.95}N$ barriers, while optical confinement was achieved with an $Al_{0.15}Ga_{0.85}N$/GaN waveguide. Because the laser was grown on *c*-plane sapphire, which cannot easily be cleaved, mirror facets were formed by reactive ion etching, not a process to be welcomed by anyone charged with commercial exploitation (!) but more than adequate to prove an existence theorem.

This dramatic development, only a few years after the first observation of *p*-type conductivity in GaN, set Nichia apart from almost all its rivals—only Akasaki's group at the Nagoya Institute of Technology were in a position to compete, which they did to a limited extent by reporting laser action in a rather unconventional vertical structure towards the end of 1995. Nevertheless, some significant improvements were obviously needed. Continuous operation was essential for the DVD application, as was a considerable reduction in operating voltage, cleaved mirror facets and at least a modest reduction in threshold current were also highly desirable. All these were tackled within the next couple of years, with remarkable success—by the end of 1997 Nakamura was able to report room temperature CW operation of lasers with operating lives greater than 3000 h and predicted lifetimes in excess of 10,000 h (though, not surprisingly, it took another year and a half to prove it!). Prove it they did and, by January 1999, Nichia announced that it was shipping sample quantities of violet lasers with guaranteed lifetimes of 10,000 h under continuous operation, a truly remarkable achievement.

These commercial samples differed considerably from the above prototype—they employed only two quantum wells, they relied on cleaved mirrors, by the expedient of changing the crystal orientation of the sapphire substrate from *c*-plane to *a*-plane, they operated with threshold currents of $1.2\ \text{kA cm}^{-2}$ and voltages of 4–5 V. As much as 400 mW of power could be obtained from each laser facet which was more than sufficient for both reading and writing DVD disks and for application to laser printing. Two key innovations in technology were responsible for these improvements, one involving a novel method of circumventing the problems associated with heteroepitaxy and the other aimed at improving the electrical conductivity of the structure. Growth directly on sapphire substrates inevitably resulted in high dislocation densities ($10^{14}\ \text{m}^{-2}$) in the nitride layers and, though these appeared to have negligible effect on LED operation, Nakamura found them to be significant in reducing laser lifetimes. Their density could be reduced to about $10^{10}\ \text{m}^{-2}$, however, by the use of a novel growth method known alternatively as epitaxial lateral overgrowth (ELOG) or lateral epitaxial overgrowth (LEO) in which a GaN layer grown on sapphire was patterned photolithographically with SiO_2 stripes before regrowing a second layer of GaN. Figure 7.16 shows how the regrown GaN grows first between the stripes with high defect density but then grows laterally over the stripe with very much lower density. If lasers are then made on the material over the stripes, they show much improved lifetimes compared with those grown on unpatterned substrates. The second innovation was to replace the AlGaN cladding layers with AlGaN/GaN superlattices which improved the doping efficiency and therefore the electrical conductivity. (As the Al content in AlGaN increases it becomes more difficult to incorporate dopants and, what is more, their ionization

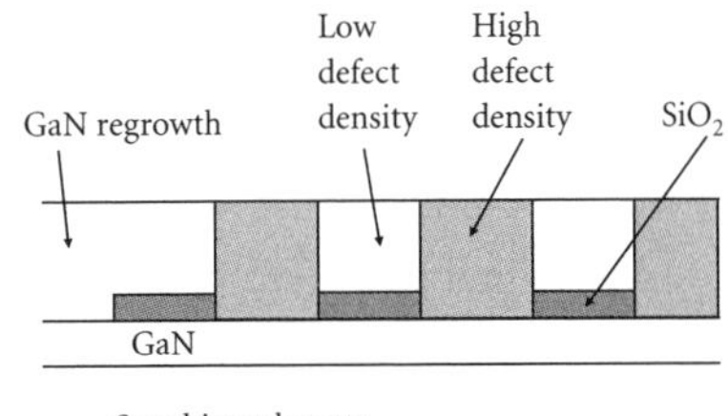

Figure 7.16. Schematic diagram to illustrate the phenomenon of lateral overgrowth (ELOG). An initial layer of GaN grown on a sapphire substrate contains a high density of dislocations. It is then patterned with SiO_2 stripes and a second GaN layer grown over them. Initial growth occurs only between the stripes and the material is highly defective but, when the GaN thickness exceeds that of the SiO_2, the film starts to grow laterally over the stripes at a much higher rate than in the vertical direction and this lateral growth is characterized by a very much smaller dislocation density. It is this material which is used as the basis for growth of the laser structure.

energies increase, thus reducing the doping efficiency, so it is more effective to dope a superlattice containing GaN layers.) The use of superlattices also reduced the tendency for the layers to crack.

Needless to say, the rest of the world was hardly going to take Nichia's leading position lying down and a huge catching-up exercise has been under way since 1997. The number of companies reporting a nitride laser capability now includes Toyoda Gosei, Cree, NEC, Xerox, Osram, Samsung, NTT, Matsushita, Sony, Fujitsu, etc. Not all of these will survive in the battle to supply the world's DVD blue lasers but it well illustrates the importance everyone attaches to this vital new component. Nor is this the end of the line—the large band gap of AlN (6.2 eV) encourages the belief that similar devices can be developed with emission wavelengths well out into the UV. Nichia have already taken the first step, with a recent report of CW laser action at 369 nm and, though further progress will undoubtedly be difficult, there is every hope that even shorter wavelengths may be reached in the near future. In this context, it is interesting to compare the relative positions of II-VI and III-V short wavelength laser diodes. Both material systems are capable of producing shorter wavelengths, though the nitrides appear to be well ahead at present in respect of operating lifetimes, while the II-VI materials have a distinct edge in terms of threshold current. From a theoretical viewpoint, the relatively large electron effective mass in GaN ($m_e = 0.22m$) implies that threshold currents will always be much greater than those appropriate to AlGaAs devices, for example, and the present figure of about 1 $\mathrm{kA\,cm^{-2}}$ may be close to the lowest achievable, while the best ZnSe-based lasers are characterized by values as low as 200 $\mathrm{A\,cm^{-2}}$. On the other hand, the problem of degradation is far from beaten in II-VI devices, whereas the nitrides appear to be surprisingly rugged in this respect. Dislocations in GaN are quite remarkably stable and show very little tendency to move, unlike their counterparts in the II-VIs, a phenomenon related to the significantly stronger covalent bonding in GaN. In this context, it may be important that recent work on II-VI lasers at Wurzburg in Germany has demonstrated the incorporation of Be into both barrier and cladding layers, Be chalcogenides (BeS, BeSe, BeTe) have stronger covalent bonds than their Zn equivalents. It is also worth bearing in mind that similar degradation problems were experienced, and overcome in AlGaAs laser development, so the achievement of long-lived II-VI devices certainly cannot be ruled out. But all this is mere speculation—we must wait with interest to see what the future will bring.

Progress on the development of commercially viable violet lasers has been so dramatic that it is easy to overlook the fact that there is still much that is not yet understood concerning the recombination mechanisms which control light emission from InGaN quantum wells, so we shall end this section with a brief discussion of some of the relevant physics. The argument revolves mainly around the precise nature of the well material and the presence of strain. The band gap of $In_xGa_{1-x}N$ was measured as a function of composition quite a long time ago by Osamura et al. (1972) who found a distinct downward bowing in the plot of E_g vs x. This was later parametrized by Nakamura in the form:

$$E_g = xE_g(\mathrm{InN}) + (1-x)\,E_g(\mathrm{GaN}) - bx(1-x), \tag{7.6}$$

with $E_g(\mathrm{InN}) = 1.95$ eV, $E_g(\mathrm{GaN}) = 3.44$ eV, and the bowing parameter $b = 1.00$ eV (which means that the band gap of the 50% alloy is 0.25 eV smaller than the value given by a linear interpolation). These measurements were made on thick films which were therefore completely relaxed (i.e. unstrained) but it must be remembered that the material in a GaN/InGaN quantum well is under compressive strain which results in an increase in gap above the values given by equation (7.6). On the other hand, there have been recent measurements which claim to find a considerably larger bowing parameter, throwing doubt on even the basic unstrained energy gaps. The situation is further confused by the fact that it is expected for any random alloy that there will exist so-called 'tail states' extending to energies below the nominal band edge and these states are of a localized nature, unlike band states (i.e. electrons or holes in these states are no longer free to move through the crystal lattice). Finally, there is evidence to suggest that InGaN can rarely be grown as a random alloy—more often, there is segregation of InN and GaN to a greater or lesser extent, but generally to an increasing extent as the In content is increased. This tends to decrease the effective gap but also adds significantly to the density of localized tail states and this has important consequences for the nature of recombination radiation. It follows, for instance, that recombination between electrons and holes which have thermalized into tail states will occur at longer wavelengths than predicted by the alloy band gap and should be characterized by long radiative lifetimes in much the same way as found for the D–A transitions we discussed when considering GaP light emission processes. (The states are spatially separate, which reduces the overlap of the electron and hole wavefunctions.) Indeed, this is generally observed to be true in photoluminescence measurements on InGaN thin films but the interpretation is still far from clear for two reasons. First, there is evidence to suggest that the segregation of InN and GaN may proceed to the point where InN quantum dots are formed and, second, even in well-behaved wells there exists an alternative explanation for most of the experimental data. This rejoices in the name of 'the quantum confined Stark effect' (QCSE).

The reader may recall that, in our discussion of AlGaN/GaN HEMT devices, we made the point that in such nitride structures there exist

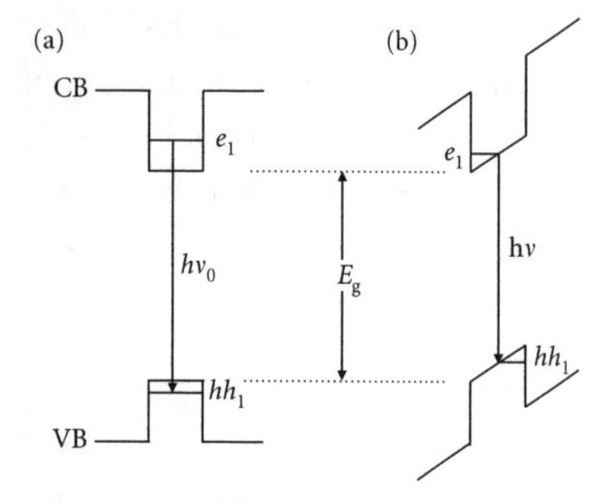

Figure 7.17. Energy bands of a single quantum well (a) in the absence of an electric field and (b) with a strong field applied. The field forces electrons towards the left-hand side and holes towards the right-hand side of the well. This reduces the overlap between electron and hole wavefunctions and also changes the energies of the confined states in such a manner that the photon energy $h\nu$ resulting from radiative recombination is significantly smaller than in the field-free case, $h\nu_0$.

strong electric fields due to both naturally occurring and to strain related polarization effects. Similar considerations apply to InGaN quantum wells and the resulting fields can strongly modify radiative recombination. Figure 7.17 illustrates the effect of an electric field on the band diagram for a quantum well and demonstrates that radiative recombination takes place between electrons and holes which have been pushed to opposite sides of the well. This has two consequences: the emitted photons have smaller energy than that corresponding to the field-free case and the electron–hole wavefunction overlap is significantly reduced—in other words, the recombination lifetime can be considerably lengthened. Experimental evidence that these effects do actually occur in InGaN quantum wells has been obtained by many researchers and makes it difficult to pin down the precise nature of the recombination process occurring in any specific situation, the effects of the QCSE and of localized tail states being closely similar. In the case of the laser the situation is even more complicated by the high free carrier densities required to reach threshold. Tail states are probably all filled so that band states become important again and electric fields in quantum wells are largely screened out, minimizing the QCSE and making an adequate description of the overall process extremely difficult to obtain. What is more, very recent reports suggest that the band gap of InN is actually about 0.7 eV (rather than the 1.95 eV previously measured) so the anticipated values of the alloy band gaps must be smaller than those previously assumed. And, if the reader is now totally confused by all this, I can only sympathize—the situation is simply very confusing! Ideally, technology follows science but there are certainly times when it runs well ahead of scientific understanding and this seems like an excellent example. Blue lasers exist and work extremely well—one day we shall probably understand just how!

Bibliography

Akasaki, I. and Amano, H. (1996) *J. Cryst. Growth*, 163, 86.

Bergh, A. A. and Dean, P. J. (1972) *Proc. IEEE*, 60, 156.

Bergh, A. A. and Dean, P. J. (1976) *Light Emitting Diodes*, Clarendon Press, Oxford.

Bhargarva, R. N. (1975) *IEEE Trans. Electron Dev.* ED-22, 691.

Bhargava, R. N. (1997) (ed.) *Wide Bandgap II–VI Semiconductors* EMIS Data Reviews No. 17, INSPEC, IEE, London.

Casey, H. C. and Panish, M. B. (1978) *Heterostructure Lasers*, Academic Press, New York.

Craford, M. G. (1977) *IEEE Trans. Electron Dev.* ED-24, 935.

Edgar, J. H., Strite, S., Akasaki, I., Amano, H., and Wetzel, C. (1999) *Gallium Nitride and Related Semiconductors*, EMIS Data Reviews No. 23, INSPEC, IEE, London.

Gil, B. (ed.) (1998) *Group III Nitride Semiconductor Compounds*, Clarendon Press, Oxford.

Gooch, C. H. (1973) *Injection Electroluminescent Devices*, John Wiley & Sons, London.
Keitz, H. A. E. (1971) *Light Calculations and Measurements*, 2nd edn, Macmillan, London.
Madelung, O. (1964) *Physics of III–V Compounds*, John Wiley & Sons, New York.
Mohamad, S. N. and Morkoc, H. (1995) *Progress in Quantum Electronics* (ed. M. Osinski) Elsevier, London.
Mohamad, S. N., Salvador, A. A., and Morkoc, H. (1995) *Proc. IEEE*, 83, 1306.
Morkoc, H., Strite, S., Gao, G. B., Lin, M. E., Sverdlov, B., and Burns, M. (1994) *J. Appl. Phys.*, 76, 1363.
Nakamura, S., Pearton, S., and Fasol, G. (2000) *The Blue Laser Diode*, 2nd edn, Springer, Berlin.
Neumayer, D. A. and Ekerdt, J. G. (1996) *Chem. Mater.*, 8, 9.
Orton, J. W. and Foxon, C. T. (1998) *Rep. Prog. Phys.*, 61, 1.
Ponce, F. A. and Bour, D. P. (1997) *Nature*, 386, 351.
Smith, S. D. (1995) *Optoelectronic Devices*, Prentice Hall, London.
Thomas, D. G., Gershenzon, M., and Trumbore, F. A. (1964) *Phys. Rev.*, 133, A269.
Thomas, D. G. (ed.) (1967) II–VI Compound Semiconductors, Proceedings of the International Conference, Benjamin, New York.
Willardson, R. K., Beer, A. C., and Weber, E. D. (eds.) (1997) 'II–VI Blue/Green Light Emitters—Device Physics and Epitaxial Growth,' in *Semiconductors and Semimetals,* Vol. 44, Academic Press, New York.
Willardson, R. K. and Weber, E. R. (eds.) (1997) 'High brightness light emitting diodes,' in *Semiconductors and Semimetals*, Vol. 48, Academic Press, New York.
Willardson, R. K. and Weber, E. R. (eds.) (1997) 'Gallium nitride,' in *Semiconductors and Semimetals*, Vol. 50, Academic Press, New York.
Willardson, R. K. and Weber, E. R. (eds.) (1998) 'SiC materials and devices,' in *Semiconductors and Semimetals*, Vol. 52, Academic Press, New York.
Willardson, R. K. and Weber, E. R. (eds.) (1999) 'Gallium nitride,' in *Semiconductors and Semimetals*, Vol. 57, Academic Press, New York.

CHAPTER 8

Communicating with light

8.1 Fibre optics

Without doubt, the introduction of light into the world telecommunications network represents one of the most important 'revolutions' in modern society, and semiconductors have played a vital role in its success. This chapter is therefore concerned to describe that contribution, but first we shall try to put the semiconductor role in context by providing a brief history of the fibre-optic revolution, itself. This is made all the easier by the availability of an admirably written overview in the book '*City of Light*' by Jeff Hecht (1999) which is strongly recommended to anyone wishing to explore the subject in greater detail.

The fibre-optic revolution occurred in the surprisingly short period 1970–90, during which fibre in one form or another took over the role of making connection between telephone switching stations, all the way from local city networks to transoceanic long haul cables. Given the generally conservative nature of the world's telecom business, this is a remarkably short timescale for such a major transformation. Copper wire had provided the mainstay of remote information exchange technology from the development of the electric telegraph in the 1840s, together with the first transatlantic telegraph cable in 1866 (ably assisted by the physicist William Thomson—later Lord Kelvin), through that of the telephone in the 1880s, until the advent of radio in the years after the First World War began to offer serious rivalry. Even with the development of microwave links after the Second World War, copper remained the medium for all local connections and, what is more, for the first transatlantic telephone cable TAT-1 (which used coax) in 1956 (this, by the way, depended on valve amplifiers at each of 51 repeaters—the first transistorized cable did not appear until 1968!). In 1963 the Telstar communication satellite was launched to herald the era of long distance satellite links which threatened to put transoceanic cables out of business but the planning of new land-based connections was still based on copper, albeit, perhaps, in the form of millimetre waveguide for large volume traffic. That optical fibres should transform this scenario almost completely in the space of just a few years was nothing short of remarkable. Today, we take it as read that local, medium haul and long haul connections should be dominated by fibre, forgetful, perhaps, that the first practical fibres did not even exist before 1970 (and, in the case of long haul, not before 1980). The only remaining debate concerns the famous 'last mile', taking data links right into the individual subscriber home. These are still largely copper on the grounds of cost, though it may only be a matter of time before pressure for interactive wide-band links forces these too to be made in glass.

The driving force for change was the ever-growing need for more bandwidth, based on two distinct demands; for increasing numbers of simple telephone connections and, at the same time, for the introduction of services requiring large data rates such as television or computer data links. Added to this was the rapidly increasing demand for efficient long distance telephone connections. It is difficult, now, to remember that even during the 1950s subscribers were frequently obliged to book long distance calls well in advance and might then have to put up with a noisy and indistinct connection. The direct dialling of crystal clear calls to all parts of the world which we now take for granted became available surprisingly recently!—and it depends, of course, on the use of digital data handling techniques which, as we saw in Chapter 4 (Box 4.1), require significantly more bandwidth than corresponding analogue methods.

The original analogue telephone system required some 3 kHz bandwidth for each transmission which was readily provided by a pair of copper wires. All was well so long as the number of calls allowed each one to be routed along a separate line but rapidly increasing demand emphasized the need to send many calls down the same line and required a more sophisticated approach. This took the form of a modulated carrier, different calls employing different carrier frequencies, initially of order, 100 kHz, and therefore needing bandwidth of 100 kHz, rather than 3 kHz. Gradually, the carrier frequency increased into the megahertz range and eventually into the gigahertz range and coaxial cable replaced the original transmission line technology in order to minimize radiation loss from the line. However, the 'skin effect' led to increasing dissipation loss in coax cable as the frequency increased and this made it impractical to transmit gigahertz frequencies over distances greater than about 1 km. The immediate answer was provided by the use of microwave links for medium haul transmission but even these were not immune from loss, which limits the ultimate distance available for any analogue transmission system. (Amplifier booster stations amplify noise as well as the required signal so lead to no improvement in signal-to-noise ratio.) Digital techniques provided the only reliable answer to long distance transmission and were eventually introduced during the 1960s (though they had been proposed as long ago as 1937 by the English inventor Alec Reeves who was the inspiration for fibre-optical communications research at the STL laboratory near Harlow, in Essex). The first serious intention to employ digital signals

(PCM—pulse code modulation) was associated with the Bell Labs proposal to set up a millimetre waveguide system for intercity connections which was under development during the late 1950s, again with the motive of obtaining another order of magnitude increase in bandwidth. However, there were serious difficulties with this—even small bends caused signal leakage and the fact that it was considerably 'over-moded' (i.e. there were very many propagation modes within the guide) led to problems of mode-hopping. Field trials, which eventually went ahead in 1975 in both the United States and England, proved that millimetre waveguide systems were, indeed, possible but extremely difficult to engineer. However, time ran out for millimetre waves—optical fibres, which offered another four orders of magnitude of bandwidth, came to fruition at the crucial moment and millimetre waveguide was shelved before it could play so much as a supporting role. On the other hand, digital methods rapidly became accepted as standard—though they demanded greater bandwidth than their analogue counterparts, there was bandwidth to spare if light was to be used as the carrier. (Let us not forget, by the way, that the original telegraph system was itself based on digital methods—Morse code providing an invaluable, if slow, communications language—and the first ever transatlantic cable was laid as early as 1866, using regularly spaced repeater stations to regenerate the signal, just as is done today. It was the coming of the telephone which introduced the analogue techniques which dominated communications for the next hundred years.)

The use of light as a medium for communications is not, of course, a new phenomenon. Indian smoke signals and the famous Armada bonfire signals are but two early examples, while a much more versatile system was the semaphore signalling chain established in France by Claude Chappe at the end of the eighteenth century. His chain of repeater stations facilitated the transfer of text messages over distances of 100 km or more (provided visibility between adjacent stations was adequate!), reducing the time for communication from days to hours and being of considerable assistance to Napoleon's armies in their numerous campaigns. The English Admiralty established a similar system between London and Portsmouth at a significantly later date. However, all these relied on free space propagation which could be frustratingly unreliable over distances greater than about 100 m—any development of a truly reliable optical communication system clearly required a method of guiding light within a secure pipeline of some kind. Quite a number were explored before the successful use of glass fibre waveguides in the 1970s.

The idea that light could, indeed, be guided appeared first in a demonstration by another Frenchman, Daniel Colladon in 1841, while the Englishman John Tyndall (working with Faraday) repeated the experiment in 1854, apparently unaware of Colladon's priority. They arranged light beams to shine along narrow jets of water and observed that, as the jets curved under the influence of gravity, the light curved with them, a phenomenon which was quite widely employed during the Victorian era to make elaborate and attractive illuminated fountains. A more utilitarian application was proposed (though never seriously implemented) by an American, William Wheeler, in 1880 with a view to guiding the light from an electric arc through a system of light pipes to illuminate the various rooms in a house. The year 1880 also saw the invention of the 'photophone' by Alexander Graham Bell, a light beam (actually, a sunbeam!) being modulated by reflecting it off a vibrating mirror and detected by a selenium photoconductor. Bell waxed lyrical about it: 'I have heard articulate speech produced by sunlight. I have heard a ray of sun laugh and cough and sing. I have been able to hear a shadow, and I have even perceived by ear the passing of a cloud across the sun's disc.' Unfortunately, the signals could only be transmitted over short distances and the device never achieved wide application. Nevertheless, it surely represented an important first step towards practical light wave communication.

The first use of glass fibres had nothing to do with light guiding but occurred as an essential part of the physicist C. V. Boys' measurement of the gravitational constant using a torsion balance. His contribution was, nonetheless, important in demonstrating how very fine, yet strong, glass fibres could be 'drawn' from a crucible of molten glass. His method was a trifle startling in so far as it involved using a crossbow to shoot an arrow down the length of a corridor with a trail of molten glass attached! This was in 1887—it was not until 1930 that Heinrich Lamm, a German medical student, demonstrated the transmission of an optical image through a bundle of glass fibres. There was considerable interest in such experiments within the medical profession with a view to developing a flexible 'gastroscope' for exploring patients' intestinal tracts, though practical success had to wait until the 1950s. A problem with the early structures was their use of unclad fibres which allowed light to leak between adjacent fibres, muddying the image. The key idea of using a cladding of glass with a smaller refractive index than the core material (see Box 8.1) had a somewhat chequered history, though it appears to have been first proposed in 1951 by an American Professor of Optics, Brian O' Brien. The first practical demonstration of glass-clad fibres was at the end of 1956 when an undergraduate student at the University of Michigan, Larry Curtiss fused a rod of high refractive index inside a tube of lower index and pulled the resulting rod into a fibre. One of the first applications of such fibres (in 1958) was in the form of fibre bundles used to couple images between a pair of image intensifier tubes—this programme, masterminded by J. Wilbur Hicks, represented the principal commercial success of the American Optical Company where O' Brien had moved in 1953. Hicks, while still at

Box 8.1. Fibre-optic waveguides

The simplest form of optical waveguide consists of a cylindrical core of glass with refractive index n_1 surrounded by a tubular cladding glass with refractive index n_2, as shown in Figure 8.1(a). The key to its use as a waveguide is the phenomenon of total internal reflection which we met in our discussion of light emission from an LED (Box 7.2). In that case, it was a nuisance to be overcome but in the case of optical fibres it is a godsend to be welcomed. *Total* reflection of a light beam at the interface between core and cladding implies zero loss by leakage and allows the possibility of long lengths of low loss fibre. However, it also implies a limit on the acceptance angle θ_i for light impinging on the open end of such a fibre (Figure 8.1(b)). (This and much more is discussed in numerous books on fibre-optic systems—see Agrawal 1997 and references therein.) From Snell's law we see that:

$$\sin\theta_r = n_1^{-1}\sin\theta_i, \quad \text{(B8.1)}$$

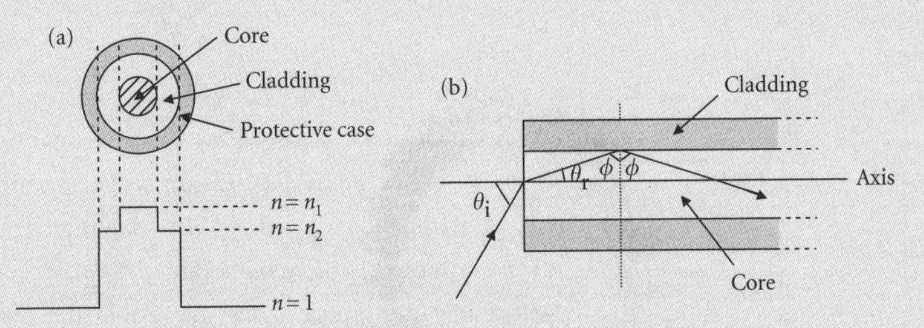

Figure 8.1. Schematic picture of light propagation in a step-index fibre based on a ray diagram. (a) The refractive index variation across a fibre diameter is shown here; (b) Shows a typical light ray entering the fibre at its centre and being refracted, then totally internally reflected at the core–cladding interface. There is a maximum value for the incident angle θ_i for total internal reflection to occur. This represents the acceptance angle for the fibre.

while at the critical angle for total internal reflection, we have:

$$\sin\varphi_c = n_2/n_1. \quad \text{(B8.2)}$$

All angles φ greater than φ_c result in total reflection and this, in turn, implies that the incident angle θ_i should be less than θ_c, where:

$$\sin\theta_c = (n_1^2 - n_2^2). \quad \text{(B8.3)}$$

(We have used the relation $\sin\varphi = \cos\theta = (1-\sin^2\theta)^{1/2}$.)
$\sin\theta_c$ is known as the numerical aperture of the fibre and given the symbol NA. In practice, n_1 and n_2 differ by only a small amount, so we may write, approximately, that:

$$\mathrm{NA} = n_1(2\Delta)^{1/2}, \quad \text{(B8.4)}$$

where $\Delta = (n_1 - n_2)/n_1$ is the fractional change in refractive index at the core-cladding interface.

In the interest of making the acceptance angle as large as possible, we should also try to make Δ as large as possible, but this has an unfortunate down-side in the form of 'modal dispersion', different rays progressing down the fibre at different rates. Let us compare two such rays, the one making the critical angle φ_c with the interface, the other travelling straight down the centre of the fibre, making no contact with the interface. The actual velocity of both rays

Box 8.1. Continued

is $v = c/n_1$ but they travel different distances in progressing a distance L along the fibre, so there is a time delay Δt between their respective arrivals at this point:

$$\Delta t = (n_1/c)[L/\sin\varphi_c - L] = (L/c)(n_1^2/n_2)\Delta. \tag{B8.5}$$

If we were to focus a light pulse onto the end of the fibre, its energy would be spread over all angles within the acceptance angle and it is apparent that, in progressing along the fibre, it would suffer a degree of broadening given by the time delay Δt. Clearly, this broadening should be less than the separation between successive pulses, so it implies a limit on the allowed bit rate B, consistent with the length of the fibre L. The relationship is:

$$BL < (c/\Delta)(n_2/n_1^2). \tag{B8.6}$$

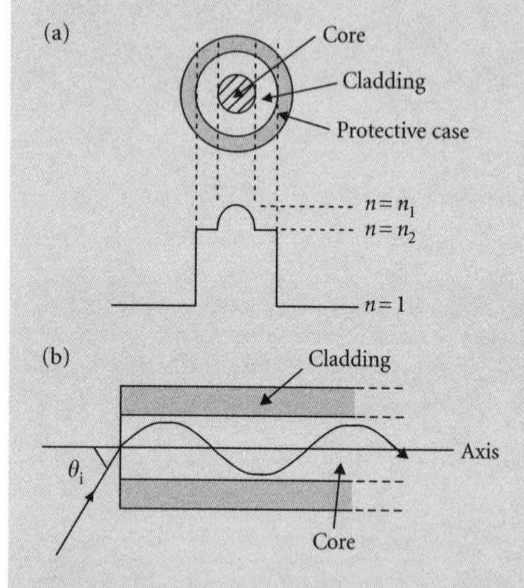

Figure 8.2. Corresponding to Figure 8.1 but for the case of the graded-index fibre in which the refractive index of the core glass varies in an approximately parabolic form. This has the effect of producing curved light rays as shown in 8.2(b) which travel along the fibre with considerably less dispersion than is the case for step-index fibres.

Given that both n_1 and n_2 are fairly close to unity, we can approximate this as $BL \sim c/\Delta$ which makes it very easy to see how the BL product (a useful figure of merit for the fibre) depends on Δ. For example, a typical value is $\Delta = 2 \times 10^{-3}$, giving $BL \sim 100$ Mb s^{-1} km—we might transmit a bit rate of 10 Mb s^{-1} over a distance of 10 km but a rate of 1 Gb s^{-1} could only travel about 100 m before becoming unreadable. Note that the corresponding value of numerical aperture is NA ~ 0.1, giving an acceptance angle of only 6°.

To avoid this rather serious limit on the performance of optical fibre, Stewart Miller of Bell labs proposed in 1965, the idea of a graded-index fibre in which the refractive index of the core is graded smoothly between centre and edge as shown in Figure 8.2(a). In this type of fibre, the ray diagram looks roughly as shown in Figure 8.2(b). The key difference between this and the simple step-index fibre is the fact that the local velocity of each ray is no longer constant but changes with its distance from the fibre axis, being larger as this distance increases (remember that $v = c/n$). Thus, the rays which travel the greater distance do so at greater velocity and this property tends to reduce the time dispersion as compared with step-index fibre. By designing the shape of the refractive index gradient appropriately, it is possible to achieve a BL product given by:

$$BL < 8c/n_1\Delta^2, \tag{B8.7}$$

from which we find that a fibre having $\Delta = 0.01$ can transmit bit rates of 100 Mb s^{-1} over distances of 100 km, while having significantly larger acceptance angle, which makes the launching efficiency correspondingly greater.

American Optical, was actually responsible for drawing the first 'single-mode' glass fibre (see Box 8.2) but immediately left to set up his own business and did not pursue it. The company did not see itself as being in the communications business so failed to pursue it either, though, later, single mode fibre came to be recognized as one of the most

Box 8.2. Waveguide modes

The ray-diagram approach to describing optical waveguide properties, which we used in Box 8.1, is helpful in providing an easily visualized picture but is limited in its capability. An accurate account must be based on the solution of Maxwell's electromagnetic equations (as is done in the standard texts, such as Agrawal 1997 and references therein). Let us merely state that Maxwell's equations can be manipulated so as to give a wave equation whose solution leads to the prediction that there are several distinct modes of propagation of energy along the fibre. The essence of these solutions depends upon the boundary conditions—for example, the step in refractive index at the core–cladding interface in a step-index fibre—and involves the radius of the fibre core. Each of these modes is characterized by a different spatial arrangement of the electric and magnetic fields within the core glass and by a different propagation rate (the velocity with which energy travels along the fibre). This so-called 'waveguide dispersion' results in broadening of a pulse of light as it travels along the fibre because the light energy is inevitably spread over a number of different modes, each travelling with its own velocity. Fortunately, the magnitude of the time dispersion given by this analysis of mode propagation corresponds closely to the result obtained in Box 8.1 using geometrical optics!

Perhaps more importantly, mode theory suggests the possibility of minimizing waveguide dispersion by reducing the diameter of the fibre core 'a'. A typical multimode fibre might have an inner radius $a = 25$ μm, supporting over a hundred modes but this number is fairly rapidly reduced by making the diameter smaller and, if $a < 4$ μm it is easy to arrive at a situation where only a single mode can propagate. Such a 'single mode fibre' no longer suffers from multimode dispersion and offers the possibility of a considerably improved BL product. However, dispersion is still not zero because of 'wavelength dispersion'—the velocity of propagation depends on the wavelength of the light being used, and the light source itself has a finite spread in wavelength $\Delta\lambda$. (We discussed this for the case of an LED, for example, in Chapter 6, Box 6.4). Thus, the energy in a pulse of light still suffers a spread in velocity and therefore it spreads in time as it travels down the fibre. This time dispersion, however, is considerably smaller than that due to multimode propagation and results in a BL product given by:

$$BL < (D \cdot \Delta\lambda)^{-1}, \qquad \text{(B8.8)}$$

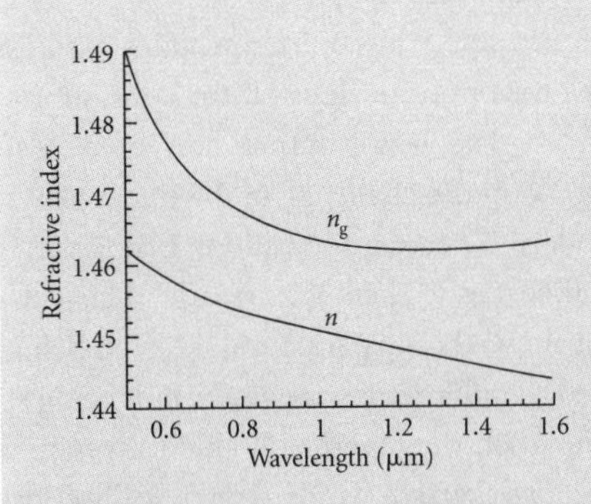

Figure 8.3. Variation of the refractive index of pure silica with wavelength between 0.5 and 1.6 μm. 'n' is the refractive index defined by the phase velocity $v = c/n$ while 'n_g' is the group index defined by the group velocity $v_g = c/n_g$. The relationship between n and n_g is given in equation (B8.9). (From Agrawal, G. P. (2002) '*Fibre Optic Communication Systems*', p. 40, John Wiley & Sons, New York) This material is used by permission of John Wiley & Sons Inc.

where the dispersion parameter D takes a value of typically $D \sim 1$ ps km^{-1} nm^{-1} and results in a BL product of 2×10^4 Mb s^{-1} km if the source is an LED with a linewidth of 50 nm or 5×10^5 Mb s^{-1} km using a multimode laser with a linewidth of 2 nm (compared with, typically, $BL = 100$ Mb s^{-1} km for a multimode fibre).

However, this is not the end of the story because there also exists a phenomenon known as 'material dispersion' which depends on the fact that the refractive index of glass (and, for that matter, other solids, too) varies with the wavelength of the light involved. The trend is for the refractive index to decrease as the wavelength increases. Figure 8.3 shows data for pure silica where the lower curve labelled 'n' represents the ordinary index $n = c/v$ (where v is the phase velocity of light within the glass), while the upper curve labelled 'n_g' corresponds to the group velocity v_g (which is the velocity with which energy travels within the medium), that is, $n_g = c/v_g$. It is easy to show (see, for example, Ditchburn 1952: section 4.29) that the relationship between n_g and n is:

$$n_g = n[1 + (\lambda/n)\,dn/d\lambda]^{-1}, \qquad \text{(B8.9)}$$

Box 8.2. Continued

which allows the curve for n_g to be derived from that for n. The important point to notice is that n_g shows a minimum value at $\lambda \approx 1.25\ \mu m$, for which case material dispersion is zero. The precise wavelength for zero material dispersion actually depends somewhat on the doping used to control the refractive index of the core or cladding glass (e.g. GeO_2 or P_2O_5 may be used to increase n in the core region or B_2O_3 to decrease n in the cladding) but it remains fairly close to $1.3\ \mu m$.

To evaluate the overall performance of a single mode fibre, it is necessary to combine waveguide and material dispersion, the net dispersion going through zero at a wavelength slightly greater than $1.3\ \mu m$ for a standard step-index fibre. However, clever design of the refractive index profile has allowed this 'zero-dispersion' wavelength to be shifted to $1.55\ \mu m$, thus combining minimum dispersion with minimum loss. However, we should note that dispersion, even then, is not quite zero, because of second-order effects, the minimum BL product being given by:

$$BL < [S(\Delta\lambda)^2]^{-1}, \tag{B8.10}$$

where the second-order dispersion parameter S takes a typical value of $S = 0.05$ ps (km^{-1} nm^{-2}) and the BL product, when using a multimode laser with $\Delta\lambda = 2$ nm, is then $BL = 5 \times 10^6$ Mb s^{-1} km (or 5 Tb s^{-1} km). This corresponds to a bit rate of 10 Gb s^{-1} being able to travel over a distance of 500 km (though the loss of 100 dB over this length of fibre would require perhaps five repeaters to regenerate the signal). Even better performance is possible using a single mode laser. However, it should be emphasized that there are practical difficulties in maintaining the laser wavelength adequately close to the zero dispersion condition and real systems rarely aspire to such performance. The best that can be anticipated, in practice, is a small value of the linear dispersion parameter D (equation (B8.8)).

important components in the whole of the telecommunications armoury. Such is often the manner of technological progress—closer to random walk than linear motion—and the reason is clear—looking into the future is a lot harder than looking into the past.

Another company which was also *not* in the communications business was Corning Glass but they took the opposite view—if the communication highway was to be made from glass, it was in their best interests to maintain an interest—so they took up the challenge in the mid-1960s at much the same time that Standard Telecommunications Laboratories (STL) and the British Post Office (as it then was) made their own commitment. Interestingly, Bell showed only peripheral interest, being, at that point, strongly committed to millimetre waveguide, but, being so well endowed with facilities and staff, they were still able to mount a small holding programme. The drive was towards reducing fibre loss which, at that time was horrendously high, of order 1 dB m^{-1} (a factor of two reduction in light intensity for each 3 m of fibre). Following a detailed theoretical study, the STL group estimated that losses should be no greater than 20 dB km^{-1}, if fibres were to be of even marginal importance, which meant reducing the absorption coefficient by approximately 50 times and this demanded a much better appreciation of the source of the loss than was then available. Many people felt that this was so improbable of success that the way forward lay in developing hollow light pipes, and a variety of ideas were explored. Copper pipes with internal silvering proved too lossy because of the very large number of reflections involved, so a system of confocal lenses (lenses separated by four times their individual focal length) was tried with a view to avoiding reflections altogether. An even more ingenious idea was to make gas lenses by developing a temperature gradient within a gas-filled pipe but it proved too difficult to stabilize the light beam. Indeed, all such ideas suffered from problems associated with bends—it might just be possible to set up a guide provided it was absolutely straight but real transmission systems were obliged to incorporate curves. Perhaps more seriously, the minute bending associated with temperature fluctuations and land settlement around buried cables was sufficient to destabilize the beam, in much the same manner as it afflicted millimetre waveguides. After much effort, the sad conclusion was reached in the early 1960s that 'The only thing left is optical fibres'! (Hecht 1999: 104)

No matter what the difficulties, it was clear what had to be done—optical fibre glass must be made less lossy, and this meant a sharp reduction in the density of impurities, particularly those from the iron transition group (see the periodic table in Chapter 1, Table 1.1). But first it was important to select the most promising glass. Up to this point, most workers had used low melting point glasses which were relatively easy to manipulate but which were also very far from pure. It was left to Bob Maurer, leader of a small research group at Corning, to make the bold decision to work on fused silica glass. Corning had considerable experience of silica and Maurer knew that it was the purest glass available so, in spite of the difficulty of working with a material whose melting point was somewhere in excess of 1600°C, he led his group out on a glass limb and concentrated on making silica fibres. Having discovered how to pull fibres, it was then necessary to develop appropriate core and cladding glasses which presented a significant problem—it required a method of doping silica so as to modify its refractive index while maintaining its inherent purity. This they achieved, using Corning's long experience, by doping the core glass with titania (TiO_2) and, by 1970, they were able to demonstrate fibre with a loss of 16 $dB\,km^{-1}$, measured at a wavelength of 633 nm. Here was the breakthrough the world of fibre optics had been waiting for—it was now clear that optical fibre could, at least in principle, meet theoretical demand for use in a communication system. Telecom engineers (and managers) sat up and noticed. There was much else to do in the way of finding suitable light sources and detectors, checking out the mechanical properties of fibre and developing ways and means for launching light efficiently into a tiny glass waveguide (Box 8.1) but there was now an 'existence theorem' which encouraged telecom companies worldwide to invest in further research. It also encouraged Corning to redouble their own efforts to improve on this early success. This they did with aplomb, reporting in 1972 a fibre having a core doped with GeO_2, rather than titania, which boasted of losses down to 4 dB km^{-1}. Equipment for drawing silica fibres appeared almost miraculously at Bell Labs, at STL, at the British Post Office Laboratory at Martlesham Heath, at the University of Southampton, at Nippon Telegraph and Telephone Company (NTT) and Fujitsu in Japan, and in many other laboratories around the world. The race to develop practical systems was now well and truly on! Somewhat bizarrely, the first of these was commissioned by the Dorset police force in south-west England in 1975. A lightning strike had wrecked the electronics in the police control room in Bournmouth and the chief constable wanted an aerial link which would insulate the replacement equipment from any further attack. Standard telephones and cables (STC) were asked to supply a fibre transmission line which answered the need in a matter of weeks and STC were handed a wonderful piece of advanced publicity. It might have been a long way from a long haul telephone system but it demonstrated possibilities better than a dozen papers at scientific conferences (not that the research laboratory, STL overlooked the possibility of presenting them, of course!).

The next major step forward was made in Japan. In 1976, Masahara Horiguchi of NTT, working with Horoshi Osanai of the Fujikura Cable Company reported silica fibres with losses of 0.47 dB km^{-1} measured at a wavelength of 1.2 μm, another major reduction which opened the way to truly long haul fibre systems. They further improved on this 2 years later, reporting losses as low as 0.2 dB km^{-1} at 1.55 μm. Previously, systems design had been based on the use of a GaAs laser working at 880 nm (the first room temperature CW laser appeared in 1970) but two developments changed everyone's thinking on this—first of all, it was discovered that silica fibre showed zero dispersion (see Box 8.2) at a wavelength of 1.2–1.3 μm and, second, the Japanese workers showed that its lowest loss occurred at a wavelength close to 1.55 μm (see Figure 8.4). (Note that, in order to reach these low loss

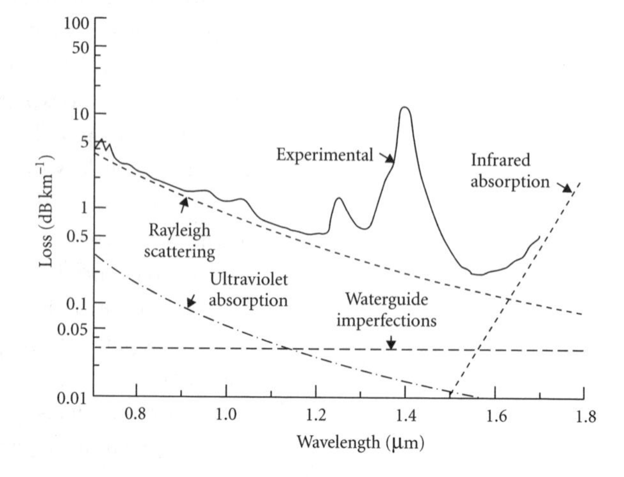

Figure 8.4. Experimentally measured loss in pure silica fibres as a function of wavelength, showing minimum loss at a wavelength of 1.55 μm. The subsidiary minimum at $\lambda = 1.3$ μm corresponds to minimum material dispersion (see Box 8.2). The loss peaks in the vicinity of 1.4, 1.25, and 1.0 μm result from residual water vapour at a level of about 10^{-8}! (From Agrawal, G. P. (2002) *'Fibre Optic Communication Systems'*, p. 56, John Wiley & Sons, New York) This material is used by permission of John Wiley & Sons Inc.

levels, it was necessary to reduce iron group impurity densities below one part in a billion and that of water vapour below ten parts per billion!) While losses at 880 nm were certainly low enough to satisfy the requirements of relatively short distance communications, it rapidly became clear that long and medium distances would be best served by using the longer wavelengths. However, there was a difficulty—no suitable light source existed at either of these wavelengths and a whole new development had to be set in train to provide it. Thus was born the InGaAsP laser which we shall describe in Section 8.2.

Real fibre systems began to appear with some degree of urgency from about 1977 onwards. Both ATT and the British Post Office used graded-index fibres to transmit telephone traffic at 880 nm wavelength in 1977 and, in 1978, they agreed to develop the first transatlantic fibre-optic cable which was to use a single mode fibre and operate at a wavelength of 1.3 μm. In the same year, NTT demonstrated a 53 km transmission system working at 1.3 μm. From 1982 single mode fibre displaced graded-index fibre and in 1984 the first working submarine cable was installed by British Telecom (now split away from the Post Office) between the Isle of Wight and the English mainland—a relatively short distance but an important demonstration of intent. In 1985 single mode fibre spread across North America, carrying long distance telephone signals at 400 $\mathrm{Mb\,s^{-1}}$ and upwards, and in 1988 the first fibre transatlantic cable TAT-8 went into service, using single mode fibre at 1.3 μm wavelength. In 1987, David Payne at Southampton University demonstrated the first erbium-doped fibre amplifier which allowed optical signals to be amplified without first transposing them into electronic form. This constituted a tremendous step forward and it is important to note that erbium amplifies at 1.55 μm rather than 1.3 μm, a major factor in favour of recent moves towards concentrating on this wavelength. The other factor is, of course, the minimum fibre loss which occurs at 1.55 μm and, now that 'dispersion-shifted' fibres also show minimum dispersion at 1.55 μm, there seems to be little point in opposing the trend. The 1990s saw the development of almost unbelievable bit rates in optical fibres, reaching 1000 $\mathrm{Gb\,s^{-1}}$ (1 $\mathrm{Tb\,s^{-1}}$) by 1996 (at Bell, NTT, and Fujitsu laboratories). Practical systems operated at rates up to 10 $\mathrm{Gb\,s^{-1}}$ by virtue of wavelength division multiplexing (WDM) (each fibre carrying four different wavelengths, each wavelength carrying signals at 2.5 $\mathrm{Gb\,s^{-1}}$) and systems operating at 640 $\mathrm{Gb\,s^{-1}}$ have been planned for operation in 2003. For future submarine cables, the sky appears to be the only limit!

So much for our potted history of optical fibres in the communications business. But what does all this say about the semiconductor devices which are designed to be used with them? We shall close this section with an attempt to summarize the crucial issues raised. These can conveniently be considered under three headings: sources, detectors, and electronics. The system clearly needed a source of light of appropriate wavelength which must also satisfy a number of other criteria, an efficient optical detector which could operate at the required data rate (speed) and associated electronics which could be used to build signal regenerators, amplifiers, wavelength discriminators, etc. Let us look at each in turn.

As we remarked above, the first fibre systems operated at wavelengths near 880 nm which were provided by GaAs light emitting diodes (LEDs) or lasers, the reason being that these were the only suitable sources available, but when it became clear in the 1970s that longer wavelengths possessed advantages in terms of low loss and dispersion, there was an urgent demand for lasers operating at 1.3 and 1.55 μm. Furthermore, they had to be stable, with long life (particularly for transoceanic cable use), and able to be modulated at high bit rates, bit rates which have shown a tendency to increase without limit! An important parameter, in this connection, was soon seen to be the temperature stability of threshold current and wavelength. Particularly in WDM systems, it was vital, not only that the wavelength could be precisely selected but that it must remain within tight limits at all times during operation (even while ambient temperature made wide excursions). Looking at the need to minimize time dispersion in long haul systems, it was also important that the laser emission linewidth should be as small as possible and this led to a demand for single mode operation. It was also desirable that these sources should work at high overall efficiency, particularly those to be used in submarine cable repeaters where power had to be provided via long supply lines.

With regard to the photodetector, it was relatively straightforward to specify the requirements. These were for sensitivity (at the appropriate wavelength), fast response and absence of noise. Particularly for use with the long wavelength lasers, the band gap of the detector material should be small enough to absorb the appropriate radiation, but should not be too small so as to minimize thermal generation of free carriers, which represent a noise background against which the photogenerated signal must compete. Long separation between repeaters in a long haul system implied low signal levels at the detector, the need to minimize the number of repeaters demanding maximum sensitivity in the detector. It must also be capable of responding at high frequencies so as not to distort the digital pulses which must be clearly recognized in the interests of minimizing error rate.

Finally, we note that there was an essential need for a range of electronic devices to function as pulse generators, modulators, amplifiers, etc. Low noise amplifiers were needed to follow the photodetectors and serve as drivers for the regenerating pulse generators used in repeaters, power amplifiers were needed to act as pulse modulators for driving lasers, mixers were required in conjunction with coherent (i.e. superheterodyne) receivers, which also demanded appropriate IF amplifiers. One of the principal issues here is that of optoelectronic integration—combining optical and electronic devices on a single chip so as to optimize performance and minimize cost. Because long wavelength lasers were built on InP substrates, this implied the development of InP-based electronic devices, though there is now pressure to develop lasers based on GaAs substrates which can be combined with standard GaAs electronic devices. This is still very much in the melting pot, but it is certain that the future of 'photonic' circuitry lies in greater integration, with development of complete modules to replace the plethora of individual devices which currently provide the various essential functions. At present, light wave systems appear to be in somewhat the same situation as transistor electronics was before the development of silicon integrated circuits. The analogy may well be imperfect (most analogies are!) but there is every indication of considerable further change before photonics reaches maturity. For the moment, though, we shall look at the various optoelectronic devices as individuals.

8.2 Long wavelength sources

Returning to the 1970s, when it first became clear that fibre optics had a future in communications, we have already noted that the AlGaAs/GaAs double heterostructure (DH) laser had reached the important milestone of CW operation at room temperature. It was natural enough, therefore, that telecoms engineers should look to incorporate it in the first fibre systems, even though it was not a single mode laser, with the consequent limitations implied by the relatively large dispersion of multimode fibre (even graded-index fibre). At that stage, it hardly mattered because fibre losses of 15–20 $\mathrm{dB\,km^{-1}}$ restricted the distance between repeaters to a mere kilometre or so and dispersion effects were relatively unimportant. The main contributions of the GaAs laser were to provide a stop-gap until long wavelength devices could be developed and to provide a relatively cheap source for short-haul links (local area networks, LANs) where distances were naturally less than a kilometre. However, its significance should certainly not be overlooked in so far as it demonstrated the important principle that practical fibre-optic communication systems could actually be made to work.

Nor, while on the subject of GaAs, should we overlook the fact that even GaAs LEDs were able to make a valuable contribution. Compared with lasers, they suffer from the drawbacks of lower power, wider beam angle, broader emission linewidth, and smaller modulation bandwidth but, again, in the context of LANs, they certainly offered a much cheaper alternative to lasers while providing a performance adequate for many small-scale systems. In the early stages of fibre communications LEDs also offered far greater reliability—GaAs lasers suffered serious degradation problems associated with 'dark line defects' (dislocations which propagated through the structure and destroyed the device) which took considerable time and research effort to eradicate—for some years LEDs could therefore be relied upon to give much longer operating lifetimes and were preferred in many situations. Two types have been employed—surface emitting and edge emitting. Typical of the former was the so-called 'Burrus diode' (invented by C. A. Burrus of Bell Labs in 1970) in which the optical fibre was bonded with epoxy cement in close proximity to the active layer, while the diode current was restricted to a region immediately below the fibre end. The epoxy had the further function of reducing the refractive index step at the point where the light was exited, thus improving light collection efficiency, close to 1% of the internally generated light being accepted by the fibre. The edge emitting device looked structurally very similar to a laser diode except that a highly reflecting coating was applied to one end of the active region and an anti-reflection coating to the emitting end. Because of light guiding along the active region, the emission angle was much smaller than that in a surface emitting device and significantly greater collection efficiencies were possible.

As we saw in Chapter 6, the linewidth of LED emission is of order 2 kT (at room temperature, approximately 50 meV) which, for a GaAs LED, corresponds to about 30 nm. Because different wavelengths travel at slightly different velocities in the fibre (wavelength dispersion), this results in a limitation to the so-called *BL* (bit rate $\times$ distance) product (see Box 8.2) but this is very much smaller than the time dispersion in a multimode optical fibre (see Box 8.1) and is therefore unimportant for the LAN application. In other words, an LED will perform quite as well as a laser diode, in terms of dispersion, when we are concerned with short propagation distances. Perhaps the more serious limitation is the modulation bandwidth for direct modulation—that is, when the source is modulated by switching its drive current. In the case of an LED, the maximum speed of modulation is determined by the electron–hole recombination lifetime which, for direct gap materials, is typically 3 ns, resulting in a maximum modulation rate of about 50 MHz. This contrasts somewhat starkly with the Gb s^{-1} bit rates which we referred to in the previous section but has been found entirely adequate for many local links (bear in mind that the modem connecting a personal computer to the telephone line operates, even today, at no more than about 50 kb s^{-1}). We shall say no more about LED sources, preferring to concentrate on the more progressive (and more exciting!) long wavelength laser diodes. However, it should not be forgotten that LEDs still have a role to play in those situations (of which there are many!) where cost considerations dominate those of technical sophistication.

The search for long wavelength laser materials began surprisingly early. In 1972, two interesting developments were reported. Sugiyama and Saito at the Musashino laboratory of NTT published an account of a DH laser based on the AlGaAsSb system which emitted at a wavelength of 980 nm. Fifteen percent antimony was included in the material to reduce the band gap from the 1.43 eV of GaAs to 1.27 eV for the alloy GaAsSb. Some 25% Al was included in the confining layers to provide a lattice-matched material with larger gap. The whole structure was grown by liquid phase epitaxy (LPE) on a GaAs substrate (to which it was *not* matched) and pulsed room temperature operation required a threshold current density of 8.5 kA cm^{-2}. At much the same time, Gerald Antypas et al. at Varian Associates in Palo Alto reported an exciting innovation in the form of a 'negative affinity' photocathode for an infrared night vision tube based on the quaternary alloy InGaAsP, lattice-matched to InP. We shall discuss this in greater detail in Chapter 9 but, for our present purpose, it is sufficient to recognize that the device was sensitive at wavelengths up to 1.1 μm, implying the possibility of light generation at this wavelength using a similar alloy. Given the fact that InP has a gap of 1.34 eV, it would clearly be possible to use this as confining layers in a DH laser, the whole structure being lattice-matched to an InP substrate. Box 8.3 explains the significance of the InGaAsP system for long wavelength devices.

Both ideas were pursued by others and in 1976 Nahory et al. at Bell labs reported on a GaAsSb/AlGaAsSb DH laser which operated CW at room temperature with a much reduced threshold current density of 2.1 kA cm^{-2} and an emission wavelength of 1.0 μm. Again, it was grown by LPE on a GaAs substrate, employing a graded GaAsSb layer to minimize the effect of lattice-mismatch. Bogatov et al. in the Soviet Union also achieved success with AlGaAsSb and then in 1975 reported the first lasers based on InGaAsP/InP, while in 1976, Jim Hsieh at Lincoln Labs, MIT also reported a room temperature CW laser based on the InGaAsP/InP system with threshold current density of 4.7 kA cm^{-2} and wavelength of 1.1 μm. There was also competition from RCA where Neuse and Olson had earlier demonstrated laser action at 1.0 μm in the alloy GaInAs. They grew their structure on a GaAs substrate, using a graded GaInP layer to take up the mismatch between GaAs and GaInAs. It was in 1978 that the low loss at 1.3 and 1.55 μm in silica fibres was first appreciated and the question then had to be asked: could any of the above material systems be used to extend laser operation to these longer wavelengths?

The answer, in principle, was 'yes' for all three but only the InGaAsP/InP system could do so while maintaining a lattice-match and this, as ever, proved to be the overriding consideration. Defects are

Box 8.3. Band gap of lattice-matched $Ga_xIn_{1-x}As_yP_{1-y}$

The importance of the quaternary alloy $Ga_xIn_{1-x}As_yP_{1-y}$ lies in the fact (see Figure 8.5) that it can be lattice-matched to an InP substrate over a range of compositions which result in band gaps lying between 1.35 and 0.75 eV (corresponding to wavelengths between 0.92 and 1.65 μm). This includes the wavelengths of interest for long distance fibre-optic communications, viz. 1.3 and 1.55 μm and also provides band gaps appropriate for both carrier and optical confinement in a DH or QW laser, the use of lattice-matched alloys ensuring minimal defect densities.

The requirement of lattice-matching implies a relationship between the atomic fraction of Ga and that of As in the alloy (i.e. between the parameters x and y). To see what this relationship must be we note that it depends on the differences between the lattice constants of the binary compounds GaP, InP, and InAs (see Figure 8.5). Suppose that we perform an imaginary experiment in which we start with InP and substitute some fraction of In atoms with Ga—this will shift the lattice constant of the alloy to the left but we can balance this by replacing an appropriate fraction of P atoms with As (which shifts the lattice parameter to the right). However, because the difference between a(InP) and a(GaP) is much larger than that between a(InAs) and a(InP), the introduction of Ga has a much greater effect than that of As, so, to maintain the required balance, we should introduce correspondingly *less* Ga than As. To put this on a quantitative basis, we assume that the lattice parameter of the ternary alloy $Ga_xIn_{1-x}P$ varies linearly with x (which is a statement of Vegard's Law), so we can write:

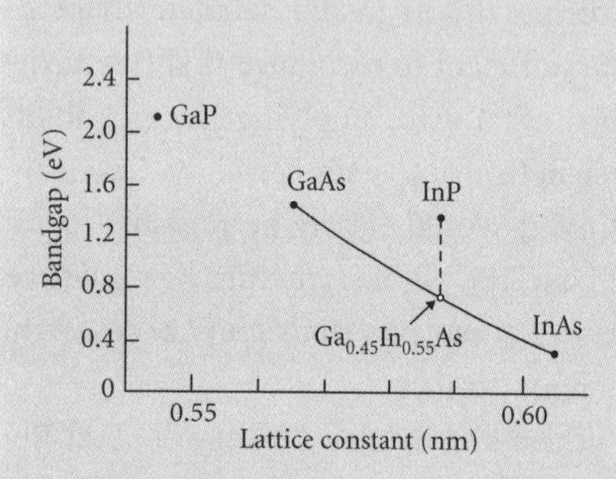

Figure 8.5. Band gap E_g vs lattice parameter 'a' for the alloy system $Ga_xIn_{1-x}As_yP_{1-y}$. By controlling the ratio $y:x$ so as to maintain $y = 2.21x$, it is possible to maintain lattice-matching to an InP substrate. The corresponding band gaps (which are all direct) trace out the region defined by the dashed line, ranging from 0.75 to 1.35 eV.

$$a(x) = a(\mathrm{InP}) - x[a(\mathrm{InP}) - a(\mathrm{GaP})] \tag{B8.11}$$

and, similarly for $InAs_yP_{1-y}$:

$$a(y) = a(\mathrm{InP}) + y[a(\mathrm{InAs}) - a(\mathrm{InP})] \tag{B8.12}$$

from which it follows that, in order to maintain lattice-match for the quaternary, we must have:

$$x[a(\mathrm{InP}) - a(\mathrm{GaP})] = y[a(\mathrm{InAs}) - a(\mathrm{InP})]$$

or

$$y/x = [a(\mathrm{InP}) - a(\mathrm{GaP})]/[a(\mathrm{InAs}) - a(\mathrm{InP})] \tag{B8.13}$$

and, inserting the values $a(\mathrm{GaP}) = 0.5451$ nm, $a(\mathrm{InP}) = 0.5869$ nm, and $a(\mathrm{InAs}) = 0.6058$ nm, we obtain the result that:

$$y/x = 0.0418/0.0189 = 2.21. \tag{B8.14}$$

Since neither x nor y can exceed unity, it follows that the lattice-matched alloy range is specified by $0 < y < 1$, for which case $0 < x < 0.453$.

Given the above relationship between x and y, we can now write an expression for the band gap of the alloy in terms of a single parameter. Fitting the results of photoluminescence and photoreflectivity measurements allows us to write this (in terms of y) as:

$$E_g(y) = 1.35 - 0.72y + 0.12y^2 \text{ (eV)}, \tag{B8.15}$$

which includes the slight downward bowing of the curve E_g vs y which, as we have seen earlier, is typical of many ternary alloy systems. Putting $y = 1$ in this expression yields a minimum band gap, corresponding to the alloy $Ga_{0.45}In_{0.55}As$, of $E_g(\mathrm{min}) = 0.75$ eV, as already stated. The values of y which correspond to wavelengths of 1.3 and 1.55 μm are $y = 0.61$ and $y = 0.90$ ($x = 0.28$ and $x = 0.41$), respectively.

anathema to semiconductor laser diodes and, if long operating lifetimes are to be achieved, a lattice-matched system always wins, hands down. So it was in this case, where reliability was, if anything, more important than in any other application. We shall therefore concentrate attention on developments in this system which has proved spectacularly successful in satisfying the demands of fibre systems. The story began, as we have already seen, in 1976 when Jim Hsieh achieved CW operation at room temperature from a laser emitting at 1.1 μm. By the following year his MIT group was able to report similar devices operating at 1.3 μm and in 1980 Hsieh helped to set up a new company Lasertron with the express goal of manufacturing long wavelength lasers for the fibre communications industry. It was an excellent decision—by the mid-1980s Lasertron employed 350 staff and enjoyed a turnover of $28 million. However, they were not alone—a brief scan of the literature in 1979 reveals that NTT and KDD in Japan, STL in England, together with RCA and Bell in America were all developing long wavelength devices, the ultimate goal of 1.55 μm operation already having been achieved.

It is interesting that, with the single exception of RCA, who used vapour phase epitaxy (VPE), all these developments were based on LPE, following the pioneering work of the Varian team which originally developed GaInAsP for long wavelength photocathodes. Whereas micron-thick layers were an essential requirement for photocathodes, and LPE was therefore an entirely appropriate growth technology, DH laser structures really demanded epilayers down to 0.1 μm in thickness and the later, but inevitable move to quantum well structures reduced this by a further order of magnitude. LPE therefore made a vital contribution to the early developments but was obviously destined to be replaced by more flexible methods such as metal-organic vapour phase epitaxy (MOVPE) and molecular beam epitaxy (MBE). (There was, in any case, a technical problem in applying LPE to the longer wavelength quaternary alloys.) In this case it was MOVPE which took the prize—by the early 1990s scarcely any other growth method is ever referred to. We may recall that GaAs was first grown by MOVPE in 1973 (Manasevit) so the extension to other III-V compounds and alloys followed naturally as demand developed. InP growth was reported by Duchemin from Thomson CSF in 1979 and the application to GaInAsP by Jean Pierre Hirtz in 1981. The first MOVPE lasers operated in pulsed mode in 1981 but progress was rapid and by 1983 the Thomson workers (Razeghi et al.) had obtained CW operation at threshold current densities below 500 $\mathrm{A\,cm^{-2}}$—from this point, MOVPE never looked back. (Anyone wishing to follow the MOVPE story in greater detail might like to refer to Razeghi 1989.)

In order to appreciate subsequent progress in laser technology, it may be helpful to reiterate the key requirements which determined research objectives during those early years. Apart from the choice of wavelength (1.3 and 1.55 μm) which demanded appropriate control of alloy composition (see Box 8.3), it was clearly desirable to minimize threshold current and its temperature-dependence, achieve long-lived stable operation, and obtain single mode operation in the interest of minimizing the emission linewidth. Remarkably enough, all these objectives were well on the way to being satisfied by 1985 when the situation was summarized in the '*Semiconductors and Semimetals*' series (Willardson and Beer 1985). Threshold current densities had been reduced by an order of magnitude from the early values of order 5 kA cm^{-2} and some understanding was emerging as to the origin of its dependence on temperature. Let it be said immediately that this latter feature has been seen all along as one of the problems associated with long wavelength lasers and one which has taken up considerable scientist-hours in attempts to improve it. It is common practice to express the temperature-dependence in terms of a parameter T_0, as follows:

$$J_{th}(T) = J_{th}(T_1)\exp\{(T - T_1)/T_0\}, \tag{8.1}$$

where T_1 is a convenient reference temperature (e.g. room temperature). This implies an exponential variation of J_{th} with device temperature. While there is little theoretical justification for using this form, there is considerable pragmatic support, even though it may sometimes be necessary to use different values of T_0 to characterize different temperature ranges. Without doubt, it is a useful single parameter which can be applied to compare different lasers. Note that a large value for T_0 implies a *small* variation of J_{th} with temperature. The problem with long wavelength quaternary lasers could be stated succinctly in terms of their having typical values of $T_0 \sim 50$K, compared with the corresponding values for GaAs lasers of $T_0 \sim 150$K. It severely restricted their operation at elevated temperatures, an important practical limitation for devices required to function at moderately high power levels in a wide variety of environments.

Explaining this behaviour in physical terms was less straightforward. A number of mechanisms was considered during the early 1980s, including the radiative recombination process, inter-valence band absorption, carrier leakage over the heterobarrier in a DH structure, and non-radiative 'Auger recombination'. Let us look briefly at each. The radiative lifetime depends on temperature through the fact that the effective density of states in the conduction and valence bands is itself temperature-dependent and there is evidence to suggest that this may play a role in determining T_0 at low temperatures. A contribution to non-radiative current is made by the optical absorption occurring when

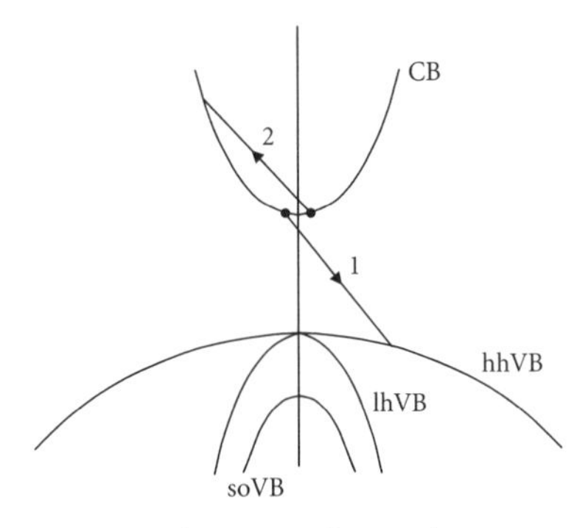

Figure 8.6. Illustration of a two-electron Auger recombination process such as that which occurs in GaInAsP long wavelength lasers. The energy produced by the recombining electron (1) is transferred to the second electron (2) which is excited high into the conduction band. This excited electron returns to the bottom of the band by emitting a sequence of acoustic phonons so the overall process is therefore non-radiative. Note that momentum is conserved between the two electrons involved. Several similar processes occur, involving transitions between the different valence bands.

a hole in the valence band near the Γ point transfers to the split-off valence band but direct measurement of the absorption coefficient showed this to be negligible in terms of the measured values of T_0. Carrier leakage certainly plays a role in determining T_0 in GaAs lasers and in AlGaInP lasers where the confining barrier heights are relatively small but, in the case of 1.3 and 1.55 μm lasers, these barriers are too large to allow significant leakage and this was not thought to be an important mechanism. Auger recombination is probably much more significant in long-wavelength devices and, given various uncertainties in calculating recombination rates, can account reasonably well for experimental observations. Proper understanding was made difficult by the fact that several different Auger processes were known to exist and it was not easy to determine the transition probabilities for each one. (An example of such a process is illustrated in Figure 8.6, involving a two-electron transition in which a conduction band electron drops into the valence band, the energy being given to a second electron which is excited higher into the conduction band.) Calculation of Auger recombination lifetimes showed the strong temperature dependences necessary to explain experimental T_0 values and, what is more, suggested that T_0 values for 1.55 μm lasers should be rather smaller than those for 1.3 μm devices, as was generally found in practice. Because the Auger process is inevitable in long wavelength lasers, the conclusion was that very little could be done about the small T_0 values of DH GaInAsP devices. It would generally be necessary to control the ambient temperature and to run the laser at modest current densities so as to minimize joule heating.

On the credit side, however, it was soon appreciated that long-wavelength devices were very much more durable than their GaAs counterparts. The two important failure mechanisms which plagued early GaAs lasers were the so-called dark line defects and catastrophic mirror damage but neither was found to be anything like so serious for quaternary devices, probably on account of the much smaller photon energies involved. Non-radiative recombination, resulting in band gap energy being dumped as heat in the active region, could lead to destructive dislocation propagation from the substrate through the active region or to overheating of the mirrors. The significantly smaller photon energies in long-wavelength devices meant much smaller heat quanta and thereby contributed to improved device stability. This simple explanation entirely ignores, of course, any detailed understanding of degradation mechanisms but it was clear, right from the word 'go', that 1.3 μm lasers showed quite remarkable durability and, by 1985, predicted operational lifetimes in excess of 25 years (10,000 h) began to look perfectly realistic. This was a practical bonus which had not been predicted but which was all the more welcome to an industry which depended on long-term investment in ultrareliable performance. Detailed understanding would follow but was by no means essential to the appreciation of Fortune's unaccustomed smile.

The quest for single mode operation, on the other hand, was very much in the hands of the laser design engineer and its achievement represented a major success in laser fabrication technology. First let us be clear what we mean by 'single mode' operation. We have to distinguish between 'lateral' and 'longitudinal' modes associated with the laser optical cavity. Ideally, the latter consists of an approximately rectangular parallelepiped (or brick) with dimensions of order: length, $L = 500\ \mu m$; width, $W = 2\ \mu m$; and thickness, $t = 0.2\ \mu m$. The length, as we have discussed in connection with GaAs lasers in Chapter 5 is defined by partially reflecting end mirrors, the thickness by epitaxially grown optical confining layers whose function depends on the introduction of relatively small steps in refractive index (see Figure 5.19), while the width requires more detailed discussion. The longitudinal modes are determined by L, being separated in wavelength by an amount $\Delta\lambda = \lambda^2/2\mu L$ (equation B5.24), and the lateral modes are determined by the width W. In order that the laser waveguide should support only the fundamental transverse mode, W must be less than a limiting value W_0 which is of order of the wavelength λ. Single mode operation implies that the laser should oscillate in the fundamental transverse mode and in a single longitudinal mode only, and we must examine these two requirements separately. First we look at the question of the lateral modes.

Two quite different mechanisms can operate in defining the width of the cavity, giving rise, on the one hand, to 'gain-guided' and, on the other, to 'index-guided' lasers. An example of a gain-guided structure is shown in Figure 8.7(a), the extent of the active region being determined by the top contact stripe. Current flow and therefore carrier injection is restricted to the region immediately below this contact, implying that optical gain is similarly confined. In principle, it seems necessary simply to define the contact stripe with a width less than W_0 in order to obtain the desired fundamental mode but unfortunately this represents an oversimplification on account of current spreading below the stripe and the effect of current injection in modifying the refractive index profile beneath the contact. It turns out that the presence of free carriers, effects a reduction in the refractive index which results in 'antiguiding' of the optical wave. The net result is that it becomes extremely difficult to stabilize the mode structure of a gain-guided laser, more or less precluding its use for single mode operation.

Figure 8.7(b) illustrates an example of the alternative, index-guided structure in which W is defined by steps in refractive index, somewhat similar to those which determine the effective thickness t. This

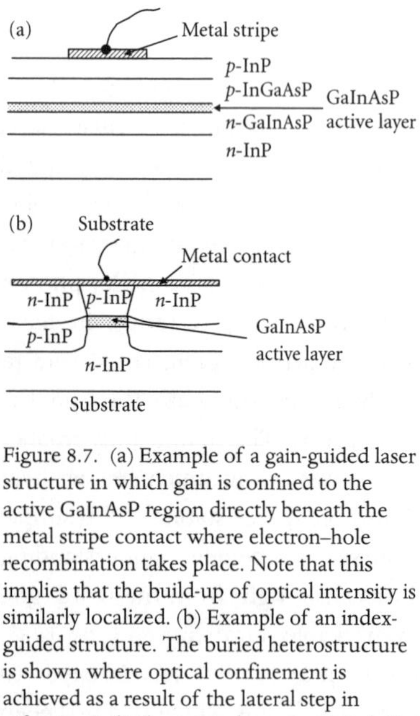

Figure 8.7. (a) Example of a gain-guided laser structure in which gain is confined to the active GaInAsP region directly beneath the metal stripe contact where electron–hole recombination takes place. Note that this implies that the build-up of optical intensity is similarly localized. (b) Example of an index-guided structure. The buried heterostructure is shown where optical confinement is achieved as a result of the lateral step in refractive index between the active GaInAsP and the p-type InP.

structure, introduced in 1980, is known as a 'buried heterostructure' (BH) laser. It is made by growing an *n*-type InP layer, followed by an undoped GaInAsP active layer, then etching a mesa to define the width W and regrowing a *p*-type InP layer which isolates the active region. A second etch is followed by the final *n*-type InP layer which provides current confinement. Its essence is the precise control of the width W by the first mesa etch and the presence of a refractive index step $\Delta n = 0.3$ between the active region and the adjacent *p*-type InP. In order to achieve a single lateral mode W should be less than W_0, where:

$$W_0 = \lambda / 4(\Delta n)^{1/2}. \tag{8.2}$$

Inserting the value $\Delta n = 0.3$ results in a value of $W < 0.5\lambda$ which represents a modest challenge to the technologist. In practice, somewhat more sophisticated structures are preferred in which the index step is reduced to about $\Delta n = 0.01$ and $W_0 = 2.5\lambda$ but, from our point of view, it is the principle which is important, rather than the technological detail. In any case, both were well established by 1985.

In order to discuss the question of longitudinal modes we refer again to Chapter 5 and recall that the mode separation given by equation (B5.24) amounts to approximately 0.35 meV when $L = 500$ μm. This compares with a typical gain distribution which spreads over some 50 meV and means that several longitudinal modes are likely to be excited once the threshold current is only slightly exceeded. Indeed, little can be done to prevent this in a simple BH laser, implying that the optical signal has an effective width of order 1 meV (or, in wavelength terms, approximately 2 nm). As we point out in Box 8.2, this gives rise to a limiting BL product of about 5×10^6 Mb s^{-1} km, even for a 'zero-dispersion' single mode optical fibre, a value significantly too large to allow the high bit rates planned for future long-haul systems (e.g. 640 Gb s^{-1}—see Section 8.1). Only a single mode laser source could be contemplated for such developments, the question being how to select just one longitudinal mode from the hundred or so which lie within the gain curve. Several approaches were investigated during the early 1980s and considerable progress made.

Examination of equation (B5.24) suggested a simple method of mode selection—reduction of the cavity length L implies a corresponding increase in mode separation which may lead to no more than one or two modes being close enough to the gain peak to be excited. In particular, if L were reduced from 500 to 5 μm, no more than two modes would lie within the gain spectrum and there would be a very high probability of just one of them being excited. Such a line of argument clearly led to the use of a vertical cavity surface emitting laser (VCSEL)

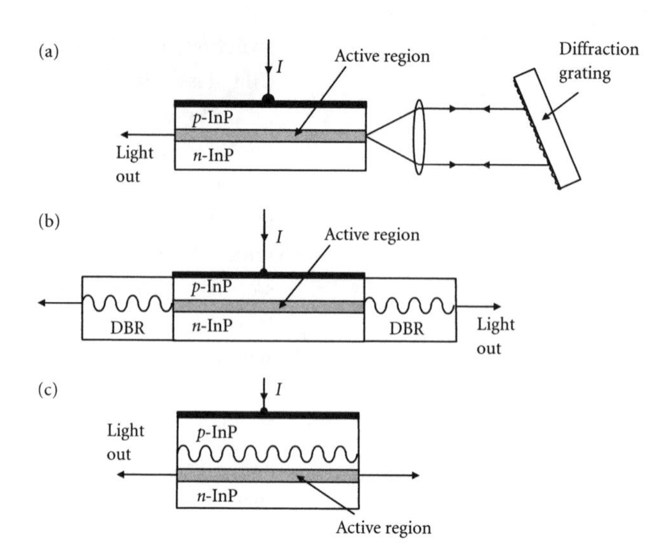

Figure 8.8. Three examples of single mode long wavelength laser structures. In (a) the mirror selectivity is obtained by the use of an external mirror in the form of a suitably orientated diffraction grating. In (b) the conventional mirror facets are replaced by a pair of distributed Bragg reflectors (DBR), while in (c) the optical feedback is distributed all through the lasing region (DFB). Note that the Bragg mirrors are made by etching a grating pattern into the appropriate layer, then depositing a second layer having different refractive index. This provides the required periodic variation in refractive index along the axis of the laser structure.

structure and, though some limited success was achieved, making long wavelength VCSELs proved extremely difficult (we shall discuss this in detail below) so some alternative method was needed. The preferred method depended on using wavelength-selective feedback to replace the standard Fabry–Perot mirrors common to nearly all lasers developed up to this point. Very simply, if one could design a mirror which reflected strongly at only one wavelength (a wavelength within the gain spectrum), then the laser would oscillate at this wavelength and at no other. The trick would be to design a mirror whose reflectivity peaked at a wavelength corresponding to one of the longitudinal modes and having a reflectivity–wavelength curve sharp enough to exclude the modes on either side.

Various means for achieving this have been investigated, three such have been illustrated in Figure 8.8. The simplest to understand involves the use of an external mirror, rather than relying on the laser facets themselves. One facet should be coated with an anti-reflection film so that all the laser light is transmitted to the mirror and, if this mirror takes the form of a diffraction grating, as shown in Figure 8.8(a), its reflectivity can be tuned to select the desired wavelength by rotating it to the appropriate angle. It has the additional advantage of allowing the laser to be readily tuned over a useful wavelength range—typically 50 nm. The disadvantage of this arrangement is its inconvenience, making the overall structure many times larger than that of the laser itself and requiring mechanical adjustment to effect tuning. A far better approach is to design mirrors based on the multilayer principle, which is used to make mirrors for VCSELs (i.e. Bragg stack mirrors—see Section 6.6, Figure 6.25), where the reflectivity peaks at a wavelength determined by the thicknesses of the layers having different refractive indices. The period of the stack Λ is related to the free space wavelength of the laser emission λ_0, and to the average refractive index of the semiconductor n by equation (8.3):

$$\Lambda = \lambda_0/2n. \tag{8.3}$$

In a lateral laser structure it is not possible to grow this alternating refractive index structure epitaxially—instead it is achieved by etching the surface of one layer (say InP) to form a corrugated surface, then growing a second layer of different refractive index (InGaAsP) over the corrugations. When looked at edge-on, such a structure presents an alternating refractive index to a wave propagating parallel to the original InP surface (see Figure 8.8(b)). As can be seen in the figure, similar mirrors are formed at either end of the laser, the reflectivity maximum occurring at a wavelength determined by the period of the etched corrugation. For an operating wavelength of 1.55 μm, this period should be about 0.2 μm which may be achieved using holographic techniques—two laser beams are caused to interfere at the semiconductor surface, coated with a layer of photoresist, the resist being developed in sympathy with the interference pattern, then used as a mask to etch the required corrugations. Such a laser is usually referred to as a distributed Bragg reflector (DBR) laser. The third alternative incorporates optical feedback within the gain region of the laser itself—in other words, the Bragg mirror corrugations are an integral part of the gain region, feedback being continuous throughout the structure (see Figure 8.8(c)). This variation is known as a distributed feedback (DFB) laser and has been widely used in fibre systems operating at high bit rates ($B > 2.5$ Gb s^{-1}). Note that the end facets of the device also act as local mirrors but the 2000 or more corrugation periods within the cavity dominate the feedback and ensure that only a single longitudinal mode oscillates.

The GaAs laser not only served as a source for many of the early fibre systems but provided a prototype for the development of long wavelength devices, both DBR and DFB structures having been developed in the AlGaAs/GaAs system during the early 1970s. The use of Bragg reflectors was first demonstrated at Bell Labs in 1971 by Kogelnik and Shank who applied it to a dye laser but their lead was soon followed in a number of other laboratories (Bell, Xerox, Caltech, Hughes, Hitachi, Toshiba) where it was applied to GaAs lasers. Initial work employed rather basic optically pumped structures but electrical injection lasers were demonstrated at Xerox and at Hitachi in 1974. Threshold currents were high and it soon became apparent that there were two reasons for this, the fact that the feedback corrugations had not been etched deeply enough and, in the case of the DFB structures, the damage introduced by ion beam etching of the feedback corrugations. The answer to this latter problem lay in removing the feedback region from the active region of the laser, as shown in 1975 by Casey et al. of Bell Labs who made use of a separate confinement double heterostructure to achieve $J_{th} = 2.2$ kA cm^{-2} in a DFB laser. In the same year, another Bell group, Reinhart et al. demonstrated a DBR GaAs laser with $J_{th} = 5$ kA cm^{-2}. An important feature of these Bragg mirror lasers was the very much smaller dependence of the lasing wavelength on temperature, compared with conventional Fabry–Perot devices. Typically, $d\lambda / dT = 0.07$ nm K^{-1} for a DFB laser (determined by the temperature-dependence of the refractive index), compared with 0.3 nm K^{-1} for a conventional device made from the same material (determined by the semiconductor band gap). They were also characterized by an emission wavelength which remained constant as the injection current was increased above threshold, unlike their Fabry–Perot counterparts which suffered from mode-hopping.

These GaAs developments were clearly encouraging for extension to long-wavelength devices and in 1979 workers at both Hitachi (DFB) and the Tokyo Institute of Technology (DBR) demonstrated GaInAsP/InP lasers operating at low temperatures. These were soon followed by room temperature operation (Tokyo—1980 and NTT—1981). By 1984 room temperature CW operation had been reported by several Japanese laboratories (Tokyo, NTT, KDD, NEC), by two groups at Bell Labs and by the BT laboratory at Martlesham Heath in England. Once again, a major issue concerned the need to achieve sufficiently deep feedback corrugations and Westbrook et al. at Martlesham employed an interesting variation on the standard growth method by combining LPE for the first part of the structure with MBE (i.e. a low temperature growth method) for the upper layers to avoid partial dissolution of the corrugations. Another important issue was that of achieving single mode operation when the laser was used to transmit the nanosecond pulses appropriate to practical fibre systems, earlier work having demonstrated that lasers were capable of single mode emission when driven, CW often failed to maintain this under short pulse operation. Suffice it to say that these DFB lasers passed the test with flying colours—bit rates of 2 Gb s^{-1} were demonstrated at Bell Labs from a 1.5 μm wavelength device with an output linewidth as small as 0.1 nm. Finally, threshold current densities had come down to 2 kA cm^{-2}, output powers had increased to about 30 mW and single mode operation could be obtained at currents up to twice the threshold value and at temperatures up to about 100°C. The source requirements for future long haul fibre systems were now assured. Not that it would not be possible to make further improvements, but the basic demands could certainly be met by the middle of the 1980s.

Interestingly, the DFB lasers involved up to this time covered various modifications of the double heterostructure but did not include quantum well devices. As we saw in Chapter 6, quantum well laser

(QW laser) development in the AlGaAs/GaAs material system dates from 1975–6, but those working on long wavelength devices probably had too much to do to worry about quantum wells much before the middle of the 1980s. While it is true that InGaAs/InP multiple quantum well (MQW) laser structures were grown by LPE as early as 1977, serious interest in such devices appears to date roughly from the time when MOVPE growth of InGaAsP became widely available around 1985. (MBE made much less impact because of the difficulty of handling phosphorous in an ultrahigh vacuum (UHV) system.) It was already well established that AlGaAs/GaAs MQW lasers offered important advantages with respect to reduced threshold current and increased output power but any possible impact on fibre communications remained yet to be recognized. Key issues here included the minimization of threshold current and output linewidth, together with the optimization of modulation bandwidth, and the decade between 1985 and 1995 saw each of these aspects thoroughly explored, with encouraging results, deriving from the high post-threshold gain characteristic of QW lasers.

As we point out in Box 8.4, the CW linewidth of a typical single mode laser varies as a^{-2}, where 'a' is the laser gain coefficient dg/dn (n being the density of free electrons—not the refractive index!), and this quantity is significantly larger in QW lasers. In fact, linewidths as small as 100 kHz were measured in several long wavelength MQW lasers during the decade 1985–95, compared with typically 10 MHz for a bulk device. However, this, in itself, is of limited benefit if the laser is to be modulated by direct variation of the drive current because high speed modulation leads to 'frequency chirping', a sympathetic oscillation in mode frequency, which effectively increases the linewidth to few tens of gigahertz and it is this figure which must be used in estimating the limits of practical system performance. Once again, MQW lasers have the advantage; as $\Delta\nu_{\text{chirp}}$ varies inversely with the gain coefficient, linewidths of about 20 GHz (corresponding to $\Delta\lambda = 0.2$ nm) are typical. This is a factor ten smaller than the effective linewidth of a multimode laser and results in a correspondingly increased BL product. The practical limit (see Agrawal 1997: 200) allows a transmission distance of about 100 km when the bit rate is 20 Gb s^{-1}, a performance which has been demonstrated using 1.55 μm MQW DFB lasers in conjunction with dispersion-shifted single mode fibre. It is worth noting, though, that the use of a CW laser together with an external modulator, to avoid chirping has potential for further improvement (we shall examine this in Section 8.4).

The modulation bandwidth B_{m} is also enhanced in QW lasers. B_{m} is effectively determined by the 'relaxation oscillation' resonant frequency Ω_{r} which varies linearly with $a^{1/2}$ (see Box 8.4). Values of B_{m} as large as 30 GHz were achieved during the 1990s, in response to ever-increasing demand from the systems designer.

Box 8.4. Laser dynamics

In our earlier discussion of laser operation we considered only steady-state aspects, CW operation being appropriate, for example, to CD and DVD systems. However, in the case of lasers for optical communications, it has always been seen as a significant advantage that they can be modulated by direct variation of the drive current. This inevitably raises questions such as: how fast can the laser be modulated—that is, what is the limiting modulation bandwidth?, what effect does modulation have on other parameters, such as the emission linewidth?, and how do these effects depend on the particular laser structure involved? More generally, how should we select a laser for high speed fibre-optic transmission? In this box we briefly consider three important aspects of laser performance: CW linewidth, modulation bandwidth, and so-called 'frequency chirp'. Throughout we refer to single longitudinal mode operation such as obtained from a long wavelength DFB laser. The subject of laser dynamics is inevitably complex, involving the solution of coupled rate equations (see Agrawal and Dutta 1986: ch. 6), often in the large signal regime. All we attempt to do here is collect a few simple results which illustrate the important features of dynamic laser behaviour, relevant to optical communications.

It is a common feature of all oscillators that, as oscillations build-up, the emission line becomes narrower so we clearly expect the laser linewidth to be significantly smaller than that of an equivalent LED (for which $\Delta\nu \sim 10^{13}$ Hz). For a multimode laser the effective linewidth is determined by the longitudinal mode separation, together with the number of modes excited and is typically about 2 nm (or about 3×10^{11} Hz) but for a single mode DFB laser we can anticipate something very much smaller. In fact, the frequency width of a single mode is given by the equation:

$$\Delta\nu = \Delta\nu_0(1 + \alpha^2), \tag{B8.16}$$

where

$$\Delta\nu_0 = R_{sp}/4\pi P_{ph}. \tag{B8.17}$$

R_{sp} is the spontaneous emission rate and P_{ph} is the total number of photons within the laser cavity. Note that the power output P is proportional to P_{ph}, so $\Delta\nu_0$ is inversely proportional to the laser power. At an output level of 1 mW, $\Delta\nu_0 \sim 1$ MHz.

The parameter α, known as the 'linewidth enhancement factor' plays an important role. For a typical bulk laser $\alpha \sim 5$ so the emission line has a width of roughly 25 MHz at $P = 1$ mW (or 1 MHz at 25 mW). It originates from a rather subtle mechanism associated with spontaneous emission noise. The random nature of spontaneous emission results in fluctuations in the free electron density within the laser cavity which, in turn, causes small variations in the refractive index, μ, and these effect small changes in the frequency of the oscillating mode. The wavelength of the mth mode is given by:

$$L = m\lambda/2, \tag{B8.18}$$

where L is the cavity length and λ the wavelength within the semiconductor. Bearing in mind that $\nu = c/\mu\lambda$, we easily arrive at:

$$\nu = mc/2\mu L, \tag{B8.19}$$

Box 8.4. Continued

showing that the mode frequency depends inversely on refractive index. From a practical viewpoint, it is important to know that α varies inversely with the laser gain coefficient $a = \partial g/\partial n$, where, for a bulk laser:

$$g = a(n - n_0), \tag{B8.20}$$

n_0 being the electron density at transparency. Thus, for minimum linewidth, it is necessary to maximize 'a' which is most readily achieved by using an MQW laser where the step density of states results in much larger values for 'a' than are appropriate to bulk devices. In practice, values of $\alpha < 1$ have been achieved in MQW lasers, resulting in CW linewidths as small as 70 kHz.

High speed modulation of a single mode laser results in a related phenomenon known as frequency chirping. Current modulation changes the free carrier density and, therefore, the refractive index which causes the mode frequency to change in sympathy. The net effect is a broadening of the emission line $\Delta\nu_{\text{chirp}}$ which depends on the amplitude of the modulation current I_m and the linewidth enhancement factor α. Thus:

$$\Delta\nu_{\text{chirp}} \propto \alpha I_m. \tag{B8.21}$$

In this case, linewidth is *increasing* with laser power. It also increases with increasing modulation rate and is roughly a factor two larger at 1.5 μm wavelength than at 1.3 μm. In practice, for 1.55 μm lasers and bit rates above 1 Gb s^{-1}, $\Delta\lambda_{\text{chirp}} \sim 0.3$ nm, at drive currents of order 40 mA or, in frequency units, $\Delta\nu_{\text{chirp}} \sim 4 \times 10^{10}$ Hz. As may be anticipated from the mechanism, this is many orders of magnitude larger than the corresponding CW linewidth (though an order less than that of a multimode laser).

Finally, we examine the limiting modulation bandwidth which is closely related to the phenomenon of 'relaxation oscillations'. If a laser is suddenly switched on, the amplitude of the light output does not immediately settle to its equilibrium value but first oscillates at a frequency of order of a few gigahertz known as the 'relaxation frequency'. The 3 dB modulation bandwidth B_m (the frequency at which the modulated output power falls to half its low frequency value) is generally found to be approximately equal to the relaxation frequency Ω_r. This frequency is given by the relation:

$$\Omega_r = \{\Gamma v_g a(I - I_{th})/Ve\}^{1/2}, \tag{B8.22}$$

where Γ is the optical filling factor, v_g the group velocity of photons within the laser cavity, 'a' the gain coefficient, and V the volume of the cavity. In order to maximize B_m, we should aim for a large gain coefficient and small cavity length. In the case of a quantum well laser which has the virtue of a large gain coefficient, it is also necessary to use several wells in order to optimize the filling factor. (Note that Ω_r represents the fundamental limit to modulation frequency—considerable care is needed to minimize parasitic resistance and capacitance which may limit B_m to even lower values.) In practice, B_m ranges from about 5 to 25 GHz.

The InGaAsP/InP material system lends itself conveniently to the requirements of a separate confinement double heterostructure laser. As we saw in Box 8.3, the band gap can be tailored between 0.75 and 1.35 eV, while maintaining lattice-match, so a typical 1.55 μm laser may use 10 nm wide $Ga_{0.47}In_{0.53}As$ wells with $Ga_{0.28}In_{0.72}As_{0.61}P_{0.39}$ barriers and InP

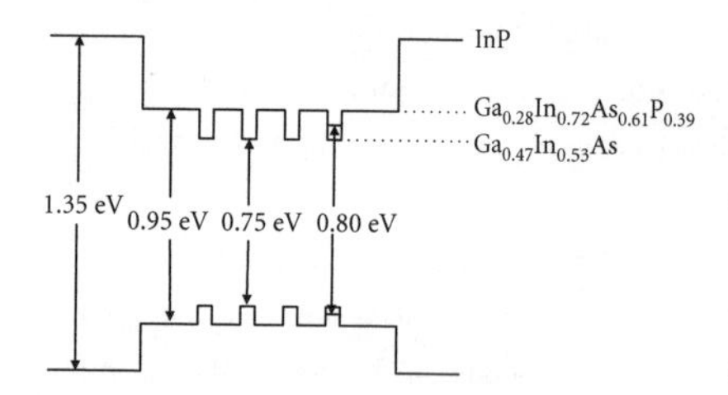

Figure 8.9. Energy diagram of a typical 1.55 μm GaInAsP/InP MQW laser. The active region is made up of four GaInAs quantum wells with $L_z = 10$ nm, separated by GaInAsP barriers whose energy gap $E_g = 0.95$ eV, corresponds to a wavelength of 1.3 μm. The optical confining layers are InP.

optical confinement layers, the appropriate energy level structure being illustrated in Figure 8.9. No sooner were such structures under development than the further advantage associated with the use of strained quantum wells was introduced, following the proposal by Alf Adams of Surrey University in 1986. We have already noted the application of strain to AlGaInP red lasers in Chapter 7—it lifts the degeneracy of the valence bands at the Γ point, leaving a light hole band uppermost whose smaller density of states results in lower threshold current and increased slope efficiency. Adams' original suggestion was made, of course, in the context of long-wavelength lasers and these desirable effects were first demonstrated in this context. When Thijs et al. (1994) reviewed the position in 1994, they were able to point to the achievement of threshold current densities of 100 A/cm^{-2} and threshold currents as low as 1 mA, which made a major contribution to overall efficiency, and to the effect of the increased gain coefficient in improving dynamic performance, as outlined in the previous paragraph. Strained well structures are now accepted as standard in almost all communication lasers. We should also note one further development—most of these devices employ several wells in order to improve the optical confinement factor Γ (the ratio of the cross-sectional area giving gain to the total cross-section occupied by the optical mode). The mode width is of order of the wavelength of light in the structure (i.e. about 0.5 μm), while L_z is no more than a few percent of this, so it is usual to employ perhaps 10 wells, thereby multiplying the total strain by a factor of 10 and leading to strain relief by the introduction of dislocations. To maintain a high level of strain in the wells *without* introducing dislocations, it is now a general practice to use barrier material which is strained in the opposite sense and, by careful choice of composition and barrier thickness, obtain an *overall* strain close to zero, a technique known as 'strain compensation'.

The degree of sophistication introduced into what began as a relatively simple device (a DH laser) can be seen to be considerable—distributed feedback, multi-quantum wells, separate optical confinement (index guiding), strain compensated active region, the need for two or more growth sequences, and using low temperature growth to avoid subsequent diffusion effects (diminution of the feedback corrugations)—but this is nothing more than standard practice in today's semiconductor industry. What matters is the ability to satisfy system requirements and, if all this is necessary, then it must be made available. And at an acceptable price, of course! Altogether, the industry can be well-pleased with its contribution (ably supported by valuable innovations from university research, of course!), though there is one aspect which falls somewhat short of ideal—the temperature coefficient of threshold current is still greater than desired (T_0 is too small, being typically about 70 K, when 150 K would be more acceptable). This applies even to MQW lasers and is the result of the Auger recombination process which tends to be significant in lasers which operate at long wavelengths (small photon energies). We shall have more to say about this in Section 8.5 when we discuss recent trends and developments—for the moment we can leave the optical transmitter in capable hands and move on to consider the receiver.

8.3 Photodetectors

The principle of the semiconductor photodetector was discovered many years before any need became apparent in the context of optical communications. We saw in Chapter 2 that both photoconductivity and photovoltage effects were discovered in selenium in the 1870s. In fact, there is an interesting link between photoconductivity and the subject of this chapter, long distance communication—when Willoughby Smith discovered the phenomenon of photoconductivity in 1873, he was in process of testing a submarine cable for which purpose he had need of a stable high value resistor. In his subsequent report, published in Nature, he described his discovery in the following terms:

> When the [*selenium*] bars were fixed in a box with a sliding cover, so as to exclude all light, their resistance was at its highest, and remained very constant, fulfilling all the conditions necessary to my requirements; but immediately the cover was removed, the conductivity increased from 15 to 20 percent, according to the intensity of light falling on the bar.

Bearing in mind the years of effort devoted to unravelling the many complexities of light-induced electrical effects in semiconductors, one may be forgiven for drawing comparison between Smith's box and its much earlier counterpart owned by Pandora! His discovery was followed by many years of empirical research to establish the complex nature of the phenomenon and the wide range of materials which evinced it. (A useful summary of this early work is contained in the classic book on photoconductivity by Richard Bube—Bube 1960—where anyone eager for more detail will find a number of further references.) As we have commented before, the lack of an acceptable quantum theory of semiconductivity meant that little progress could be made in understanding these effects before the late 1930s but a mass of experimental detail was collected. In particular, it was recognized that photoconduction was a bulk material property, while the photovoltaic effect was associated with surfaces or interfaces. The lack of good quality single crystal material also militated against proper understanding, though it certainly did not prevent these effects being put to good use—empirically designed photodetectors were made from selenium, cuprous oxide, and thallous sulfide which found application, for example, during the early years of photography and in the cinematograph industry.

As in the development of the transistor, high quality single crystals of germanium and silicon were essential to scientific progress. We saw in Chapter 3, for example, how silicon and germanium *p-n* junctions could be disclosed by shining light on them and measuring the resulting photovoltage, and this led to the development of light detectors whose design was driven by a combination of reproducible experimental data and theoretical modelling. Indeed, by the middle of the 1950s photodetectors had already been demonstrated over a wide range of wavelengths based on various semiconductors: Ge, Si, CdS, InSb, PbS, PbSe, PbTe, etc., and their operating principles established in considerable detail. Nevertheless, there was a great deal of further development to come before the high speed, highly sensitive photodetector employed in today's optical communications systems would emerge. In this section, therefore, we shall try to follow the logic of this and, in the process, gain an appreciation of the many different types of photocells which have been developed and learn something of the appropriate physics.

The photoconductor is, perhaps, the easiest type of photodetector to understand. Figure 8.10(a) illustrates its operation. A uniform 'brick' of high resistivity semiconductor has two metal contacts attached so as to allow its resistance to be determined by the application of a modest voltage while measuring the resulting small current. Shining light with a wavelength corresponding to the semiconductor band gap on top of the sample generates holes and electrons which drift to the contacts under the influence of the applied electric field, resulting in an increase in current (the 'photocurrent') which is proportional to the intensity of the light. In the case of a suitably high resistance sample, the photocurrent may be much larger than the original dark current, so a measurement of the total current (for a fixed applied voltage) gives a measure of the light intensity. Two questions immediately arise: how thick should the sample be to optimize the device sensitivity and how fast will the response be to a short pulse of light (an obvious question to ask in the context of digital communications). The first of these is easy to answer—the thickness is determined by the absorption coefficient for the light. If this be α, we can write the simple expression:

$$\Delta L = (1 - R)L_0[1 - e^{-\alpha x}] \tag{8.4}$$

for the amount of light absorbed in a thickness x of semiconductor. L_0 is the intensity of light falling on the top surface of the sample and R is the fraction of the light which is reflected from it. Ideally, we wish ΔL to represent something close to the total intensity L_0, so we require $R \sim 0$ (i.e. we should use an anti-reflection coating) and $e^{-\alpha x} \sim 0$ (which implies that αx should be at least 3—or $x \geq 3/\alpha$). For an indirect gap material, such as silicon or germanium, α is typically about $1\text{–}2 \times 10^5\,\mathrm{m}^{-1}$ so we should use a sample thickness of at least 20 μm—say 30 μm to be

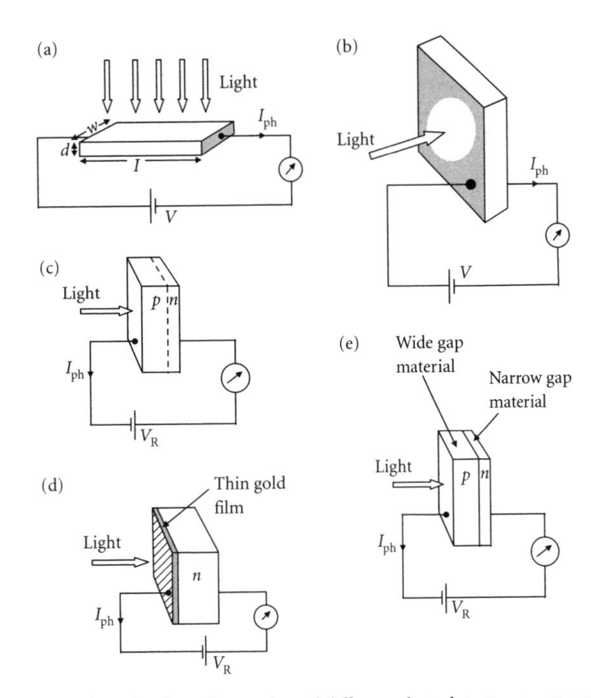

Figure 8.10. Outline sketches of a number of different photodetector structures. (a) The 'conventional' photoconductor, which is made thick enough to absorb all the light falling on it and is lightly doped so as to minimize the dark current. The light intensity is measured by recording the photocurrent I_{ph} under a small applied voltage V. (b) A modified photoconductor geometry is shown, in which the photocurrent is measured through the same face as that through which the light enters: (c), (d), and (e) Examples of *p-n* junction photodiodes which operate under reverse bias V_R. The high junction field rapidly separates holes and electrons, producing a photocurrent in the external circuit, as shown. The surface barrier diode in (d) utilizes a thin film of gold to form a Schottky barrier, the thickness being of order 15 nm, thin enough to transmit a high proportion of the incident light. The heterojunction device shown in (e) employs a wide gap material as a window which absorbs very little light, allowing most of it to be absorbed in the *n*-type material close to the junction.

safe. What is more, this argument clearly implies that the light should have a photon energy somewhat greater than the semiconductor band gap and implies a long wavelength cut-off in the sensitivity—light with wavelength longer than $\lambda_g = c/\nu_g = hc/E_g$ will not be absorbed. (Remember that, if λ is measured in microns and E in electron volts, the relation is $\lambda_g = 1.240/E_g$.) This represents an important property of a semiconductor to be used as a photodetector and means that we must choose a material appropriate to each and every application. For detecting visible light, a band gap of 1.5–2.0 eV would be suitable, whereas detecting thermal radiation with wavelength of 10 μm demands a material with band gap of about 0.1 eV.

Response speed is a more complicated question. It depends on the interplay between a number of different characteristic times, including the dielectric relaxation time t_d which we discussed in Section 3.3 (see the definition in equation (3.10)), the minority and majority carrier recombination lifetimes, and the transit time for a free carrier to move from the place where it is injected to reach a collection point such as a metal contact. Let us first ask a supplementary question. Suppose we suddenly inject a packet of electric charge into a semiconductor sample—how long will it be before its effect is detected in the measuring circuit? It is here that the dielectric relaxation time plays a fundamental role, representing, as it does, the timescale on which electrical neutrality is re-established, following a disturbance. Suppose that we were to inject a small quantity of extra charge into a metal wire and ask the question, how long will it take before this creates an effect at the end of the wire (i.e. in the external circuit)?, the answer is: a very short time indeed, because the injected charge is neutralized locally by rapid movement of nearby electrons, and this movement is transferred along the wire by similar local disturbances until electrons are 'pushed' out of the end of the wire. The time it takes for this to happen is of the order of the dielectric relaxation time which, for a metal is extremely short. Thus $t_d \sim \varepsilon_0 \rho \sim 10^{-11} \times 10^{-7} = 10^{-18}$ s. For most practical purposes, the response is instantaneous. (Notice that it is *not* necessary, in this case, to wait until the injected charge moves down the wire to reach its end.) However, in the case of a high resistance semiconductor sample whose resistivity $\rho \sim 100\ \Omega\,\text{m}$, we find $t_d \sim 10 \times 10^{-11} \times 100 = 10^{-8}$ s and the response time now is certainly finite. It might, in fact, be comparable to either the recombination lifetime or an appropriate transit time. Recombination times in III–V compounds are often of order 10 ns, though in silicon and germanium we might expect times of the order of microseconds or longer (in which case this will be the response time—see Box 8.5). What about a typical transit time? Consider, first, the case of the photoconductor in Figure 8.10(a) where the separation L of the metal contacts may be of order 1 mm or more. A free electron moves with a velocity $v = \mu_e V/L$, where V is the voltage between the contacts and μ_e is the electron mobility; so the transit time is easily seen to be given by:

$$t_t = L/v = L^2/\mu V \tag{8.5}$$

and, using typical values of $L = 1$ mm, $\mu = 0.1\,\text{m}^2\text{V}^{-1}\text{s}^{-1}$, and $V = 1$ V, we obtain $t_t \sim 10^{-5}$ s. This is a relatively long time but, considering another plausible situation, that of a reverse biased *p-n* junction depletion width $d = 10^{-7}$ m and the high field within it, which implies a 'saturation drift velocity' for electrons of $v_s = 10^5\,\text{m}\,\text{s}^{-1}$, we obtain $t_t \sim 10^{-12}$ s. This is actually much shorter than the dielectric relaxation time for the intrinsic depletion region, so here it *is* necessary to wait for the injected charge to drift across the depletion region before any response is registered in the measuring circuit. Clearly, we have to examine each particular case carefully before pronouncing on the nature of its performance.

If all this is rather confusing, the situation will probably become clearer by considering some specific cases. As we show in Box 8.5, the

Box 8.5. Photodetector response time

Consider the photoconductor illustrated in Figure 8.10(a) and assume the semiconductor to be silicon. Suppose that, at time $t = 0$, the light beam is switched on—how rapidly will the effect be registered in the external circuit? We should attempt to answer this question in two parts, first: how rapidly will the photoinduced free carrier densities build-up within the semiconductor? and second: how long will it take this build-up to be registered? Let us try to answer the second question first. Assuming the semiconductor is very lightly doped, so that its resistivity is as high as 100 Ω m, we can calculate the dielectric relaxation time $t_d \sim 10^{-5}$ s (see main text). However, as the photoinduced carrier density increases, this time will diminish (making it an undesirably complex situation!). The transit time for electrons to reach the electrodes may also be about 10^{-5} s, making it even more complex! What we can say is that no matter how fast free carriers are generated, there will be a delay of some microseconds before this is registered in the external circuit. Now for the first question.

The light, being absorbed by the semiconductor, will generate both holes and electrons in equal numbers. Let us simplify the problem by assuming that free carriers decay only by direct, bandto-band recombination and define a recombination lifetime τ. If, at some time t the total number of electrons is $N(t)$ and the (constant) generation rate of electrons is G (where G is proportional to the light intensity), we can write the net rate of increase in N as:

$$dN/dt = G - N/\tau. \tag{B8.23}$$

Making the assumption that $N = 0$ at time $t = 0$, this differential equation is easily solved to give the following expression for $N(t)$:

$$N(t) = N_0[1 - \exp(-t/\tau)], \tag{B8.24}$$

showing that the build-up of free carriers proceeds at a rate determined by the recombination lifetime τ. N_0 is the steady-state value of N which is given by:

$$N_0 = G\tau, \tag{B8.25}$$

as can easily be seen by putting $dN/dt = 0$ in equation (B8.23). Notice that a large value for τ is desirable for high sensitivity, whereas rapid response demands a small value!

A typical value of τ for silicon might be 10^{-6} s, in which case N would reach its steady-state value in a few microseconds, though, bearing in mind the dielectric relaxation effect, the overall time before this was registered might be slightly longer. The important point to notice is that, even if the light pulse had a rise-time of 1 ns, the detector would take several microseconds to respond. It is even easier to demonstrate that the decay of free carrier density when the light is switched-off is given by:

$$N(t) = N_0 \exp(-t/\tau), \tag{B8.26}$$

showing that a similar response time is appropriate to the signal decay on switch-off.

In order to make the photodetector faster, it would be necessary to reduce τ by doping the silicon with precious metal atoms such as gold or platinum (which act as recombination centres and 'kill' the lifetime) but it would also be necessary to reduce t_d by doping with donors (or acceptors) to reduce the silicon resistivity. Both, however, would reduce the sensitivity. We saw above that sensitivity is proportional to τ—increasing the background carrier concentration would also increase the so-called 'dark current', making it more difficult to detect the photocurrent against this background. All in all, photoconductors are not well-suited to operation as fast light detectors, so let us look at an alternative.

Box 8.5. Continued

Suppose we arrange to absorb the light within the depletion region of a *p-n* junction. This is a high field region which has the property of separating holes and electrons very rapidly, sweeping electrons into the *n*-side and holes into the *p*-side of the junction, where they become majority carriers (see Figure 8.11). In fact, the junction field is large enough so that carriers are quickly accelerated to reach their 'saturation drift velocity' of about 10^5 m s^{-1}, so, if the silicon doping is chosen to make the depletion width equal to about 1 μm (in order to absorb a reasonable fraction of the light), the transit time for carriers to cross it is approximately 10^{-11} s. This is much shorter than the dielectric relaxation time t_d for the depleted junction region, so t_t will, in this case, determine the response time of the detector. In other words, t_t is now the time which determines how rapidly carriers are lost from the active material and therefore should replace τ in the equations above. Electrons and holes are swept across the depletion region so rapidly that they have no time to recombine—τ, like t_d, is no longer a relevant parameter. However, the dielectric relaxation time of the doped *n*- and *p*-type contact regions of the silicon will be of order 10^{-13} s, so they will effectively transfer the signal to the external circuit in a time much less than the transit time t_t. Clearly, this provides a much more promising approach to making detectors fast enough to respond to Gb s^{-1} bit rates. Even so, there is another compromise to consider. If we wish to make a highly sensitive detector, we should increase the thickness of the depletion region with the aim of absorbing a larger fraction of the light but this inevitably increases the transit time and makes the device slower.

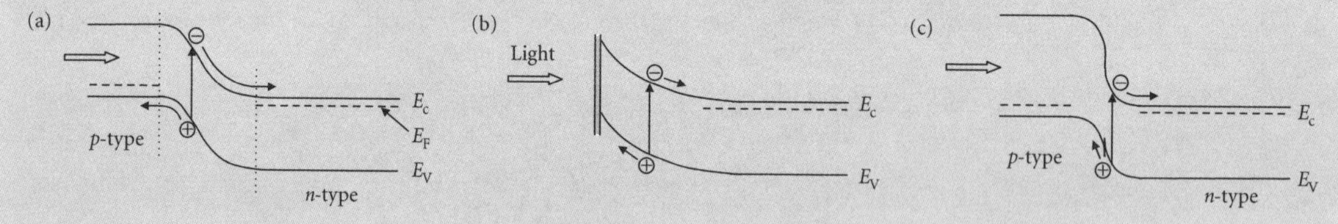

Figure 8.11. Energy band diagrams of reverse biased junctions, showing their use as photodetectors. The high junction field separates holes and electrons, sweeping them rapidly into the doped contact regions, where they become majority carriers. This flow of holes and electrons can be measured as a photocurrent in the external circuit. (a) A conventional *p-n* homojunction is shown. (b) A surface barrier detector, based on an ultrathin metal film which transmits most of the light. (c) The use of a *p-n* heterojunction is illustrated. The wide gap *p*-type material acts as a transparent window for the incident light. Note the presence of a small spike in the valence band which may restrict hole flown across the junction.

photoconductive detector illustrated in Figure 8.10(a) is not well suited to achieving fast response so we shall concentrate attention on various types of photodiode in which the light impinges in a direction normal to the junction plane. The *photoconductor* shown in Figure 8.10(b) illustrates this point, the absorbed light being detected by measuring a photocurrent flowing in the same direction as the incident light. It was a relatively small modification to include a *p-n* junction near the back of the structure to collect minority carriers and generate a similar photocurrent via the junction field and this was the geometry adopted in many of the early experiments performed in the 1950s. For example, in 1949 Shive at Bell Labs used a point contact diode on the back of a thin Ge sample to collect carriers generated by light shining on the opposite face. The device was sensitive only over a small area (about 0.2 mm diameter) but proved capable of detecting light at wavelengths up to 1.6 μm with a frequency response 'flat to 200 kc s^{-1}' (the Hertz having not yet been widely adopted). It is interesting to us today, to note just how slow this appears but we should remember that the junction transistors of this time were struggling to achieve speeds of 20 kHz (see Table 3.2). In 1951, Pietenpol, another Bell Labs worker, reported on a modified structure in which the collector took the form of a *p-n* junction formed by impurity diffusion from the back of a thin Ge wafer, though the response speed in this case was no more than 20 kHz. (Note that, though the use of Ge provided a response up to 1.6 μm wavelength, this was purely coincidental. No one at that time could possibly have predicted the later need for a response at 1.6 μm—Ge just happened to be the best developed and most widely available semiconductor.)

It is instructive to think about the modus operandi of these devices. Light incident on the front surface is absorbed in the relatively thick *n*-type Ge which forms one side of the junction, so here we are concerned with a photocurrent of minority carrier holes which must *diffuse* to the junction (there being only a very small electric field in the *n*-type material means that drift would be negligible) before being swept through the depletion region by the junction field. As we argued in Chapter 3, diffusion of minority carriers proceeds together with an equivalent majority carrier flow which ensures electrical neutrality—it is only at the junction that holes and electrons are separated so as to form a photocurrent in the external circuit. But more important for our interest in response speed is the fact that diffusion is a slow process—it proceeds at a rate determined by the 'diffusion velocity' $v_D = L/\tau$ (L = diffusion length, τ = recombination lifetime), which, for Ge, takes a value of about 10 m s^{-1} (or 10^7 μm s^{-1}). The Ge slice should be at least 10 μm thick in order to absorb a reasonable fraction of the light so this suggests a limiting response time of 10^{-6} s. At that time, however, there were problems with producing such thin semiconductor layers (transistor base widths, for example, were of order 500 μm) and the practical response time was probably closer to 10^{-5} s, with a corresponding frequency response of about 30 kHz. The achievement of 200 kHz in Shive's experiments implies a Ge sample thickness of about 50 μm which represented the result of careful etching.

There was yet another problem with this structure, concerning the ubiquitous surface states which so frustrated Shockley's attempts to make a field effect transistor (FET). Light was incident on a Ge surface, at which surface states were waiting to capture minority carriers (holes) and encourage them to recombine with electrons from the metal contact. In other words, the 'surface recombination lifetime', τ_s, was very much shorter than that in the bulk τ_r so most of the holes generated near the front surface were drawn in the 'wrong' direction, clearly undesirable from the viewpoint of sensitivity! It led, in fact, to the early development of 'surface barrier photodiodes' (see Figure 8.10(d)) which utilized a gold film evaporated onto the semiconductor surface to form a Schottky barrier. Its thickness of about 15 nm was thin enough to allow most of the light through. Holes generated in the bulk Ge diffused back to the surface where they were swept into the contact by the barrier field (see Figure 8.11(b)). Such structures had the advantage of not needing particularly thin semiconductor slices—the whole photodetection process took place near the incident surface while the rest of the semiconductor served only as an electrical contact. It had, therefore, to be moderately well doped, which limited the size of the reverse bias that could be applied, before breakdown occurred. However, the use of a surface barrier effectively avoided competition between two collector junctions—all the minority carriers could, in principle, be collected at the surface. 'In principle' because some fraction recombined in the bulk semiconductor before reaching the depletion region, representing a form of competition between diffusion and recombination. If the diffusion length L were long, compared with the absorption length α^{-1}, then most of the carriers would be collected at the surface—if not, relatively few would make the journey, and sensitivity would suffer in consequence. While Ge was characterized by long diffusion lengths ($L \sim 50\ \mu m$ compared with $\alpha^{-1} \sim 20\ \mu m$), this was not the case for all semiconductors. In Si, for example $L \sim 5\ \mu m$, which tilted the balance somewhat in the opposite direction.

The concept of the surface barrier photodiode was demonstrated very early in the history of single crystal semiconductors—Benzer, of Purdue University, described Ge photodiodes made either using a metal point contact or by evaporating a thin film of Au onto a high resistance Ge crystal in 1949. The reader may recall that we described in Chapter 2 the Purdue interest in Ge crystals in connection with their work on cat's whisker microwave detectors and the photodiode emerged as a by-product of their study of high back voltage rectifiers. It was followed up by others, including Pantchecknikoff at RCA Laboratories who reported making sensitive surface barrier photodiodes on Ge in 1952. The natural extension to silicon crystals took a surprisingly long time—it was only in 1962 that Ahlston and Gartner of the US Army Signal Labs at Fort Monmouth reported similar diodes made on lightly doped *n*-type silicon.

Perhaps one reason for this was a tendency to concentrate on *p-n* junction photodiodes, such as those described in 1958 by Sawyer and Rediker at Lincoln Labs, using a thin base Ge diode where the Ge wafer was etched down to a thickness of only a few microns. This had the benefit of a faster response, with a 3 dB cut-off frequency as high as 2 MHz. They developed a detailed theoretical model of their diodes which suggested response rates as high as 20 MHz should be possible but it is notable that they based this entirely on the diffusion of minority carriers to the junction region and did not include the contribution of carriers generated *within* the depletion region. In the case of a thin base device, the amount of light absorbed within the depletion region becomes quite comparable to that absorbed in the *n*-type base, a point which, incidentally, had been made by Gartner 4 years earlier! As we pointed out in Box 8.5, it also results in photocurrents with a very much faster reponse time. To design a fast photodiode, one should aim to absorb as much as possible of the light in the depletion region.

The 1960s was the decade of the heterostructure, witnessing, in particular, the birth of the DH laser but also much work was done towards the understanding of heterostructure properties and their application to photodetection (see the book by Milnes and Feucht 1972, for example). In 1962 Rediker was again involved in what turned out to be a significant step forward when he proposed using a heterojunction photodiode to avoid the problem of surface recombination in a *p-n* junction device. The idea was to use a wide gap *n*-type GaAs window in the form of an epitaxial layer deposited on a *p*-type Ge crystal (see Figure 8.10(e)). Such a detector would show a pass-band wavelength response, the limits of the response curve being fixed by the band gaps of the GaAs (1.43 eV) and the Ge (0.67 eV)—that is, the detector would be sensitive to all wavelengths between 870 nm and 1.85 μm. The Lincoln Labs workers proposed using this particular heterostructure because (as we saw in Section 5.2), the lattice constants of GaAs and Ge are extremely well matched, a feature which promised a low density of interface states and would hopefully provide an ideal form for the *p-n* junction detector. In practice, the GaAs/Ge interface proved less amenable than they hoped—the GaAs could never decide whether to grow from an As layer or from a Ga layer, which produced monolayer steps in the GaAs, acting like grain boundaries—but the principle was to prove invaluable in other material systems, as we shall see further.

Two other important developments date to the 1960s, those concerned with *p-i-n* and avalanche diodes. Let us examine them in turn. The *p-i-n* diode represents a simple (with the advantage of hindsight clever ideas always seem simple!) extension of the *p-n* junction diode in which the depletion region is effectively replaced by a very lightly doped near-intrinsic layer, sandwiched between heavily doped *p*- and *n*-type contact regions. The energy band diagram of the device under reverse bias is illustrated in Figure 8.12, showing how electron–hole pairs generated in the *i*-region are swept to the appropriate contacts in exactly the same way as it happens in a *p-n* junction. The big advantage of the *p-i-n* structure is that it allows the thickness of the high field region to be controlled independently of the *p* and *n* doping levels and tailored to maximize the amount of light absorbed within it. In the *p-n* junction, of course, the depletion thickness is determined by the doping levels on either side, leading to a conflict between the need for a wide depletion region (which demands *low* doping) and highly conducting contact regions (which demand *high* doping). Thus, in the case of a silicon device, the *i*-layer may be some 30–50 μm thick to obtain close to 100% light absorption, while the contacts can be doped

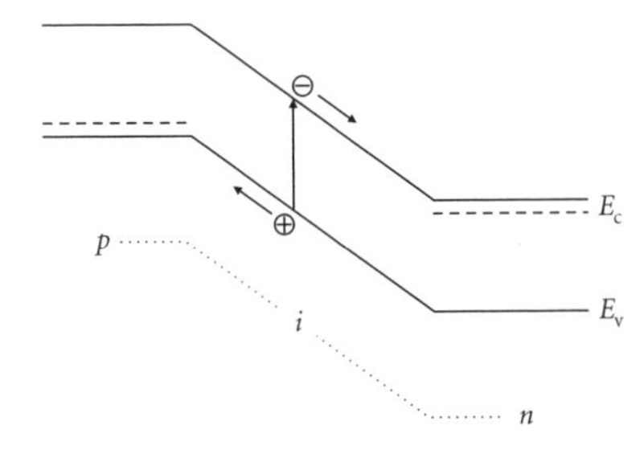

Figure 8.12. Energy band diagram of a typical *p-i-n* diode, showing how electrons and holes generated within the central *i*-region are swept towards the contact regions where they become majority carriers. The advantage of the *p-i-n* structure over a simple *p-n* junction is that the width of the *i*-region can be chosen independently of the doping levels in the contacts.

at levels of order 10^{25} m^{-3}. By making the top contact layer much thinner than the *i*-layer, it is possible to minimize the diffusion current into the junction and obtain a response time which is largely determined by the transit time for carriers to drift across the *i*-layer—in this case $2\text{–}3 \times 10^{-10}$ s, which corresponds to a frequency of ~1 GHz. Silicon *p-i-n* diodes were widely used in the early fibre-optic systems together with GaAs emitters, their high efficiency, reliability, and moderate cost making them ideal for this wavelength (880 nm).

The rather modest response time of these silicon diodes was quite acceptable in the first generation of fibre systems (*c*.1980) but, as we have already seen, later requirements for higher bit rates demanded significantly greater speeds, which, in turn, demanded narrower *i*-regions. However, to maintain adequate light absorption implied the use of materials with greater absorption coefficients and this, of course, meant direct band gap materials. At the same time, the trend towards long wavelength systems implied the use of a narrow gap material such as $Ga_{0.47}In_{0.53}As$, lattice-matched to InP, the band gap of this alloy ($E_g = 0.75$ eV), allowing detector sensitivity out to about $\lambda = 1.65$ μm. It also led to the introduction of heterostructures in which an InP window served to transmit light to the absorbing *i*-layer so as to eliminate the diffusion component of the photocurrent altogether. At a wavelength of 1.5 μm, the absorption coefficient of the alloy is $\alpha \sim 10^6$ m^{-1}, suggesting a layer thickness of 3 μm, which allows operation at frequencies up to 10 GHz. Even faster response could be obtained, at the expense of quantum efficiency, by reducing the *i*-layer to a micron or less but this inevitably led to problems with diffusion current generated in the contact material *behind* the i-layer, so it became necessary to make the back contact out of InP as well, only the very thin *i*-layer being made from GaInAs. This double heterostructure, which looked not unlike a DH laser structure, could then be designed for response speeds up to 100 GHz.

There is, however, a 'but'! An *i*-layer, sandwiched between two contact layers constitutes a capacitor *C* which appears in series with the load resistor R_L (used to measure the photocurrent) and this *RC* time constant can easily dominate the system response, no matter how fast the diode itself may be. Suppose we are aiming for an operating frequency of 100 GHz, which requires the transit time to be of order 1 ps, we must make the *i*-layer no greater than 0.1 μm thick. This implies $C = \varepsilon\varepsilon_0 A/d \sim 10^{-3}$ A (where *A* is the diode area in m^2). If $R_L = 100\ \Omega$, and we demand that $R_L C - 10^{-12}$ s, this requires $A < 10^{-11}$ m^2, or the diode diameter must be no greater than about 4 μm! Note the amazing coincidence that this corresponds to the core diameter of a single mode optic fibre (!) which implies that we can stick the detector directly on the end of the fibre and hope to collect all the light while keeping the time constant adequately short. 'All very well', one might say, 'but what about stray capacitance?' Fortunately, it is not our responsibility to design the necessary circuitry but the above simple calculation demonstrates just how difficult such design can be, and helps us to remain humble in the face of those who actually do it—successfully! Suffice it to say that success depended on integrating a photodiode with a low noise FET on the same semiconductor chip.

After that little diversion into the realms of the nearly impossible, we should return to the second important development of the 1960s, that of the avalanche photodiode (APD). A well-designed *p-n* junction or *p-i-n* diode operates with quantum efficiency close to unity—that is to say, each photon incident on the front surface results in one electron being collected in the external circuit. The photocurrent I_{ph} is given by:

$$I_{ph} = \eta e N_{ph}, \tag{8.6}$$

where N_{ph} is the number of photons incident on the diode per second and η is the quantum efficiency ($\eta \leq 1$). The interesting feature of the APD is that it has built-in multiplication—the photocurrent in this case being given by:

$$I_{ph} = M\eta e N_{ph}, \tag{8.7}$$

where the multiplication factor M may be as large as several hundred times. The mechanism involved is avalanche multiplication within the reverse biased depletion region when the bias is close to its breakdown value. The very high field existing within the depletion region accelerates photogenerated carriers to velocities high enough to cause impact ionization of the semiconductor—that is, to raise further electrons into the conduction band. This process then repeats over and over, building up a free carrier density many times larger than that occurring in an ordinary *p-n* or *p-i-n* diode. If we regard the product $M\eta$ as an effective quantum efficiency, it is clear that this can take values very much greater than unity and the advantage, particularly when detecting weak optical signals, is obvious. Considerable work went into developing the APD in silicon and germanium during the (1960s) and Si APDs were frequently used in first generation fibre systems. In practice, the structure of an APD differs from that of a *p-i-n* diode by having an extra layer in which the multiplication takes place, carriers being generated by absorption in an *i*-layer but then transferred to a high field multiplication layer in order for the photocurrent to be enhanced.

The subject of overall receiver sensitivity is complex and we make no attempt to discuss it here (see Agrawal 1997: ch. 4) but there are three comments we should make about APD performance. The first is that the avalanche process is inherently noisy and it follows from a detailed examination of APD noise that the performance depends on certain properties of the material in which the avalanche occurs. From the nature of impact ionization, it is apparent that it may be driven either by electrons or by holes and it is customary to define two ionization coefficients, one for electrons α_e and the other for holes α_h. The parameter of importance for receiver performance is the ratio of these two ionization coefficients—if it is close to unity it is bad news, while if the ratio is either large or small, it is good. Silicon is characterized by $\alpha_e/\alpha_h \sim 300$ and makes excellent APDs, while Ge has a ratio close to 1 and has a much poorer noise performance. Unfortunately, the ratio in GaInAs is rather similar to that in Ge and is far from ideal as an APD material. Second, it is essential, in making an APD, to obtain an avalanche which is spatially uniform across the whole area of the diode and this is often difficult to achieve. It reflects, of course, the uniformity of the material in question and, again, silicon behaves well, while InGaAs has proved more difficult. Third, it should not be thought that a large value of M is necessarily desirable under all circumstances because multiplication has a detrimental effect on receiver bandwidth. This is simply expressed as:

$$M(\omega) = M(0)[1 + (\omega t_e M(0))^2]^{-1/2}, \qquad (8.8)$$

where t_e is an effective transit time (which is greater than t_t as a result of the time taken to build-up the avalanche) and $M(0)$ is the multiplication factor at low frequencies. This equation implies a receiver bandwidth B which is given approximately by:

$$B = [2\pi t_e M(0)]^{-1}, \qquad (8.9)$$

illustrating the trade-off between bandwidth and sensitivity.

But, to return to our story, the early success with silicon diodes led naturally to attempts to develop APDs in the long-wavelength region of the fibre-optic spectrum though, initially, with rather limited success. Not only was it difficult to obtain an avalanche in GaInAs due to tunnelling breakdown preceding avalanche breakdown, but the adverse effects of the nearly unity ratio of ionization coefficients made noise performance less than ideal. However, there was considerable effort in the decade 1970–80 aimed at improving the situation and, eventually, success was achieved by separating the absorption and multiplication regions so that multiplication occurred in InP, rather than InGaAs. (Such diodes were affectionately known as SAM—which stood for, separate avalanche and multiplication.) In InP, $\alpha_h > \alpha_e$, so the device was designed so that holes initiated the avalanche in n-type InP, resulting in the structure shown in Figure 8.13 where absorption takes place in the thin GaInAs film, the rest of the structure being made from InP. The only difficulty with this concerned the large band gap step between GaInAs and InP which made it difficult to transfer

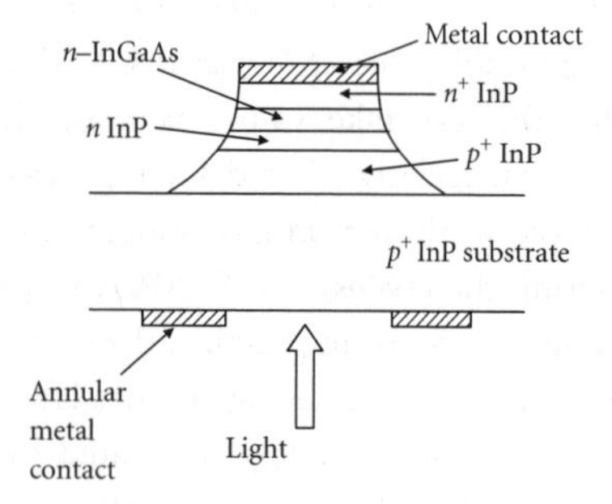

Figure 8.13. Structure of a typical InGaAs APD. Light is absorbed in the thin, lightly doped InGaAs layer, while multiplication takes place in the adjacent n-InP layer. Electrical contact is made through the heavily doped n^+-InP and p^+-InP contact regions. The diode area is determined by the etched mesa whose diameter is typically about 10 μm to minimize diode capacitance.

holes across the interface and led to disappointing response speeds. Finally, towards the end of the 1980s, this was solved by grading the interface using a lattice-matched, graded GaInAsP layer some few tens of nanometres thick. Multiplication factors of about 10 and gain-bandwidth products of order 100 GHz were obtained in the early 1990s with structures like this. Even better performance was obtained when the multiplication region was made from a short period AlGaInAs/AlInAs superlattice, though, of course, this only became possible when MOVPE growth replaced the long-favoured LPE process during the 1980s.

One might have imagined that the amount of information to be transmitted would reach some sort of plateau but there is no sign of it yet! Which means that photodiodes must work faster and faster, while still maintaining high quantum efficiencies, and that means new and better designs must materialize each decade. What, then, did the 1990s produce? Three things: Fabry–Perot cavities to improve light absorption efficiency, semiconductor waveguides to aid in the integration of the optical receiver and metal–semiconductor–metal (MSM) photodiodes for improved frequency response and ease of integration. Let us look at each in turn. We have already commented upon the trade-off between bandwidth and sensitivity which results from the need for an adequate absorption length for the light (to obtain high quantum efficiency) while pressing for a minimum depletion width (to minimize transit time) and, as the demand for higher and higher bit rates grew inexorably, this compromise became more and more stressful. Various solutions were sought which, in their several ways, decoupled the absorption length from the drift length. First of these was the idea of placing the absorbing film within a Fabry–Perot optical cavity so that the light made multiple transits through the absorbing layer. Thus, even though the layer might be as thin as 0.1 μm, in order to reduce t_t to the order of 1 ps, the quantum efficiency could still take values of order 50% or more. How was this done? By epitaxial growth of a Bragg mirror stack immediately beneath the absorbing layer and by topping off the structure with a dielectric mirror made by depositing alternate films of CaF_2 and ZnSe. Such structures were first developed for diodes designed to detect radiation at 880 nm from a GaAs laser. The absorbing layer was composed of roughly 0.1 μm of $In_{0.1}Ga_{0.9}As$, the Bragg stack of AlAs/GaAs pairs, and the top dielectric mirror was transparent at the wavelength of interest, the whole being grown on a GaAs substrate. The long wavelength equivalent might have employed an InP/GaInAsP Bragg stack on an InP substrate but, unfortunately, the refractive index difference in this material system is rather small so the preferred method utilized wafer bonding of an InP-based photodiode with a GaAs-based Bragg stack. The top mirror demanded no modification.

The idea of using a semiconductor waveguide to feed light signals to appropriate points in a receiver circuit led to their use with edge-illuminated detector diodes. If, rather than illuminating a diode from above (i.e. at right angles to the junction plane) as we have implied so far, one were to illuminate it from the edge (i.e. in the junction plane), the absorption length could easily be several microns, while permitting the transit width to be of submicron dimensions. The waveguide took a form similar to that inherent in an InGaAsP laser, consisting of an InP/InGaAsP double heterostructure, while the photodiode might be a conventional InP/InGaAs *p-i-n* structure with an absorbing layer 0.2 μm thick and lateral dimension of order 10 μm, which would be more than sufficient to absorb all the light, while implying an overall device area small enough to reduce the diode capacitance to very acceptable proportions.

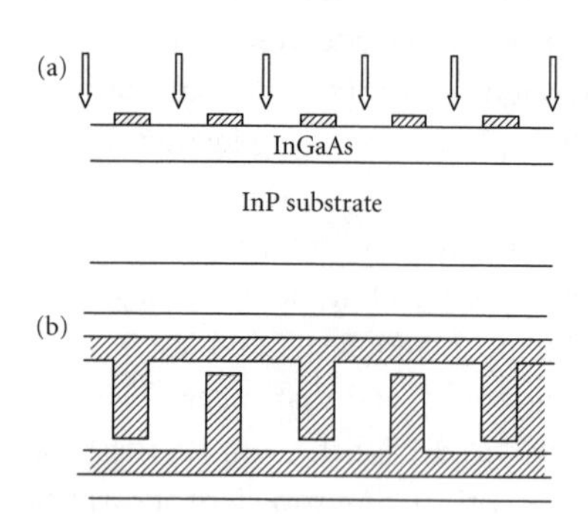

Figure 8.14. Schematic drawing of an MSM photodetector (a) in cross-section and (b) in plan. Typical dimensions are: finger width 1 μm, finger separation 2 μm, overall area 30 × 30 μm. Alternate fingers of the interdigitated structure are connected together and biased oppositely. Illumination from above generates free electrons and holes in the regions between the fingers, which drift to opposite electrodes.

Finally, perhaps the most significant development of the 1990s was the MSM photodiode, an example of which is illustrated in Figure 8.14. It was a very simple structure, consisting of a set of interdigitated metal fingers deposited by evaporation onto a semiconductor surface, the metal being chosen so as to form a Schottky barrier with the semiconductor in question. Illumination from above resulted in free carrier generation in the regions between the fingers, carriers which were swept to the electrodes by the field between them. The structure had the advantage of a very low interelectrode capacitance (partly because of the very small effective area of the capacitor) while the transit time was determined by the electrode separation. Typical dimensions were: overall area 30 × 30 μm, finger width 1 μm, finger separation 2 μm. The absorption length could be relatively large, allowing nearly 100% light absorption, though the shadowing effect of the electrodes reduced the effective quantum efficiency to 50–60%. Though the transit time was no longer determined by the absorption length, it was difficult to make devices with interelectrode spacing much less than 0.5 μm so the ultimate value of t_t was no less than 5 ps. In practice, the absorption length did influence t_t (measured values of t_t being typically greater than 10 ps) so it was desirable to minimize it as far as possible and this, again, was achieved by employing a Fabry–Perot cavity to reflect the light beam back and forth through the absorbing film. The first efficient detectors were made for the 880 nm wavelength region, and were based on the AlGaAs/GaAs material system which was relatively straightforward, GaAs making a good Schottky barrier (barrier height of order 0.8 eV) with many metals such as Au or Pt.

The extension to long wavelengths was less than straightforward, because the absorbing layer in $In_{0.53}Ga_{0.47}As$, due to its small band gap, exhibited only a small barrier height (about 0.2 eV). A solution was eventually found in the form of a thin barrier layer of AlInAs between the metal and the InGaAs absorber. AlInAs has a much larger gap and forms good barriers with metals such as Au. It had to be thick enough to prevent tunnelling, and thicknesses of about 80 nm were typical. Devices with areas of 50×50 μm showed RC time constants of about 2 ps and transit times of a few tens of picoseconds, yielding detector bandwidths of 3 GHz. Generally, bandwidths tended to be significantly smaller than was anticipated (10 GHz), a problem traced to the build-up of charge at the interface between the AlInAs and the GaInAs layers because of the large difference in their band gaps. This was later overcome by introducing a short period AlInAs/GaInAs superlattice (period 6 nm) to grade the band gap between the two alloys, subsequent layers being graded in thickness from 9 : 1 ratio through to 1 : 9 ratio. (As an aside, we note that this solution was described by Zhang et al., working in Shanghai—more and more such contributions are appearing from the Chinese mainland and will, no doubt, continue to do so.) The net result was a bandwidth of 10 GHz, as predicted. This performance, while certainly useful, was not earth-shaking—the principal advantage of the MSM detector was actually the ease with which it could be integrated with an associated low noise FET amplifier, its planar structure being compatible with that of the FET. As we remarked earlier, the development of optoelectronic integrated circuits (OEICs) is now seen as having paramount importance for the future of fibre-optic communications.

8.4 Optical modulators

The facility with which a semiconductor laser diode could be modulated by simply varying its drive current represented an undoubted convenience. On the other hand, the resulting frequency chirping certainly constituted an unfortunate side-effect and, as we noted earlier, there was also a limit to the modulation bandwidth set by the phenomenon of relaxation oscillations. This obviously raised the question as to whether it might not be better to run the laser under CW conditions while achieving the desired modulation of the light in a separate device, potentially, offering higher modulation bandwidth, while maintaining the extremely narrow linewidth obtainable in a single mode laser. These were both desirable objectives in the context of higher bit rates and longer transmission distances, so we devote a short section here to considering some aspects of this approach. It has implications for the design of the laser as well as introducing us to semiconductor modulators, per se.

The concept of optical modulators based either on the electro-optic effect or the magneto-optic effect has a long history. Rotation of the plane of polarization of a light beam by a magnetic field, was discovered by Faraday in 1845 and the influence of an electric field on the refractive index of a dielectric solid was studied by Kerr in 1876. It is this latter, the electro-optic effect, which has been applied to the development of fast light modulators for communication purposes so we shall ignore the wide range of alternative phenomena and concentrate simply on this. Kerr, himself, observed that an isotropic solid subjected to an electric field F displayed birefringence (the refractive index in a direction parallel to the field differing from that perpendicular to the field), the refractive index *difference* being proportional to F^2. Later studies showed that certain anisotropic crystalline materials showed a *linear* electro-optic effect (sometimes known as the Pockels effect) which was significantly larger than the original Kerr effect and much effort was devoted to finding materials with large electro-optic coefficients. Examples include potassium di-hydrogen phosphate (KDP), ammonium di-hydrogen phosphate (ADP), lithium niobate ($LiNbO_3$), lithium tantalate ($LiTaO_3$) and a number of polymers. Early attempts to develop modulators for the 1.3–1.55 μm spectral regions, made in the early 1980s, following the discovery of these 'windows' in the period 1976–8, were based on $LiNbO_3$. More recently, polymer materials have come into prominence, though ultimately, it would appear, semiconductor modulators are likely to dominate the commercial telecoms scene on account of their greater integratability.

In its simplest form (for more detail see, for example, Smith 1995: ch. 7), the electro-optic modulator consists of a thin slab of material with electrodes top and bottom to provide the necessary field, while the light propagates along the length of the slab, parallel to the electrode plane (in this configuration the modulator is known as a 'Pockels cell'). In effect, the modulator rotates the plane of polarization as the light travels through the crystal so, in order to obtain amplitude modulation, a pair of crossed polarizers is employed, as illustrated in Figure 8.15. In the absence of a voltage between the electrodes, no light is transmitted through the overall structure—application of a voltage (typically 2–5 V) rotates the polarization and allows some fraction of the light to be

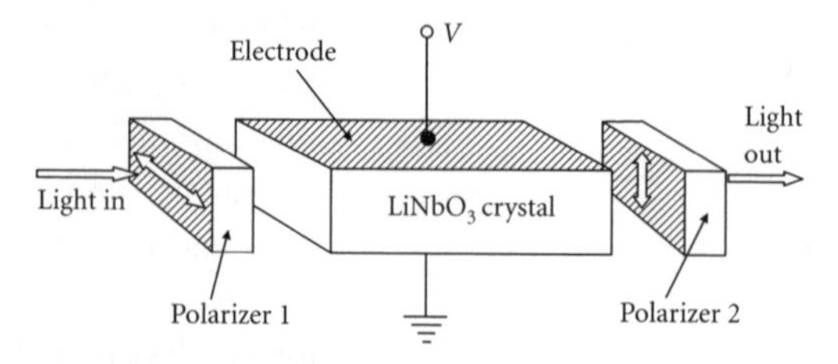

Figure 8.15. Outline of an electro-optic modulator based on a single crystal of $LiNbO_3$. The incoming light is polarized horizontally by polarizer 1, while polarizer 2 is crossed with respect to 1. In the absence of a bias on the modulator crystal, no light can be transmitted through the system but when a bias is applied, which rotates the plane of polarization of the light in the $LiNbO_3$, some fraction of the light is transmitted. In particular, when the rotation angle is $\pi/2$, all the light is transmitted so it is possible to produce rectangular pulses of light by switching on and off the appropriate voltage $V(\pi/2)$.

transmitted. Rotation through an angle of $\pi/2$ results in all the light being transmitted (save for a small insertion loss) while electric pulses of the appropriate amplitude produce rectangular light pulses, as required for digital signal transmission.

The question arises, of course, of just how fast the modulator response is. To obtain speeds appropriate to the high bit rates now typical of fibre systems implies that the designer must take account of the transit time of light through the modulator—for a length of 1 cm (required to obtain a sufficiently large rotation) this time is of order 100 ps, comparable with the length of a single bit at 10 $\mathrm{Gb\,s^{-1}}$. In fact, it is impossible to resolve individual pulses at bit rates greater than this unless the applied microwave field travels along the modulator at the same velocity as the light, implying that a specific part of the light field sees the same applied field at all times during transit. In other words, the electrode system must act as a transmission line—and it turns out to be more than a little difficult, in practice, to match the two velocities. This problem certainly exercised many of the protagonists during the 1980s, a whole host of different solutions being tried with gradually improving degrees of success. Nothing could be done about the velocity of light propagation so it was necessary to modify that of the electrical signal by constructing some form of slow wave structure. Suffice it to say that modulation bandwidths increased gradually from the order of 10 GHz in 1985 to almost 100 GHz in 1995, representing a really significant achievement and keeping roughly in line with the demands of the system engineers.

Some of the best results were obtained using polymer materials, rather than the much longer established, but expensive $LiNbO_3$. It was also common practice to employ a Mach–Zehnder interferometer, rather than the simple arrangement shown in Figure 8.15. In this configuration, the light beam propagates along an optical waveguide which contains a 'Y' junction where the energy is split equally between two identical arms, followed by a second junction where the two waves are recombined. One of the arms is provided with electrodes, as in Figure 8.15, which serve to control the phase of the light in that arm. In the absence of any applied voltage, the two beams interfere constructively but if the wave in one arm is slowed down by a field-induced change of refractive index, so that it acquires a phase difference of π with respect to the unchanged wave, the two interfere destructively and the recombined beam has an intensity very close to zero. In this way, 'extinction coefficients' of 30 dB were achieved in both the 1.3 and 1.55 μm bands with drive voltages of 2–3 V, a highly acceptable performance, indeed, apart from the size of the overall structure which was measured in centimetres, rather than microns.

It was this latter aspect which provided the spur to those developing the alternative modulator based on the property of electro-absorption

(E-A) in semiconductors, in particular that due to the quantum confined Stark effect (QCSE) in semiconductor quantum wells. We have already described this effect in Chapter 6 (Section 6.6 and, in particular, Figure 6.27)—the excitonic absorption associated with the $n = 1$ confined states in a quantum well shifts to lower photon energy (longer wavelength) as a function of an electric field applied normal to the plane of the well. If, at zero bias, the absorption band is arranged to lie at slightly higher energy than the photon energy $h\nu$ of the light we wish to modulate, very little absorption takes place. Application of an appropriate bias then shifts the absorption band to an energy just below $h\nu$ and the light is then strongly absorbed. In practice, it is necessary to use several wells to obtain sufficient attenuation but figures of 10–15 dB have been reported for 1.55 μm modulators based on this principle.

The effect was first discovered in AlAs/GaAs MQW structures in the early 1980s and was also applied to the longer wavelength, 1.3 and 1.55 μm bands once their importance became clear. In 1989 Kotaka et al. reported a 1.55 μm MQW modulator based on InGaAlAs wells sandwiched between InAlAs barriers, light travelling in the plane of the wells over a distance of 100 μm. The on/off ratio was 10 dB for a 4 V bias and the bandwidth $B > 20$ GHz which clearly demonstrated the potential of the method. During the early years of the 1990s many workers in America, Europe, and Japan studied similar systems, including, in particular, InP/InGaAsP structures. By 1996, Weinmann et al. were able to report a modulator based on InGaAs/InGaAsP tensile strained quantum wells (the strain enhanced the strength of the absorption) with a 42 GHz bandwidth and an operating voltage of only 1.8 V. It was 120 μm in length and used a ridge waveguide to guide the light. They also pointed out an advantage of the E-A approach, namely that the modulation depth was insensitive to the polarization state of the light.

Such devices established the principle of the E-A modulator but it was clear to all that another important hurdle had yet to be surmounted—that of integration. By this time (mid-1990s) there was little doubt in anyone's mind that the future of fibre-optic systems must depend on the development of the optical equivalent of the silicon integrated circuit, that is, the OEIC (nowadays perhaps better known as the photonic integrated circuit). The emergence of the semiconductor planar optical waveguide in the 1960s and 1970s already offered the possibility of optical integrated circuits but very little further progress was in evidence and it was not before the 1990s that integration acquired a serious image, driven by the continuing remarkable growth in demand for greater bit rates. Combining a modulator with a DFB laser was always possible but when done by hand, it was both tricky and time consuming (i.e. costly!). If fibre-optic systems were to reach their obvious commercial potential, components must be integrated together whenever possible and the laser–modulator combination was a case in point. The first requirement was clear—they should both be constructed from the same materials, that is, InGaAsP/InP—but therein lay a problem. They both depended on MQW structures which, at first, sounded ideal but, when considered at the nitty-gritty level, proved frustratingly unaccommodating. Assuming that the two sets of wells and barriers were to be formed in the same epitaxial growth stage, the absorption transition in the modulator section would coincide exactly with the laser emission which would therefore be strongly attenuated. Application of a bias could only shift the absorption edge *downwards* in energy which made almost no difference to the overall effect! What was needed was an absorption edge which, at zero bias lay slightly *above* the laser transition, so as to minimize the absorption—application of a bias would then shift it downwards to produce a strong attenuation. But this looked like an impossibility. Impasse? Well, nearly so—one should never reckon without the ingenuity of the R&D scientist when seriously challenged.

First attempts to find a way round this little difficulty employed complicated sequences of epitaxy and selective etching which did work—but only occasionally! Imagine the difficulty of growing a modulator structure which had to match in height with that of the laser, so that they would be optically coupled. Something subtler was clearly necessary and not one but two solutions emerged during the course of the 1990s. First came Aoki et al. in 1991 with the observation that the epitaxial growth rate of InGaAs could be controlled by depositing it in the space between two stripes of SiO_2—the narrower the gap between them, the faster was the growth (for gaps in the range 5–400 μm). This gave them the possibility of growing two sets of quantum wells on the same substrate, with different well widths, simply by depositing two lots of SiO_2 with different gaps between them. Different well widths meant different confinement energies so it was now possible to separate the $n = 1$ exciton energies even though there was only a single growth stage. All that was required was to make the laser from the wider set of wells and the modulator from the narrower. In 1993 they demonstrated a working laser–modulator combination with 12.6 dB attenuation for a 1 V bias and having a bandwidth of 14 GHz.

The second solution made use of a phenomenon first explored in AlAs/GaAs QWs during the 1980s—that of quantum well disordering, based on the interdiffusion of the well and barrier materials. In other words, aluminium atoms diffused into the wells, while Ga atoms diffused into the barriers, changing the basic band gap of the well material and, at the same time, smearing out the sharp well–barrier interfaces so that the wells were no longer 'square' but tended to become roughly parabolic (in energy). The overall result was to cause a blue-shift in the exciton transition—that is, a shift to higher photon energy. To effect such interdiffusion required elevated temperatures but the key to the integration problem lay in the discovery that it could also be influenced by the deposition of a film of SiO_2 on top of the MQW structure. In fact, the presence of the oxide stimulated much more rapid diffusion and this provided another mechanism for differentiating two regions of an MQW sample. All that was required was a stripe of oxide over one area, while leaving the rest uncovered. The modulator could then be made on the disordered region and the laser on the un-disordered region. What could be simpler? Miyazawa et al. demonstrated the application to the InGaAsP system in 1991, using a Si_3N_4 film, rather than SiO_2 and in 1995 Ramdane et al. described a successful 1.55 μm laser–modulator combination, using SiO_2 as the stripe material, with a 10 dB attenuation for a 3 V bias. In the same year Tanbun-Ek et al. also described an integrated DFB laser-Mach–Zehnder modulator using the same technique. The way was clear, at last—integrated laser–modulator combinations could now be regarded as standard units. It was a relatively simple matter to integrate a photodiode to monitor laser output power as well and use it to control the output via an appropriate feedback circuit. All of which gave the semiconductor modulator a distinct edge over the electro-optic modulator. While it was true that the bandwidths available with the E-A modulator still lagged somewhat behind those appropriate to electro-optic devices, the convenience of monolithic integration more than compensated—and, in any case, progress in this direction is still continuing.

8.5 Recent developments

The demand for faster and more powerful communications systems shows no sign of slowing—more and more people are discovering the wonders of the Internet, visual information is going 'digital', casual interperson communication via e-mail is as likely to involve pictures as words, interactive systems are increasing in popularity, etc. No sooner is one demand for bandwidth satisfied than two others lay claim for attention and the huge bandwidth made available by the move from microwave to optical carrier frequencies, which at one time seemed almost infinite, now appears likely to be used up within the foreseeable future. How are the suppliers of the essential communication links reacting? Two trends are apparent—techniques for more effective use of the available bandwidth and methods of satisfying the demand at more competitive prices. Examples of the first of these are provided by the increased emphasis on wavelength division multiplexing (WDM) and the development of coherent detection systems. Examples of the second include the rapid development of optical integrated circuitry, to which we have already referred on several occasions, attempts to replace the standard InP-based technology with one based on GaAs and the introduction of long wavelength VCSELs. Needless to say, such developments make corresponding demands on the semiconductor technologist so, in this final section, we briefly outline some of the relevant semiconductor device developments. It is, perhaps, worth pointing out that the associated semiconductor device market, alone is predicted to reach $25 billion by the year 2004, so his efforts (where sucessful!) will scarcely go unrewarded.

The drive to transmit more and more information through each individual fibre rapidly led to the introduction of time division multiplexing (TDM) in which different bit streams were interlaced within the time domain. This involved the use of shorter digital pulses which demanded higher frequency response from both source and detector, the practical limit being set by the performance of the individual components, as we have already discussed. The next step along the multiplexing avenue (which, incidentally, was not restricted in this way) involved making use of different wavelengths to carry different sets of bit streams. In principle, this opened the whole optical spectrum, though the need for low fibre loss and low dispersion limited long distance communication to wavelengths around 1.3 and 1.55 μm. An obvious question was: how close together could the various channels be? The answer depended on how fast the individual channels were to operate. In other words, if the TDM bit rate was of order 10 $\mathrm{Gb\,s^{-1}}$, this implied a bandwidth for each channel of 10 GHz which limited the smallest separation between channels to 40 GHz (in order to avoid cross-talk between channels) and this, in turn, limited the total number of channels available within the allowed frequency bands. For example, if we assume the band near 1.55 μm to be limited to the wavelength range 1.48–1.60 μm, this corresponds to just over 15 THz (1.5×10^{13} Hz) and allows approximately 400 separate channels. Future trends are towards faster bit rates which imply greater channel separation but this may be compensated by the use of even greater spectral bandwidth.

What did all this have to say to those long-suffering mortals responsible for laser design? One very important requirement became clear—a WDM system demanded large numbers of laser sources with accurately controlled emission wavelengths, each of which should be stable to within a few gigahertz (roughly 0.03 nm, or about 3 parts in 10^5) in spite of inevitable ambient temperature fluctuations. As we saw earlier, a typical DBR laser has a temperature coefficient of wavelength $d\lambda / dT = 0.07\ \mathrm{nm\,K^{-1}}$ so this implied that the laser temperature should be kept stable to about 0.5K! But this was not the end of the matter. Any acceptable communication system must provide a reliable service with minimal down-time and this implied the availability of spare lasers to cover the whole waveband—in the example above no less than

400 different devices! Clearly, a more efficient method of cover would utilize tuneable lasers, thus reducing the number of spares to a much smaller number. It demanded, of course, a simple, convenient method of wavelength tuning.

We saw in Section 8.2 that tuning over some 50 nm could be achieved by selecting the operating wavelength with an external diffraction grating (Figure 8.8(a)) but this could scarcely be described as 'simple and convenient'. A better approach might be to make use of the temperature-dependence of laser wavelength. If temperature had, in any case, to be stabilized, it should be straightforward to control it over an appropriate range, 1 nm variation in wavelength requiring approximately 15°C change in temperature. This had the advantage of providing a smooth tuning characteristic but was obviously limited to a relatively small tuning range—to cover even 10 nm required an unreasonably large variation in temperature. It was also slow to switch between wavelengths, while some modern applications demand nanosecond switching times! A purely electronic method of wavelength control would obviously be preferable and much work has gone into developing suitable structures. Two successful approaches are illustrated in Figure 8.16, one based on

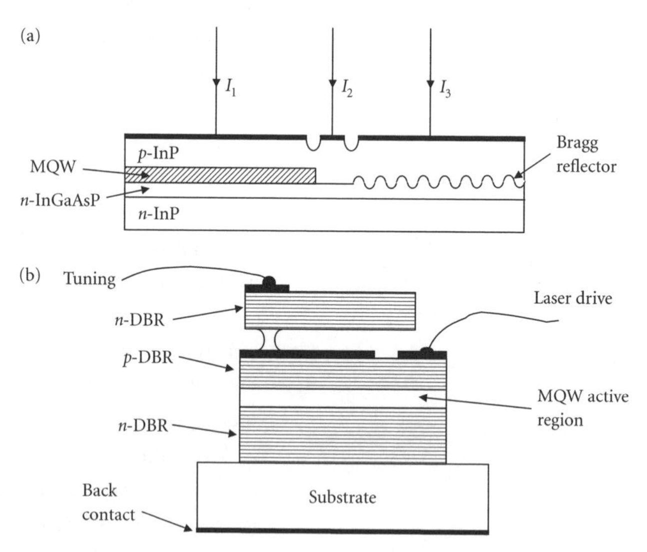

Figure 8.16. This figure illustrates two very different methods of manufacturing electrically tuned laser diodes. (a) In this case, tuning is achieved by injecting current I_3 into the Bragg reflector section of a three-section laser structure. The free carriers generated result in changes to the refractive index which modify the resonant frequency of the Bragg reflector. I_1 represents the drive current to a conventional MQW, in-plane laser, while I_2 is used to adjust the refractive index of the central, phase-matching region which effectively alters the cavity length of the laser. (b) In this diagram is shown an example of an electromechanically tuned VCSEL. The top mirror is cantilevered so that a voltage applied to the tuning contact bends the cantilever and moves the position of the mirror with respect to the active region, thus effecting a change in effective cavity length. Smooth variation of wavelength with applied voltage can be obtained over large tuning ranges.

a multisection laser structure, the other using an electromechanical principle.

The multisection DBR laser shown in Figure 8.16(a) consists of three electrically isolated regions, each of which can be driven by separate current sources. The left-hand section constitutes a conventional MQW laser structure which provides gain when pumped above threshold, the right-hand section acts as a variable wavelength Bragg reflector, while the middle section is a phase-matching section which effects a degree of fine tuning. The essential tuning property of the reflector depends on the variation of the refractive index of the materials on either side of the *p-n* junction with free carrier density. (Remember that the relation between the free-space wavelength for maximum reflectivity, λ_0 and the wavelength of the Bragg corrugations, Λ is given by equation (8.3) ($\lambda_0 = 2n\Lambda$), n being the average refractive index of the two constituent materials.) Changing the free carrier density by changing I_3 in Figure 8.16(a), shifts the peak reflectivity smoothly as a function of current. However, the laser emission wavelength tends to jump discontinuously from one Fabry–Perot mode to the next as the Bragg peak coincides with each mode in turn, and such mode-hops are typically of order 0.5–1.0 nm, depending on the length of the active region. As it is desired that WDM wavelengths differ by only 0.3 nm, there is clearly a need to smooth this mode-hopping behaviour, hence the need for the central section which can be thought of as introducing a controlled variation in active region length. A further complication arises when large wavelength shifts are required, leading to the use of arrays of different period Bragg gratings, rather than the single one shown in Figure (8.16), together with a fourth section which acts as a course tuner to select a particular grating from the array. We shall not attempt to go into further detail but it is important to appreciate something of the sophistication of design which WDM demands from its laser sources.

The alternative approach to wavelength tuning, introduced during the early years of the new millennium, is based on the use of a VCSEL in which the top mirror can be moved electromechanically so as to vary its separation from the active region. VCSELs are, as we noted in Section 8.2, very short cavity devices and, in consequence, are essentially single wavelength emitters (the Fabry–Perot modes being a long way apart in wavelength). By varying the effective cavity length slightly, it is possible to tune these devices smoothly over a wide range of wavelengths and one practical method of achieving this is shown in Figure 8.16(b). The top mirror is cantilevered over the laser aperture, its angle being controlled by applying a voltage to the tuning contact. Because the III–V compounds are piezoelectric, the resulting field applies a force to the cantilever which effects the required movement. Tuning ranges as large as 40 nm have been demonstrated with similar electromechanically tuned structures.

For some time, VCSELs had been seen to offer advantages in fibre communications on account of their single mode operation, circular emission pattern (well matched to a cylindrical fibre), and ease of packaging. The ability to check device yield on the chip before separation of individual devices was yet another appealing feature, while the possibility of their being used in parallel optical data processing offered further encouragement. These advantages were turned to practical effect during the late 1990s in the form of 880 nm GaAs-based VCSELs which found wide application in short haul systems using multimode fibres. However, the desired extension to long wavelength devices proved frustratingly difficult on account of the relatively small refractive index differences available within the InGaAsP material system—which adversely affected the performance of multiple Bragg reflectors. Whereas the maximum difference in refractive index in the AlAs/GaAs system is approximately 0.6, in the quaternary material it is limited to about 0.3 which requires an impractically large number of mirror layers to achieve the necessary high mirror reflectivity for VCSEL operation. The alternative of employing InGaAlAs/InGaAs mirrors, lattice-matched to an InP substrate proved only marginally better. This apparent impasse was eventually circumvented by incorporating AlAs/GaAs mirrors into InGaAsP/InP structures (the cantilever mirror in Figure 8.16(b), for example, was made from AlGaAs/GaAs multilayers in order to provide adequate reflectivity with relatively few layers) but not without additional complexity. The first sucessful attempts depended on wafer bonding techniques in which mirrors and active regions were separately grown, then fused together under moderate temperature and pressure. Somewhat later, it was found possible to grow the complete structure epitaxially by incorporating suitable strain relieving interlayers but neither method could be regarded as entirely acceptable when viewed with a manufacturer's eye. Some radical rethinking became a matter of considerable urgency.

This emerged in 1996 from the Hitachi company in Japan and depended on the development of a new alloy system InGaAsN which could be lattice-matched to GaAs substrates, while allowing laser operation at wavelengths close to 1.3 μm. Attempts to incorporate nitrogen into GaAs or GaP structures followed naturally, once a successful technology for growing GaN films had been established at the beginning of the 1990s. Just as the ternary alloy GaAsP provided a range of band gaps for use in red and yellow LEDs, it seemed reasonable to anticipate an even wider range from the alloy GaAsN (1.43–3.43 eV) which would include the whole of the visible spectrum. However, in the event, such hopes proved groundless—not only was it difficult to incorporate more than a few per cent of N atoms into GaAs (or GaP) but the effect on band gap turned out to be the opposite of that anticipated. Because of an unusually large bowing parameter in the GaAsN system, adding

N to GaAs actually *reduced* the gap, rather than increasing it, but what was disappointment for the visible LED protagonist gave hope to the long wavelength VCSEL enthusiast. The question, in this case, was how *small* a gap might be achieved—could it be reduced from 1.43 to 0.95 eV ($\lambda = 1.3$ μm) or even to 0.80 eV ($\lambda = 1.55$ μm)?—and the crucial contribution of the Hitachi scientists was their recognition of the further advantage to be gained from adding indium to the brew. It was well known, of course, that substitution of Ga by In atoms had the effect of reducing the band gap but, if it were possible to incorporate both In and N together, the resulting quaternary alloy could be lattice-matched to a GaAs substrate. The lattice parameters of (cubic) GaN, GaAs, and InAs are 0.45, 0.565, and 0.606 nm, respectively, giving values for the differences $[a_0(\text{GaAs})—a_0(\text{GaN})] = 0.115$ nm and $[a_0(\text{InAs})–a_0(\text{GaAs})] = 0.041$ nm. Following an argument similar to that propounded for the GaInAsP quaternary system in Box (8.3), this suggests that, if the ratio of In : N is made equal to 2.8, the lattice parameter of the resulting alloy will approximate to that of GaAs, thus opening the possibility of VCSEL structures lattice-matched to GaAs.

The Hitachi workers demonstrated a VCSEL based on this approach which operated at a wavelength of 1.2 μm in 1996—more recently the running has been taken up by the Infineon company in Munich (an offshoot of Siemens) and at Sandia in the United States but all these groups found that luminescence efficiencies degraded rapidly when more than about 2% N was incorporated, leading to the use of strained InGaAsN quantum wells in the active layer. The present position is that promising performance has been obtained at 1.3 μm wavelength, with improved thermal properties (compared with standard InGaAsP DBR lasers)—T_0 values of around 100 K have been reported, whereas values of about 60K are typical of standard devices. The difficulty of incorporating nitrogen, however, has so far militated against the development of lasers operating in the 1.55 μm band. Future progress in this direction probably awaits a better understanding of the material growth process, a feature of which appears to be the need to grow at surprisingly low temperatures (typically 500°C), and all the successful work, so far, has been based on MBE, rather than on MOVPE.

In conclusion, we might refer to yet another aspect of the struggle to develop acceptable GaAs-based lasers—the question (once again!) of integration. The ready availability of high speed GaAs electronic devices manufactured on large (6 in.) substrates offers the tempting possibility of cheap, monolithic photonic integrated circuits—a holy grail for the optical communications industry—which makes it all the more frustrating that 1.55 μm 'GaAs' lasers are currently unobtainable, when seen in the context of a strong move towards this wavelength for long haul systems. It is also pertinent to note that the ultimate speed restriction on electronic devices may actually be reached with InP-based circuits, rather than with those based on GaAs. Exactly how the balance of conflicting demands and possibilities will be reconciled is yet to emerge but we can be sure of many interesting future developments—this remains one of the most exciting and rapidly moving fields in semiconductor R&D.

Bibliography

Agrawal, G. P. (1997) *Fibre Optic Communication Systems*, 2nd edn, John Wiley & Sons, New York.

Agrawal, G. P. and Dutta, N. K. (1986) *Long Wavelength Semiconductor Lasers*, Van Nostrand Reinhold, New York.

Bube, R. H. (1960) *Photoconductivity of Solids*, John Wiley & Sons, New York.

Ditchburn, R. W. (1952) *Light*, Blackie & Sons Ltd, London & Glasgow.

Hecht, J. (1999) *City of Light*, Oxford University Press.

Milnes, A. G. and Feucht, D. L. (1972) *Heterojunctions and Metal Semiconductor Junctions*, Academic Press, New York.

Pearsall, T. P. (e.d.) (1982) *GaInAsP Alloy Semiconductors*, John Wiley & Sons, Chichester.

Razeghi, M. (1989) *The MOCVD Challenge*, Vol. 1, IOP Publications, Bristol.

Smith, S. D. (1995) *Optoelectronic Devices*, Prentice Hall, London.

Thijs, P. G. A., Tiemeijer, L. F., Binsma, J. J. M., and van Dongen, T. (1994) *IEEE J. Quant. Electron*, QE 30, 477.

Willardson, R. K., and Beer, A. C. (eds.) (1985) *Semiconductors and Semimetals*, Vol. 22 (parts A, B, C, and D), Academic Press, San Diego.

CHAPTER 9

Semiconductors in the infrared

9.1 The infrared spectral region

The existence of radiation in the infrared (IR) spectral region was first discovered two centuries ago. In 1800, Sir William Herschel used a simple mercury-in-glass thermometer to demonstrate the heating effect of radiation with wavelengths greater than that of red light, having dispersed it with a glass prism. Since his pioneering work, an enormous amount of development has led to wide-ranging applications based on infrared radiation, including systems for night vision, thermal imaging, burglar alarms, fire alarms, temperature measurement, infrared spectroscopy, infrared lasers, missile guidance, infrared photography, detection of breast cancer, industrial process control, remote control of TV sets, etc. and semiconductors have played a vital part in many of these developments both as infrared sources and as detectors, hence this chapter. Nor should we overlook the tremendous importance of infrared measurements in furthering the basic understanding of the nature of electromagnetic radiation—it was largely on this basis that Planck, in 1901, came to propound his famous radiation law, incorporating the revolutionary concept of radiation quanta, or photons. This, in turn, led to Einstein's quantum theory of electron photoemission (1905), the Bohr theory of the atom (1913), and quantum/wave mechanics by Heisenberg and Schroedinger (1925–6). Needless to say, many of the technological developments referred to depended heavily on this improved understanding, while it is equally true that an inverse relationship also existed. For example, the observation of complex atomic and molecular infrared spectra provided an important stimulus to progress in theoretical quantum chemistry. This introductory section provides a brief outline of the significant features of infrared radiation as utilized in imaging and control systems—for a comprehensive account, the reader should consult the book by Smith, Jones, and Chasmar (1968), while the important subject of infrared spectroscopy is dealt with in Herzberg's (1945) classic treatment (for a recent account see Duxbury 2000).

But first, as the French Michelin Guide writers are fond of saying, 'un peu d'histoire'. Following Herschel's initial discovery, progress was stately, rather than vigorous. The first significant advance was the development of the thermopile (an array of thermocouples), which provided considerably improved sensitivity compared with the simple mercury thermometer and led, in 1847, to the discovery of optical interference effects in the infrared. This was important because it proved IR radiation to be a wave motion like that of visible light and, for the first time, allowed accurate measurement of its wavelength. For a long time, measurements were confined to near-infrared wavelengths but by 1880 the range had been extended to 7 μm and by the turn of the century to well beyond 100 μm—in fact, it gradually came to be realized that there was effectively no limit. Indeed, the gap between short radio waves and long infrared waves was finally closed in the 1930s. However, other significant discoveries were made well before that—in 1843 Becquerel discovered the photographic effect of near-infrared radiation and in 1880, Langley developed the use of a diffraction grating to disperse IR radiation, aided by his use of the newly invented bolometer, a thermal detector which was much more sensitive than available thermopiles. The First World War stimulated considerable effort, on both sides of the conflict, to develop infrared systems for military use, including the first application of the 'internal photoelectric effect' (i.e. photoconductivity). This was reported in 1917 by T. W. Case (of the Case Research Laboratory in the Unites States) who found that thin films of thallium sulfide could be suitably activated to form highly sensitive and fast detectors at wavelengths out to approximately 1.1 μm. The development of sensitive IR film followed shortly afterwards in 1919. The Second World War stimulated further developments in photoconductive detection and imaging (note that IR imaging systems offered much improved resolution compared with microwave radar because of the shorter wavelengths employed), PbS films showing sensitivity out to 4 μm, followed shortly afterwards by the other lead chalcogenides, PbSe and PbTe, which extended the range to nearly 10 μm. These semiconductor photon detectors were two orders of magnitude more sensitive than the best thermal detectors and showed much shorter response times (microseconds, rather than milliseconds), of crucial importance to the later development of missile guidance systems based on the detection of heat radiation from aeroengines. They also provided an essential spur to the development of infrared spectroscopy for the study of molecular vibrational and rotational spectra which has proved, and continues to prove indispensable both in the laboratory and in a wide range of industrial applications, including automatic process control. In particular, the application to synthetic rubber manufacture in the United States in 1943 (based on thallium sulfide detectors) provided a major stimulus. Of almost equal importance was the follow-up, during the 1960s, to the measurement of phonon spectra in a wide range of solids (including, of course, semiconductors), which furthered much improved understanding of solid-state physics. Other wartime developments were based on the application of the infrared photoelectric effect (electron emission from an irradiated solid surface) in the form of the S1 (Ag—O—Cs) photocathode, first reported from RCA in 1930 and sensitive out to a wavelength of 1.3 μm. This was used in the manufacture of night vision tubes as an aid to night driving (with infrared headlights), also in the aptly named 'sniperscope' rifle sight and 'snooperscope' surveillance equipment. Greater infrared sensitivity was later achieved using the multi-alkali (Na, K, Cs, Sb) S20 photocathode, followed in the 1970s by the even better, negative-electron-affinity semiconductor cathode, based on GaAs and its alloys InGaAs and InGaAsP. Combined with the multi-channel electron multiplier developed during the 1960s, these infrared photocathodes provided remarkable low-light-level performance in image converter tubes, initially for military use but more recently as essential aids to the study of nocturnal wildlife! The 1950s and 1960s saw the development of many more detector materials, such as InSb, impurity-doped Ge, and the ternary alloys HgCdTe and PbSnTe, the latter of which also led to the first far-infrared laser in 1966, while during the 1980s and 1990s quantum well (QW) structures in (e.g.) AlGaAs/GaAs made an impact on both detector and laser development, using 'inter-sub-band', rather than 'inter-band' transitions. This flourish of material activity was, of course, stimulated by a rapid increase in system developments, centred largely on long wavelength thermal imaging, one of the principal aims of recent work being the achievement of room temperature operation (thermal radiation detectors had to be cooled to obtain adequate sensitivity). Another is the integration of detector arrays with charge-coupled device (CCD) image processors, thus achieving 'all-solid-state' imaging systems in the far-infrared. Finally, we note that semiconductor lasers now exist, covering all infrared wavelengths out to and beyond 100 μm, further enhancing the range of infrared techniques available. And if this seems like an awesome amount of information to be condensed into a single paragraph, it should leave the reader in no doubt as to the major importance of the infrared spectral region, both for fundamental studies and for technological applications. Furthermore, he or she may yet again recognize the significance of the interaction between these two activities—a recurring theme in our overall discourse! But now (at last!) we can return to the mainstream of our discussion.

One of the most significant features of thermal radiation from a 'black body' (i.e. a body which absorbs all radiation falling upon it), when at temperature T, greater than absolute zero, concerns the distribution of emitted energy as a function of wavelength. Classical theory was clearly in error in predicting that the distribution function tended to infinity as the wavelength λ approached zero (the so-called ultraviolet (UV) catastrophe), leading Planck to his alternative formulation whose mathematical form is given by equation (B9.1) in Box 9.1. The essentials are apparent from the plots of emitted power against wavelength for

Box 9.1. The Planck distribution function

A 'black body' is an idealization, having the property that it absorbs all radiation falling on it at all wavelengths (as is well known, a close approximation to this ideal can be realized by employing an enclosure at uniform temperature T, with a small hole in it through which the appropriate radiation can pass). It follows from this definition that a black body also emits radiation at all wavelengths with maximum efficiency while all real bodies emit less efficiently. This is usually expressed mathematically by multiplying the black body emission intensity by an emissivity factor $\varepsilon(\lambda) \leq 1$. Note that the emissivity ε is, itself, usually a function of wavelength which significantly complicates the relation between real bodies and the ideal. Nevertheless, it is useful to keep in mind the fundamental properties of a black body because these can be derived from the application of thermodynamics and serve as limiting values to which real bodies may, as it were, aspire.

An important aspect of black body radiation is its distribution as a function of wavelength $P(\lambda)$, a function which was central to the development of the quantum theory of radiation. Classical theory predicted a distribution which approached infinity as the wavelength approached zero and it was this anomaly that encouraged Planck to introduce the concept of radiation quanta (or photons) and which led him to his alternative distribution function:

$$P(\lambda,T)\,d\lambda = (2\pi hc^2/\lambda^5)[d\lambda/\{\exp(ch/\lambda kT) - 1\}] \tag{B9.1}$$

where h is Planck's constant, c the velocity of light *in vacuo*, and T the temperature of the black body in degrees absolute. In this form, $Pd\lambda$ represents the amount of energy radiated per second by unit area of the black body between wavelengths λ and $(\lambda + d\lambda)$. By differentiating equation (B9.1), it is possible to show that $P(\lambda)$ has a maximum at $\lambda = \lambda_m$, where:

$$hc/\lambda_m = 4.965kT \tag{B9.2}$$

and, expressing λ_m in μm, this approximates fairly closely to:

$$\lambda_m = 3000/T. \tag{B9.3}$$

The peak value $P(\lambda_m)$ is given by:

$$P(\lambda_m) = 1.286 \times 10^{-11}\, T^5\, \mathrm{W\,m^{-2}}(\mu\mathrm{m})^{-1}, \tag{B9.4}$$

while the total power emitted (obtained by integrating over all wavelengths) is:

$$P_T(T) = 5.669 \times 10^{-8}\, T^4\, \mathrm{W\,m^{-2}}. \tag{B9.5}$$

Equation (B9.5) is familiar as Stefan's law. From it, we see that a black body at room temperature ($T = 300$K) radiates approximately 460 W from each square metre of its surface. On the other hand, the sun at a black body temperature of 6000K radiates 73.6 MW m^{-2}. Similarly, equation (B9.3) tells us that the peak wavelengths are approximately 10 and 0.5 μm, respectively, explaining why evolution has resulted in our developing maximum visual sensitivity close to this wavelength (see Figure 7.1). It also suggests that thermal imaging systems might be expected to operate with maximum efficiency at wavelengths in the region of 10 μm. Figure 9.1 provides examples of black body radiation profiles for a number of practical thermal sources. Note that in Figure 9.1(a) the function $F(\lambda, 300\text{K})$, plotted on a linear scale, is a dimensionless ratio $P(\lambda)/P(\lambda_m)$ appropriate to bodies at room temperature, $T = 300$K, whereas $P(\lambda)$ in Figure 9.1(b) has units of W m$^{-2}(\mu$m$)^{-1}$.

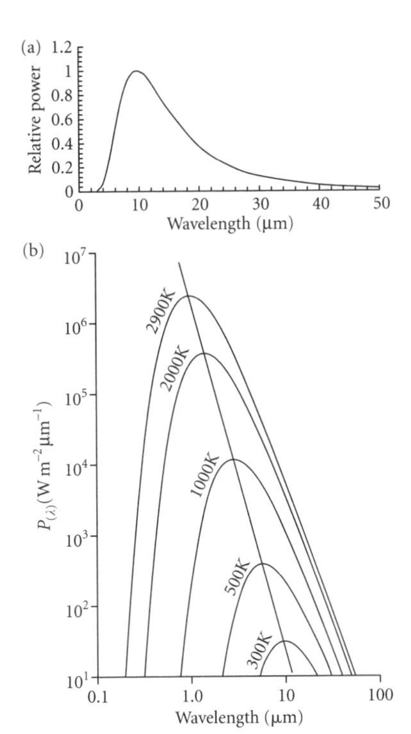

Figure 9.1. Plots of the Planck distribution function (equation (B9.1)) corresponding to the radiant power emitted from a black body at absolute temperature T, as a function of wavelength λ. In (a) we show a linear plot of the normalized distribution function for a black body temperature of 300K, while in (b) we show logarithmic plots of the actual power emitted from black bodies at selected temperatures. $T = 2900$K corresponds to a typical tungsten filament lamp (though attention must be paid to the envelope material if the radiation at wavelengths longer than 3 μm is required), while typical glowbars may operate in the temperature range 1000–2000K. This plot demonstrates clearly the linear shift of the peak intensity towards shorter wavelengths as the temperature is increased—see equation (B9.3).

black bodies at various temperatures shown in Figure 9.1. As can be seen, the radiant power peaks at a wavelength λ_m which decreases inversely with temperature, the peak power also being a strong function of temperature (which explains the use of the logarithmic plot in Figure 9.1(b)). Two special cases are clearly of importance, those of solar radiation, characterized by a temperature $T = 6000$K, and of thermal radiation from bodies at room temperature, $T = 300$K. In particular, the peak wavelengths occur at $\lambda_m = 0.5$ μm (500 nm) and $\lambda_m = 10$ μm, respectively, which explains why evolution has led to our having maximum visual sensitivity in the vicinity of 500 nm (see Figure 7.1) and why thermal imaging systems are designed to be sensitive at wavelengths in the 'mid-infrared'. (Very roughly, we may define 'near-infrared', 'mid-infrared', and 'far-infrared' as corresponding to wavelengths near 1, 10, and 100 μm, respectively.)

In any system designed to detect or form images of objects, two critical parameters are the intensity of radiation reaching the detector from the object (the signal) and the intensity of background radiation which may serve to confuse (i.e. interfere with) the signal. On the other hand, background radiation frequently plays an essential role in *generating* the signal, which points up the distinction between systems where the signal takes the form of thermal radiation (radiation *emitted* by the object) and those where it consists of radiation *reflected* by the object. The latter rely, of course, on the presence of background radiation from an appropriate source. (In visual imaging, with which we are most familiar, this may be the sun, moon, or suitable artificial lighting, in night vision systems it may take the form of night sky radiation in the near-infrared or, at long wavelengths, a laser beam.) In the first case, the object is characterized by its (wavelength-dependent) emissivity and in the second by its (wavelength-dependent) reflectivity. We often learn as much about the object from its 'colour' as from its shape. Infrared systems may be of either type. Thus, near-infrared radiation from the night sky is used in reflective mode by night vision tubes, while thermal radiation emitted by hot (or more often merely warm) bodies is used at longer wavelengths. However, it is probably true to say that, thermal systems are of considerably greater importance, which is one reason why we began this section with a discussion of thermal radiation.

A third important feature of any detection system concerns the transmission properties of the medium in which the system has to operate. In practice, background radiation may not contain the smooth distribution of wavelengths indicated in Figure 9.1, due to wavelength-selective atmospheric absorption or scattering—similarly, thermal radiation from an object may fail to reach the detector if its wavelength coincides with that of an atmospheric absorption band. Clearly, it is important that system designers should be familiar with the transmission properties, not only of the earth's atmosphere but also of any solid media forming lenses or windows through which the radiation must pass before reaching the actual detector. For example, a glass dewar, used to cool a detector, will absorb radiation at wavelengths greater than 3 μm and must therefore incorporate an infrared window if the system is required to operate at longer wavelengths. This may consist, for example, of alumina, fused quartz, rock salt, or one of a variety of fluorides (e.g. CaF_2) and bromides (e.g. KBr), according to the wavelength range of importance (see Smith et al. 1968, ch. IX for details).

The study of atmospheric transmission has, itself, a long history, dating to the middle of the nineteenth century, and by 1886, excellent data were already available over the wavelength range 1–5 μm which showed the existence of various 'windows' where radiation losses were extremely small. Later work in the 1930s and 1940s extended this to wavelengths of 20 μm and beyond and comparison with careful absorption measurements indicated that the principle absorption bands resulted from the presence of water vapour and carbon dioxide in the atmosphere. A typical example of atmospheric transmission between 1 and 14 μm is shown in Figure 9.2, revealing the existence of windows in the regions of 1–2, 3–5, and 8–14 μm, which are clearly important for the design of infrared imaging systems. In practice, all three bands have been utilized, as we shall see in Section 9.3.

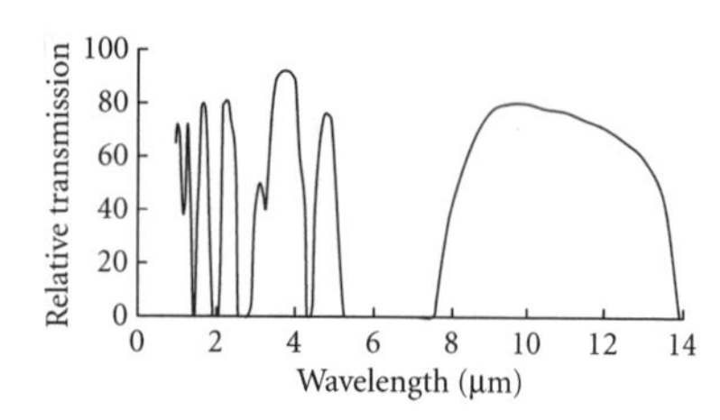

Figure 9.2. Measured atmospheric transmission over the wavelength range of 1–14 μm. The ordinate represents the percentage transmission through approximately one mile of air near sea-level. Note the regions of high transmission in the wavelength ranges 1–2, 3–5, and 8–14 μm. The plot represents a smoothed version of data originally reported by H. A. Gebbie, W. R. Harding, C. Hilsum, A. W. Pryce, and V. Roberts, *Proc. Roy. Soc. A206*, 87 (1951).

The importance of atmospheric transmission is well illustrated in the visible part of the spectrum where, as we know all too well from personal experience, it can be seriously affected by the presence of fine water drops or smoke particles which scatter visible wavelengths. However, it is frequently found that the degree of scattering decreases with increasing wavelength so infrared transmission is less affected than that in the visible, though this depends in detail on the sizes of the scatterers. Provided their diameters are much less than the wavelengths involved, the Rayleigh scattering law applies and the scattering cross-section varies like λ^{-4}, implying a rapid improvement in imaging performance at longer wavelengths. Smoke particles tend to have diameters less than 1 μm, which has led to the successful use of mid-infrared imagers in firefighting, for example. On the other hand, fog particles tend to be larger, in the range 1–10 μm, and significantly longer wavelengths would be required to obtain a similar advantage. However, there is less thermal radiation available in the far-infrared which militates against this approach unless suitably bright long wavelength sources are available (e.g. far-infrared lasers).

Having, I hope, successfully set the infrared scene, it is now appropriate to consider in more detail the nature of the essential components in any infrared system—principally those of source and detector. The following section serves to provide a broad overview of the appropriate devices, while subsequent sections consider the various semiconductor materials and devices individually.

9.2 Infrared components

In general, we may anticipate that any practical infrared system will include a suitable source of energy, an optical system, and an appropriate detector. In the case of passive detection or imaging systems, the source may be a natural one, such as the night sky or even the object to be detected, itself. In active systems, it may take the form of a laser or a high temperature thermal source such as a suitably filtered tungsten lamp. We discuss infrared lasers in Section 9.6 and so will say no more about them here. Similarly, because thermal sources are little more than heated bodies of various kinds, we simply refer the reader to Chapter 8 of Smith, Jones, and Chasmar (1968). This section takes a broad look at detectors, the emphasis being on semiconductor photon detectors which constitute our principal interest but also includes some thermal devices for comparison. We shall examine the various different types of detector which have contributed to the exploitation of the infrared spectrum, leaving accounts of specific materials to the sections which follow.

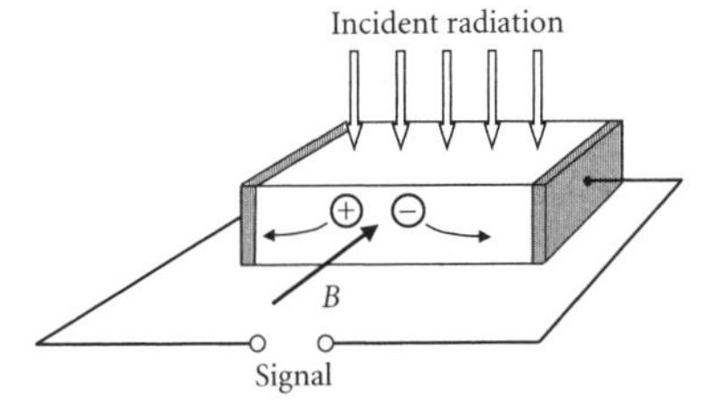

Figure 9.3. Sketch of a semiconductor PEM photodetector. Free carriers generated by absorption of the incoming radiation diffuse downwards into the sample and are deflected by the presence of the magnetic inductance B. Holes and electrons move in opposite directions, setting up a photovoltage between the end contacts.

It is convenient to make two broad divisions between the various photon detectors employed in infrared spectroscopy and imaging applications. In the first instance, one can distinguish three detector geometries—the photoconductor (PC), the photodiode (PD) and the photoelectromagnetic (PEM) detector—in the second, two types of semiconductor material—the so-called 'intrinsic' and 'extrinsic' detectors. We have already made the acquaintance of photoconductors and photodiodes in the previous chapter (Section 8.3) and the reader is referred to the accounts given there. The PEM device was demonstrated by Hilsum and Ross in England and by Kurnick and Zitter in the United States towards the end of the 1950s. Its operation is illustrated in Figure 9.3. Free carriers generated by the absorption of radiation near the upper surface of the sample diffuse away from the surface into the bulk and are deflected sideways by an applied magnetic field. As can be seen, electrons and holes move in opposite directions and set up a photovoltage between the end contacts which constitutes the output signal. (This is effectively a Hall signal produced by diffusing, rather than the more usual drifting, free carriers.) In certain circumstances, it can provide a larger signal than is available from a simple photoconductor but the requirement for a magnet makes it less convenient and it has never been commercialized. The distinction between intrinsic and extrinsic materials has been made on several occasions—an intrinsic photoconductor employs a highly pure semiconductor sample, photocarriers being generated by optically induced transitions across the fundamental energy gap while, in an extrinsic device the transition involved is between an impurity level and either the valence or the conduction band (see Figure 9.4). We shall discuss the importance of these transitions below.

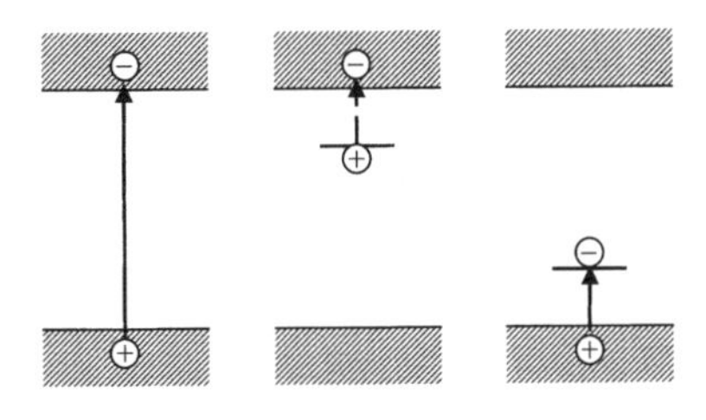

Figure 9.4. Semiconductor band diagrams to illustrate the distinction between intrinsic and extrinsic photoconductors. In the former, radiation is absorbed as a result of transitions between valence and conduction bands, in the latter between an impurity level and one of the bands.

It is also convenient in this section to discuss an appropriate figure of merit, known as the 'normalized detectivity' $D^{\star}$, which can be used to compare the sensitivities of different detectors, the other important characteristic being their respective response times τ. For a proper appreciation of detectivity we shall be obliged to consider the various noise sources which affect detector performance but we do this in an essentially qualitative fashion, eschewing most of the mathematics which feature in the usual textbook treatments (for a detailed discussion see the article by Tom Elliott and Neil Gordon in *Handbook on Semiconductors*, 1993).

The concept of detectivity is a general one applicable to any radiation detector. It depends on the fact that the smallest signal which the detector can hope to record is limited by random noise. Assuming that the detector output, in the form of a signal voltage, is fed through an amplifier to an appropriate display, the parameter of importance is the signal : noise ratio S/N (i.e. signal power/noise power) at the amplifier output. In practice, for most infrared systems, noise generated in the amplifier is negligible, so we need only consider the value of S/N at the detector output terminals and detectivity is defined in these terms. It is usual to define the 'noise equivalent power' NEP as being the signal power at the detector input which gives rise to a S/N ratio of unity at the detector output. This, then, is the minimum detectable signal and represents an important indicator of detector performance. In order to calculate it for any particular detector, it is necessary to evaluate all contributions to the noise power at the detector output and these are of two types—those occurring within the detector itself and those inherent in the signal. For example, in a thermal imaging system, the radiation impinging on the detector consists of a component from the object (at, let us say, an effective temperature of 500K), together with a component from the thermal background at a temperature of 300K (see Figure 9.5). Clearly, background radiation will interfere with the signal and therefore represents a noise source, but a little thought suggests that, with all the wonders of signal processing available to the modern electronics expert, it should be possible to 'back off' any steady background level and leave the desired signal completely untrammelled. This would, indeed, be the case if the background were really constant but we should remember that it consists of a stream of photons emitted randomly and it is the random nature of the emission which gives rise to noise. The radiation power impinging on unit area of the detector fluctuates in time and it is the amplitude of these fluctuations which has to be compared with the signal power in evaluating the S/N ratio. This is discussed in greater detail in Box 9.2 where we define the commonly used figure of merit $D^\star$ which can be used to compare the performance of a detector with that of any other. Briefly, $D^\star$ is the reciprocal of the 'noise equivalent power' NEP (i.e. $D^\star = NEP^{-1}$) where NEP is defined for unit receiver bandwidth and unit detector area.

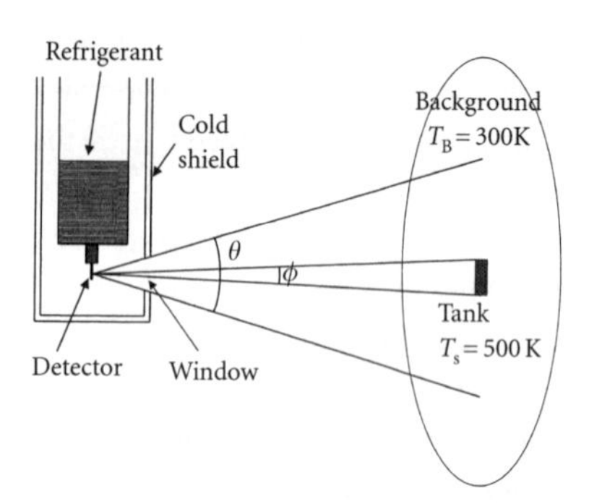

Figure 9.5. Schematic diagram of a cooled infrared detector sampling thermal radiation from a scene, consisting of a tank (a black body at temperature $T_S = 500K$) against a background at temperature $T_B = 300K$. The effective viewing aperture is defined by the angle θ while the object subtends an angle φ. In order to minimise the amount of background radiation reaching the detector, it is necessary to cool all the material surrounding the detector.

Let us be clear what we have said in the above paragraph. Background noise in the detector sets an ultimate limit to the performance

Box 9.2. Background limited detection

To appreciate the concept of background limited detection we can consider a typical problem such as the attempt to use a detector, sensitive at wavelengths close to 4 μm, to detect the presence of a tank (at an effective temperature of 500K) against a background at a temperature of 300K. Suppose the optical system in front of the detector receives radiation over a cone angle of θ radians and suppose that the tank subtends a much smaller cone angle of φ radians (see Figure 9.5). Even though the intensity of 4-μm radiation from unit area of the tank is nearly one hundred times greater than that from the same area of background (see Figure 9.2(b)), the integrated background 'signal' reaching the detector may well be comparable to that from the tank. The problem of detection can then be stated in the form: 'is the detector capable of recording the presence of the tank signal against that of the background?'

Before trying to answer this question, we must be clear about the nature of background noise. The background radiation itself does *not* represent the noise power—indeed, it should be a straightforward matter to 'back-off' a steady background level, leaving the signal completely unaffected. The fact is, though, that the background is *not* steady because it consists of a random (in time) emission of a stream of photons—the number of background photons per second N_B impinging on unit area of detector (integrated over all wavelengths to which the detector is sensitive) therefore fluctuates in time with an amplitude which is proportional to $N_B^{1/2}$ and the total noise power is therefore proportional to $(AN_B)^{1/2}$, where A is the detector area. The question we should be asking is: 'how does the signal from the tank compare with this background noise fluctuation?' and this leads to the concept of NEP, defined as the signal power at the detector input which results in a S/N (i.e. signal power/noise power) of unity. Detailed considerations result in the expression:

$$\mathrm{NEP} = h\nu(2AN_BB/\eta)^{1/2} \tag{B9.6}$$

where ν is the frequency of the radiation, N_B is the background photon flux impinging on the detector, and η is the detector quantum efficiency (electrons per incident photon). In this definition, the bandwidth of the receiver B, is taken as 1 Hz.

This quantity NEP is obviously an important measure of detector performance because it represents the smallest signal power which can be detected. It is, perhaps, slightly less than ideal as a figure of merit because *good* performance is characterized by a *small* number—for preference a figure of merit should increase with improved performance—so it was soon replaced by its reciprocal. The 'detectivity' is defined as: $D = (\mathrm{NEP})^{-1}$ and a slightly more general parameter obtained by removing the detector area from the expression (i.e. referring the detectivity to unit area and unit bandwidth). This results in the widely used 'normalized detectivity' $D^\star = A^{1/2}D$, thus

$$D^\star = (h\nu)^{-1}(\eta/N_B)^{1/2}, \tag{B9.7}$$

which has the rather strange units of $\mathrm{m(Hz)^{1/2}W^{-1}}$. (Frequently in the literature one finds $\mathrm{cm(Hz)^{1/2}W^{-1}}$—reader beware.)

Clearly, the value of N_B to be used in equation (B9.7) will depend on background temperature and on the angle of view (θ in Figure 9.5). It will also depend on the range of wavelengths to which the detector is sensitive—remember that N_B is the photon flux integrated over all detected wavelengths. $D^\star$ also depends on the wavelength through the first term in (B9.7) so it is not possible to quote any one definitive value for $D^\star$ appropriate to background limited detection. Nevertheless, $D^\star_{\mathrm{BLIP}}$ is a useful parameter in so far as it represents the ultimate performance one can expect from a perfect detector—in certain circumstances real detectors may actually approach this performance quite closely.

(i.e. detectivity) of any photodetector—any other sources of noise in the detector itself can only degrade performance below this background limited level. Note that, if we assume a detector quantum efficiency of unity, the expression given in equation (B9.7) for $D^\star$ represents the best (i.e. largest) value possible and it depends only on the nature of the background and the optical aperture (i.e. the angle θ in Figure 9.5). However, we need to be careful in interpreting Figure 9.5 because the detector 'sees' not only the background radiation within the cone of angle θ but also its immediate surroundings, including the window frame through which it peers into the outside world. All these will therefore radiate energy to the detector and contribute to the background noise, a contribution which may only be minimized if they are cooled well below room temperature. In the case of the cooled detector shown in Figure 9.5, this can conveniently be managed by ensuring that all the surrounding structural items are at the temperature of the refrigerating bath, thus ensuring that the only significant background radiation comes from the scene. Then, and only then, can we achieve strictly background limited performance—usually abbreviated to 'BLIP' (background limited infrared photodetector). The additional condition for ensuring BLIP performance is that all other noise sources in the detector are negligible compared with that from the background, a condition which clearly becomes more difficult to satisfy as the background contribution is reduced by making the detector aperture (i.e. θ) smaller. Because $D_{BLIP}^\star$ depends on θ and on the range of wavelengths to which the detector is sensitive, it is impossible to quote a precise value but a typical value for detectors operating over the 5–10 μm waveband might be $D_{BLIP}^\star = 10^9$ m(Hz)$^{1/2}$W^{-1}. (The strange units appear because $D^\star$ refers to a detector of unit area (1 m^2) operating into an amplifier with a bandwidth $B = 1$ Hz,—see Box 9.2.) What this means is that the minimum signal detectable has a magnitude of 10^{-9} W at the detector input. However, if the detector area is actually 1 cm^2, rather than 1 m^2, this figure reduces to 10^{-11} W, while if the receiver bandwidth is increased to 100 Hz, both figures should be increased by a factor ten (NEP $\propto B^{1/2}$).

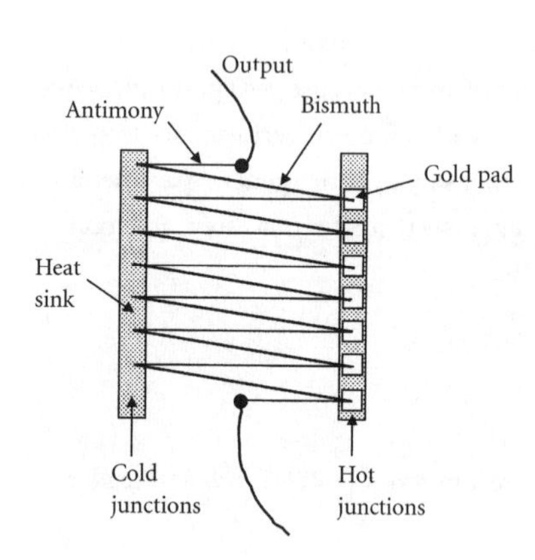

Figure 9.6. Arrangement of thermocouples in a typical infrared thermopile detector. Each of the hot junctions, which see the radiation, is covered by a 'black' gold pad to optimize the absorption of radiation. Each cold junction is buried in a heat sink to maintain a constant reference temperature. The output signal is fed to a sensitive galvanometer or low noise amplifier.

But that is quite enough about background. It is time to talk about detectors. First, we shall briefly discuss thermal detectors and then provide a general outline of photon detectors. As pointed out in Section 9.1, the first reasonably sensitive detector to be applied to infrared measurements was the thermopile, an array of thermocouples arranged as shown in Figure 9.6 for the case of Sb/Bi pairs. (This series connection was adopted with the object of increasing sensitivity but it turned out, on closer theoretical analysis, that a single thermocouple was equally effective.) The hot junctions were blackened using gold films to enhance the absorption of radiation, while the cold junctions were held in contact with a heat sink. To complete the detector, a sensitive galvanometer was used to obtain a tangible output signal. Indeed, much of the development effort devoted to improving performance went into galvanometer design. One interesting example concerned the C. V. Boys radiomicrometer (1888), based on his success in making fine quartz fibre suspensions (we became familiar with his innovative method of making them in the previous chapter, Section 8.1). However, sensitive as some of these designs were, they lacked robustness and later developments relied on electronic amplifiers. Because of the problems in designing stable DC amplifiers, the preferred procedure lay in chopping the signal at a low frequency (a few Hz) and employing an AC amplifier tuned to the chopping frequency. The other principal development has been concerned with thermocouple materials which have included numerous metallic alloys and some semiconductors. Recent devices made by evaporation of metal films are both rugged and *relatively* rapid in response—$D^\star$ may be as high as 10^7 m(Hz)$^{1/2}$W^{-1} and response times as short as 1 ms (see Stevens 1970). A feature of these thermopiles, as, indeed, of all thermal detectors, is a response which remains independent of wavelength out to 100 μm and beyond.

The next innovation in thermal detector technology came in 1880 when Langley introduced the bolometer in the form of a strip of platinum whose resistance changed with its temperature. In practice, he used two identical strips, only one of which was exposed to the radiation. The two platinum resistors were included in a Wheatstone bridge which could be balanced with a variable resistor in the absence of any radiation. The unbalance resulting from radiation reaching the first platinum strip represented the desired signal voltage. Later development made use of chopped radiation and a low frequency amplifier in much the same way as described above for the thermopile. Semiconductor materials, which have much larger temperature coefficients of resistance than metals and therefore provide greater responsivity, were first introduced in 1946 by W. H. Brattain (of Ge point contact transistor fame) and J. A. Becker at Bell Labs. The best $D^\star$ values and response times appear to be very comparable with those for thermopiles mentioned above. However, an interesting variation on the bolometer theme (also dating from the 1940s) made use of a superconductor close to its transition temperature where the change of resistance with temperature is extremely large, resulting in considerably improved performance—$D^\star$ values approaching 10^9 m(Hz)$^{1/2}$W^{-1} have been reported—though the need for liquid helium cooling and very careful temperature control has largely confined the superconducting bolometer to laboratory use where it has been found valuable as a detector in the very far-infrared region.

The Golay cell is yet another product of 1940s innovation. It consists of a small gas-filled cell with an infrared absorbing film at one side and a very thin flexible mirror at the other. Heat from radiation absorbed in the film raises the gas temperature which, in turn, increases the pressure, distorts the mirror, and deflects a light beam which is reflected from it. The deflection is recorded via a highly sensitive light amplifier, the resulting signal being fed to a conventional electronic amplifier. The best detectivity D^* is about 3×10^7 m(Hz)$^{1/2}$W^{-1} with typical response time of about 10 ms.

Another thermal detector which was first proposed in 1938 but not seriously developed until the 1960s is the pyroelectric detector (described in detail by Putley 1970). It utilizes the fact that certain non-centrosymmetric crystalline materials such as $LiNbO_3$, $BaTiO_3$, and triglycine sulfate (TGS) show a change in electrical polarization when their temperature is changed, they are said to be 'pyrolectric', and this polarization change may be detected electronically. Electrodes placed either side of a thin sheet of the material form a capacitor whose state of charge varies with the temperature of the crystal. If the incoming radiation is modulated at frequency ω, an AC signal at this frequency appears across the capacitor, which may then be amplified. The performance of today's pyroelectric detectors is similar to that of the best bolometers but with the added advantage of faster response. At the expense of significantly reduced detectivity, devices have even been made to respond at frequencies up to 1 MHz. Commercial versions are widely available and have been incorporated in many burglar alarm systems.

All the above devices depend for their function on sensing a change in temperature resulting from absorption of the incident radiation. We now turn to the semiconductor devices known as 'photon detectors'. Absorption of infrared photons leads to a direct change in their electrical properties which has nothing to do with any possible change in temperature. For example, in a photoconductor, the device resistance decreases as a result of optically generated electrons and holes and in a photodiode, an electric current is produced by the separation of such electrons and holes at a *p-n* junction. Three advantages are immediately apparent: there is no need to coat the device with a 'black' film to encourage efficient absorption of the incoming radiation (band-to-band absorption in a direct gap semiconductor being efficient in its own right), nor is there any need to make the device fragile in attempting to minimize its thermal mass and, finally, the response time of a photon detector is usually very much faster than that of a thermal detector because it is limited by an electronic rather than a thermal time constant, typically microseconds, rather than milliseconds. Another feature, which is not quite so obvious, relates to the fact that the number of electrons generated by the radiation is proportional to the number of photons per second impinging on the detector N_{ph}—that is, $n \propto N_{ph}$. Thus, the signal voltage $V_S \propto N_{ph}$ and, if we define the responsivity of the detector (in volts per watt) as being:

$$R_\lambda = V_S / P_{ph}, \tag{9.1}$$

where P_{ph} ($= N_{ph} \cdot h\nu$) is the radiant signal power at the detector, we see that $R_\lambda \propto (h\nu)^{-1} \propto \lambda$. The longer the wavelength of the incident radiation, the greater the responsivity, a result which is applicable to any kind of photon detector.

In attempting to match a photon detector to a particular application, the distinction between intrinsic and extrinsic devices (depending on whether absorption of photons takes place as a result of valence band-to-conduction band or impurity level-to-band transitions) becomes important. Concentrating first on the former type, the principal requirement is that the semiconductor should have an energy gap small enough to absorb the desired radiation, a condition which can be written as:

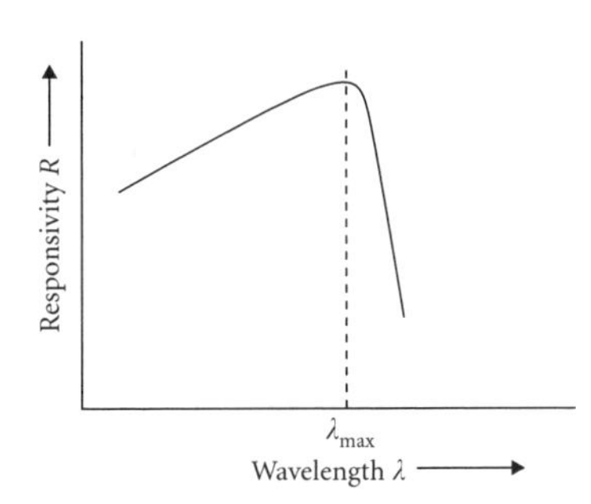

Figure 9.7. Illustration of the way in which the responsivity R of a typical photon detector varies with the wavelength λ of the incident radiation. R increases linearly at short wavelengths, falling more or less sharply to zero beyond a cut-off wavelength λ_{max} defined in equation (9.2). The steepness of the cut-off depends on the nature of the semiconductor absorption edge.

$$E_g\,(\text{eV}) \leq 1.240/\lambda(\mu\text{m}). \tag{9.2}$$

Alternatively, equation (9.2) can be interpreted as defining the longest wavelength which can be absorbed by a specific semiconductor with band gap E_g. Very roughly, we can expect the responsivity to increase with increasing wavelength up to a cut-off wavelength λ_{max} where it drops more or less rapidly to zero, as shown in Figure 9.7. The fact that the cut off is not infinitely sharp depends on the shape of the absorption edge (see, for example, Figure 3.11), which differs significantly between direct and indirect gap materials. This distinction is also important in its implication for the absorption length (the distance over which the incident radiation intensity is reduced to $(1/e)$ of its incident value), typical values near the band edge being 1 μm for direct and 10 μm for indirect energy gaps. This, in turn, determines the thickness of material required for effective absorption of the radiation, 3 and 30 μm, respectively.

In the case of an extrinsic detector, absorption tends to be at much longer wavelengths than appropriate to transitions across the fundamental energy gap. For example, a donor impurity in silicon may have an ionization energy of about 50 meV, corresponding to a cut-off wavelength of 25 μm, whereas the narrow gap semiconductor, InAs has a band gap of 0.35 eV and a cut-off wavelength of 3.5 μm. Another important difference is that the absorption coefficient depends on the density of impurity atoms (e.g. donors) which makes it difficult to obtain adequate absorption in extrinsic detectors. In some cases the solubility of impurities is inadequate and, even when this is not a limiting factor, problems arise in relation to the interaction between impurity atoms when present in densities of order $10^{26}\,\text{m}^{-3}$ or more, where their spacing is of order 1 nm.

This brings us to an important aspect of how to make the best choice of material for any particular application. At first thought, it would appear that a material with a very small band gap would be suitable for detecting all wavelengths but this overlooks the fact that, the smaller E_g, the larger the intrinsic carrier density n_i (equation 1.3) and it is desirable to minimize n_i in the interest of reducing detector noise. As we discussed in Chapter 1, n_i results from thermal generation of electrons and holes—in thermal equilibrium a dynamic balance is established between generation and recombination—and, while the effects of a steady value of n_i might be backed off electronically, the generation and recombination processes are random in nature, just like that of the thermally generated background which we discussed above. Thus, the larger n_i, the greater the generation–recombination (G–R) noise and the poorer the detectivity, which implies that we should choose a semiconductor with band gap just small enough to detect the longest wavelength of interest, but no smaller. In addition, because $n_i \propto (m_e m_h)^{3/2}$, it is desirable, where possible, to select a semiconductor with small effective masses in both valence and conduction bands. But, of much greater significance, is the choice of operating temperature—the exponential dependence of n_i on $(-E_g/2kT)$ results in a strong variation of G–R noise with temperature, and cooling the detector is generally necessary if BLIP performance is to be achieved. While thinking about G–R noise, it is well to appreciate yet another subtlety in the design of a photoconductive detector. Because GR noise is proportional to the total number of intrinsic free carriers in the sample, it can be reduced by reduction of the sample thickness. On the other hand, if this is taken too far, the signal current will also be reduced because only a fraction of the incident photons are absorbed. Optimum design requires that the thickness is roughly three times the absorption length defined above, that is about 3 μm for a direct gap material. Indirect gap materials suffer, by comparison, because they must be made many times thicker in order to absorb an acceptable fraction of the signal photons.

Finally, in addition to all these considerations, it is necessary to take account of the transmission properties of the earth's atmosphere referred to in Section 9.1 (see, in particular, Figure 9.2). The presence of transmitting 'windows' in the regions of 8–14, 3–5, and 1–2 μm has led to these being widely selected for infrared systems, the first two for thermal imaging, the third for night vision tubes, and, as we shall see in the following sections, photon detector band gaps have been chosen to match these requirements.

9.3 Two world wars—and after

Though infrared devices have now found many applications in ordinary life, it is clear that the main thrust of development was stimulated by military needs. The first semiconductor photodetector to demonstrate useful infrared performance was studied at the T W Case Research Laboratory in Auburn, New York, during the First World War. This was the 'Thalofide cell', a photoconductor made by fusing thallium sulfide in the presence of oxygen, followed by a vacuum treatment to improve its sensitivity. The latter reached a maximum at a wavelength of 1 μm, dropping to zero for wavelengths beyond 1.2 μm, characteristic of a material with a band gap of about 1.1 eV (not, of course, that such an explanation was available at the time—band gaps only came to life with the quantum theory of solids in the 1930s). The best devices showed a 50% lowering of dark resistance when subjected to light of intensity 0.06 fc from a tungsten filament lamp. Response times of 10 ms were recorded. Though they saw some degree of service during the latter stages of the war, being copied in several other countries (Italy, Germany, USSR, and the United Kingdom), these detectors suffered from instability and were easily damaged by exposure to visible or UV radiation. These problems were eventually overcome in the early stages of the Second World War but, being sensitive only in the near-infrared, where they were useful as detectors of night sky radiation, they came into competition with infrared image tubes which, as we shall see, rapidly came to dominate this spectral region.

Following the first of the two great wars, there was a lull in work on infrared devices until Germany began rearming during the 1930s which saw important developments in the use of lead sulfide (PbS) as an infrared detector. A photovoltaic device had, in fact, been demonstrated as early as 1904 by Bose, using natural galena crystals but these were of very variable quality and lacked reliability—as we now know, they contained uncontrolled amounts of impurities. The important breakthrough in the 1930s resulted from the controlled deposition of thin films of lead sulfide, either by vacuum evaporation onto glass substrates or by chemical reaction of lead acetate with thiourea in the presence of sodium hydroxide. As with the Thalofide cell, oxygen was found to play a significant role in improving sensitivity and was deliberately incorporated (by means of a small vapour pressure of O_2 in the evaporator or by a post-deposition oxygen bake in the case of chemically deposited films). Interestingly, German research made much use of the photovoltaic effect associated with a point-contact diode—our old friend the cat's whisker which played such an important role as a high frequency rectifier in the early days of radio and (based on Ge and Si) as a microwave radar detector in Second World War (see Chapter 2).

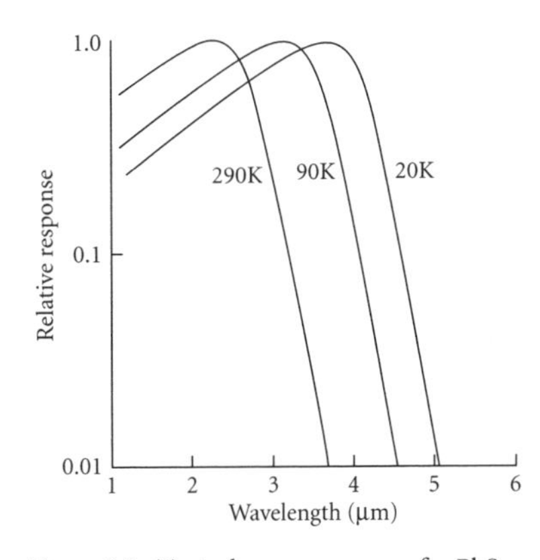

Figure 9.8. Typical response curves for PbS photoconductors, showing the way in which the long wavelength cut-off moves to longer wavelengths as the temperature is lowered. This allows a degree of 'tuning' of the response peak to match system requirements, though cooling also improves the detectivity.

An important property of PbS was its response to significantly longer wavelengths, compared with Tl_2S_3 (see Figure 9.8), showing a cut-off at about 3 μm (at room temperature), shifting to 4 μm when the device was cooled to 90K with liquid oxygen (today's health and safety protagonists would be up in arms, of course, but the much safer liquid nitrogen was not then available!). Thus, lead sulfide became the first semiconductor sensitive to thermal radiation (from bodies at modest temperatures) and attracted widespread interest as the second great war duly imposed itself on an unwilling world. Considerable development effort was devoted to lead sulfide on both sides of the Atlantic, to a significant extent independently, on account of the military secrecy which surrounded it. This separatism has remained true to a large extent even today, as exemplified by the fact that American publications such as '*Semiconductors and Semimetals*' (see Vols. 5, 15, and 18) or *Proc. IRE* (Vol. 47) contain articles from American authors, while European counterparts such as *Advances in Physics* (Vol. 2) or *Handbook on Semiconductors* (Vol. 4) rely largely on European contributors.

The result of this work was the development of the (then) most sensitive detector available for the wavelength region 1–4 μm. Values of detectivity $D^* = 3 \times 10^9$ m(Hz)$^{1/2}$ W^{-1} for detectors cooled to a temperature of 195K (obtained with solid CO_2) were measured, which was close to 300K BLIP performance. Cooling the detector to lower temperatures led to an increased cut-off wavelength, as shown in Figure 9.8, though the ultimate performance could only be attained when care was taken to screen room temperature background radiation. As can be seen in Figure 9.8, PbS detectors, cooled by liquid oxygen (or nitrogen) can be used at wavelengths out to just beyond 4 μm but this does not quite cover the whole of the 3–5 μm atmospheric window—some extension would obviously improve the performance when detecting 500K thermal radiation (which peaks at about 6 μm—see Figure 9.1). What is more, the 8–14 μm window is well beyond the reach of PbS and some major modification in material would obviously be required to exploit this region. In fact, the first step towards this came quite early, the wartime work in Germany exploring the use of lead selenide and telluride, PbSe and PbTe, though apparently with very limited success. It was only after the war that further developments in England showed how an excellent performance could be obtained from these materials, and, importantly, with significant extension in cut-off wavelength (see, for example, Smith 1953). Figure 9.9 shows measured response curves for all three of the lead salts which provided encouragement in that a cooled PbSe detector is seen to be sensitive at wavelengths out to just beyond 8 μm, while adding a touch of mystery in that the PbTe cut-off lies *between* those of PbS and PbSe, rather than, as we should expect from its chemistry, at longer wavelength still. Though, sadly, the 8–14 μm band was still out of range, both PbSe and PbTe were well able to cover the 3–5 μm band, and with excellent sensitivity.

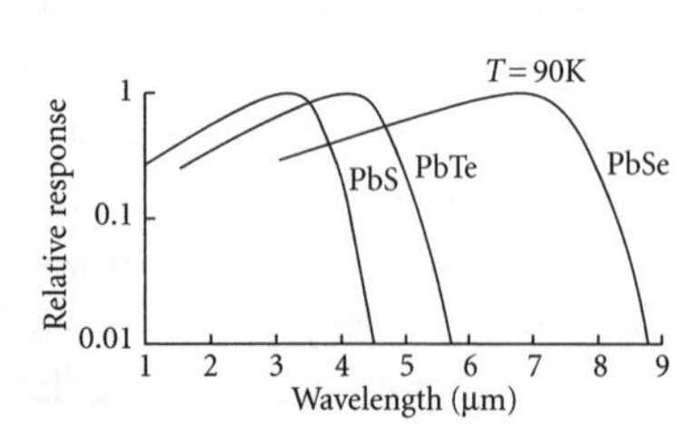

Figure 9.9. Relative response curves for the three lead salts at a temperature of 90K. Notice the surprising fact that the PbTe curve lies between the other two, rather than at longer wavelengths, as might be expected from their chemical properties.

All these devices consisted of thin (approximately 1-μm thick) photoconducting films deposited on glass or quartz substrates—quartz being preferred because impurities in glass were found to diffuse into the film and degrade its performance. Not surprisingly, therefore, they were polycrystalline, with grain size about 0.1 μm, and this made for uncertainty concerning their precise mode of operation. We shall be taking up the detailed discussion of polycrystalline materials in Chapter 10, when we meet numerous other examples, including the technologically important polysilicon, but it would be reprehensible not to raise one particular question here. What role was played by the grain boundaries between each small crystallite, making up the film? Very briefly, they were likely to act as sinks for electric charge which caused band-bending within the grains, resulting in the development of potential barriers between each grain. Two aspects may influence photoconductivity: photoinduced carriers must surmount these barriers in order to progress towards the end contacts (effectively increasing the sample resistance), while trapping of optically generated minority carriers at the grain boundaries results in a reduction of barrier height (thus, reducing the resistance again), complications which made proper understanding of film properties far from easy (it was thought, incidentally, that oxygen played a role in determining the height of these inter-grain barriers). As we shall see in the next chapter, there are tradeoffs between grain size, barrier height, and semiconductor doping which lead to some unexpected behaviour in polycrystalline photoconductors but these were little understood in the 1940s and 1950s when the exploration of the lead chalcogenides was under way. It was perhaps ironic that the move from single crystals to polycrystalline films had achieved undoubted technological success, while, at the same time raising significant barriers to the scientific understanding of performance limits. The growth of artificial single crystals with controlled impurity levels would surely provide answers? Or would they?

Successful growth of artificial single crystals was first achieved by W. D. Lawson at the Royal Radar Establishment (RRE), Malvern, England in 1951 (PbTe) and 1952 (PbS and PbSe). He used a version of the Bridgman method (growth from the melt), the respective melting temperatures being 1127°C, 1081°C, and 924°C for PbS, PbSe, and PbTe. There were problems with strain introduced during cooling but the Czochalski growth method, which later solved this problem, could not be applied because of the high vapour pressures of the chalcogenides (the liquid encapsulation method introduced to alleviate this problem in GaAs growth did not appear until 1965, see Chapter 5). The availability of single crystals allowed a wide range of optical and electrical measurements to be made in efforts to establish the fundamental properties of these materials. The results of optical absorption measurements supported the interpretation of photodetector cut-off wavelengths as corresponding to band gap energies of (at room temperature) $E_g = 0.30$ eV (PbS), 0.22 eV (PbSe), and 0.25 eV (PbTe), thus confirming the unexpected ordering. Temperature coefficients of E_g were closely similar for all three salts at approximately $+2.5 \times 10^{-4}\,\mathrm{eV\,K^{-1}}$, again, consistent with the behaviour of the detector response curves. (Note that this positive value is opposite from that found for most semiconductors, and is probably related to crystal structure—the lead salts crystallize in the NaCl face-centred cubic form, having octahedral symmetry, rather than the tetrahedral symmetry of the Groups IV, III-V, and II-VI materials.) Hall and conductivity measurements revealed high free carrier mobilities but led to thermal band gap energies (i.e. derived from the temperature-dependence of the intrinsic free carrier density) several times greater than the optical values, a discrepancy for which, initially, there seemed to be no explanation. The problem was probably associated with the rather high extrinsic carrier densities in most of the crystals examined—in fact, later work has led to better agreement, with both thermal and optical band gaps being revised. The accepted values are now $E_g = 0.42$, 0.27, and 0.31 eV, while the temperature coefficient is $+4.7 \times 10^{-4}$ eV K^{-1} but this only became clear in the 1970s, some two decades later! (Yet another illustration of the considerable difficulty of mastering semiconductor materials.) Early attempts to calculate band structures pointed to the lead salts' all having indirect energy gaps though, again, later work has shown this to be in error. It is now established that these chalcogenide semiconductors have direct gaps, though with the respective band extrema at the L point of the Brillouin zone, that is, lying along (1 1 1) type directions (there being four equivalent such extrema as we saw for the conduction band minima in Ge in Chapter 3). A more detailed account of the lead salt properties is given by Lovett (1977)—we shall say no more about it here.

Difficulties with the physics detracted not at all from the efficacy of the lead chalcogenides as infrared detectors but there did gradually emerge one significant practical problem, that of response time. Measured minority carrier lifetimes in thin film samples were found to lie in the vicinity of 0.1–1.0 ms (probably as a result of carrier trapping at grain boundaries), thus limiting response times. While these values were still less than the response time of available thermal detectors, they came to be seen as too long for the important application to thermal imaging, which was a rapidly emerging technology in the late 1950s. Any detailed discussion of imaging systems would take us too far from our brief—readers are referred to the book by Lloyd (1975) and to the articles by Morten (1971) and Jervis and Lockett (1971)—but some appreciation of the importance of detector response time is, nevertheless, desirable.

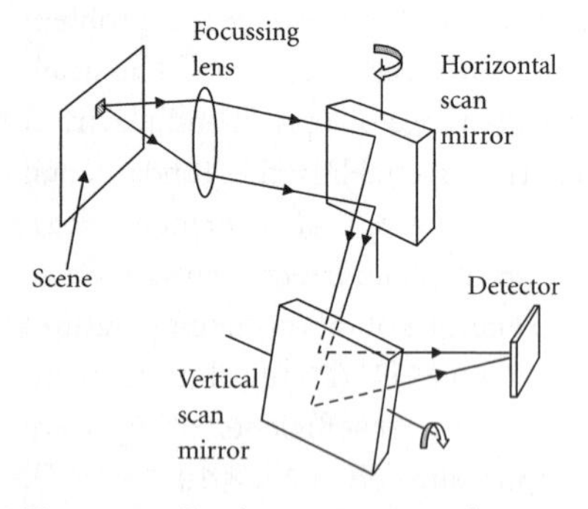

Figure 9.10. Simple scanning system for an infrared imager based on a single radiation detector. The first rotating mirror serves as a line scanner (i.e. to provide a horizontal scan) while the second provides the frame scan (i.e. the vertical scan). Note that the two mirrors rotate about mutually perpendicular axes.

Though later work led to the use of one- and two-dimensional arrays of detectors, in the first instance, infrared imaging systems were based on a single detector element, the scene being scanned mechanically over the element by rotating mirrors, prisms, or a suitable chopping disc. An example is shown in Figure 9.10 where a pseudo-TV type raster is effected with two mirrors rotating about mutually perpendicular axes, the line scan mirror rotating considerably faster than the frame scan (a whole line must be scanned in the time taken by the frame scan to move between one line and the next). However, a crucial requirement is the need to complete a whole frame in a time less than the integration time of the human eye, namely about 1/25th of a second and a simple calculation suggests that, in this case, each (notional) scene pixel is incident on the detector for a time not greater than $(1/25N)$s, where N is the total number of pixels. For even a crude image made up of $20 \times 20 = 400$ pixels, this implies a response time for the detector of better than 10^{-4} s. Millisecond response times were simply not acceptable, an appropriate target level being more like 1 to 10 μs. What was to be done?

Two possibilities presented themselves—one was to use photodiodes rather than photoconductors (remember from Chapter 8 that PDs are very much faster than PCs), the other was to change material. The first option demanded controlled *n*- and *p*-type doping of the lead salts in single crystal form which, though not by any means impossible, was still far from straightforward. The fact that background doping levels in single crystals were relatively high would not represent a disadvantage because, in a simple *p-n* diode, the *n*- and *p*-sides should be fairly heavily doped to minimize series resistance, however, it did imply a more complex technology, photodiodes always being trickier to make than photoconductors. The second option was more a matter of chance but, as chance would have it, a promising candidate was on hand in the shape of a III-V material, InSb. The Pandora's box of III-V compounds had been opened in the early 1950s by Professor Welker and the Siemens company in Germany, leading, as we saw in earlier chapters, to the emergence of GaAs, GaP, InP, and a host of alloys which continue to grab headlines wherever silicon finds it difficult to venture. Because of its very small energy gap $E_g = 0.17$ eV, we have so far ignored InSb but it clearly came into its own in the context of infrared detection. With a cut-off wavelength of about 7 μm at room temperature, reducing to 5.5 μm at 77K, it was an ideal candidate to challenge the lead salts. Single crystals were relatively easy to grow with high purity and, what was more, minority carrier lifetimes were measured in the range between 1 ns and 1 μs. All this was established well before the end of the 1950s and, by 1957, InSb already had a pedigree in both photovoltaic and photoconductive detector applications (see, for example, Kruse 1970). Interestingly enough, the Americans preferred to use diodes, while the British preferred photoconductors but there was little difference in overall performance. In either event, it was clear that InSb could not be overlooked in the context of thermal imaging.

The key feature of InSb was its availability in well-controlled single crystal form, and, because of its relatively low melting point (530°C) and the low vapour pressure of antimony at this temperature, crystal growth was relatively straightforward. The material was first purified by zone refining, mainly to remove Te (a donor) and Zn (an acceptor). Crystals were then grown by either the horizontal Bridgman or the Czochralski process at atmospheric pressure, typically in a hydrogen atmosphere. Excellent summaries of the early electrical and optical data on InSb are given in the books by Hilsum and Rose-Innes (1961) and by Madelung (1964). In common with other III-V compounds with modest energy gaps, InSb is a direct gap material, as is desirable for a photodetector, and it is also characterized by a small electron effective mass ($m_e = 0.012$ m at room temperature). The purest crystals obtained had net donor concentrations of order 10^{19} m^{-3} (compared with a room temperature intrinsic free carrier concentration of $n_i = 1.5 \times 10^{22}$ m^{-3}), though 10^{20}–10^{21} m^{-3} was more usual, and electron mobilities at room temperature close to 8 $m^2 V^{-1} s^{-1}$, an encouragingly large value for the application to *p*-type photoconductive detectors (compare, for example, the value of $\mu_n =$ 0.9 $m^2 V^{-1} s^{-1}$ for GaAs which was regarded as giving it an important advantage over silicon). Peak electron mobilities close to 100 $m^2 V^{-1} s^{-1}$ at reduced temperatures were similarly mouth-watering when cooled detector operation was implied by the need to reduce GR noise.

It is well worth noting that, because of the availability of high quality crystals having unusually high free carrier mobilities, InSb was the first compound semiconductor to be investigated in depth. Detailed studies of both majority and minority carrier properties were made in the 1950s and 1960s together with theoretical analyses which largely preceded those done on GaAs. Indeed, many of the techniques which were used so successfully in analysing GaAs behaviour were derived from the early InSb work. For a comprehensive account of the Hall and conductivity measurements the reader should consult Earnest Putley's (1968) seminal book *The Hall Effect and Semiconductor Physics*.

Measured recombination lifetimes showed interesting variation with temperature and doping level, lying typically within the range 10^{-7}–10^{-9} s, values which are clearly short enough for scanned imaging applications but, perhaps, a trifle too short for optimum detector sensitivity (detectivity D is proportional to τ_n in a *p*-type photoconductor, see Box 9.3). To illustrate the complexities of the physics of recombination we should note that several possible recombination mechanisms may play a role. The simplest of these is direct radiative recombination but in many cases non-radiative processes are dominant. In *p*-type InSb lifetimes were found to be controlled either by Auger recombination (in which a recombining free carrier gives its energy to another free carrier, exciting it high into the appropriate energy band) or by the so-called 'Shockley–Read' (S–R) mechanism in which electrons and holes recombine via deep impurity levels within the forbidden energy gap (first an electron is captured from the conduction band by the centre, then a hole from the valence band). This latter process has one particularly interesting feature—depending on whether the density of recombination centres is less than or greater than the majority carrier concentration, the electron and hole lifetimes may be equal or unequal to one another, which has important consequences for device performance. In *p*-type InSb, τ_n was found to be equal to τ_p at room temperature but $\tau_n \ll \tau_p$ at 77K, requiring a significantly different theoretical approach to the analysis of device behaviour at the two temperatures. In particular, the photoconductive

Box 9.3. Photoconductive detector sensitivity

The literature abounds with theoretical derivations of photodetector performance, taking account of the various noise mechanisms which may limit it. These are often complex, depend on specific models, and include various approximations. It is beyond our brief to look in detail at any of these calculations—the interested reader can find them in several of the references cited in the references, such as Kruse (1970) or Elliott and Gordon (1993)—but, in this box, we present an overview, as it were, of one such derivation, examining the assumptions made and the conclusions to be drawn from it. To be specific, we choose the example of an InSb photoconductor operating at room temperature. We refer to Figure 9.11 for the appropriate geometry.

It is straightforward to derive an expression for the short-circuit photocurrent I_S:

$$\begin{aligned} I_S &= ewd(n\mu_n + p\mu_p)F_x \\ &= ewN\tau_n\mu_nF_x \end{aligned} \tag{B9.8}$$

where N is the number of *absorbed* photons per unit surface area per second ($N = \eta N_{ph}$), τ_n is the electron recombination time, and F_x is the applied electric field. We have also assumed, as appropriate for InSb, that $\mu_n \gg \mu_p$. It follows that the open circuit voltage V_S, which appears at the input terminals of the pre-amplifier is given by $V_S = I_S R$, where R is the sample resistance. If we assume that the sample is intrinsic (which is readily achieved in InSb at room temperature where $n_i = 1.5 \times 10^{22}\ \text{m}^{-3}$), we can write the sample resistance as $R = 1/wdn_i e\mu_n$ so

$$V_S = 1N\tau_n F_x/n_i d. \tag{B9.9}$$

It follows from this that we can write the detector responsivity R_λ (in volts per watt) as: $R_\lambda = V_S/P_{ph}$, where P_{ph} is the radiant power falling on the total detector area, that is, $P_{ph} = N_{ph}h\nu.\text{lw}$ and, assuming unity quantum efficiency, so that $N = N_{ph}$, we have:

$$R_\lambda = \tau_n F_x/wdn_i h\nu. \tag{B9.10}$$

To obtain an expression for the minimum detectable power NEP we need to relate the signal power P_S to the incident power P_{ph}. Thus, $P_S = V_S{}^2/R$ where $V_S = R_\lambda P_{ph}$, so we have:

$$P_{ph} = (RP_S)^{1/2}/R_\lambda \tag{B9.11}$$

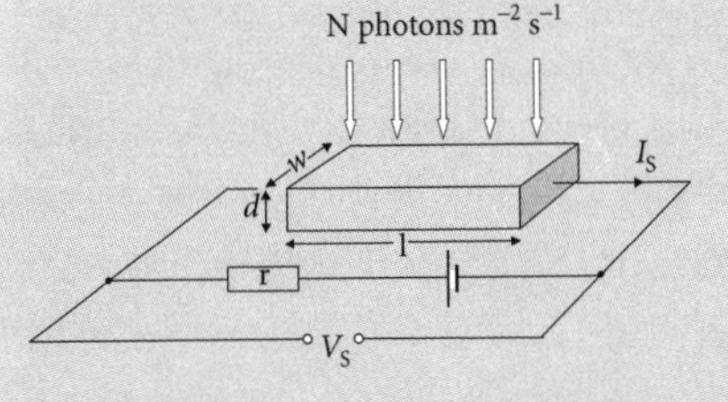

Figure 9.11. Sample geometry of a typical photoconductive detector, showing sample dimensions and arrangement for irradiation. I_S is the photocurrent which depends on the applied DC voltage. The voltage V_S which appears across the device terminals represents the signal applied to the pre-amplifier input terminals.

and, therefore:

$$\text{NEP} = (RP_N)^{1/2}/R_\lambda \tag{B9.12}$$

where we have put $P_S = P_N$ (P_N being the noise power) as the condition for minimum detectable signal. Note that $(RP_N)^{1/2} = V_N$ (the noise voltage), so, finally, we can write the detectivity D as:

$$D = R_\lambda/V_N. \tag{B9.13}$$

(Note that, in calculating the noise voltage V_N, we assume a receiver bandwidth of $B = 1$ Hz.)

Box 9.3. Continued

This result is quite general but, in order to proceed further, we need to know the nature of the noise which limits performance. In this case, Johnson noise in the sample generally dominates other noise sources and we can write $V_N = (4kTRB)^{1/2}$ with $B = 1$ Hz. Note that Johnson noise is independent of the sample area so no advantage accrues in this case from defining $D^\star$.

It may now be helpful to calculate numerical values for the important parameters in order to feel a little more secure in our understanding. To do so we take appropriate values for the input parameters as follows: $\mu_n = 8\ \text{m}^2\,\text{V}^{-1}\,\text{s}^{-1}$, $\tau_n = 3 \times 10^{-8}$ s, $\text{l} = \text{w} = 1$ mm, $d = 3\ \mu\text{m}$, $n_i = 1.5 \times 10^{22}\,\text{m}^{-3}$, $\nu = 6 \times 10^{13}$ Hz ($\lambda = 5\ \mu\text{m}$), $F_x = 3 \times 10^3\ \text{V}\,\text{m}^{-1}$, $B = 1$ Hz, from which we obtain:

$R = 1/wd(n_i e\mu_n) = 17\ \Omega$, $R_\lambda = \tau_n F_x/wdn_i h\nu = 50\ \text{V}\,\text{W}^{-1}$, $P_N = 4kTB = 1.7 \times 10^{-20}$ W, $V_N = (P_N R)^{1/2} = 0.5$ nV, $\text{NEP} = V_N/R_\lambda = 1.0 \times 10^{-11}\ (\text{Hz})^{-1/2}\,\text{W}$, $D = (\text{NEP})^{-1} = 1.0 \times 10^{11}\ (\text{Hz})^{1/2}\,\text{W}^{-1}$, ($D^\star = D(\text{lw})^{1/2} = 1.0 \times 10^8\,\text{m}(\text{Hz})^{1/2}\,\text{W}^{-1}$), $N_{\text{NEP}} = \text{NEP}/\text{l}wh\nu = 2.7 \times 10^{14}\ \text{m}^{-2}\,\text{s}^{-1}$, $n_{\text{NEP}} = N\tau_n/d = 2.7 \times 10^{12}\ \text{m}^{-3}$. Note how small is this value of n compared with n_i—the optically induced change of sample resistance is a minute fraction of the steady state value. Note, too, the somewhat surprising result that R_λ is independent of the electron mobility μ_n, while it is proportional to the recombination lifetime τ_n. Any reduction in lifetime in the interest of improved response rate can only be achieved at the expense of reduced responsivity (and, hence, detectivity).

One further comment is in order. In equation (B9.10), we see that the responsivity is proportional to the applied DC field F_x and it might seem that R_λ could be increased without limit by increasing the applied voltage between the end contacts. There are two reasons why this is not possible in practice—first, dissipation of DC power in the sample, causing it to overheat, and, second, free carrier sweep-out. In the former case, a limit of about $5\text{–}10 \times 10^4\ \text{W}\,\text{m}^{-2}$ is imposed by the need to minimize temperature rise and in our example this implies that the value $F_x = 3 \times 10^3\ \text{V}\,\text{m}^{-1}$ is close to the maximum allowed. 'Sweep-out' refers to the possibility that carriers are swept to the end contacts before they can recombine in the bulk. It is a central assumption of our model that electron–hole recombination takes place only in the bulk and not at any of the surfaces (i.e. why we use the bulk recombination lifetime τ_n in equation (B.9.8)). In the first instance, this implies that the surface recombination velocities of the top and bottom surfaces must be small (a requirement which can be met by careful surface treatments) and, in the second, that the applied field is insufficient for sweep-out effects. This condition demands that the drift length L_d between bulk recombination events be smaller than the sample length l. Thus:

$$L_d = \mu_n F_x \tau_n < 1. \tag{B9.14}$$

In our example this sets a limit on F_x of about $3 \times 10^4\ \text{V}\,\text{m}^{-1}$, which is much greater than the thermal dissipation limit and therefore irrelevant. However, if the detector dimensions were to be reduced to $10 \times 10\ \mu\text{m}$, sweep-out would then set a limit of $300\ \text{V}\,\text{m}^{-1}$, compared with the value of $3 \times 10^3\ \text{V}\,\text{m}^{-1}$ set by dissipation (the dissipation limit to F_x is independent of device area, the relationship being $F_x < (RP_D\text{w}/\text{l})^{1/2}$, where P_D is the dissipation per unit area of the detector).

response time τ_{PC} is no longer equal to τ_n as we assumed, for instance, in Box 9.3. While any attempt to pursue this further would take us far too deeply into mathematical obscurities, it is worth just noting the complexities involved in a proper understanding of device behaviour.

All three types of detector, PC, PEM, and PD were developed during the late 1950s and the first and last have since been successfully employed in a wide range of applications. While PEM detectors demonstrated performances comparable to PC devices, they were never seriously commercialized so we shall say no more about them here. As an example of detector fabrication technology, consider the processing of an InSb photoconductor. A high purity sample was first sliced, then polished, and etched to yield surfaces with low recombination velocity, the slice was then stuck down onto a quartz substrate and diced (i.e. sawed into rectangles), followed by further etching to obtain an optimum thickness (typically 5 μm), then contacted by indium soldering of platinum wires at opposite ends of each device. Alternatively, linear arrays of detectors were made using photolithography to define small rectangles with one common electrical contact and separate individual contacts for each element. Performance figures for PC devices are presented in Figure 9.12, where it can be seen that detectivities of devices cooled to 77K approach very closely 290K (2π) BLIP performance, with maximum D^* values of $5 \times 10^8\,\mathrm{m(Hz)^{1/2}\,W^{-1}}$. InSb diode detectors were made by diffusing Zn into an n-type slice, dicing and contacting in the usual way. Detectivities as high as $5 \times 10^9\,\mathrm{m(Hz)^{1/2}\,W^{-1}}$ were obtained by restricting the field of view to an angle of 15°.

So much for intrinsic detectors—the development of extrinsic devices went on in parallel, based largely on doped Ge crystals. As already intimated, the various intrinsic detectors available in the postwar period were well able to cover the 3–5 μm wavelength atmospheric window but

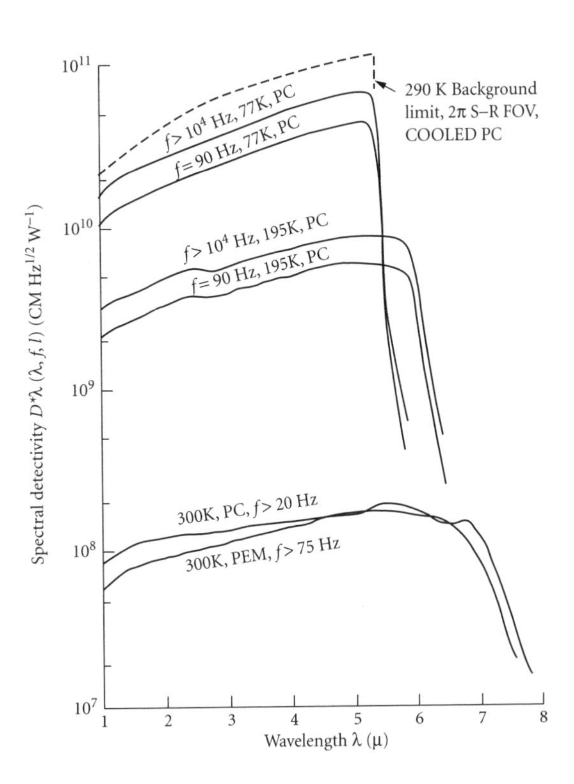

Figure 9.12. Experimental values of normalized detectivity D^* for practical InSb photoconductive detectors at three temperatures. Note the reduction of cut-off wavelength as the temperature is lowered. The performance at 77K is close to the room temperature background limit for a 2π-radian field of view. (From P. W. Kruse (1970) '*Semiconductors and Semimetals*', (eds. R. K. Willardson and A. C. Beer, Vol. 5, p. 15, Academic Press, Figure 45). Reprinted with permission from Elsevier.

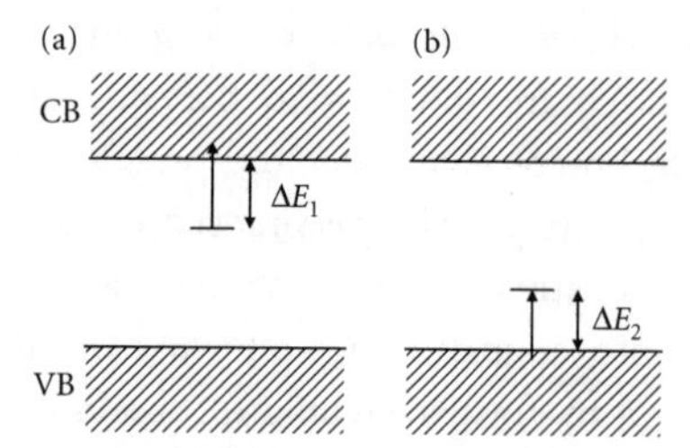

Figure 9.13. Band diagrams to illustrate the optical transitions involved in extrinsic photoconductors. In (a) an electron is excited from an impurity level just below the conduction band into the band, while in (b) an electron is excited from the valence band into a suitable impurity level. The absorption coefficients depend on the density of impurities which can be achieved and are generally much smaller than those associated with intrinsic (band-to-band) transitions.

not that in the 8–14 μm band, while this latter was of particular interest for detectors designed to register objects with temperatures close to room temperature (see Figure 9.1). In the absence of other more suitable materials, attempts were made (largely in the United States, during the period 1955–62) to service this important band using extrinsic detectors, in which the desired optical absorption corresponded to transferring an electron from a shallow impurity level into the conduction band or from the valence band into an appropriate impurity level, as shown in Figure 9.13. Photoconduction occurred by free electrons or free holes, respectively. Cut-off wavelengths corresponded to the energies ΔE_1 and ΔE_2 so, by choice of suitable impurities, it was possible to adjust the response curve to match the desired range. An essential requirement was that the impurity level in 9.13(a) actually contained an electron (or in (b) that the level was empty) and this implied an operating temperature low enough to satisfy the condition $\Delta E \gg kT$. For example, if a wavelength of $\lambda = 10$ μm were involved, this required a temperature $T < 0.1(hc/k\lambda) = 140$K. However, an even more stringent requirement was set by the need to minimize thermal noise so, in practice, considerably lower temperatures were needed. Figure 9.14 shows a number of response curves measured for the impurities Au, Hg, Cu, and Zn, in Ge, together with appropriate operating temperatures. Wavelengths out to 100 μm could be accessed using shallow donor levels (due to Sb) or shallow acceptor levels (due to B), while even longer wavelengths could be reached using impurity levels in InSb.

Though such materials were valuable in extending the available wavelength range, they suffered from poor optical absorption, the absorption coefficient being proportional to the density of impurity atoms which could be incorporated and this was frequently limited to rather modest values (e.g. 10^{22}–10^{23} m^{-3}). To improve sensitivity, it was therefore necessary to site the detector within an optical integrating sphere which ensured many passes of the radiation through the detector.

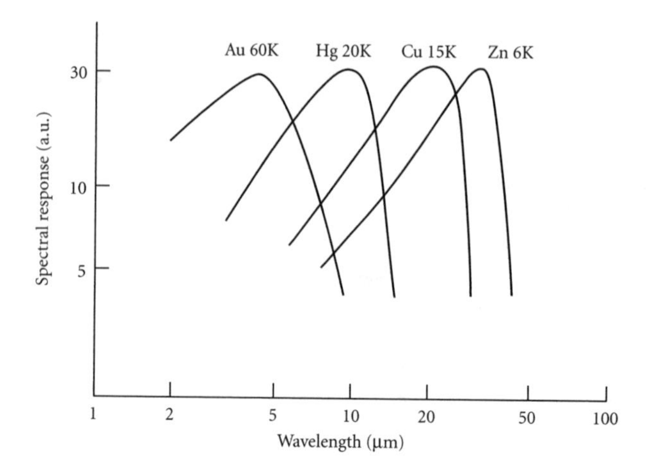

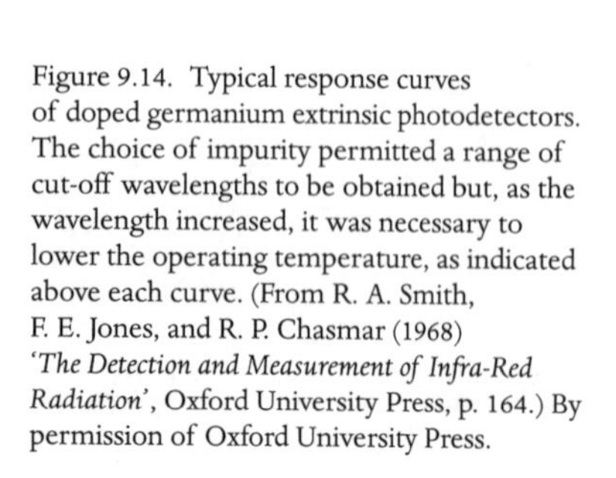

Figure 9.14. Typical response curves of doped germanium extrinsic photodetectors. The choice of impurity permitted a range of cut-off wavelengths to be obtained but, as the wavelength increased, it was necessary to lower the operating temperature, as indicated above each curve. (From R. A. Smith, F. E. Jones, and R. P. Chasmar (1968) *'The Detection and Measurement of Infra-Red Radiation'*, Oxford University Press, p. 164.) By permission of Oxford University Press.

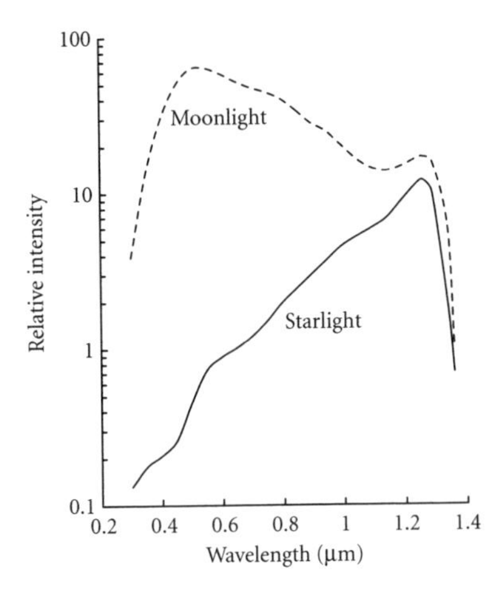

Figure 9.15. Night sky radiation intensity for a moonlit and a starlit night. As can be seen, effective night vision on starlit nights requires the utilisation of the spectral region between approximately 0.8 μm and 1.4 μm. (From Richards, E. A. (1969), *Advances in Electronics and Electron Physics* Vol. 28B, p. 661, Figure 2). Reprinted with permission from Elsevier.

This, together with the low operating temperatures, meant that such detectors were not widely used outside the laboratory. At the end of the 1950s, then, there was a clear need for new intrinsic materials to meet the demand for detectors in the 8–14 μm band but, before returning to this important problem, we must give attention to the shorter wavelength region between 1 and 2 μm, which was of interest on account of the spectral distribution of radiation from the night sky. The plot of radiation intensity in Figure 9.15 makes it clear that, in the absence of moonlight, effective night vision (based on reflected radiation) depends on our ability to utilize the 'starlight' band of wavelengths between about 0.8 and 1.4 μm which peaks at about 1.25 μm. In practice, this region has been exploited by vacuum tubes, which takes us from the realms of photoconductors to that of the photocathode.

As we remarked earlier, any infrared system can be expected to include a detector, followed by a signal amplifier. The image tube or photomultiplier is no exception to this rule, though both detector and amplifier function somewhat differently from those considered so far. Absorption of light by the photocathode results in electrons being ejected into vacuum where they may be accelerated to an anode held at a significantly large positive potential. The energy gained through this potential difference constitutes power gain and, in the simplest type of tube, which consists only of a vacuum diode, this energy may be used to stimulate visible light emission from a phosphor screen. It is apparent that such an arrangement may be used as an image convertor, an infrared image focussed on the cathode, being transferred to the anode in the form of a visible image. Alternatively, the electron density may be very considerably increased by the use of a set of secondary-emitting dynodes—such is the basis of the photomultiplier. Our interest here is in the properties of the photocathode which, in many cases, takes the form of a semiconducting thin film. An excellent account of the early development of photocathode materials is given in Sommer (1968) while a brief discussion of the appropriate tubes can be found in Chapter 4 of Smith et al. (1968).

The first demonstration of photoemission can be attributed to Hertz in 1887. In studying the radio waves generated by passing a spark between a pair of metal electrodes (see Chapter 2), he observed that the spark would pass across a larger gap if the negative electrode were illuminated with ultraviolet light. This was soon followed by the discovery that alkali metals showed a photoelectric effect when irradiated with visible light. Any hope of understanding these phenomena had to await the discovery of the electron by J. J. Thomson in 1897, the postulation of radiation quanta by Planck in 1901, and Einstein's quantum theory of photoemission in 1905. On this basis, electrons were emitted into vacuum with energy E given by:

$$E = h\nu - E_{\mathrm{th}}, \tag{9.3}$$

where $h\nu$ was the photon energy of the radiation and E_{th} was a threshold energy which was a property of the material in question. Photoemission was only possible when $h\nu$ exceeded this threshold.

In somewhat the same fashion as the early studies of radio waves, initial studies of photoemission concentrated on understanding the process, rather than on seeking useful applications but this was partly due to the very low quantum efficiencies of the materials under investigation. However, this situation changed dramatically in 1929 when Koller at RCA discovered the Ag—O—Cs photocathode (later designated as S1) which had a quantum efficiency approaching 10^{-2} in the visible and UV parts of the spectrum and, of even greater interest from our present point of view, useful sensitivity at wavelengths out to about 1.1 μm. Clearly, it should be possible to use infrared sensitive cathodes in image convertor tubes as outlined above and, indeed, the S1 cathode formed the basis of image converter tubes developed during the Second World War both in the United States and in England. (Because of their modest IR sensitivity, they were used in conjunction with searchlights or headlights, suitably filtered to remove the visible radiation—they were not yet sensitive enough to utilize the night sky.) Nevertheless, the main thrust of cathode development lay in the direction of photomultipliers and image tubes for visible light, the TV camera tube being a case in point.

One of the more significant results of the Ag—O—Cs discovery was its effect in stimulating research to find other materials with even higher quantum efficiencies, though, because the photoemission process was far from being well understood, the search was largely empirical and characterized by numerous blind alleys. Nevertheless, it did produce several successful materials, including a number of alkali metal antimonides, for example, Cs_3Sb (1936), $Na_2KSb(Cs)$ (1955), and K_2CsSb (1963). The Cs_3Sb cathode showed excellent performance in the visible spectral region, having a quantum efficiency of about 10^{-1} but showed no significant response in the near-infrared, so it was of no interest for night vision tubes. However, in 1955, Sommer (also at RCA), while studying an Li_3Sb cathode, accidentally discovered the 'multi-alkali' cathode. His source of Li was contaminated by both Na and K which have much higher vapour pressures than Li, so, when he attempted to evaporate Li, he succeeded only in evaporating the impurities, making a material which approximated to the composition Na_2KSb. Treatment of this with Cs resulted in a cathode with high visible response ($\eta > 10^{-1}$) and a long wavelength response which extended as far as 0.9 μm. This came to be known as the S20 cathode which was widely adopted for infrared image tubes, sensitive enough for application to passive viewing. Because of its military importance, it was studied in many laboratories in attempts to extend the IR response even further, eventually leading to the S25 version with a cut-off at about 1 μm. There was little understanding of the mechanism of photoemission so all this work was done empirically and many of the 'recipes' developed were obscured by either military or commercial classification (or both!). Broadly speaking, cathodes were formed by depositing a film of antimony to a controlled thickness then treating it with potassium to form K_3Sb, followed by sodium in excess of stoichiometric proportion. At this point photoemission quantum efficiency was low and had to be recovered by careful addition of alternate layers of potassium and antimony. Finally, small quantities of caesium and antimony were added until maximum photoemission was obtained. Needless to say, there were many minor variations on this basic theme, aimed at squeezing a little more sensitivity or obtaining a wider spectral response, though no proper understanding of these process steps was (or is!) available. A variety of IR image tubes based on the S20/25 formula has been available since the 1970s.

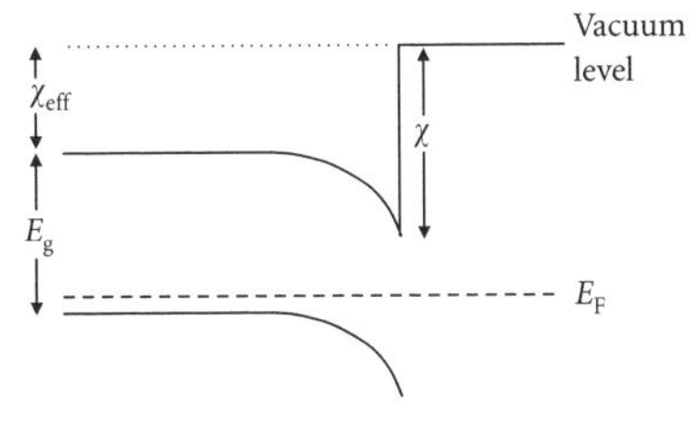

Figure 9.16. Band diagram to illustrate the photoemission process for a *p*-type semiconductor. The effect of surface band-bending is to reduce the effective electron affinity from χ to χ_{eff}. The minimum photon energy for electron emission is $E_{th} = E_g + \chi_{eff}$.

Though detailed understanding of the emission process was lacking, it was soon well established that these antimonide materials were semiconductors and a tentative model developed for their band structure. Figure 9.16 provides an approximate representation of the situation in an S20 type cathode with $E_g = 1.0$ eV and $\chi = 0.5$ eV. Band-bending near the semiconductor surface has the effect of reducing the effective electron affinity so that the emission threshold E_{th} has a value close to 1.2 eV. It is clear from the figure that:

$$E_{th} = E_g + \chi_{eff} = E_g + \chi - eV_B \tag{9.4}$$

where eV_B is the band-bending at the semiconductor surface. Note that a *p*-type semiconductor has the advantage of a very high density of electrons in the valence band which enhances the probability for photon absorption.

This, then, was the situation roughly 15 years after the end of the Second World War. Much progress had been made but there were obvious gaps in the technology such as the lack of convenient photoconductors working in the 8–14 μm window and a lack of uniform detector arrays for imaging applications. The crude mechanical scanning techniques in use were little more than a stopgap solution—what was clearly needed were detector arrays combined with some form of electronic scanning. Another desirable goal was that of obtaining adequate detector performance at higher temperatures—in many applications the use of liquid nitrogen cooling was far from convenient. As far as the near-infrared region was concerned, image tubes were certainly available but their performance was only just adequate—some improvement in photocathode technology would still be very welcome.

9.4 Growing sophistication—the 1960s and 1970s

As we have made clear in several earlier chapters, the key to success in semiconductor-based activities is the correct choice of material and the mastering of its technology—the exploitation of the infrared spectral region was no exception. There was an obvious need for new detector materials to cover the 8–14 μm window, which demanded energy gaps of approximately 0.1 eV, and within the space of a few years two suitable material systems had not only been proposed but also demonstrated in single crystal form. First in 1959 came the II-VI ternary alloy $Hg_{1-x}Cd_xTe$, or MCT (often referred to in Europe as CMT!), followed a few years later by the IV-VI ternary $Pb_{1-x}Sn_xTe$ (not, so far as I am aware, ever referred to as PST!). MCT was an English invention due to W. D. Lawson and co-workers at RRE Malvern, while (PbSn)Te was American, being the child of Ivars Melngailis and co-workers at MIT Lincoln Labs. There are many similarities between the two materials, in particular, the possibility of obtaining any desired band gap in the infrared region by appropriate choice of 'x' but, for pedagogical convenience, we shall deal with them separately.

The Malvern group were able to demonstrate the important features of the HgTe–CdTe system, namely that there was a continuous distribution of compounds all the way from HgTe, which is a semimetal with an effective energy gap (at 77K) of $E_g = -0.3$ eV, through to CdTe, a semiconductor with $E_g = 1.6$ eV. They also found an approximately linear variation of band gap as a function of the parameter 'x' which implied that a value of $x = 0.2$ would result in the desired gap of 0.1 eV for 8–14 μm operation. What was more, all these alloys were characterized by direct energy gaps, as required for efficient optical absorption, showed minority carrier lifetimes of order 0.1–1.0 μs as appropriate to the demands of imaging systems and possessed extremely small electron effective masses as needed for minimum G–R noise generation. All that had to be done was to make uniform, thin samples of the correct alloy composition, connect wires at either end, and cool the resulting photoconductor to 77K—what could be simpler?

Perhaps the first reservation should be concerned with the nature of the three constituent elements. Never before had anyone proposed an important semiconductor material composed of such a malicious cocktail of poisons! It was problematic which of the three constituted the greatest danger! Added to which was the physical fact that the vapour pressure of mercury at the melting point of the compound was sufficient to cause serious explosion hazard (a hazard which was translated into violent eruption in more than one research laboratory en route to the answering of infrared imaging prayers). And, needless to say, all three constituents were to be prepared with impurity levels well below the 6–9 s limit, before being handed over to the crystal grower to produce large, uniform, strain-free and defect-free boules of material with precisely controlled composition, and residual free carrier concentration of 10^{20} m^{-3} or less (the intrinsic free carrier concentration in $Hg_{0.8}Cd_{0.2}Te$ at 77K is approximately 3×10^{19} m^{-3}). It is so easy for the systems engineer and even the device physicist to overlook these material challenges that the articles by Hirsch et al. (1981) and Micklethwaite (1981) in Vol. 18 of *'Semiconductors and Semimetals'* should probably be required reading before anyone be allowed to set hand to MCT in whichsoever manner.

Mercury was easy enough to purify by vacuum distillation, provided that due care was taken over the choice of container material and that reactive metal impurities were first removed by oxidation and precipitation. However, noble metals could only be removed by a final, careful distillation process which proceeded more slowly than that used as standard for non-semiconductor applications. Cadmium could also be purified by vacuum distillation to achieve 6–9s quality but experience has shown that the desired low carrier concentration in the finished semiconductor requires further purification by double zone refining under a hydrogen atmosphere. Tellurium was first purified by a number of chemical treatments, then zone refined up to four times, again in a hydrogen atmosphere. The three constituents must then be combined to form the required alloy material, care being required to minimise oxygen incorporation—oxygen being a donor in MCT. Then began the task of growing suitable single crystal material.

Acceptable crystals were first produced by one or other of several versions of melt-growth. Important features were found to be: careful homogenization of the melt by annealing for up to 24 h at a temperature of about 850°C, while gently rocking the crucible; very slow freezing of the melt to produce a still homogeneous solid; or, alternatively, very rapid freezing followed by a further high temperature anneal to improve crystal quality. The development of suitable bulk crystals was thus achieved at a number of American laboratories (in particular, Honeywell) and at the Southampton factory of the (then) Mullard Radio Valve Company (picking up from the initial work at RRE Malvern). Epitaxial growth was pioneered in France at the CNRS, Bellevue laboratory by M. Rodot and co-workers, during the second half of the 1960s. Their technique was that known as 'close-spaced epitaxy' in which a CdTe substrate and a polycrystalline wafer of (HgCd)Te were placed side by side in a vacuum furnace. At typical temperatures of 550°C, MCT evaporated and was transported across the 'space', to be deposited as a single crystal film on the substrate. An important consideration here was the fact that the lattice-mismatch between $Hg_{0.8}Cd_{0.2}Te$ and CdTe is less than 3×10^{-3} (0.3%), allowing reasonably defect-free films to be grown on what was effectively a transparent substrate, while interface quality was adequate to maintain the desired low interface recombination rate for optimum photoconductive detector performance. It was also convenient that CdTe wafers were available in semi-insulating form so as not to short-circuit the electrical conductance of the film.

By the end of the 1960s, then, the growth of both bulk and epitaxial single crystals of MCT with a range of compositions was well proven and most of the basic electronic properties of the material had been measured but, as ever, the demands from the applications end of the business had not stood still. It was clear that larger and larger slices would be required as the development of detector arrays for advanced imaging systems proceeded apace and, what was more, such slices would have to be highly uniform in composition and in minority carrier lifetime. This because each element in the array should provide the same performance in terms of long wavelength cut-off and detectivity if image quality was to be maintained, as the number of elements increased. As an example, we might consider a two-dimensional array of 500×500 elements, each detector being 50 μm square—an overall wafer size of 25×25 mm^2 which was well beyond the range of boule diameters available in bulk-grown crystals of MCT. What was more, the uniformity in composition over an area of 10×10 mm^2 from a typical Bridgman-grown slice was little better than 5%, resulting in a comparable spread in cut-off wavelength. In addition, minority carrier lifetimes were strongly dependent on the presence of defects in the crystal structure which were extremely difficult to control in melt-grown samples. The availability of good quality CdTe crystals of up to 30-mm diameter was therefore a powerful stimulus to the development of improved epitaxial techniques and this rather dominated the material scene during the second half of our period.

A serious difficulty with epitaxial films grown by the close-spaced method lay in their being non-uniform along the growth direction, a result of Hg diffusion from the film into the substrate during the extended growth period (a typical growth run might last up to 100 h!). Some more practical method was clearly necessary and effort was therefore devoted to vapour phase epitaxy (VPE), from about the beginning of the 1970s, then, from 1975, to liquid phase epitaxy (LPE) as well. Various modifications of the VPE process were proposed but most were based on some form of hydrogen transport using separate sources for each constituent element. A major problem was concerned with making sure the desired chemical reactions occurred only on the substrate and not in the gas phase. To this end, the substrate was held at a significantly lower temperature than the various gases involved, which required thorough cooling of the substrate holder. Typical substrate temperatures were in the range 350–550°C which were still high enough to encourage unwanted interdiffusion of Hg and Cd at the film–substrate interface. As an example, epilayers might be reasonably uniform over the top few microns but with a 5-μm thick interface region of graded composition.

LPE was performed either from mercury or from tellurium solution at melt temperatures in the region of 500°C but, once again, there were serious problems with compositional uniformity through the layers and further difficulty in achieving the low free carrier densities required for photoconductive devices. Indeed, in spite of all these efforts, it appears that up to 1980 nearly all device work was still based on the use of bulk material—there was clearly a need for epitaxial deposition at considerably lower substrate temperatures.

The state of development of both photoconductive and photovoltaic MCT detectors had reached quite an advanced stage by the end of the 1970s, as described in considerable detail in the articles by Broudy and Mazurczyck (1981), Reine et al. (1981) (both sets of authors being from the heavily involved Honeywell laboratories) and by Elliott (1981) of RRE Malvern. In both cases, there was generally very good agreement between device performance and theoretical prediction based on measured material parameters, provided attention was paid to reducing surface recombination velocities. This could be achieved by suitable etching with Br-methanol solution followed by the deposition of a film of ZnS or by the rather subtle method of introducing an accumulation layer near the surface by deliberate doping. For example, an n^+ layer on an n-type sample results in band-bending near the surface which repels minority holes back into the bulk of the sample. We have already said quite a lot about photoconductors but perhaps the subject of diodes requires some further comment.

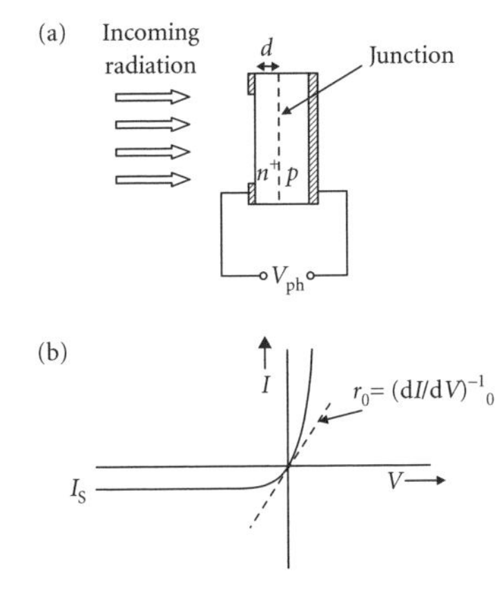

Figure 9.17. Structure (a) of a typical n^+-p infrared photodiode, together with an appropriate current–voltage characteristic (b) where I_S is the reverse saturation current. The diode might be made by diffusing excess Hg into a p-type MCT substrate to produce an n^+ region which is sufficiently heavily doped to allow most of the radiation through to the junction region where it is absorbed. The separation of electrons and holes by the zero-bias junction field generates a photocurrent I_{ph} which sets up an open circuit voltage $V_{ph} = r_0 I_{ph}$.

Various methods are available for making p-n junction diodes in MCT. Most of the early work made use of the fact that conductivity type could be controlled by stoichiometry—Hg deficiency results in p-type conduction, while excess Hg allows conversion to n-type. A typical diode would therefore be made by diffusing Hg into a p-type substrate to convert the surface region to n^+. Later work (second half of the 1970s) used ion implantation to achieve type conversion, success being achieved both with Hg and with the donors B and Al. A typical device structure is shown in Figure 9.17. Radiation is absorbed largely in the n-region to the left of the junction and minority carriers must diffuse to the junction where the high junction field (even at zero applied bias) separates holes and electrons, generating an open circuit photovoltage V_{ph} proportional to the intensity of the incoming radiation. High quantum efficiency requires that the thickness d of the n-region be matched to the absorption length for the radiation (say 2–3 μm) and that the carrier diffusion length be at least as large. Also, because minority carriers are generated near the diode front surface, it is important to reduce the surface recombination velocity by suitable surface treatment (see above). In Box 9.4, we show that a useful figure of merit for a diode detector is the product r_0A of the diode slope impedance ($r_0 = dV/dI$) and the diode area A. This parameter is independent of area and

Box 9.4. Simple theory for photodiode behaviour

The theory of photodiode behaviour (see Figure 9.17) has something in common with that for the photoconductor discussed in Box 9.3. It depends on the absorption of a stream of N_{ph} photons per unit area per second, which effectively generates a small photocurrent $I_{ph} = \eta e N_{ph} A$, where η is the quantum efficiency and A is the diode area. In practice, the diode may be operated either in reverse bias or under zero bias but we shall consider only the latter, for which case the open circuit photovoltage is given by:

$$V_{ph} = (dV/dI)_0 I_{ph} = r_0 I_{ph}, \tag{B9.15}$$

where r_0 is the slope impedance of the diode I–V characteristic at the origin (i.e. at zero bias—see Figure 9.17(b)). Assuming the diode current is dominated by diffusion (rather than G–R in the depletion region or tunnelling), we can write the I–V relation as:

$$I = I_S\{\exp(eV/kT) - 1\}, \tag{B9.16}$$

and obtain:

$$r_0 = kT/eI_S. \tag{B9.17}$$

Hence, the responsivity R_λ (in volts per watt) is given by:

$$R_\lambda = V_{ph}/N_{ph}Ah\nu = r_0\eta e/h\nu = \eta kT/h\nu I_S. \tag{B9.18}$$

To obtain an expression for the detectivity, we need to know the nature of the noise which limits the detector sensitivity. For an ideal diode, the noise current is given by:

$$\langle I_N{}^2\rangle = 2eI_S\{\exp(eV/kT) + 1\}B, \tag{B9.19}$$

which, for zero bias, becomes:

$$\langle I_N{}^2\rangle = 4eI_SB = 4kTB/r_0 \tag{B9.20}$$

Note that this is equivalent to Johnson noise in a resistor of magnitude r_0. Using equation (B9.13) in Box 9.3, we can write the detectivity D as:

$$D = R_\lambda/V_N = (r_0\eta e/h\nu)/r_0(4kT/r_0)^{1/2} = (\eta e/h\nu)(r_0/4kT)^{1/2} \tag{B9.21}$$

and, finally, the normalized detectivity, $D^\star$ is given by:

$$D^\star = (\eta e/h\nu)(r_0A/4kT)^{1/2}. \tag{B9.22}$$

Note that we have put $B = 1$ Hz in equations (B9.21) and (B9.22).

Box 9.4. Continued

This result demonstrates that the product r_0A is a useful figure of merit for a photovoltaic detector—in the interest of high sensitivity, it should obviously be made as large as possible. But note that this cannot be achieved by making A as large as possible—in fact, the product r_0A is independent of A, because r_0 varies as A^{-1}. From (B9.17) we see that:

$$r_0A = (kT/e)(A/I_S) = kT/eJ_S, \tag{B9.23}$$

where $J_S = I_S/A$ is the diode saturation current density. Standard diffusion theory for J_S allows us to write this in terms of material parameters. For an n^+-p junction, the result is:

$$r_0A = (N_A/en_i^2)(\tau_n kT/e\mu_n)^{1/2} \tag{B9.24}$$

where N_A is the acceptor doping on the p-side of the junction and τ_n and μ_n are the recombination lifetime and mobility of electrons in the p-type material (note that, as for the corresponding photoconductor, a long lifetime is required for optimum sensitivity). If we insert the following parameter values for MCT at 77K: $N_A = 10^{22}\ \text{m}^{-3}$, $n_i = 3 \times 10^{19}\ \text{m}^{-3}$, $\tau_n = 10^{-7}$ s, and $\mu_n = 10\ \text{m}^2\text{V}^{-1}\text{s}^{-1}$, we obtain $r_0A = 5 \times 10^{-4}\ \Omega\,\text{m}^2$ (i.e. $5\ \Omega\,\text{cm}^2$, to use the units frequently found in the literature). Inserting this value in equation (B9.22), together with $\nu = 3 \times 10^{13}$ Hz ($\lambda = 10\ \mu$m) we find $D^* = 2.8 \times 10^9\ \text{m(Hz)}^{1/2}\text{W}^{-1}$, which is very much better than BLIP performance for a 180° field of view. Note that $D^* \propto T^{-1/2}$ which implies reduced performance with increasing temperature.

depends on the nature of the current flow through the diode—pure diffusion gives optimum performance, while G–R within the depletion region, quantum mechanical tunnelling, or surface leakage lead to progressively degraded performance. Again, various noise mechanisms play a role—in Box 9.4 we examine the case of a diffusion limited diode to show that D^* at $\lambda = 10\ \mu$m for a diode operating at 77K can easily exceed that appropriate to BLIP performance.

In addition to the important application to thermal imaging, MCT diodes have also been used in conjunction with CO_2 gas lasers, emitting at a wavelength of 10.6 μm. In particular, as we saw in Chapter 8, diodes offer much faster response times than photoconductors and MCT diodes were operated at modulation frequencies in excess of 1 GHz in a heterodyne detection system by the French group as early as 1972. It is worth noting an interesting design feature of such diodes when applied to the detection of a single narrow band of wavelengths. The high n-type doping near the surface results in a layer which is fairly transparent to the incoming radiation (this depends on the so-called Moss–Burnstein shift of the absorption edge which occurs when the Fermi level is pushed into the conduction band by heavy doping) so that absorption occurs in the depletion region and in the adjacent p-region. Photocarriers are therefore generated within the depletion region or close to it so that they can readily diffuse to it, thus making for a high speed response. However, accurate control of composition is essential if such a mechanism is to function reliably.

Though the principal application of MCT detectors has been to thermal imaging in the 8–14 μm window, they have been used over a wide range of wavelengths and have gradually come to dominate the whole field of mid-infrared detection—though not without a fight! Lead–tin telluride came on the scene in the mid-1960s (see Melngailis and Harman 1970) and promised to mount a serious challenge. It too, offered a range of direct band gaps in the range $E_g = 0$–0.3 eV which covers wavelengths of interest for thermal imaging. It too, was readily available in single crystal form, being grown by Bridgman, Czochralski, or vapour phase processes and measured minority carrier lifetimes were typically in the range 10^{-7}–10^{-8} s. Though the band structure differed in detail from that of MCT, in so far as both PbTe and SnTe were semiconductors (whereas HgTe is a semimetal), the alloy band gap appeared to go through zero for values of x (the mole fraction of Sn) between 0.3 and 0.4 (the precise value depending on temperature) as a result of a band structure inversion—that is, the conduction band in PbTe becoming the valence band in SnTe (and vice versa). As in MCT, there is an approximately linear variation of band gap with composition, the important composition range, for $T = 77$K, being $x = 0.05$ ($E_g = 0.2$ eV) to $x = 0.20$ ($E_g = 0.1$ eV). However, PbSnTe does suffer from one disadvantage compared with MCT—its dielectric constant is dramatically large. Whereas MCT is characterized by a typical semiconductor value of $\epsilon = 12\epsilon_0$ (for $x = 0.2$), the corresponding figure for PbSnTe is of order $400\epsilon_0$, which implies very large values of depletion layer capacitance in *p-n* junction diodes and results in RC limited, rather than transit time limited, response times. The GHz operation obtained from MCT diodes appears extremely unlikely in the case of PbSnTe devices.

The PbSnTe crystals grown by the vertical Bridgman process were typically about 7 cm in length and 2 cm in diameter. Lateral compositional uniformity was better than ± 0.005 mole fraction while the variation along the growth direction was perhaps 10% over a distance of 5 cm (which corresponds to about 0.001 mole fraction through a 0.5 mm slice). The principal problem lay in the background impurity density (or non-stoichiometry) which led to free carrier densities of order 10^{26} m^{-3}. Careful annealing under an argon atmosphere for several days at 650°C resulted in material with free carrier density of order 10^{23} m^{-3} or slightly less but this lengthy processing clearly represented a disadvantage for widespread application to commercially acceptable devices. Nevertheless, respectable devices were obtained during the 1960s—both photoconductive and photovoltaic detectors were made for the 8–14 μm window with $D^\star$ values of order 10^7 m(Hz)$^{1/2}$ W^{-1} at 77K. However, as so often seems to be the case, the advent of epitaxy improved these figures quite substantially. LPE was introduced in the mid-1970s and helped to improve detectivity to the order of 10^9 m(Hz)$^{1/2}$W^{-1}. A primitive form of Molecular beam epitaxy (MBE) was also applied to the growth of hetero-epitaxial diodes (employing a window of wider gap material PbTe on a PbSnTe substrate) by Hohnke and co-workers at the Ford Laboratory in Dearborn, Michigan (Ford's involvement provides yet another evidence of the increasingly wide range of applications for IR devices). They obtained detectivities as high as 5×10^8 m(Hz)$^{1/2}$ W^{-1} at 10 μm wavelength and 5×10^9 m(Hz)$^{1/2}$ W^{-1} at 6 μm. More recently, PbSnTe devices have been fabricated on silicon substrates with a view to establishing an all-solid-state imaging technology but, in spite of these worthy efforts, there can be no denying the ascendancy of MCT—for example, a review published in 1991 listed no fewer than 20 laboratories around the world which had invested in the MBE growth of MCT for infrared devices. And this is not to mention a probable further 20 which had chosen metal-organic vapour phase epitaxy (MOVPE) as their preferred technology. All of which clearly demonstrates both the worldwide importance of infrared devices and of MCT as the preferred material. But this is running well ahead of our present time period. We have yet another important development to record in the run up to 1980—the 'negative affinity' photocathode.

The search for photocathode materials having useful sensitivity in the near-infrared region appeared to have reached its essentially pragmatic limit with the development of the S20 cathode, when, in 1965, Scheer and van Laar at the Philips Nat Lab in Eindhoven reported their observation of a qualitatively new phenomenon, the negative affinity semiconductor surface. In fact, they were probably more interested in understanding semiconductor surface physics but there was no doubting the practical possibilities of their discovery. The first point of interest was its reference to a single crystal of GaAs, rather than to an ill-defined mixture of polycrystalline alkali antimonides. Perhaps this might open the way to a more scientific approach to photocathode research? And, to a significant, though by no means total extent, so it did. Their discovery can be understood in terms of the band diagram shown in Figure 9.18 which represents a *p*-type GaAs sample whose vacuum-cleaved surface has been treated with evaporated caesium to reduce its electron affinity. The exciting feature here is that this effect, together with an appropriate amount of surface band-bending, was sufficient to reduce the effective electron affinity χ_{eff} to a value close to zero! Electrons were free to leave the GaAs surface without let or hindrance! (provided only that they could cross the band-bending region without loss of energy). An important consequence was the large increase in escape depth for emitted electrons, compared with available cathode materials. Until this moment, photogenerated electrons could escape only from a depth determined by the hot electron mean free path which was of order 10 nm. With the advent of zero, or negative electron affinity, the escape depth was determined by the (thermalized)

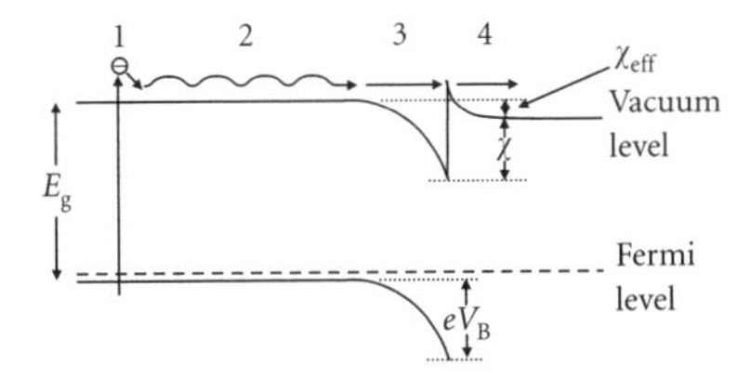

Figure 9.18. Band diagram of *p*-type GaAs which has been treated with Cs and O to produce a negative affinity surface (i.e. $\chi_{\text{eff}} < 0$). The process of photoemission involves four stages: (1) represents photoabsorption, (2) represents electron thermal diffusion, (3) represents electron drift through the band-bending region, and (4) represents escape from the surface.

electron diffusion length, of order 1 μm. Compared with Figure 9.16, this really represented a qualitative change, a veritable quantum leap into space (as, indeed, for electrons, it certainly was!).

Bearing in mind the history of electron emission from metal surfaces which had been considerably enhanced by co-treatment with both caesium and oxygen (rather than caesium alone), it was hardly surprising when the corresponding enhancement was also observed for the semiconductor GaAs. Nevertheless, it was, perhaps, ironic that the observation was made quite independently in the Philips sister laboratory in Redhill, England (at the time, still known as Mullard Research Laboratories). Andrew Turnbull and Geoff Evans were members of a team studying photocathodes for infrared image tubes under production in the Mullard Mitcham factory and were therefore in a favourable position to perform the necessary experiment. It worked famously, the results were published in 1968 and the negative electron affinity cathode was well and truly launched. Quantum efficiencies close to 100% were very much in prospect. However, the GaAs band gap of 1.43 eV implied a long wavelength cut-off of 870 nm which was far from ideal when set against the requirement for a night vision tube sensitive out to 1.2 or 1.3 μm—the question whether the negative affinity principle would be compatible with narrower gap materials had yet to be addressed. So too had the task of making a transmission cathode appropriate to a practical image tube. But there were, of course, plenty of people keen to try.

The story of the quest for practical infrared negative affinity cathodes has been well reviewed in the book by Bell (1973) and the review by his Varian colleague Escher (1981), reflecting the major contribution made by the team at Varian in Palo Alto (but see also the review by Rougeot and Baud of Thomson CSF in 1979). Needless to say, the account given here can do no more than pick out some of the highlights. Perhaps we might begin by looking a little more closely at what Scheer and van Laar actually did. Their interest in the atomic and molecular understanding of surfaces had led them to work under ultrahigh vacuum (UHV) conditions (similar to those used in MBE, for example) where the background pressure in their vacuum system was low enough (10^{-10} torr or better) that an atomically clean surface would not be accidentally contaminated, even at the monolayer level, during the course of an experiment. The experiment in question involved cleaving a bulk crystal of GaAs inside the vacuum system so as to produce a perfectly clean surface (GaAs cleaves along (1 1 0) planes so this was a (1 1 0) surface), then measuring photoemission as a function of sequential treatments with caesium atoms. After approximately one monolayer had been deposited the emission went through a maximum, corresponding to the effective electron affinity reaching a value very close to zero. Notice in Figure 9.18 that, when χ_{eff} becomes zero, the emission threshold coincides with the GaAs energy gap E_g, as also made clear by equation (9.4). Measuring this threshold, therefore, provides a method of estimating χ_{eff}. Two features of the experiment are worth emphasizing—first, it must be performed in UHV conditions to ensure surface cleanliness and second, there is an optimum thickness of Cs, any excess resulting in reduction of the emission intensity.

The next few years added a good deal of further information—it became clear, for instance, that emission from the GaAs–Cs surface was not very stable and showed significant variation across a single face and from sample to sample, a 'good' cleavage being an important requirement for high quantum yield. Indeed, the work already referred to by Turnbull and Evans showed that activation with caesium and oxygen achieved yields at least as high as those obtained with caesium alone but with considerably improved stability, though once again there was an optimum coverage. (Later work showed that more or less complex activation procedures might result in even greater yields, though with little associated understanding of their mechanisms.) Perhaps of greater significance, these same workers also studied the behaviour of samples whose surfaces had been contaminated by exposure to air. Not surprisingly, initial yields were seriously reduced compared with those from vacuum-cleaved surfaces but of particular interest was their observation that, using Cs + O, it was possible to improve yields to within about a factor 3 of vacuum-cleaved samples, following suitable in situ cleaning procedures. This was highly encouraging when examined in the context of possible future vacuum tubes—there was little likelihood of production teams enthusing at the prospect of achieving 'good' cleavage faces over diameters of 3 cm or more within the tube environment! What is more, it paved the way for the introduction of epitaxy into tube technology, giving greater flexibility in the choice of crystal orientation, much of the later work being based on the (1 0 0) surface, rather than the (1 1 0). As we shall see, it was also of fundamental importance in the development of transmission mode cathodes.

Figure 9.18 indicates that emission involves four distinct steps, which we shall examine individually. First is the absorption process in which an electron is excited from the valence band into the conduction band. The important parameter here is the absorption length α^{-1} which, for a direct gap material such as GaAs takes a value close to 1 μm—as we have said previously, something like 3 μm of material are required to absorb most of the incident radiation (actually 95%). The second step involves thermalization to the bottom of the conduction band and thermal diffusion of these minority electrons to the semiconductor surface (as we pointed out above, this step represents a qualitative difference, compared with the emission process in conventional cathode materials where only hot electrons can hope to escape). If most of them are to reach the edge of the depletion region before they recombine with majority holes in the bulk, the distance they travel should be less than a diffusion length L_n—in other words, there is a requirement for $L_n > 3\alpha^{-1}$. In practice, L_n depends on the *p*-type doping level but for $N_A = 3 \times 10^{24}$ m^{-3} L_n is typically about 3 μm which is only marginally adequate. In the third stage, electrons are accelerated through the depletion layer by the depletion electric field. To minimize recombination within the width of the depletion layer this should be significantly less than L_n, which is easy to achieve. If we assume $eV_B = E_g/3$, we can calculate the depletion width according to:

$$W = (2\varepsilon\varepsilon_0 E_g / 3e^2 N_A)^{1/2}, \tag{9.5}$$

where *e* is the relative dielectric constant of the semiconductor and N_A is the doping level. For GaAs doped at $N_A = 10^{25}$ m^{-3} we obtain $W = 7.5$ nm. Much more significant is the probability of energy loss by emission of optical phonons (having energy 35 meV in GaAs) for which the mean free path is 10 nm. This implies the need for the relatively heavy doping assumed above. However, the diffusion length L_n decreases with increasing doping level so the choice of doping must be a compromise. In practice, a value of $N_A = 3 \times 10^{24}$ m^{-3} is common, as suggested above. The final stage is electron emission through the surface layer of Cs—O, the probability of which is determined by the electron affinity, controlled in practice by the details of the activation process. We shall simply assume that this has been done in an optimal manner and leave it at that.

Once a qualitative understanding of how a negative affinity cathode would function, two important questions naturally arose: how might it be possible to make a transmission cathode suitable for an image tube? and how could the response be extended further into the infrared to make better use of the night sky radiation? Of course, both problems were pursued in parallel but we must obviously deal with them separately. The importance of developing a transmission cathode was appreciated almost immediately and attempts were made to deposit thin polycrystalline films of GaAs on quartz or glass substrates in much the same way as was done for the S20 cathode. Needless to say, results were disappointing. Reflection mode devices gave emission yields some 10 times less than had been measured on single crystal surfaces, while transmission yields were as much as 10 times down even on these. Doubtless, the main problem concerned the small diffusion lengths and high interface recombination velocities associated with poor quality material and there was little hope of improving them. An alternative approach using single crystals was imperative.

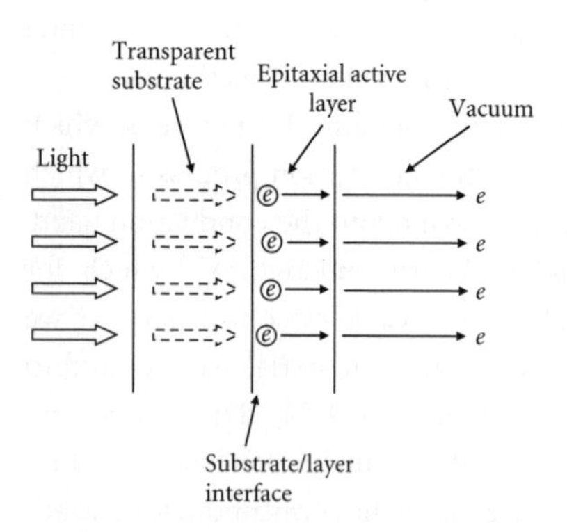

Figure 9.19. Schematic structure of a NEA transmission photocathode, showing the method of irradiation. It is important that the substrate/layer interface is of high quality, otherwise electrons generated in the active layer may recombine at the interface, rather than diffusing to the vacuum surface.

First attempts were based on the use of a GaP substrate which, with its wider band gap, was transparent to the near-infrared radiation of interest but once again results were disappointing. Let us look at the problems. Figure 9.19 shows the type of structure involved, an epitaxial GaAs layer grown on a transparent single crystal substrate, irradiated through the substrate. The first problem was that some 30% of the radiation was reflected at the substrate back surface (though this could be largely overcome with a suitable anti-reflection coating). Second, many of the electrons generated within the GaAs film recombined at the substrate/GaAs interface (which was characterized by a high interface recombination velocity), particularly when there was a significant lattice-mismatch, as is true for the GaP/GaAs interface. Third, because of this mismatch, the quality of the GaAs was poor, as reflected in low values of diffusion coefficient. Little wonder that the overall emission yield was unsatisfactory. The next step involved growing a graded interface layer of GaAsP to smooth the transition from substrate to active layer but still results were disappointing—as found in work on many other devices, such as lasers, LEDs, and bipolar transistors, good quality films could only be grown on closely lattice-matched substrates. GaAs epitaxial films grown on GaAs substrates gave excellent photoemission yields in reflection mode (once the art of in situ surface cleaning had been mastered) but these were, of course, no use for transmission cathodes. Might it be possible to use an AlGaAs substrate in some way? Bulk crystals were not available but it was possible to grow fairly thick layers by LPE. Attempts to grow AlGaAs on GaP substrates proved only moderately successful and it was not until 1975 that a technologically satisfactory solution was found by the Varian team, Antypas and Edgcombe reporting success with a GaAs/AlGaAs/glass structure, illuminated through the glass. To achieve this, they grew a structure GaAs(substrate)/AlGaAs/GaAs/AlGaAs, which was stuck down on a glass plate, followed by removal of the GaAs substrate, and the first AlGaAs layer by appropriate etching techniques, made possible by the existence of selective etches which would dissolve either GaAs or AlGaAs (but not both). This piece of technological magic eventually resulted in a practical photomultiplier tube with a spectral response which was almost flat from 1.4 to 3.0 eV, with nearly 20% quantum yield. But what of the need for better infrared response?

Needless to say, this required a change of active material from GaAs to a semiconductor with somewhat narrower band gap. For example, to achieve a response at a wavelength of 1.2 μm required a band gap of just over 1 eV, considerably less than that of GaAs. First thoughts were in the direction of InGaAs which has a gap of 1.0 eV at the composition $In_{0.4}Ga_{0.6}As$ but the familiar problem of lattice-mismatch meant that it could not be grown successfully on GaAs. The solution, as we discovered in Chapter 8, was to employ the quaternary InGaAsP, lattice-matched to InP, yet another feather in the Varian cap. The band gap of InP, being 1.34 eV, meant that a structure of $InP/In_{0.77}Ga_{0.23}As_{0.5}P_{0.5}$ could be expected to respond to wavelengths between 0.93 and 1.2 μm. The important question was: could such an alloy be activated to negative electron affinity? Figure 9.18 makes it clear that, as E_g is reduced, it must become more difficult to achieve negative values for χ_{eff}. The practical answer was that emission yield in the region of 1.2 μm was reduced to a value of order 10^{-3} electrons/ incident photon, compared with values of over 10% at 0.8 μm wavelength. In fact, there was a progressive reduction in yield as the band gap was narrowed but the improvement over GaAs for night vision tubes was certainly well worth having. Figure 9.20 gives an impression of the results available in 1980, including both transmission and reflection cathodes, compared with the best available S1 and S20 cathodes. Clearly, the effort invested in 'new' devices had been well worthwhile, both from the viewpoint of highly efficient visible sensitivity (typified by the GaAsP cathode in Figure 9.20) and in the desired extension into the infrared.

An indication of the increased sensitivity of the negative affinity devices is provided by the measured sensitivities in $\mu A\,lm^{-1}$ quoted for each type of cathode. It has been customary to measure the photocurrent produced by exposure to a standard tungsten lamp running at a temperature of 2856K and, though most of the energy from such a lamp is in the near-infrared, while the lumen is a unit related to the eye response, this method of specifying sensitivity has been widely adopted. The best result for an S20 cathode lies in the region of 500 $\mu A\,lm^{-1}$, whereas the corresponding figure for GaAs (in reflection) is 2100 $\mu A\,lm^{-1}$. For comparison, InGaAsP cathodes show sensitivities of up to 1600 $\mu A\,lm^{-1}$.

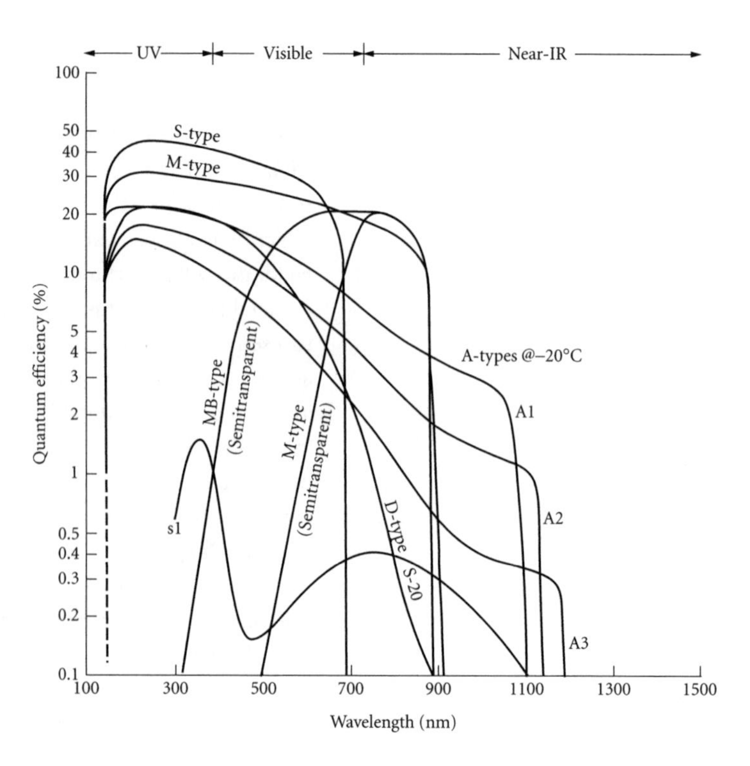

Figure 9.20. Summary of spectral response curves for a range of photocathodes. The curves marked 'A' are InGaAsP reflection mode devices, those marked 'S' are GaAsP, reflection mode and those marked 'M' are GaAs, reflection and transmission mode. Curves for S1 and S20 transmission cathodes are shown for comparison. (From Escher, J. S. (1981), *Semiconductors and Semimetals*, Vol. 15, p. 195 (eds. R. K. Willardson and A. C. Beer) Academic Press, Figure 42). Reprinted with permission from Elsevier.

The concept of negative electron affinity has found application in a number of other ways, such as secondary emitters (dynodes) and cold cathodes—negative affinity silicon surfaces have also been reported—but one area of interest, particularly relevant to the theme of interaction between pure and applied science has been the demonstration of spin polarized electron emission from a GaAs NEA cathode. The essential point here is that, in contrast to conventional photocathodes, the emitted electrons have thermal energies and it follows from the band structure of GaAs that three times as many electrons with spin $s = -1/2$ are emitted as with spin $s = +1/2$ and this high degree of polarization has been used in studies of electron scattering phenomena in atomic and molecular physics. It is quite impossible to predict just how any newly discovered phenomenon is likely to affect other branches of science—here we have yet another unexpected success.

9.5 Quantum wells, superlattices, and other modern wonders

A fair degree of sophistication in infrared detection and imaging had obviously been achieved by the end of the 1970s but such was the military importance of this area that efforts to improve performance still further were, if anything, increased. Even when war was not actually under way, it was usually under contemplation and what looks to us now as the remarkably relaxed attitude to armament characteristic of the 1920s and 1930s seems unlikely ever to be repeated. From this point of view, there were two principal issues, those of high temperature detector operation and of all-solid-state imaging, neither the use of liquid nitrogen cooling nor of mechanical scanning being regarded as ideal in systems to be employed in harsh military environments. From the scientific viewpoint, on the other hand, there was the usual drive to think new thoughts and introduce new concepts both of which have seen expression in the past two decades. Where military urgency controlled the purse strings, research funding for interesting new ideas was rarely hard to find. (This is rather well illustrated by the *Proceedings of the 1989 United States Workshop on the Physics and Chemistry of MCT and Related Compounds* (published in *J. Vac. Sci. Technol.*, Vol. A8) in which papers were presented by (at a quick estimate!) no less than 40 different and distinct research organizations, 13 industrial labs, 18 universities, and 9 Government labs, representing the United States, Canada, Australia, Japan, England, France, and Israel. Such was seen to be the worldwide importance of MCT at the beginning of the 1990s.)

Scientifically, the principal innovations during this period were concerned with the introduction of quantum wells and superlattices—remember from Chapter 6 that the semiconductor quantum well began

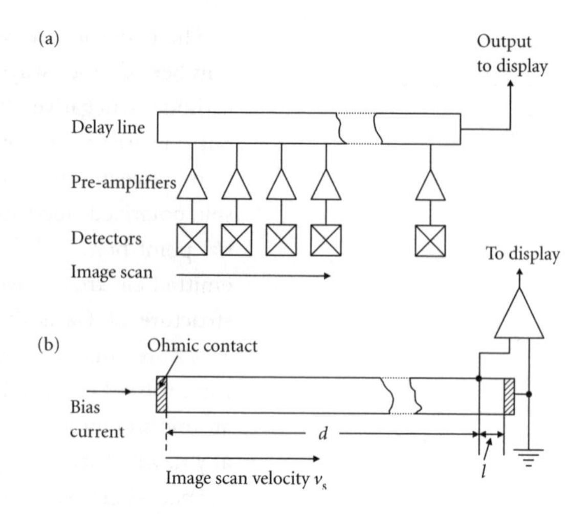

Figure 9.21. Illustration of how the SPRITE detector functions in the same way as a whole row of individual detectors. In (a) is shown a row of detector elements, each with its own pre-amplifier, being used to generate an image, while (b) shows the SPRITE which consists of a long single element with a read-out region at the right-hand end, requiring only a single amplifier. The image is scanned along the length of the detector at a velocity v_S which is equal to the velocity with which electrons travel along the semiconductor bar. In this way, each image pixel generates signals (i.e. increases in conductivity) which add together at the read-out.

to make its presence felt in the mid-1970s—and with the development of new epitaxial growth methods but we shall first look at a number of developments in the modus operandi of conventional detectors which also had a major influence on infrared systems. Perhaps the first of these was an ingenious idea from Tom Elliott of RSRE Malvern which he proposed in 1981. This was the so-called SPRITE (signal processing in the element) detector which, as illustrated in Figure 9.21, performed the same function as a whole row of individual elements. The detector consisted of a long bar (typically 1 mm in length) of *n*-type MCT with a short read-out region at its end. The image was scanned in a raster fashion so that each line travelled over the detector at a velocity v_S. By varying the applied DC bias along the bar so as to make the drift velocity v_d of photogenerated holes equal to v_S, each image pixel generated signals (in the form of regions of increased conductivity) which added together at the read-out. It was, of course, necessary that the hole transit time $t_t = d/v_d$ should be shorter than (or, at least, comparable with) the hole lifetime τ_h in the MCT material which, in the example quoted in Elliott's paper took the form of a Bridgman-grown bulk sample. For $d = 1$ mm and a bias field of $6800\,\mathrm{V\,m^{-1}}$, $v_S = 300\,\mathrm{m\ s^{-1}}$ ($\mu_h = 0.045\ \mathrm{m^2V^{-1}s^{-1}}$) and $t_t = 3\ \mu\mathrm{s}$. For a MCT alloy with $x = 0.2$ (cut-off wavelength $\lambda = 12\ \mu\mathrm{m}$) hole lifetimes of $\tau_h = 2\,\mu\mathrm{s}$ were readily obtained at the operating temperature of 80K and the measured detectivity $D^\star$ was a very satisfactory $2 \times 10^9\ \mathrm{m(Hz)^{1/2}W^{-1}}$. The advantages of the SPRITE detector were considerably easier manufacturing technology and less demand placed on material uniformity, together with the need for only a single amplifier. The name SPRITE is derived from the essential property of the device: signal processing in the element.

The second of the modifications to device structures which was designed to improve detector performance also resulted from innovation at RSRE Malvern during the latter half of the 1980s and depended on an even more ingenious idea, that of 'carrier exclusion'. In order to explain it, we need first to look in greater detail at why the use of cooling improved detectivity. We must remind ourselves that the objective behind the design of any quantum detector of thermal radiation was to obtain a sensitivity as close as possible to that represented by BLIP operation, which meant reducing internal detector noise below the level of background radiation noise. It was therefore essential that the important sources of detector noise should be understood. In fact, in the majority of cases the limits were set by G–R noise but it was necessary to be clear about mechanisms. G–R noise could arise from at least three processes: direct radiative transitions between valence and conduction bands, S–R (phonon) transitions via a deep impurity (or defect) level, or as a result of an Auger process—and the immediate question was which of these was likely to be of greatest significance. The answer lay between Auger and S–R and it is interesting to examine the parameters which control their relative importance. There were no less than 10 possible Auger generation processes which might contribute to G–R noise but one, in particular, could be expected to dominate—this was the so-called Auger 1 process in which a hot electron in the conduction band lost energy, while a valence band electron gained sufficient energy to excite it into the conduction band. Clearly, the probability of such a process depended on the density of free electrons in the conduction band, while detailed calculations showed that it also depended on the factor $\exp\{-E_g/kT\}$. It was this latter term which determined the importance of Auger processes in narrow gap materials—at room temperature E_g/kT had a value of about four in 8–14 μm MCT, whereas in GaAs (for example) it was about 55, giving a probability ratio of something like 10^{20} between the two materials! And, of course, it was this factor which led to the use of cooling to improve performance—for the MCT, the ratio between room temperature and 77K amounted to a factor of about 5×10^4. (Note that for 3–5 μm detectors, where E_g is twice as large, the exponential factor is much smaller and it was possible, in this case to use temperatures of about 180K, rather than 80K.) On the other hand, S–R generation obviously depended on the density of deep levels available in the material and could, in principal, be made negligible by reducing this. However, the corresponding reduction of Auger generation by reducing free electron concentration appeared fundamentally impossible—however pure the material might be, it was impossible to beat the intrinsic carrier density. The only hope, therefore, was to resort to cooling.

Or was it? Tom Elliott, Mike White, and co-workers thought they could see a way round this apparent impasse. Perhaps it might be possible to reduce the free carrier density below the limit set by the intrinsic level. The secret lay in recognizing that the concept of n_i was based on thermodynamics which applies only to systems in thermal equilibrium—if it were possible to run a detector under non-equilibrium conditions, this apparently fundamental limit might not be so fundamental, after all. How to achieve it? By making use of the properties of contacts. An appropriate contact could be used literally to 'suck' free carriers from a sample of detector material, thus reducing their density well below the intrinsic value. Consider, for example, the case of a very lightly n-doped sample of 12 μm MCT with $N_D = 10^{20}\ m^{-3}$, having an n^+ MCT contact on one end. The intrinsic free carrier density is about $3 \times 10^{22}\ m^{-3}$ at room temperature so such a sample is clearly very close to being purely intrinsic. On the other hand, the contact region will contain (typically) $n = 10^{24}\ m^{-3}$ electrons and $p_n = n_i^2/n = 10^{45}/10^{24} = 10^{21}\ m^{-3}$ holes. If, now, the contact is biased positively, electrons will flow easily from the intrinsic region into the contact, but very few holes will flow in the opposite direction because of the very low value of hole density in the contact. The immediate effect is to reduce the electron density in the device very considerably, while scarcely changing the hole density. In fact, the electron density will fall to a value close to that determined by the donor density $N_D = 10^{20}\ m^{-3}$ and, in order to maintain electrical neutrality within the device, as many holes must be lost as electrons so the hole density does, in fact, reduce very strongly, too. But, the important feature is that the free electron density is reduced by about 300 times, which means that the Auger 1 generation process is also reduced by a factor 300. The net effect is that such an MCT sample might be used to detect 12 μm radiation at an operating temperature close to 200K (obtainable with a piezoelectric cooler), rather than the 80K previously thought necessary. Similar arguments suggest that a 4-μm detector should work with close to BLIP performance at room temperature, which is even better! Experimental progress in pursuit of these ideas is described in Elliott and Gordon 1991—already a reduction of 30 times in free electron concentration has been observed, with a corresponding improvement in D^* of about a factor five.

While it would be inappropriate here to delve too deeply into the fine details of what is a rather complex subject, one or two further comments are probably in order. First, we should note that, in addition to the use of contact layers, as described above, heterojunctions have also been employed in order to create barriers to the flow of minority carriers, second, these ideas have been applied to improve performance of diodes as well as photoconductors and, third, because such structures must be grown epitaxially, InSb made a comeback in the 1990s on the premise that it was rather easier to grow high quality epitaxial layers of this material than of MCT. The drawback to InSb, of course, is its fixed band gap which is roughly, though not perfectly, aligned with the 3–5 μm window. Nor is it possible to find a lattice-matched wide gap material—this implies the use of strained layers which must be thin enough to avoid relaxation effects because the resulting dislocations would act as S–R recombination centres. Life is never straightforward for the

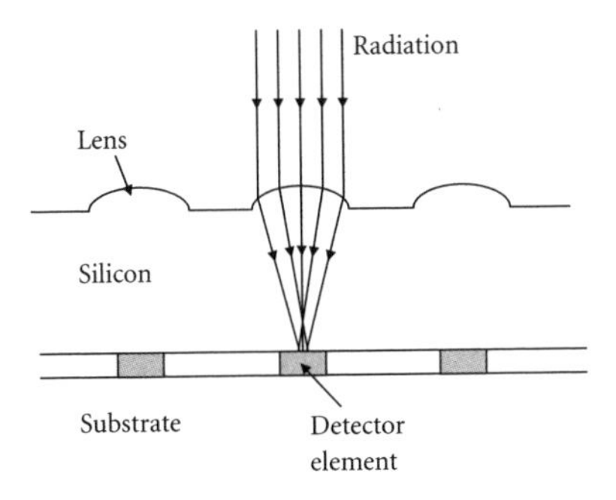

Figure 9.22. Arrangement of a microlens array designed to focus incoming radiation onto a corresponding array of small detector elements. The pitch of the array is typically 50 μm while the detector diameter is 10 μm. Lenses are formed by photolithography and etching.

device engineer! However (see Elliott and Gordon 1991), InSb/AlInSb diodes have been made which show room temperature values of the R_0A product at least 10^3 times greater than those measured on conventional diodes. What is more, such devices can be seen to have application in areas quite remote from that of infrared detection—cross fertilization operates in the semiconductor device field every bit as well as in botanical fields.

Finally, with regard to clever new device structures, further very significant improvements in performance were demonstrated at the beginning of the 1990s by the introduction of optical microlens arrays in conjunction with MCT diode arrays (yet another innovation from RSRE). The lenses were fabricated in silicon by photolithographic techniques and aligned with the diodes, as shown in Figure 9.22 so as to concentrate the incoming radiation onto a small area detector. The improved performance is a consequence of the fact that detector noise is proportional to the square root of the detector area while the signal current remains unaffected by the concentration process. Bearing in mind that reduction in detector noise is also the aim of carrier exclusion methods, it is apparent that, in the same way, optical concentration can be traded-off against operating temperature. Typical geometries involved a lens pitch of about 50 μm, focal length also of 50 μm, and diode diameter of about 10 μm—a concentration factor of 25 and improvement in S/N ratio of five times. All in all, there appears to be no fundamental reason why BLIP performance cannot be achieved at room temperature, though it will be necessary to achieve still lower background doping levels than are available at present. Epitaxial growth clearly provides the best hope.

This reference to epitaxy suggests we should now return to the important question of MCT epitaxy. During the 1970s both VPE and LPE were developed as alternatives to the difficult bulk growth methods on which MCT initially depended. However, it was still far from easy to maintain adequate uniformity in the resulting material, if only because of the strong tendency for Hg to diffuse at the growth temperatures being used (typically, 700°C for LPE, 500°C for VPE). There was certainly a need for techniques which might exploit significantly lower growth temperatures and the first of these was MOVPE, introduced in 1981 by Irvine and Mullin at RSRE Malvern, soon followed by MBE in 1982 (Faurie et al., University of Illinois). Both techniques allowed growth at lower temperatures, initially 400°C for MOVPE (though, later, below 300°C) and 200°C for MBE. They also brought with them two further advantages, the first of these being the ability to dope with impurities, rather than relying on the less well-controlled variation of stoichiometry which had been used previously, and this came with yet another advantage because stoichiometric doping was associated with the introduction of Hg-related S–R recombination levels—impurity doping allowed the density of these unwanted centres to be minimized. The second advantage was greater ease of diode fabrication. *In situ*, junction formation by simply changing dopant species offered far greater control and the low growth temperatures ensured that these grown junctions stayed where they were put. At last, MCT technology began to look more like those associated with well-established materials such as GaAs. Both epitaxy methods have the ability to grow accurately defined diode structures in which the junction region is carefully graded but there is still a need to improve background doping levels to the order of 10^{19} m^{-3} (*n*-type) and 10^{20} m^{-3} (*p*-type) in order to reduce Auger generation to levels which will allow acceptable performance at room temperature.

Control of composition is good, usually achieved in the case of MOVPE by the use of the so-called interdiffused multilayer process (IMP) technique in which individual thin layers of CdTe and HgTe are deposited, then interdiffused at the growth temperature. Uniformity of composition across a 2 in. wafer is probably adequate but the development of large staring arrays depends on the availability of larger substrates and progress in increasing the size of bulk CdTe (or CdZnTe) appears to have been minimal. This has led to numerous attempts to grow high quality layers on Si or GaAs substrates, though there is a tendency for Ga to diffuse into the layer, when growing on GaAs. (Though this problem can be minimized by growing a 5-μm layer of CdTe before starting to deposit MCT.) Growth of a polar material such as MCT on a non-polar substrate such as Si has always proved difficult and this is no exception. However, there is clearly considerable advantage to be won in the shape of Si CCD arrays for signal processing if a satisfactory method can be developed. In fact, progress using MBE growth has been encouraging. Even though defect levels in the MCT layers are still higher than when grown on CdZnTe substrates, this appears not to degrade detector performance and reasonable uniformity has been achieved in arrays with up to 640×480 pixels. This undoubtedly seems to be the way forward and it is also worth noting recent efforts to grow lead chalcogenide layers on silicon with the same object in view. Arrays of 128×128 pixels have been demonstrated successfully, though with somewhat poorer temperature resolution than obtained from MCT arrays. (Thermal resolution is expressed in terms of the 'Noise Equivalent Temperature Difference', NETD, which represents the smallest detectable temperature difference in the scene—for the MCT arrays this was better than 20 mK, while for PbTe arrays it was of order 100 mK.) Clearly, MCT on silicon shows excellent promise for ultimate imaging performance. However, the 1980s saw the appearance of yet another approach to infrared imaging which now offers strong competition to MCT, particularly with regard to the important question of uniformity over an array of detector elements.

As we remarked earlier, the development of low-dimensional structures (LDS) in the late 1970s gradually influenced an extremely wide range of semiconductor applications, and IR detection came within this sphere of influence from about 1985 onwards. The idea, which was first proposed by Esaki and Sasaki from IBM, was to make use of radiative transitions between different sub-bands in a quantum well, rather than those between valence and conduction bands (i.e. 'intra-band', rather than 'inter-band' transitions). This immediately offered an important advantage in that it depended not on 'difficult' materials like MCT but on well-established systems such as AlGaAs / GaAs where quantum well structures were already under excellent control. It also offered an extremely easy method of tuning the transition energy to match the requirement of the infrared system in question. As we saw in Chapter 6, the energy differences between different sub-bands depends on both the depth and width of the quantum well—either of which may be varied in the epitaxial growth process. The application of quantum wells to infrared detection in the form of QWIPs (quantum well infrared photodetectors) was well reviewed in 1993 by Levine of Bell Labs (Levine 1993) who was also responsible for the first practical detector, reported in 1987.

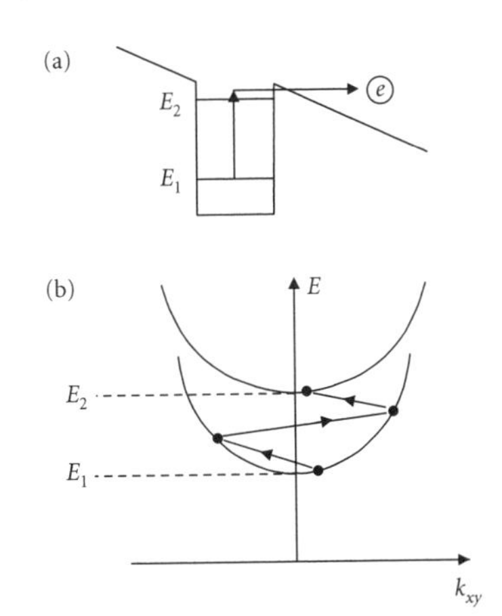

Figure 9.23. A single quantum well in the conduction band of a QWIP structure, with an electric field applied. The well is heavily doped to provide a supply of free electrons in the ground state level E_1 which implies very little potential drop across the well, itself. Incoming radiation is absorbed by optical transitions between the two confined levels, electrons tunnelling out of the well to form a photocurrent in the external circuit. In (a) is shown the spatial variation of energy in the z-direction, in (b) the variation of E with k in the xy plane (i.e. in the plane of the well). The arrows in (b) represent typical phonon transitions which give rise to dark current.

The early work was based largely on GaAs quantum wells within AlGaAs barriers. The principle is made clear in Figure 9.23 which shows an electron well containing two bound states, the upper being close to the top of the well. By correct choice of well width, it was straightforward to adjust the separation ΔE between the two energy levels to correspond to the desired detection wavelength (say 10 μm, corresponding to $\Delta E = 0.12$ eV). Provided the lower state contained an adequate density of free electrons, absorption of 10 μm photons would occur, with excitation of electrons into the (initially empty) upper level and, provided this upper level was close enough to the top of the well, some of these electrons could escape into the continuum of states which existed above the barriers. Application of an electric field across the sample resulted in their being swept out to an electrical contact, thus producing a photocurrent. In this sense, the quantum well behaved very much like a conventional photoconductor. However, an obvious difference lay in the fact that it was necessary to dope the well heavily n-type in order to provide the source of electrons in the lower confined state (this explains why the field in Figure 9.23 is low within the well). Typical doping levels were 10^{24} m^{-3}, though it turned out that the precise value was not very critical. This contrasts with the near-intrinsic behaviour of conventional photoconductors.

A little more thought soon leads one to a number of other conclusions. First, the well depth must be somewhat greater than the photon energy of the radiation to be detected and this has obvious implications for the fraction of aluminium in the barriers. We can do a simple sum to illustrate this point. To first approximation, the difference in the direct band gaps of $Al_xGa_{1-x}As$ and GaAs is given by:

$$\Delta E_g = 1.25\,x\,(\text{eV}) \tag{9.6}$$

and this difference is split in the ratio 2 : 1 between conduction band and valence band. In other words, the well depth $\Delta E_c = 0.83\,x\,(\text{eV})$. Allowing for the confinement energy of the ground state to be typically about 50 meV, we require that $\Delta E_c > (0.12 + 0.05)$ eV $= 0.17$ eV, and this implies $x > 0.2$, that is, there should be at least 20% aluminium in the barriers. On the other hand, too much Al would result in the upper state lying too low in the well for the carriers to escape. Notice that, in Figure 9.23, they leave the well by tunnelling through the top of the right-hand barrier, and tunnelling probability depends strongly on the height of the barrier through which this tunnelling takes place. Bearing in mind that the energy levels also depend on the well width L_z, it is apparent that there must be some trade-off between well depth and well width—a practical solution of the problem has $x = 0.25$ ($\Delta E_c =$ 0.21 eV) and $L_z = 6.5$ nm. (The reader may find it helpful to turn back to Figure 6.14 which illustrates the manner in which the confined energy levels depend on L_z for the case of $x = 0.3$. Here, the desired situation corresponds approximately to that of the 5-nm well, in the middle of the diagram.)

A second, relatively straightforward feature concerns the possibility of thermally induced transitions between the two levels. Note that an energy difference of 0.12 eV corresponds to no more than $4kT$ at room temperature which implies significant dark current and suggests the need for cooling to reduce it, though, as in our earlier discussions of noise in infrared detectors, it is not so much the magnitude of the current but rather the noise fluctuation which is important. In any event, most of the research and most actual device work has involved temperatures in the range 20–70K in order to minimize this noise contribution, lower, indeed, than the temperatures generally used for conventional infrared photoconductors. The reason for this is not hard to see and is made clear in Figure 9.23(b). In contrast to the situation in a narrow gap semiconductor, the 2 two-dimensional bands involved in QWIP operation both curve upwards in an E–k diagram which makes possible phonon transitions of the type illustrated in the figure, strongly increasing the probability of phonon-induced band-to-band transitions. (One way of looking at the situation would be to think in terms of the excited states in the lower band playing the same role as deep impurity states in S–R G–R in the narrow gap semiconductor.) This was, and still is a serious argument against the widespread application of QWIPs, no matter what advantages they may offer. The use of carrier exclusion techniques to reduce noise is not appropriate here, as the dominant noise contribution involves phonon, rather than Auger processes, though the use of microlenses to concentrate the radiation flux is certainly of value.

The advantages seen for QWIPs were (1) the ability to use a well-developed material technology, (2) the possibility of multi-spectral response (different well widths in the same structure), (3) ease of monolithic integration, and (4) good prospects for detectors operating in the wavelength range beyond 14 μm. One feature, in particular, which stemmed from the first point above concerned improved uniformity in large arrays—usually referred to as 'staring arrays' because they stared uninterruptedly at the infrared scene. This had become of considerable importance when it became apparent that poor uniformity could seriously affect overall performance and there was clearly a case for using QWIPs even if they had to be cooled below 77K—though there was always the hope that liquid nitrogen cooling might just be sufficient. In any event, there was little doubt during the 1990s that AlGaAs/GaAs QWIPs were capable of considerably better uniformity than available MCT arrays and this apparently led to a divergence of approach between friends and allies on opposite sides of the Atlantic Ocean. America went wholeheartedly for QWIPs, Europe put most of its eggs into the high temperature MCT basket, and it still remains a matter of debate which bet is more likely to prove successful.

The QWIP could be thought of as an advanced version of an impurity doped extrinsic detector such as silicon (which again had the advantage of a well-developed technology) but having a very much larger 'oscillator strength' (i.e. a much stronger optical absorption) and a much greater freedom to select any desired wavelength. The large oscillator strength came from the much larger physical dimension of the well, compared to the size of an individual impurity atom and the tuning ability from the ease with which the well width could be varied. There was, however, a problem with the absorption strength in that it only applied to radiation with an electric field vector parallel to the z-direction (i.e. normal to the plane of the well). Thus, radiation shining onto a QWIP normal to the top surface would not be absorbed—it was necessary to employ some special geometry to obtain a component of the electric vector along the growth direction. In early work this was achieved by cutting a 45° angle through the substrate and using this as the plane of incidence, a procedure not consistent with the demands of imaging systems. A better solution was to use an optical diffraction grating deposited on top of the structure which effectively launched waves into the semiconductor at an angle to the normal. By suitable choice of grating dimensions it was possible to increase the absorption coefficient by about two and a half times compared with the much cruder angle facet. Even better was to use a so-called 'random scattering surface' which consisted of a random arrangement of half and quarter wavelength depressions etched into the surface—a further factor of two improvement resulted, with a net responsivity close to 1 amp/watt for a typical 10 μm QWIP. Finally, we should note that a single quantum well of typical thickness 5 nm represented a very thin absorbing layer, in spite of the large oscillator strength and it was common practice to use a stack of at least 50 wells in order to absorb an acceptable fraction of the incident radiation. This raised complications for the collection of the photocurrent but certainly led to much improved device performance.

Apart from cooling, the best hope for improved detectivity lay in improving the signal strength, an aim which was in part met by making an important change in the choice of energy levels. Once again, it was a Bell initiative (in 1990) to use an upper level which lay within the continuum of states above the barriers, rather than a confined state within the well. This had the advantage that excited electrons were immediately free, strongly reducing the probability of their being recaptured into the ground state and failing to contribute to the photocurrent. Levine and co-workers were able to demonstrate a 128×128 array of QWIP detectors based on this principle. Operating temperatures were below 60K but the arrays showed excellent resolution and uniformity, together with a very high yield of working pixels. Individual device detectivity was 10^7 m(Hz)$^{1/2}$ W^{-1} at 77K, increasing to 10^{10} m(Hz)$^{1/2}$ W^{-1} at 33K. There can be no doubt that such arrays offer the major advantage of excellent uniformity, essential for large imaging systems, provided system specifications are consistent with these low operating temperatures. It remains to be seen whether significant improvements in performance can be obtained to allow the use of higher operating temperatures.

While, for obvious reasons, much of the work on QWIPs has made use of the AlGaAs/GaAs material system, there may be reasons to use other combinations. For example, because of its deeper wells, InGaAs/InAlAs, lattice-matched to InP has been used to make QWIPs for the 3–5 μm window, as have GaAs wells with AlInP lattice-matched barriers. Long wavelength devices have also been demonstrated in InGaAsP/InP, GaAs/GaInP, and InGaAs/GaAs. This latter system is interesting in using a binary compound for the barriers which facilitates transport of the photogenerated electrons. Long wavelength devices have also been made using *p*-type QW samples in which the photocurrent consists of holes but there seems to be little advantage. Perhaps the most interesting recent development in this area is the move to replace quantum wells with quantum dots. Dots have an immediate advantage in so far as they make no demand on the required polarization of the detected radiation and this simplifies pixel design. They are also characterized by atom-like energy levels, rather than bands, so the phonon process illustrated in Figure 9.23(b) is no longer applicable, making high temperature operation so much more probable. It is still possible, of course, to employ multiple stacks of dots in order to obtain adequate absorption strength.

Finally, we should not close this section without making mention of yet another interesting material innovation, in the shape of the semiconductor superlattice. The subject of superlattices is too complex for detailed discussion here but, since it is possible to design materials with effective band gaps which lie in the range appropriate to infrared detection, we must, at least, provide a brief outline. First, perhaps, it is necessary to emphasize the difference between a system of quantum wells and that of a superlattice. In the QWIP structures discussed above the barriers were always thick enough (5–10 nm or more) that the wells could be regarded as separate (i.e. not coupled together) so that the properties of the 50 wells were identical with that of a single well (apart from the increased absorption coefficient). If, on the other hand, the barriers are made gradually thinner, inter-well coupling increases and the properties of the overall structure become significantly different from those of the individual wells. In particular, short period superlattices (those made up of layers of no more than a few monolayers of each constituent) behave rather like semiconducting materials with band gaps which can be controlled by the choice of individual layer thicknesses. For example, a superlattice consisting of alternating single monolayers (ML) of AlAs and GaAs behaves in a manner very similar to that of a random $Al_{0.5}Ga_{0.5}As$ alloy but different properties can be obtained by varying the thicknesses of the two layers, while still maintaining the same ratio—for example (2ML Al plus 2ML Ga), (3ML Al plus 3ML Ga), etc. Needless to say, the AlAs/GaAs system is important because of its excellent lattice-match but many other possibilities exist.

If we wish to 'design' a material with an effective band gap in the infrared spectral region, we must choose different starting compounds—in particular $InSb/InAs_xSb_{1-x}$ and $Ga_{1-x}In_xSb/InAs$ superlattices may be made with band gaps appropriate to the 8–14 μm window. There is, however, a problem in these systems with finding a suitable high quality substrate, various different ones (GaAs, InP, and GaSb) having been tried. Possibly the most promising is the combination of the $Ga_{1-x}In_xSb/InAs$ superlattice grown on GaSb because there is a good match between the 'average' lattice constant of the superlattice and that of GaSb. At the same time, it is possible to obtain the desired band gap with acceptably small strains in the GaInSb and InAs layers. Whether such sophisticated material architecture can win out over the other approaches to long wavelength detection remains yet to be seen, but that people are seriously trying says much for the continuing vitality of the field.

9.6 Long wavelength lasers

As we saw in Chapter 5, the first semiconductor lasers to startle the physics and electrical engineering communities appeared in 1962 in the form of GaAs and GaAsP homojunction devices, operating at low temperatures in pulsed mode, and this was followed in short order by the demonstration of lasing action in various other materials. Once it had been established that a principal requirement was that for a direct band gap, it was natural that other direct gap III-V compounds be considered, including InAs and InSb, both of which could be doped *n*- and *p*-type. Reports of injection laser action in both these materials were published in 1963, again at low temperatures (see Horikoshi 1985). Melngailis et al. (1970) of Lincoln Labs (from which one of the pioneering laser groups hailed) achieved laser emission from an InAs homojunction device at a wavelength of 3.1 μm with threshold current densities of $1300\,\mathrm{A\,cm^{-2}}$ and $1.6 \times 10^4\,\mathrm{A\,cm^{-2}}$ at 4.2K and 77K, respectively (entirely comparable with the earlier GaAs values), while Phelan et al. (from the same stable) obtained laser action in InSb at $\lambda = 5.2$ μm with $J_{th} \sim 10^3\,\mathrm{A\,cm^{-2}}$ at temperatures below 2K. Both groups made use of bulk crystals with Zn-diffused junctions and cleaved mirror facets, much as the GaAs pioneers had done.

Thus was born the infrared laser diode (LD)—long before anyone quite knew what to do with it! Time, however, soon rectified this deficiency and today there are numerous applications for such lasers in the field of pollution monitoring, many of the gases involved in destroying our environment showing characteristic infrared absorption spectra whose measurement provides a convenient method of identification. However, the low levels of contamination which are significant demand highly sensitive detection systems consisting of a suitable laser diode, together with an infrared detector, a key feature being the need to match the laser wavelength with a specific absorption line. In other words, just any old laser will not do—it is essential that some method of selecting the wavelength be available. In this sense, InAs or InSb were not very attractive. Resort had to be made to material systems which offered greater flexibility—such as PbSnTe or HgCdTe, or, more recently, InGaAlAsSb! What was more, for devices to function effectively in the field, it was also desirable to achieve room temperature CW operation and this called for the sophistication of well-matched heterostructures. It took the GaAs teams until 1970 so we could hardly anticipate more rapid progress in the mid-infrared. But progress there certainly was.

The first step towards the PbSnTe (or PbSnSe) injection laser was made in 1964 when Butler et al. of Lincoln Labs reported laser action in PbTe ($\lambda = 6.5$ μm) and PbSe ($\lambda = 8.5$ μm) at a temperature of 12K. They used *p*-type Bridgman-grown bulk crystals and formed the

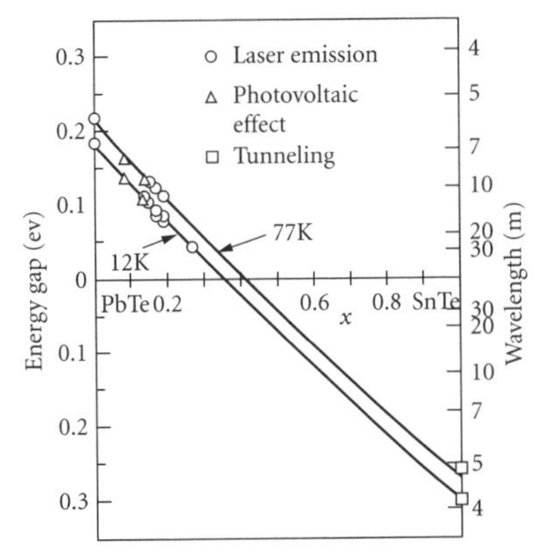

Figure 9.24. Plot of energy gap E_g for the alloy $Pb_{1-x}Sn_xTe$ as a function of x at two temperatures $T = 12K$ and $T = 77K$. For $x < 0.4$, E_g increases with increasing temperature but, for $x > 0.4$, it decreases. This is a consequence of the band inversion which occurs at about $x = 0.4$. (From Melngailis, I. and Harmon, T. C. (1970), *Semiconductors and Semimetals*, (eds. R. K. Willardson and A. C. Beer, Vol. 5, p. 111, Figure 2, Academic Press). Reprinted with permission from Elsevier.

p-n junction by annealing in a Pb-rich atmosphere (excess Pb was known to result in *n*-type doping). The rock salt structure of the group IV chalcogenides leads to (1 0 0) cleavage planes which provided convenient mirror facets. While threshold current densities were in excess of 10^3 A cm^{-2}, this was only to be expected for homojunction devices—the important point was that laser action in these materials was also possible in spite of their very different band structures (i.e. having both valence band maxima and conduction band minima at the L point in *k*-space, rather than at the Γ point). In fact, the chalcogenides are well suited to produce laser action because of the small effective masses in *both* bands (because the band extrema lie at the L point, two masses must be specified, m_L and m_T, as for Si and Ge in Chapter 3, but the average mass in PbTe is about 0.04 m). They followed this up 2 years later with the first PbSnTe and PbSnSe laser diodes which operated in the wavelength range of (roughly) 9–14 μm with $J_{th} \sim 600$ A cm^{-2} at 12K. The smaller energy gap involved implies smaller effective masses which lead to reduced threshold current. These devices even operated at 77K, though with considerably larger threshold currents ($J_{th} \sim 3000$ A cm^{-2}). In 1968, the same group reported laser action out to $\lambda = 28$ μm with even smaller threshold current ($J_{th} = 125$ A cm^{-2}), illustrating the flexibility available with this material system. They also demonstrated some interesting consequences of the band inversion which occurred as the mole fraction of tin was increased from zero through the cross-over point. They observed that, for $Pb_{1-x}Sn_xTe$ samples with $x < 0.4$, the temperature coefficient of band gap was positive, while, for $x > 0.4$, it was negative. Figure 9.24, which shows plots of E_g vs x for two temperatures $T = 77$ and $T = 12K$, makes clear how this came about. It was of practical significance because temperature control offered a method of wavelength tuning.

Following experience with GaAs lasers, it was clear that any hope of obtaining CW operation and raising operating temperatures would probably depend on the development of heterostructures to facilitate both carrier and optical confinement. Though one or two attempts were made using a method involving out-diffusion of tin from PbSnTe or PbSnSe to increase the band gap, the practical way ahead involved some form of epitaxy. VPE was first used to grow PbSnTe on PbTe substrates in 1970 (though it later appeared to fall out of favour), while MBE and LPE began to impact the laser scene in 1972 and 1974, respectively and have been widely used. The first successful MBE laser was reported by Holoway et al. of Ford Motors, who grew PbTe on BaF_2 substrates and obtained emission at 12K with a wavelength of 6.5 μm. The running was then taken up in 1973 by Walpole et al. of Lincoln Labs with a single heterostructure device made by growing *n*-type PbTe on a *p*-type $Pb_{0.88}Sn_{0.12}Te$ substrate. This 10-μm emitter worked CW up to 65K and showed significantly lower pulsed threshold current—$J_{th} = 45\ \text{A cm}^{-2}$ at 4.2K and $780\ \text{A cm}^{-2}$ at 77K. While a crude double heterostructure (DH) laser was reported as early as 1971, the first successful device came from the prolific Lincoln Labs in 1974, in the shape of an LPE-grown $PbTe/Pb_{0.88}Sn_{0.12}Te/PbTe$ structure on a PbTe substrate. This, at 12K, produced 10 mW of output power at 10 μm wavelength and ran CW up to 77K. MBE, however, bounced back with a still better device, using a $Pb_{0.78}Sn_{0.22}Te$ active layer sandwiched between $Pb_{0.88}Sn_{0.12}Te$ barriers, which operated CW up to 114K—and so the battle rolled!

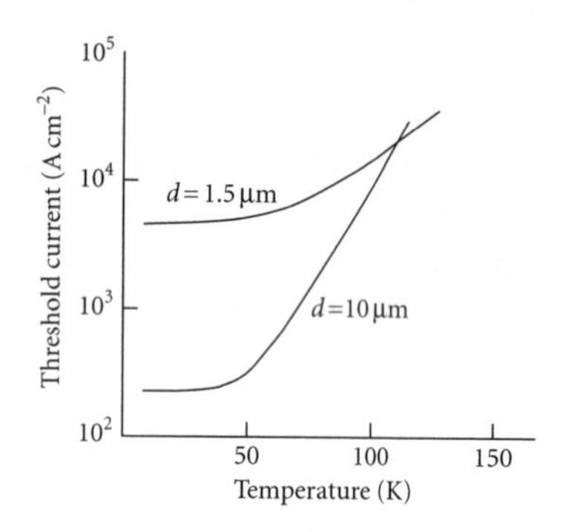

Figure 9.25. Experimental plots of threshold current density for two PbSnTe/PbTe DH lasers with different active layer thickness, $d = 1.5$ μm and $d = 10$ μm. Note the large decrease in low temperature threshold current density for the wider active region where interface recombination plays a much smaller role. The steep increase in J_{th} at higher temperatures reflects the importance of Auger recombination. Following: Horikoshi, Y., Kawashima, M., and Saito, H. (1982) *Jap Jnl Appl Phys* 21, 77 (Fig 13).

All these devices were crude in the sense that little attempt was made to design lattice-matched structures. The mismatch between PbTe and SnTe is approximately 2% so the structures used here suffered something like 0.3% mismatch which was probably sufficient to introduce significant numbers of defects at the interfaces. This suggestion was supported by the measured temperature-dependence of the threshold current, an example of which is shown in Figure 9.25. At low temperatures the curves are flat, indicating recombination lifetimes independent of temperature and probably the result of interface recombination (the shorter the lifetime, the harder one has to pump to reach the injected free carrier levels necessary for the onset of laser action). Notice that, when the active layer thickness is increased, so that interface recombination becomes less important, J_{th} shows a sharp decrease. At elevated temperatures the sharp increase of slope indicates the effect of our old friend Auger recombination. In fact, Auger effects were largely responsible for the difficulty in raising operating temperatures—the strong temperature-dependence of the Auger lifetime made it inevitable that threshold current should increase rapidly with increasing temperature (they also increased with increasing wavelength). There was, of course, no possibility of overcoming it by reducing free carrier density—high carrier densities are an essential requirement for laser action. The other factor likely to increase high temperature threshold current was carrier leakage out of the active layer due to the barriers' being too low and this indicated the need for barriers with greater band gap. This, together with the need for lattice-matched combinations led to some wide-ranging material studies which unearthed some surprising configurations. However, before discussing these, we should point out that the inclusion of selenium in PbSnTe material does allow lattice-matching and considerable success was obtained with PbSeTe/PbSnTe/PbSeTe and PbSeTe/PbSnSeTe/PbSeTe structures. In particular, low threshold currents were obtained at low temperatures, indicating much improved interface quality.

The next phase of the story is well illustrated by the work of Dale Partin and colleagues at the General Motors Laboratory in Warren, Michigan (Partin 1991). They took up the challenge of MBE growth at the beginning of the 1980s with a (then) modern ion-pumped machine which reached background pressures of 10^{-10} Torr, implying considerably higher purity levels than had been available in the earlier work of Holloway et al. at the Ford Laboratory in Dearborn (just the other side of Detroit) who worked with background pressures in the region of 10^{-6} Torr. Ford, as it were, laid the foundations with their Model T, while GM countered with a Cadilac! Not that their contribution was mere glitter—indeed, they played a serious role in the introduction of real sophistication into the IV-VI materials system. What was needed was a range of chalcogenides which crystalized in the rock salt structure (to facilitate ready alloying with PbTe, PbSe, SnTe, etc.) and which posessed both larger band gaps and lattice parameters suitable for designing lattice-matched double hetero (DH) structures. The solution they came up with can be understood from Figure 9.26 which provides the relationship between band gap and lattice parameter for a range of chalcogenides, including the wide gap materials YbTe, EuTe, SrTe, and CaTe. In particular, it is apparent that alloys of PbTe with EuTe and YbTe (or SrTe and CaTe) can readily be lattice-matched to PbTe, while having significantly larger band gaps than PbTe itself. (Similar remarks apply to PbSnTe alloys appropriate to the wavelength region between 8 and 14 μm.) Partin et al. showed that many of these alloys could be reliably grown by MBE at relatively low temperatures (e.g. 300–400°C) and the way was clear to develop, for the IV-VI compounds, the sophisticated heterostructures and quantum well structures which resulted in major improvements to AlGaAs/GaAs lasers (see Chapter 6). (The situation is similar to the one we met in Chapter 7 (Section 7.5) when discussing II-VI visible laser diodes—a similarly wide range of alloy materials was required then to obtain lattice-matched DH structures and a number of new components was drafted in to satisfy these demands.)

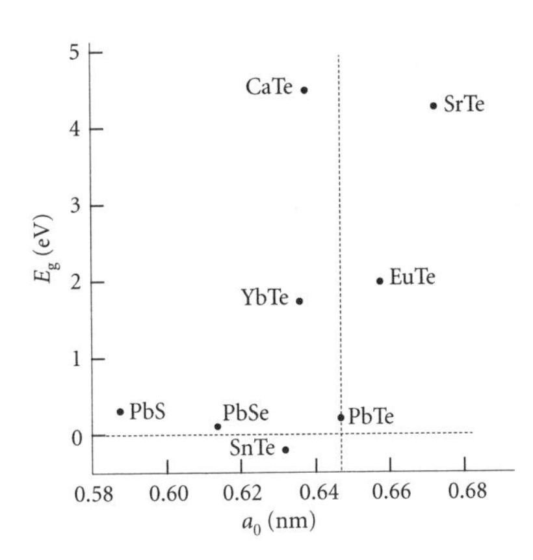

Figure 9.26. Band gaps and lattice parameters for a range of chalcogenides which crystalize in the rock salt structure. As can be seen, alloys containing EuTe and YbTe (or SrTe and CaTe) can be lattice-matched to PbTe or to $Pb_{1-x}Sn_xTe$ (for small x).

Two important aspects of these chalcogenide material developments lay in the possibility of developing QW lasers and the possibility of making laser diodes for somewhat shorter wavelengths (in answer to a likely requirement to match very low loss fluoride glass optical fibres in the 2–3 μm range). As an example of an early QW laser, we can quote the work reported by Partin in 1988, using a 30-nm PbTe well with barriers and optical confining layers both of PbEuSeTe, the barriers containing 1.8% Eu, the waveguide layers 4% Eu, while the Se fraction, in each case, was adjusted to maintain lattice-match to PbTe. Devices operated CW up to temperatures of 140K, threshold current densities being 150 A cm^{-2} at 4.2K and 2.5 kA cm^{-2} at 140K. The emission wavelength at 140K was 4.8 μm. The difference in energy gap between well and barrier was approximately 100 meV but there was no clear evidence as to how this was shared between conduction and valence bands. The likelihood was that carriers were being lost from the well at the upper temperature range, contributing to the problem of obtaining higher operating temperatures and suggesting the need for higher

Eu concentrations in the barriers. However, measurement of electron mobility in alloys of PbEuSeTe showed a surprisingly steep reduction as a function of Eu concentration, leading to significant series resistance in laser structures. There was therefore an impasse—and it turned out that this could not be remedied by resorting to Ca or Sr-containing alloys because these suffered in precisely the same manner. There was to be no easy solution in the quest for room temperature operation and, what was worse, the longer the wavelength, the worse the problem became (both Auger recombination and free carrier absorption losses increase as the wavelength increases).

The state of play in IV-VI laser diode development was reviewed in 1995 by Maurus Tacke of the Fraunhofer Institute in Freiburg (Tacke 1995). Progress seemed rather slow—though both DH and BH (buried heterostructure) lasers, grown by MBE, were fairly well established, the maximum temperature for CW operation was still no more than 200K. In attempting to satisfy the need for single mode lasers for high resolution spectroscopy in gas analysis, some effort had also been devoted to making distributed Bragg reflector and distributed feedback structures and, though these looked promising, they were still very much at the laboratory prototype stage. It was still far from clear whether the holy grail of room temperature operation was likely to be met, either soon or even ever! Part of the problem lay in the diversity of effort over a number of different material systems, it being unclear which offered the best hope for success.

The problems associated with lattice matching are, as we saw in our earlier discussion of infrared detectors, significantly reduced when MCT is used as the active material, so it seems reasonable to look for applications in laser diode fabrication. The first MCT diode laser was reported in 1991 by the group at the Rockwell International Science Center in California whose principal interest was in IR detectors. They used MBE growth at low temperatures ($\sim$200°C) to make a stripe geometry, DH laser which emitted at 2.86-μm wavelength. The operating temperature ranged between 40 and 90K and the pulsed threshold current density at 77K was a respectable 625 A cm^{-2}. The output power could not be measured but was probably of order a few milliwatts. The active region consisted of undoped $Hg_{0.605}Cd_{0.395}Te$ with confining layers of p^+-$Hg_{0.51}Cd_{0.49}Te$ and n^+-$Hg_{0.52}Cd_{0.48}Te$, the whole structure being grown on a semi-insulating $Cd_{0.96}Zn_{0.04}Te$ substrate. Doping was achieved using In for *n*-type and As for *p*-type material, though because As has a tendency to be amphoteric (depending on which lattice site it occupies), it was necessary to grow the *p*-type confining layer in the form of a CdTe/HgTe superlattice which was subsequently interdiffused, the As being incorporated only into the CdTe layers, grown (perhaps surprisingly) under high Hg pressure (this encourages the As atoms to occupy Te sites on which they act as acceptors).

Such problems with *p*-type doping, together with the need for greater output powers led to an alternative approach to the generation of laser power in the 2–4 μm region, namely the use of optical, rather than electrical pumping. In this case, the structure can be grown from undoped material throughout and can be more easily designed to include light-guiding layers. Pump light is absorbed in the outer, waveguide layers while the resulting free carriers thermalize into the central active region, where they recombine. Using a high power InGaAs/AlGaAs 0.94-μm laser diode array to pump a multi-quantum well MCT structure, emitting at 3.2 μm, the same group was able to demonstrate pulsed output powers greater than 1 W at 88K operating temperature, with 100 mW average power levels. Such performance allows much greater flexibility in the application to pollution monitoring over wide areas, whereas milliwatt devices must be used with carefully designed absorption cells employing multi-pass light paths. A disadvantage of optical pumping is the inefficiency resulting from the fact that the energy of the pump photons is inevitably larger than that of the output photons—in this case by a factor of 3.4—though this may sometimes be a price worth paying for the considerable simplification in material structure which accrues.

However, the lead chalcogenides and MCT were not the only materials capable of generating stimulated infrared emission. As we saw in Chapter 8 (Section 8.2), the antimony-based system GaAsSb/AlGaAsSb had yielded laser diodes emitting at about 1-μm wavelength in the early 1970s, and, in principle, it would be possible to reduce the band gap of the active material by including a larger fraction of Sb and thus extending the output wavelength further into the infrared. The problem here (yet again!) was one of lattice-match—these devices were grown on GaAs substrates to which they were not matched and any increase in Sb content would merely exacerbate the problem ($a_0 =$ 0.60955 nm for GaSb, 0.56535 nm for GaAs). In fact, two innovative steps were required before this could be achieved: first, the substrate had to be changed to GaSb and, second, In had to be included in the active layer, resulting in a quaternary compound $Ga_xIn_{1-x}As_ySb_{1-y}$ which could be lattice-matched to the GaSb substrate. Notice that we have here another system similar to InGaAsP which could be matched to InP (see Box 8.3) by satisfying a specific relation between the parameters x and y. In this case, the relation can be written as:

$$y = 0.91(1 - x)/(1 + 0.05x) \tag{9.7}$$

and the band gap of the resulting quaternary material is found to vary between limits of 0.28 and 0.73 eV ($\lambda = 4.4$–1.7 μm). This happens to fit rather neatly alongside the InGaAsP system whose minimum gap is 0.75 eV ($\lambda = 1.65$ μm).

The problem of finding a suitable cladding material might appear to be easily solved by using GaSb but this is not a valid solution because GaSb has a refractive index greater than that of the active material and cannot, therefore, act as a waveguide layer. Fortunately, two other possibilities were found: AlGaAsSb and InAsPSb, both of which can be lattice-matched while providing an appropriate refractive index for waveguiding and large enough band gaps for carrier confinement. The first successful lasers based on the GaInAsSb/AlGaAsSb system were reported by Kobayashi et al. of NTT in 1980. They demonstrated room temperature pulsed emission at a wavelength of 1.8 μm with a threshold current density of 5 kA cm^{-2}—later work in other laboratories has brought this down to about 1.5 kA cm^{-2} and Russian workers achieved room temperature CW performance at wavelengths close to 2 μm. Nearly all this early work made use of LPE, which limited the range of wavelengths available because there exists a 'miscibility gap' in the GaInAsSb system for wavelengths between 2.4 and 4-μm (a region of compositions where it is impossible to grow an alloy material due to segregation of the constituents into separate spatial regions). This is a serious problem for any growth technique which employs thermal equilibrium conditions (such as LPE) but can sometimes be overcome by employing alternative methods which function far from thermal equilibrium. MBE and MOVPE are examples, and they have been used to extend the wavelength range as far as 4 μm, though at operating temperatures well below room temperature. Work at MIT, for example, (reported in 1994) resulted in 4-μm lasers which operated CW at temperatures up to 170K with $J_{th} = 8.5$ kA cm^{-2}. At 80K, $J_{th} = 100$ A cm^{-2} and the maximum output power was 24 mW. QW lasers have also been demonstrated with similar performance and optically pumped devices have yielded as much as 2-W peak power in pulsed operation at 92K.

An alternative approach to growing laser structures for the 3–4 μm wavelength region is based on the use of InAs substrates (though note that neither GaSb nor InAs substrates are available in anything like the quality or size appropriate to GaAs). Structures have made use of InAs active layers with AlAsSb cladding and InAsSb/InGaAs superlattice active regions with InPSb cladding. It is an interesting field and may develop further but we shall not attempt any detailed discussion because of its relatively youthful state of development. Without doubt, it provides a wonderful opportunity for interesting material studies but there is, so far, only limited progress towards practical laser devices.

The perceptive reader may, at this point, be wondering why, if quantum well intra-band transitions could be used successfully to make infrared photodetectors, they might not also form the basis of infrared lasers. And he would be right so to wonder—not only might they be used, indeed they are. In 1994, in an article in 'Science', Federico Capasso and his colleagues at Bell Labs, Murray Hill described their pioneering work to demonstrate what has become widely known as the 'Quantum Cascade Laser' (Faist et al. 1994), a device which bids well to revolutionize the whole field of infrared semiconductor lasers. It depended on a remarkably complex set of coupled quantum wells to set up the desired free carrier inversion between a pair of confined conduction band energy levels but went one important step beyond this by using each electron to stimulate photon emission not only in a single well but in a large sequence of wells (typically 50). Whereas each electron injected into the active region of a conventional diode laser recombines, with the emission of a single photon, in the new invention it produced a cascade of photons from the sequence of wells, thereby multiplying by as much as 100 times the amount of light power generated by the passage of each electron through the structure. (Though it should not be overlooked that the applied voltage was, of necessity, correspondingly greater.)

Figure 9.27 provides a much oversimplified illustration of how the device works. It represents the active region as a single conduction band quantum well having three confined energy states, $E_3 > E_2 > E_1$, the various wells being separated by an intermediate region which operates to inject electrons into the upper level in the following well. Electrons in level three fall into level two with the emission of a photon of energy $h\nu = (E_3 - E_2)$ (typically 80–400 meV, corresponding to wavelengths $\lambda = 15$–$3\ \mu$m), then drop into level one by emitting an optical phonon of energy approximately 35 meV, and finally, tunnel out again into level three of the next well, the sequence being repeated N times, where N is the number of sections in the whole structure. Notice that there is no *p-n* junction involved—the device is said to be 'unipolar' because it employs only one type of charge carrier. Current is carried entirely by

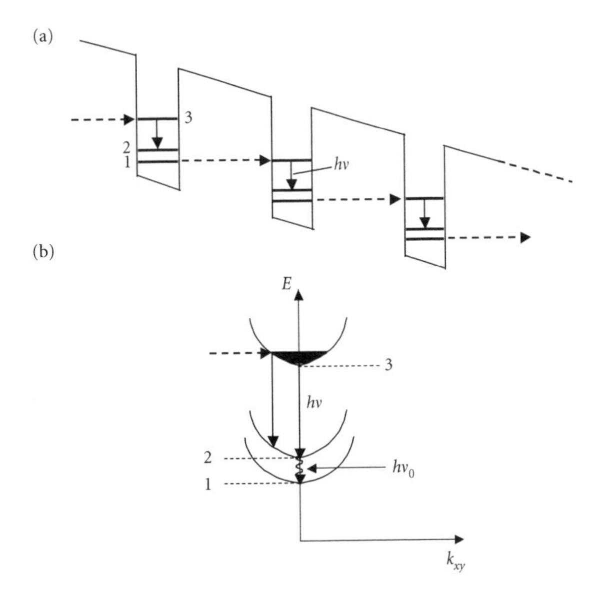

Figure 9.27. Simplified conduction band energy diagram to illustrate the operation of a quantum cascade laser. In part (a) is shown a section of the device structure containing three active regions and corresponding injection layers. Electrons tunnel into the upper level (level three), fall into level two, with the emission of a photon of energy $h\nu = (E_3 - E_2)$, then drop into level one by emitting an optical phonon and, finally tunnel out into the next active region. Part (b) shows an expanded energy diagram for the active regions, illustrating the fact that all emission processes correspond to the same photon energy as a result of the energy bands' having the same upward curvature in *k*-space.

electrons. Early in 1994 Capasso et al. described an LED based on this type of structure—to turn it into a laser involved the addition of a pair of thick outer waveguide layers which guided the emitted radiation between a pair of perpendicular cleaved window facets in the conventional way. This they did within a matter of months, announcing the first quantum cascade laser later that same year. Operated at 10K, it produced up to 8 mW of pulsed power at a wavelength of 4.3 μm with a threshold current density of about 15 kA cm^{-2}.

All this may sound relatively straightforward but that impression depends on the oversimplification introduced here as an aid to comprehension. The effective operation of the laser depends on the specific arrangement of the three energy levels involved. It is essential that electrons relax very rapidly from level two to ensure that level two remains essentially empty, thus facilitating the inversion of population between levels two and three. To achieve this demands that $(E_2 - E_1) = h\nu_0$ (where $h\nu_0$ is the optic phonon energy at the centre of the Brillouin zone in the material of the well). As we saw above, this energy is much smaller than the required photon energy and this implies that $(E_3 - E_2) > (E_2 - E_1)$, a condition which is incompatible with the physics of quantum confinement within a single well. In square wells (as we saw in Chapter 6) the opposite inequality holds, so some clever 'band structure engineering' was needed to obtain the desired arrangement of levels. Instead of using a single well in the active region, it was necessary to use three—a narrow well to obtain level three, two wider wells to provide levels one and two. By arranging that the barriers were thin enough to provide the appropriate coupling between these wells, the trio was persuaded to play together in harmony to produce the desired scale of energies. Similarly, the injection regions between the active regions were fashioned out of carefully designed superlattices, having six wells each, making the overall structure even more complicated. Given that there were nine wells and nine barriers per unit and, say, 50 units plus waveguiding layers, it becomes clear that there was a grand total of something approaching one thousand separate layers in each laser structure, each of which must be accurately tailored to meet device needs! Could the demands made on the long-suffering crystal grower possibly be racked up any further?

Not surprisingly, the Bell workers chose to use MBE to grow these improbable structures (Al Cho's name was not on the papers for nothing) and opted for an InP-based material system—$Ga_{0.47}In_{0.53}As$ wells with $Al_{0.48}In_{0.52}As$ barriers, grown on an InP substrate (though more recently other workers have employed AlGaAs/GaAs to achieve similar results). Computer-controlled shutter operation was, of course essential—no human operator could hope to keep track of all the switching steps involved! Transmission electron micrographs showed just how successful was the control and the fact that the devices worked at all, never mind so well, added further confirmation. Altogether, this represented a quite remarkable achievement, flowing from the integration of excellent physics, brilliant device design, and magnificent material technology—surely the stuff from which Nobel Prizes are fashioned? We can only wait and see!

To give a flavour of just how immediately successful these endeavours have been we need only to refer to the review article by Capasso, written for Physics World in 1999 (Capasso 1999), just 5 years after the initial breakthrough. In this he reports devices which operate at wavelengths from 3.4 to 17 μm, room temperature pulsed powers of hundreds of milliwatts per facet, CW performance at temperatures up to 175K, CW powers of up to 700 mW at 30K, while the lowest threshold current densities have been 750 $A\,cm^{-2}$ at 77K and 5.2 $kA\,cm^{-2}$ at room temperature. Already such devices had been used in gas monitoring systems—Whittaker et al. in the Steven Institute, New Jersey demonstrated the detection of traces of nitrous oxide (N_2O) in nitrogen at the level of 250 parts per billion while Paldus et al. at Stanford University detected ammonia in nitrogen at the 100 parts per billion level. For much of this pollution monitoring work it is necessary to use a single mode laser so the Capasso group has, of course (!), made one, employing a distributed Bragg mirror to select just one mode from the 20 or more emitted from a standard Fabry–Perrot cavity laser. It is clear now that numerous other groups around the world are making important contributions to laser development and the next decade will doubtless see considerable further progress. For example, recent work in Switzerland has led to the first room temperature CW devices and further development is aimed towards lasers in the 50–100 μm range, while several medical applications are under preliminary investigation. Again, we must wait with interest.

Bibliography

Bell, R. L. (1973) *Negative Electron Affinity Devices*, Oxford University Press.

Broudy, R. M. and Mazurczyck, V. J. (1981) *Semiconductors and Semimetals*, Vol. 18, p. 157 (eds. R. K. Willardson, and A. C. Beer) Academic Press, New York.

Capasso, F. (1999) *Phys. World*, Vol. 12, June, p. 27.

Faist, J., Capasso, F., Sivco, D. L., Sitori, C., Hutchinson, A., and Cho, A. Y. (1994) *Science*, 264, 553.

Duxbury, G. (2000) *Infrared Vibration-Rotation Spectroscopy*, John Wiley, Chichester.

Elliott, C. T. (1981) *Handbook on Semiconductors* 1st edn, Vol. 4, ch. 6B, p. 727, North Holland, Amsterdam.

Elliott, C. T. and Gordon, N. (1991) *Handbook on Semiconductors* 2nd edn, Vol. 4, Ch. 10, p. 841, North Holland, Amsterdam.

Escher, J. S. (1981) *Semiconductors and Semimetals*, Vol. 15, p. 195 (eds. R. K. Willardson and A. C. Beer) Academic Press, New York.

Herzberg, G. (1945) *Infra-red and Raman spectra*, Van Nostrand, New York.

Hilsum, C. and Rose-Innes, C. A. (1961) *Semiconducting III-V Compounds,* Pergamon Press, NY.

Hirsch, H. E., Liang, S. C., and White, A. G. (1981) *Semiconductors and Semimetals*, Vol. 18, p. 21 (eds. R. K. Willardson and A. C. Beer) Academic Press, New York.

Horikoshi, Y. (1985) *Semiconductors and Semimetals*, Vol. 22C, p. 93, (eds. R. K. Willardson and A. C. Beer) Academic Press, New York.

Jervis, M. H. and Lockett, R. A. (1971) *Applications of Infrared Detectors*, ch. 9, p. 107, Mullard Ltd, London.

Kruse, P. W. (1970) *Semiconductors and Semimetals*, Vol. 5, p. 15 (eds. R. K. Willardson and A. C. Beer) Academic Press, New York.

Levine, B. F. (1993) *J. Appl. Phys.*, 74, R1.

Lloyd, J. M. (1975) *Thermal Imaging Systems*, Plenum Press, NY.

Lovett, D. R. (1977) *Semimetals and Narrow Gap Semiconductors*, Pion Ltd, London.

Melngailis, I. and Harman, T. C. (1970) *Semiconductors and Semimetals*, Vol. 5, p. 111 (eds. R. K. Willardson and A. C. Beer) Academic Press, New York.

Madelung, O. (1964) *Physics of III-V Compounds*, John Wiley and Sons, New York.

Micklethwaite, W. F. H. (1981) *Semiconductors and Semimetals*, Vol. 18, p. 47 (eds. R. K. Willardson and A. C. Beer) Academic press, New York.

Morten, F. D. (1971) *Applications of Infrared Detectors*, chap. 1, p. 1, Mullard Ltd, London.

Partin, D. L. (1991) *Semiconductors and Semimetals*, Vol. 33, p. 311 (eds. R. K. Willardson and A. C. Beer) Academic Press, New York.

Putley, E. H. (1968) *The Hall Effect and Semiconductor Physics*, Dover, New York.

Putley, E. H. (1970) *Semiconductors and Semimetals*, Vol. 5, p. 259 (eds. R. K. Willardson and A. C. Beer) Academic Press, New York.

Reine, M. B., Sood, A. K., and Tredwell, T. J. (1981) *Semiconductors and Semimetals*, Vol. 18, p. 201 (eds. R. K. Willardson and A. C. Beer) Academic Press, New York.

Rougeot, H. and Baud, C. (1979) *Adv. Electron Electron Phys*. 48, 1.

Smith, R. A. (1953) *Adv. Phys*. (Suppl. Phil Mag) 2, 321.

Smith, R. A., Jones, F. E., and Chasmar, R. P. (1968) *The Detection and Measurement of Infrared Radiation*, 2nd edn, Oxford University Press.

Sommer, A. H. (1968) *Photoemissive Materials*, John Wiley, New York.

Stevens, N. B. (1970) *Semiconductors and Semimetals*, Vol. 5, p. 287 (eds. R. K. Willardson and A. C. Beer), Academic Press, New York.

Tacke, M. (1995) *Infrared Phys. and Technology* 36, 447.

CHAPTER 10

Polycrystalline and amorphous semiconductors

10.1 Introduction

The reader will hardly need reminding that one of the principal emphases in our discussion throughout the book has been on the importance of high quality single crystal materials. Progress in both physical understanding and device development appears always to depend crucially on the availability of high purity, defect-free single crystals and we have followed in numerous cases a progression from tiny (perhaps naturally occurring) crystallites through bulk single crystal boules to lattice-matched epitaxial growth. Frequently, it is only at this later stage that has it been possible to obtain performance from devices which approaches theoretical expectation. Indeed, only under these circumstances can we expect simplified theoretical models to offer a reasonable approximation to practical working structures. On the other hand, the growth of such high quality crystalline material is not achieved without considerable investment in both time and effort (i.e. money!) which inevitably stimulates the thought that it would be 'nice' if we did not always have to do it! It is also important to recognize that situations abound where the sheer size of an electronic or optoelectronic device makes it virtually impossible to base it on single crystal material. Display systems are an obvious case in point—one could not seriously imagine making a 26 in. TV display screen from expensive single crystals. Even after decades of development, silicon single crystal boules have only now achieved 12 in. diameters and no other semiconductor is even within range of this. Which, at least in part, explains why the cathode ray tube has for so long dominated the display field. High power solar cell electricity generators represent another obvious example where the large areas involved, at least, militate against the use of single crystal material, stimulating the thought that it might be possible to satisfy at least some system demands with non-single crystal material.

In Chapter 9 we instanced an example of quite useful performance from lead chalcogenide infrared photoconductors in the form of polycrystalline thin films, deposited on glass or vitreous quartz substrates, a completely non-single crystal approach; as far back as Chapter 3, we noted the use of polycrystalline silicon in forming gates in metal oxide silicon (MOS) transistors and, even earlier, in Chapter 2, made the acquaintance of the cat's whisker radio detector which relied on a polycrystalline nugget of PbS—so there certainly exist precedents. It is far from obvious just how widely similar techniques may be applied to other problems, but the fact that, in some cases, they *can* provides the *raison d'etre* for the present chapter. We shall be concerned to examine the physics and some of the applications of semiconductor materials in either polycrystalline or amorphous form. Needless to say, such materials behave in ways which differ considerably from their single crystal counterparts and we need to examine their properties in a certain amount of detail before we can hope to understand the applications. This we do in the next two sections, followed in Sections 10.4 and 10.5 by discussion of two major applications of amorphous and polycrystalline semiconductors; those of solar cell electricity generation and liquid crystal TV (LCTV) display screens. Finally, in Section 10.6, we take up the question of what has come to be called 'porous silicon', yet another form of matter with exciting properties which promises to find important applications in optoelectronics. While it has to be admitted that these examples represent only a small fraction of the overall semiconductor device portfolio, they are certainly of considerable practical importance—perhaps one chapter out of ten provides a reasonably fair indication of their significance.

10.2 Polycrystalline semiconductors

As indicated in section 10.1, polycrystalline semiconductor materials have a long history. Prior to the concerted effort put into developing the growth of artificial single crystals, which took off in the early 1950s with Ge and Si (see Chapter 3), they represented the only option available. Thus, cat's whisker detectors were based on polycrystalline PbS, semiconductor rectifiers in the 1930s used polycrystalline CuO_2 or Se and, early infrared detectors depended on thin films of Th_2S_3, PbS, PbSe, or PbTe. Remember, too, that much of the pioneering work on the semiconductor laser, described in Chapter 5, relied on cutting small crystallites from polycrystalline masses, GaAs crystal growth still being in its infancy in 1962. In most cases, of course, the achievement of optimum performance depended on introducing single crystal material, usually in the form of epitaxial thin films and frequently in the form of lattice-matched multilayer structures. However, in a few cases, polycrystalline films did provide surprisingly good device results—the example of infrared photoconductors is a case in point—so it behoves us well to look at their modus operandi. What is it about such polycrystal material that distinguishes it electronically from its single crystal counterpart?

As made clear by numerous electron microscope studies, an important structural characteristic of a polycrystal film is the presence of 'grain boundaries' which separate small single crystal regions within the film. In other words, the film consists of a conglomeration of tiny single crystals with sizes which lie typically within the range 10 nm to 10 μm (depending on the material and on the mode of preparation) which are in intimate contact with each other but where the contact regions are not, in a crystallographic sense, properly aligned. Generally, the individual crystallites are slightly misaligned with respect to each other so that the boundaries contain high densities of dislocations and so-called 'dangling bonds' (atoms not properly chemically bonded). We shall not attempt to delve too deeply into the precise nature of these interface regions but simply note the very important fact that they contain a high density of electron states which trap electronic charge and result in band-bending within the crystal grains. It is this property which gives rise to the peculiar behaviour of such films, thereby distinguishing them from corresponding bulk single crystals.

Let us look in more detail at the situation in an *n*-type film, containing N donors per unit volume (a corresponding model can equally well be constructed for *p*-type material but involves thinking in an inverted band context which most of us find less easy!). At normal temperatures where the shallow donor atoms are thermally ionized, free electrons exist in the conduction band of the grain material and are able to move freely within the grains. Some of these will reach the grain boundaries where they may be captured by interface states, becoming spatially localized in the process. This fixed negative charge has the property of repelling free electrons from the region of the grain close to the interface and gives rise to a depletion region similar to that associated with a Schottky barrier contact to a semiconductor (which we discussed briefly in Chapter 3). In terms of the conduction band energy, the depleted region is characterized by band-bending, as illustrated in Figure 10.1 (the separation between the conduction band E_C and the Fermi level E_F increases, consistent with the reduction in free carrier density in this region—remember that $n = N_C\{\exp-(E_C - E_F)/kT\}$). In Box 10.1 we show that there is a relationship between the amount of band-bending φ_b and the densities of donors N and interface trap states N_t which allows us also to calculate the extent of the depletion region W.

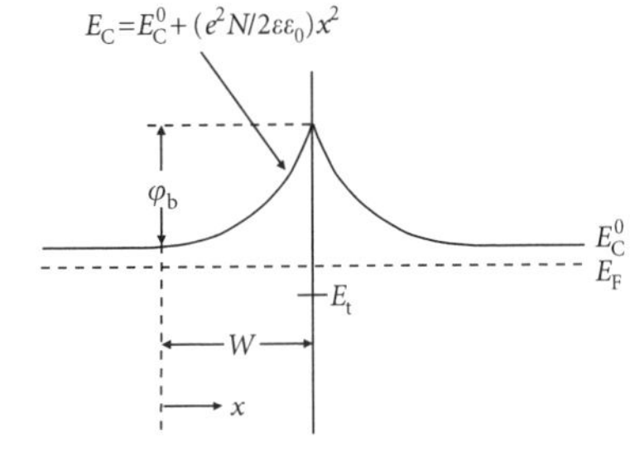

Figure 10.1. Schematic energy diagram to illustrate the bending of the conduction band in the vicinity of a grain boundary. The interface traps at energy E_t lie below the Fermi level and are therefore full of electrons. Total band-bending φ_b and depletion width W depend on the trap density N_t and doping level N in the semiconductor grains. This diagram corresponds to the case where the depletion width is significantly less than half the grain size l—that is, $N > N_t/l$.

One of the principal conclusions of Box 10.1 is the importance of the relative sizes of the depletion width W and the grain diameter l. For a certain combination of interface trap density and bulk doping level, the depletion regions from opposite sides of the grain just meet at the grain centre, so that the grain is almost depleted of free electrons (see Figure 10.2b). The appropriate relationship is $Nl = N_t$, for which case the band-bending φ_b shows a sharp maximum. Figure 10.3 provides an example of such a theoretical plot of φ_b against N, for a trap density

Box 10.1. The grain boundary depletion region

An understanding of depletion region parameters associated with grain boundaries in a polycrystalline semiconductor is crucial to any understanding of film behaviour. This box therefore provides an elementary account of this basic theory in terms of the model illustrated in Figure 10.1. We assume that the semiconductor grains contain a uniform distribution of donors with a volume density N (m^{-3}) and that the grain boundaries are characterized by a surface density N_t (m^{-2}) of interface traps with an energy E_t. (Note that there is no good reason to assume that all these traps occur at the same energy E_t—we could equally well assume some distribution of trap energies throughout the band gap but that makes the mathematics correspondingly more difficult! Let that be our justification!) Electron charge trapped at the interface results in a degree of band-bending φ_b with corresponding depletion width W which can be calculated from standard depletion layer theory. Both φ_b and W depend on the respective densities N and N_t, and on the trap energy E_t—it is this relationship which we now want to determine.

Assuming a uniform distribution of donors within the grains, the conduction band energy in the depletion region varies quadratically with position x (see Figure 10.1), according to:

$$E_C = E_C{}^0 + (e^2N/2\varepsilon\varepsilon_0)x^2, \tag{B10.1}$$

where ε is the relative dielectric constant of the semiconductor (in the calculations below we take a nominal value of $\varepsilon = 10$). This provides us with a relation between the total band-bending φ_b and the width of the depletion region W:

$$\varphi_b = e^2NW^2/2\varepsilon\varepsilon_0. \tag{B10.2}$$

As is certainly true in many cases, the trap energy E_t lies well below the Fermi level and the traps are, therefore, all filled with electrons. We can then equate the total interface charge eN_t to the charge within the two depletion regions (one either side of the interface), thus:

$$N_t = 2NW. \tag{B10.3}$$

Now we can eliminate W and obtain an expression for the band-bending in terms of N and N_t:

$$\varphi_b = e^2N_t^2/8\varepsilon\varepsilon_0N. \tag{B10.4}$$

Note that, for any particular value of trap density, φ_b varies like N^{-1}, becoming small at high doping levels but increasing as N falls. If we take a typical value of $N_t = 10^{16}$ m^{-2}, it follows that $\varphi_b = 0.023$ eV at $N = 10^{24}$ m^{-3} but increases to 0.23 eV at $N = 10^{23}$ m^{-3}, and 2.3 eV at $N = 10^{22}$ m^{-3}! This last figure is obviously unreasonable (being larger than the band gap in many cases!) and implies that something must happen to prevent it. In fact, two things may happen—first the trap level E_t may move above the Fermi level so the assumption of filled traps no longer holds, but, second, and more importantly, the depletion width increases (see equation (B10.3)) until the depletion regions fill the whole grain and, again, our simple theory no longer applies.

Let us look at this latter case in more detail. If we take the grain size to be $l = 100$ nm, we can use equation (B10.3) to obtain a value for the critical doping level at which the two depletion regions from either side of the grain just meet at the centre. This is $N_{max} = N_t/l = 10^{23}$ m^{-3} and the corresponding value of band-bending is:

$$\varphi_{b\,max} = e^2lN_t/8\varepsilon\varepsilon_0, \tag{B10.5}$$

Box 10.1. Continued

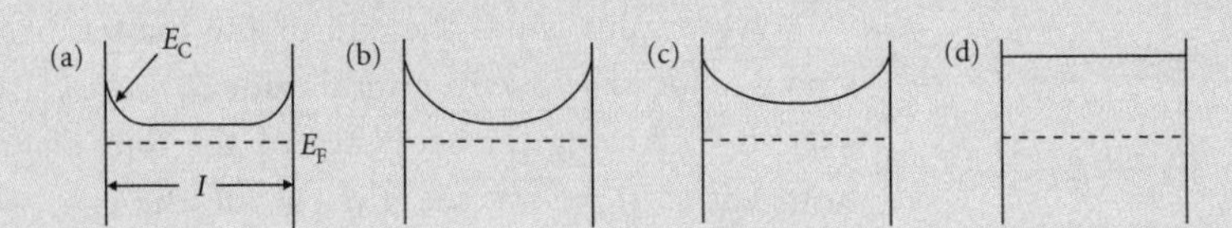

Figure 10.2. Schematic to illustrate the different regimes of band-bending within a crystal grain of dimension l. In (a) is shown the situation where $N > N_t/l$ and the depletion width W is much less than $l/2$. (b) This represents the case $N = N_t/l$, where the two depletion layers just meet at the centre of the grain; (c) applies to the case where $N < N_t/l$ but where band-bending is still finite—note that the lowest point in the conduction band has moved further away from the Fermi level; (d) corresponds to low doping levels where the band-bending is now negligibly small and the conduction band lies far above the Fermi level. This last situation is characterized by the inequality $\lambda_D > l$, where λ_D is the Debye screening length.

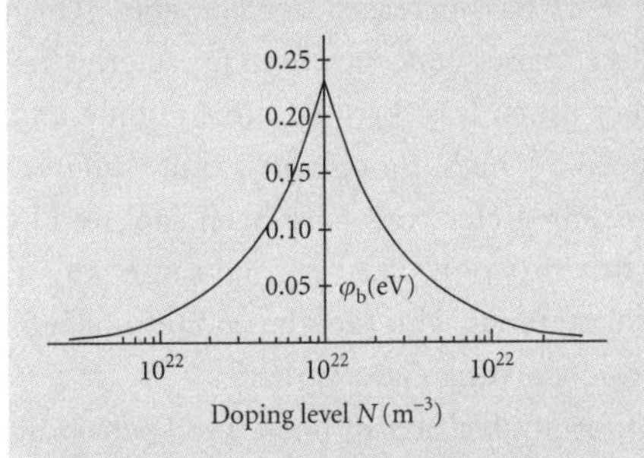

Figure 10.3. This figure shows an example of the variation in band-bending φ_b as a function of doping level in the semiconductor grains, showing the sharp maximum which occurs at $N = 10^{23}$ m^{-3}. The figure is drawn for the case where $N_t = 10^{16}$ m^{-2} and $l = 100$ nm. In plotting it, we have assumed that the interface traps all lie below the Fermi level.

which, for our case, gives a value of 0.23 eV.

We now need to ask what happens if N is decreased below this critical value. In fact, the mathematics becomes more difficult so we shall merely quote the result which is that the value of E_C^0 (equation (B10.1)) starts to increase (i.e. it moves away from the Fermi level—see Figure 10.2(c)) and the value of n within the grains decreases—n is no longer equal to N but becomes rapidly smaller than N as N decreases below N_t/l (in fact it is thermally activated with an activation energy equal to $(E_C^0 - E_t)$). At the same time, the band-bending also becomes smaller, according to the relation:

$$\varphi_b = e^2 l^2 N / 8\varepsilon\varepsilon_0, \tag{B10.6}$$

so, in this regime, φ_b varies linearly with N, tending to zero as N becomes small—in our case $\varphi_b = 0.0023$ eV when $N = 10^{21}$ m^{-3}, which is negligibly small (i.e. $\varphi_b \ll kT$) near room temperature. This implies that the conduction band is essentially flat right across the grain (see Figure 10.2(d)).

Comparing equations (B10.4) and (B10.6) illustrates the fact that φ_b goes through a sharp maximum at the critical doping density $N_{max} = N_t/l$, which is why we referred to it as $\varphi_{b\,max}$ in equation (B10.5). Using the values $N_t = 10^{16}$ m^{-2} and $l = 100$ nm, we have plotted φ_b as a function of doping level in Figure 10.3.

$N_t = 10^{16}$ m^{-2} and grain size $l = 100$ nm. This band-bending is important in its effect on electrical transport properties because free electrons have to surmount the barriers in order to move through the film, though it is clear from Figure 10.3 that, at both high and low doping levels, the barriers will have rather a small effect. If φ_b is less than kT ($kT = 0.026$ eV at room temperature), free carriers with thermal energies will easily surmount them. On the other hand, when $Nl \sim N_t$ and φ_b is near its maximum value, they may represent a major impediment to current flow (the appropriate relationship has the form $I = I_0 \exp\{-\varphi_b/kT\}$ so, if, for example, $\varphi_b = 10\,kT$, current may be reduced by a factor of order $e^{10} = 2 \times 10^4$). It now becomes clear why polycrystalline films with moderate doping levels can have very high effective resistivities, differing markedly in this respect from single crystal films.

However, this is not the end of the matter, because, if the doping level is reduced below the critical value $N_{max} = N_t/l$, even stranger things happen. Following the sequence of the four diagrams in Figure 10.2 from left to right, we see, first of all, the case where $N \gg N_t/l$ and the barriers are rather small, then the case $N = N_t/l$, discussed above, then an example where $N < N_t/l$ and the band-bending has become somewhat smaller than $\varphi_{b\,max}$ and, finally the case where $N \ll N_t/l$ and the conduction band is essentially flat throughout the grain. There is no longer any barrier to current flow but notice that the separation of the conduction band and Fermi level has increased considerably, which implies a strong reduction in free carrier concentration. In fact, the free electron concentration n is very much less than N in this regime and, again, film resistivity is unexpectedly high, though for a quite different reason, namely that most of the free electrons have been captured by interface traps which act, in this situation, in much the same way as deep traps in bulk single crystal material. This (i.e. the addition of bulk traps), for example, is how semi-insulating GaAs is made.

Having established these important features of polycrystal behaviour in an essentially theoretical manner, it is instructive to think about possible experimental techniques which may (or may not!) provide supporting evidence. In single crystal material, for example, we saw in Chapter 5 (Section 5.4) that the standard measurement of Hall effect and resistivity could be of crucial importance in separating the effects of free carrier density and of carrier mobility. This followed from the relation between the Hall coefficient R_H and carrier density ($R_H = 1/ne$) which, together with a resistivity measurement, allowed the carrier Hall mobility to be determined as $\mu_H = R_H/\rho$. (We shall ignore the subtlety introduced by the Hall scattering factor as of little relevance to our present concerns.) How, one may ask, would similar measurements impinge on the interpretation of polycrystal behaviour? In fact, the answer has been very largely a positive one—Hall and resistivity measurements have helped enormously in confirming the broad brush picture painted by the above theoretical artistry, though not, let it be said, without a certain amount of agonizing over detail. Indeed, the more pedantic critic may well be heard muttering about artistic licence in the interpretation of the Hall coefficient in non-uniform materials but we shall gloss over such niceties in our present account, simply referring the more adventurous student to the review article which Martin Powell and I wrote some years ago—*The Hall Effect in Polycrystalline and Powdered Semiconductors* (Orton and Powell 1980).

To get ourselves started, it is advantageous to consider the case of a polycrystalline sample which is fairly heavily doped so that $N \gg N_t/l$ and the inter-grain barriers are of only modest height (Figure 10.2a).

Suppose we perform a Hall measurement on such a sample and work out a Hall coefficient in the usual way. What does it tell us? The consensus of a number of different theoretical approaches, favours the suggestion that it actually measures the free carrier concentration in the bulk of the grain, which we treat here as a direct measurement of the doping density N (i.e. we assume that the donors are shallow enough that, in the bulk of the grain, they are all ionized). This implies that, in the vicinity of room temperature, the carrier density (and, therefore, the Hall coefficient) do not depend on temperature. This, however, cannot be true of the corresponding resistivity because, as we have seen, electrons are obliged to clamber over the inter-grain barriers in order to progress to the outside world. We must expect then that both conductivity σ and Hall mobility μ_H will be thermally activated, for example,

$$\mu_H = \mu_0 \exp\{-\varphi_b/kT\}, \tag{10.1}$$

which implies that we might obtain an experimental value for φ_b by measuring μ_H as a function of temperature—that is, by measuring both R_H and ρ against T. (This assumes that current flow is dominated by thermal emission over the barriers and not by tunnelling through them—care is therefore required but, suffice it to say, it can be done with a fair degree of reliability.)

This deals with the temperature-dependence of mobility but what about its magnitude? In other words, how can we interpret the pre-factor μ_0? A first thought on the subject might suggest that μ_0 should be identified with μ_1 (the electron mobility in the bulk of the grain) but more structured analysis suggests that, in most cases, this is not so. In Box 10.2 we show, on the basis of a simple, one-dimensional model of current flow through a grain, and thermionic emission of electrons over the intergrain barriers, that μ_0 actually depends on grain size l and on temperature as $T^{-1/2}$ (see equation (B10.16)). In fact, μ_0 appears to be the mobility corresponding to grain boundary scattering—that is, when the mean free path for electron scattering is equal to the grain size l. It is easy, then, to calculate appropriate values for μ_H which turn out to be in the region of 0.1 $m^2V^{-1}s^{-1}$ at high doping levels (where φ_b is less than kT), decreasing to very much smaller values at lower doping levels, when φ_b approaches its maximum value. A typical mobility might be $\mu_H = 10^{-5}$ $m^2V^{-1}s^{-1}$ when $\varphi_b = \varphi_{b\,max}$. (Note, by the way, that such low mobilities are extremely difficult to measure.)

This, however, is only half the story. We still need to consider the case of even lower doping levels where $N < N_t/l$. Following the sequence of band pictures in Figure 10.2, we should expect the measured mobility to increase again as N decreases below N_{max} and φ_b becomes correspondingly smaller. Indeed, when $N \ll N_{max}$, the conduction band becomes essentially flat throughout the film and we

Box 10.2. The Hall mobility in polycrystalline films

The interpretation of Hall measurements in polycrystalline semiconductor films implies that the Hall coefficient R_H provides a measure of the free carrier density n in the bulk of the grain. Referring, now, to the case where the doping level N is greater than the critical value N_{max} and the band picture is that shown in Figure 10.2(a), this assumption further implies that the effective carrier mobility in the film is thermally activated—that is, given by:

$$\mu_H = \mu_0 \exp\{-\varphi_b/kT\}, \tag{B10.7}$$

where φ_b is the band-bending in the vicinity of the grain boundaries. This box is concerned to derive this result from a basic model and to interpret the prefactor μ_0 in terms of this model. To derive an expression for μ_H, we must be clear about the mode of current flow through the film, which can be divided into two distinct regions, the bulk of the grain and the barrier region. We begin by writing expressions for the current density in the two regions and for the overall grain. Thus, in the bulk of the grain we have ohmic conduction, for which:

$$J_g = ne\mu_1 V_g/l, \tag{B10.8}$$

where V_g is the voltage drop across the bulk of the grain. In the region of the barrier, we write the current density in terms of thermionic emission of electrons over the barrier:

$$J_b = J_S[\exp(eV_b/kT) - 1] \approx eV_b J_S/kT, \tag{B10.9}$$

where we have taken $eV_b \ll kT$. This is justified because V_b is a small fraction of the voltage applied across the film V_A. For grains of size 100 nm and an experimental set-up in which the film is 1 cm long, $V_b/V_A \sim 10^{-5}$ and $V_b \sim 10^{-4}$V.

For the grain as a whole, we can write:

$$J = ne\mu_H(V_g + 2V_b)/l. \tag{B10.10}$$

Note that we are here assuming the depletion width W to be small compared with the grain size l so we can take l to represent both the overall grain size and the distance between the edges of the depletion regions. (The problem can easily be treated rigorously but making this assumption does simplify the mathematics a little.)

Bearing in mind that these three current densities must all be equal (in the one-dimensional model which we are using), we can eliminate the various voltages from equations (B10.8)–(B10.10) and obtain an expression for the Hall mobility μ_H which can be written in the form:

$$\mu_H^{-1} = \mu_1^{-1} + \mu_b^{-1}, \tag{B10.11}$$

where

$$\mu_b = lJ_S/2nkT. \tag{B10.12}$$

J_S is the reverse saturation current density for thermionic emission over the barriers and can be written as:

$$J_S = J_0 \exp\{-\varphi_b/kT\}, \tag{B10.13}$$

Box 10.2. Continued

which allows us to write μ_b as:

$$\mu_b = \mu_0 \exp\{-\varphi_b/kT\}, \tag{B10.14}$$

where

$$\mu_0 = lJ_0/2nkT. \tag{B10.15}$$

To evaluate μ_0 we need an expression for J_0 which can be obtained from the discussion of thermionic emission over a Schottky barrier given in Sze (1969). The final result is that:

$$\mu_0 = elv_{th}/8kT = el(8\pi m^\star kT)^{-1/2}, \tag{B10.16}$$

where v_{th} is the thermal velocity of electrons, given by $v_{th} = (8kT/\pi m^\star)^{1/2}$, $m^\star$ being the effective mass of the carriers. This allows us to make an estimate of μ_0 by assuming appropriate values for the various parameters. Taking $l = 100$ nm and $m^\star = 0.2\,m$, we arrive at a value of $\mu_0 = 0.12\ \mathrm{m^2V^{-1}s^{-1}}$ and, if we also assume $\varphi_b = 2kT$, we obtain $\mu_b = 0.016\ \mathrm{m^2V^{-1}s^{-1}}$. This can be compared with a typical value of μ_1, the mobility in the bulk of the grain, of about $0.1\ \mathrm{m^2V^{-1}s^{-1}}$. So, returning to equation (B10.11), we see that, to a good approximation, the measured Hall mobility is given simply as:

$$\mu_H = \mu_b = \mu_0 \exp\{-\varphi_b/kT\} \tag{B10.17}$$

as expected. This is a rather lengthy argument but it is important to appreciate where this final result comes from and to be able to estimate the effective mobility we can expect to measure in practice.

One final comment might be in order, while indulging in this spate of equations. We can define a mobility μ_g by supposing that free carriers will be scattered at every grain boundary and writing their mean free path λ as equal to the grain dimension l. (Note that this assumes the bulk mobility μ_1 to be high enough so that scattering within the grain is negligible.) Thus:

$$\lambda = v_{th}\tau = l \tag{B10.18}$$

from which it follows that:

$$\mu_g = e\tau/m^\star = el(\pi/8m^\star kT)^{1/2} \tag{B10.19}$$

If we neglect the small factor π, this is identical with μ_0 (equation (B10.16)) which is, at least, reasonable because, when φ_b approaches zero, we expect (see equation (B10.11)) to measure a mobility equal to μ_0 (always assuming that $\mu_0 \ll \mu_1$).

no longer expect μ_H to be thermally activated, but to approximate to μ_0, the same value as measured at high doping levels. In other words, we expect a plot of μ_H vs N to show a sharp minimum at $N = N_{max}$, rising to an asymptote of μ_0 at both high and low doping extremes. From Figure 10.3, we see that a typical value of N_{max} is likely to be about

10^{23} m^{-3} but it may vary considerably, depending on interface trap density and grain size. However, the chief lesson to learn from all this is the totally different behaviour of mobility in a polycrystalline semiconductor from that in a single crystal sample of the same material. Not only must we expect it to be thermally activated, but, when measured as a function of doping level, to show a strong minimum at some characteristic value of doping density N_{max} (the subscript 'max' refers, of course, to the maximum in band-bending which gives rise to the minimum in mobility). This, combined with the unusual behaviour of the free carrier density at low doping levels, provides a signature which allows one to characterize a particular set of polycrystalline samples.

So much for the background theory—what happened experimentally? As we have already pointed out, polycrystalline semiconductors were in practical use from a fairly early stage in the history of the subject, though, as one might expect, applied in a largely empirical manner. Perhaps the first serious attempts to understand their optoelectronic behaviour were associated with the infrared photoconductors such as PbS. It had been established as early as the 1920s, in relation to thalium sulfide photocells, that oxygen played an important role in optimizing performance and similar observations were made for the lead chalcogenides during the 1940s. It was found that oxygen could even cause type conversion from *n*-type to *p*-type conductivity and, in the early 1950s, this led to the idea that the sensitivity of these cells was related to the presence of *p-n* junctions within the film, an idea given credence by the observation that certain small areas of the film showed much higher sensitivity to a small light spot than the rest of the film (see Bube 1960: ch. 11). This was followed by detailed measurements of photoconductivity which were interpreted in terms of mobility modulation (rather than the free carrier modulation which one would expect in single crystal photoconductors). The mechanism for this involved minority carriers being trapped at grain boundaries, neutralizing some of the interface charge and, thereby, reducing the amount of band-bending. It was even discovered that nominally single crystal samples actually contained barriers, strengthening the argument that they must also occur in thin films. Various theoretical treatments of barrier photoconductivity were developed, including one by Petritz in 1956 which followed the model outlined at the beginning of this section. However, there were voices raised in defence of the modulated free carrier model of photoconductivity, some of which denied even the existence of barriers in their owners' samples. The problem was that it was very difficult to think of a definitive experiment which would settle the argument, not least because it was impossible to control doping levels in these thin film samples (as we saw above, very different behaviour can be expected depending on the relative values of doping level, grain size, and trap density and, if none of these parameters was known, it was impossible to know what to expect!). So the matter rested—not that the argument showed any sign of abatement!—until the 1970s when definitive experimental data finally became available.

In the meantime, applications of thin films grew apace. Following the demonstration of the first silicon MOS transistor at Bell Labs in 1960, Weimer invented the thin film transistor (TFT) in 1961 and a spate of similar devices was reported using a range of semiconductors—Te, PbTe, InAs, InSb, GaAs, CdS, and CdSe, to name but a few. It was necessary to employ some form of deposited gate insulator because, unlike silicon, none of these materials formed satisfactory oxides. However, there could be no denying that they showed considerable promise as amplifiers or switches suitable for use in large area circuits. We shall say no more about them here (see Sze 1969: ch. 11) but press on into the 1970s.

Many measurements of Hall effect and resistivity on polycrystalline films were reported in the 1960s and 1970s but it was not until the period 1971–5 that convincing evidence to support the barrier model was obtained. It came essentially from a rapidly developing interest in polysilicon (not only as a MOS gate but also for TFTs) films of which could be deposited in a rather more controlled manner and, more importantly, could be doped to produce known donor or acceptor levels, either by gas phase doping or by ion implantation. In 1971 Kamins (of Fairchild R&D Labs) reported on both *n*- and *p*-type films with doping levels in the range 10^{23}–10^{25} m^{-3}. He observed a marked drop in Hall mobility with decreasing free carrier density, as predicted by the barrier theory (for the case $N > N_{max}$). In 1972, Cowher and Sedgwick (IBM T. J. Watson Research Laboratory) measured similar films which covered a wider range of doping levels and observed, not only the drop in mobility, but also a sharp decrease in free carrier density when the doping level fell below 10^{24} m^{-3}, as predicted for the case $N < N_{max}$. Unfortunately, neither set of workers made measurements as a function of temperature so it was impossible to make any direct estimate of barrier height or trap depth, though indirect evidence suggested reasonable values.

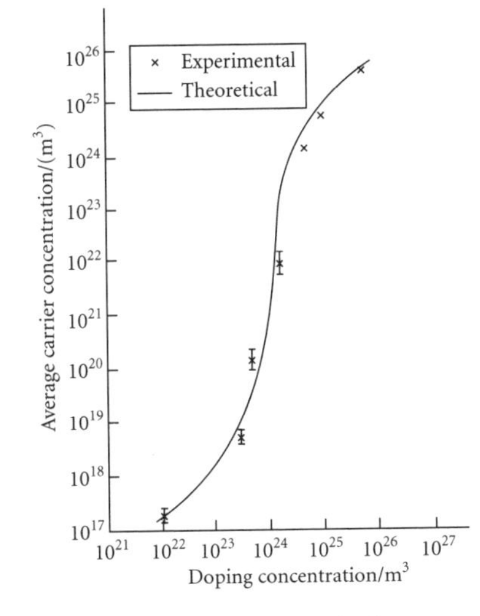

Figure 10.4. Experimental data for the variation of free hole density p as a function of doping level N for seven *p*-type silicon thin film samples. The solid line is calculated on the basis of the barrier theory of conductivity—the crosses represent experimental data. The free hole densities in samples for which $p \ll N$ are characterized by thermal activation energies. (From Seto J. Y. W. (1975) *J. Appl. Phys.*, 46, 5247, fig. 1). Reprinted with permission from American Institute of Physics.

The most complete and convincing evidence came in 1975 from the work of Seto (of the General Motors Research Labs, Warren, MI) who was also concerned with silicon films, this time doped *p*-type by ion implantation. To illustrate the significance of his work, we show his results for both free carrier density and mobility as functions of doping level in Figures 10.4 and 10.5. The sharp drop in hole density for values of N less than about 10^{24} m^{-3} coincides with the corresponding drop in mobility which also occurs at $N = 10^{24}$ m^{-3}, as predicted by theory. The solid lines in both figures were calculated from the barrier model and show excellent agreement with the experimental data. Seto estimated barrier heights φ_b from the temperature-dependence of mobility

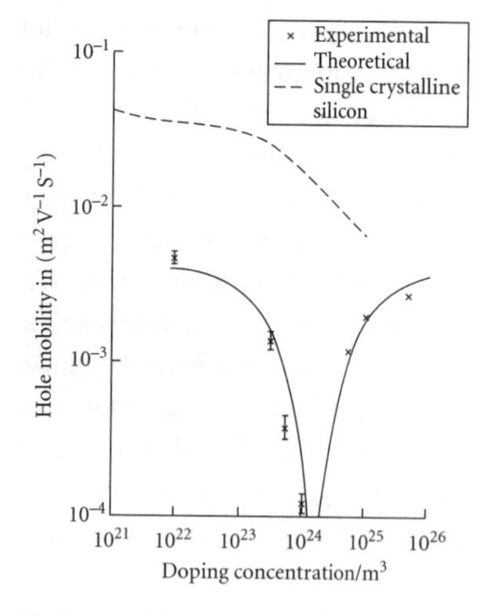

Figure 10.5. Experimental Hall mobilities measured on the same samples as shown in Figure 10.4 (from Seto 1975). The solid curve is calculated on the basis of the barrier model—crosses represent experimental data. The dashed curve represents the variation of hole mobility in single crystal silicon, for comparison. The mobility thermal activation energies show a maximum value of approximately 0.15 eV, corresponding to the minimum in mobility. (From Seto 1975: fig. 2—see Figure 10.4). Reprinted with permission from American Institute of Physics.

measured on the four samples with $N = 1 \times 10^{24}$ to 5×10^{25} m^{-3} and found the activation energies to vary roughly as N^{-1}, as predicted. The corresponding values of μ_0 were also in reasonable agreement with those calculated from the measured grain size $l = 20$ nm and the value of trap density N_t was found to be consistently about 3×10^{16} m^{-2}. Finally, it was possible to estimate a value of trap energy from the measured ratio of p/N from the lowest doped sample—the result being $(E_C - E_t) \sim 0.35$ eV. While there were one or two detailed imperfections in the fit of experimental data and theoretical prediction, there could be little doubt that this work provided strong support for the barrier theory, which could no longer be dismissed by even the most biased of protagonists. Many more reports of Hall effect and resistivity appeared, covering a wide range of materials and methods of deposition, many of these being summarized in Orton and Powell (1980). They show fair agreement with the barrier theory but are less detailed than Seto's results and so provide only general support for the theory. Seto's work stands out.

While making a major contribution to our understanding of electrical conduction in polycrystalline films, Seto did not study photoconductivity. The controversy over photoconductivity mechanisms remained, therefore, in an unsatisfactory state, even after Seto, and it was clear that something needed to be done about it. Not only was it necessary that a detailed study of dark conductivity be made on a set of known samples, but that it be extended to include photoconductivity. It was probably better, too, that it should be performed on a semiconductor, such as CdS, which had been widely used as a photoconductor. The difficulty, here, lay in the fact that CdS films could not readily be doped in a controlled manner, so it would be necessary to derive doping levels from Hall effect and resistivity measurements on the basis of the barrier model, and then use this analysis to interpret photoeffects. This, at least, was what my (then) Group at Philips Research Laboratories in Redhill did in 1980 (see Orton et al. 1982), in an attempt to pin down the mechanisms of photoconductivity at least once—if not once and for all!

A large number of *n*-type samples was deposited by spray pyrolysis—an aqueous solution of $CdCl_2$–thiourea was sprayed onto a silica glass substrate at 400°C—and the doping level varied by controlling the stoichiometry of the mix (i.e. the ratio of Cd to S ions in the solution). This allowed us to vary the doping over quite a wide range but not with any hope of knowing in advance what it would turn out to be—this had to be inferred from the Hall and resistivity data. The grain size was measured by electron microscopy and turned out to be about 300 nm (though there was quite a significant spread). Electrical measurements were made as a function of temperature over the range 70–300K, either in the dark or under illumination from a quartz–iodine lamp. Dark measurements allowed us to analyse all samples according to the barrier model, from which the estimated range of doping levels was roughly 10^{20}–10^{24} m^{-3}. Barrier heights, obtained from mobility thermal activation energies, peaked at $\varphi_{b\,max} = 0.2$ eV when $N = N_{max} = 2 \times 10^{22}$ m^{-3} and mobilities showed a corresponding minimum at this doping level. The trap density N_t was inferred to be about 5×10^{15} m^{-2}. For doping levels below N_{max}, the free carrier density dropped sharply and was thermally activated exactly as predicted by theory, suggesting a trap energy $(E_C - E_t)$ of roughly 0.4 eV. This gave us confidence in believing that we understood the behaviour of our samples in the dark (in terms of the barrier theory) and encouraged us to study photoconductivity from the same vantage point.

The details of the comparison between theory and experiment in the case of photoconductivity involved various subtleties but two clear conclusions stood out. First, samples could be readily separated into two groups, those with low doping where $N < N_{max}$, for which photoconductivity resulted almost entirely from an increase in carrier density (while mobility remained nearly constant) and those having higher doping $(N \geq N_{max})$ for which both carrier density and mobility increased, this being in good agreement with theoretical prediction. Second, for the second group of samples, there was a relation between Δn and $\Delta \mu$ (the respective changes in carrier density and mobility in response to illumination) which provided a good test of the barrier theory. In summary, we could claim that we now understood rather well the circumstances under which photoconductivity was dominated by a change in carrier density and those under which it was dominated by a change in mobility. The answer to the original controversy between those who believed in mobility modulation and those who defended free carrier modulation was now clear—they were both right!

This surely represented a happy outcome and a considerable vindication of the barrier theory but we should, perhaps, make one final comment. The theory, as we have presented it, can be no more than a first approximation to reality. In any real semiconductor film, there exists a spread in grain sizes (which can be demonstrated very clearly by electron microscopy) and, probably, a spread in grain boundary parameters (especially trap densities) which implies a spread in barrier heights and values for μ_0. A more complete theory should obviously take account of these factors and involve an element of 'percolation theory' (current percolates through a film by selecting those paths which offer least resistance—see Orton and Powell 1980). This makes the theoretical challenge significantly greater but also makes one wonder a little at the undoubted success of the basic (oversimplified) account we have given here. Science is rarely quite as straightforward as it sometimes seems.

10.3 Amorphous semiconductors

In a sense, the polycrystalline semiconductors discussed in Section 10.2 represent a half-way house between single crystal and amorphous semiconductors. The characteristic which polycrystal and amorphous materials have in common is the element of disorder in their structure. While the ideal crystal is characterized by complete long-range order, polycrystal materials show long-range order only within the volume of the crystallites—if we examine them on a length scale large enough to include many crystallites, it is clear that this degree of order no longer exists. In a perfect single crystal, if we know the coordinates of any one specific atom, we effectively know the coordinates of all the other atoms in the sample—this being a fundamental property of a crystal lattice. For the polycrystal sample, this is no longer true and the lack of order results, as we have seen in considerably modified electronic behaviour. However, it was nevertheless possible to use standard semiconductor theory to treat electrons within the crystallites, while the grain boundaries could be regarded simply as introducing high densities of localized defect states. Amorphous semiconductors, on the other hand, are characterized by a complete absence of long-range order, down to the scale of the spacing between atoms and we can no longer hope to describe their electronic properties in standard semiconductor terms. In this case a totally new theoretical approach is required which we can do little more than hint at here. Mott and Davis wrote a book of nearly 600 pages on the subject: *Electronic Processes in Non-Crystalline Materials* (Mott and Davis 1971), though the reader preferring a rather shorter account might like to consult the introductory chapter of Bob Street's excellent book on amorphous silicon (Street 1991). Volume 21 of *Semiconductors and Semimetals* (Pankove 1984) also contains a comprehensive account of the properties and applications of amorphous silicon. In this section we hope simply to provide a brief outline of the essential properties of amorphous semiconductors to serve as a basis for later discussion of some of their applications.

The most familiar example of an amorphous material is that of common window glass which, being transparent and a good electrical insulator, illustrates the fact that amorphous materials may have wide band gaps just as crystalline insulators do. And, just as there exist many crystalline semiconductors and insulators, there exist many glasses, some of which are insulators and some semiconductors. Today, by far the best-known amorphous semiconductor is amorphous silicon which, during the past two decades, has found several important applications in the field of semiconductor optoelectronics but we should not overlook the fact that the study of the electronic properties of glassy materials dates back to a time, during the 1950s, when silicon was not widely known even as a crystalline semiconductor. Much of the early work was concerned with the glassy forms of arsenic, selenium, and tellurium, and with a variety of chalcogenide glasses such as As_2Se_3 and a rather bewildering range of more complex compounds such as $As_{0.3}Te_{0.48}Si_{0.12}Ge_{0.10}$ which showed interesting switching behaviour. Meanwhile, the first successful xerographic copier in 1956 made use of a selenium photoconductor. The theoretical basis for the understanding of amorphous semiconductors was also laid down during the 1950s, though, stimulated to a considerable extent by the growing commercial promise of amorphous silicon, the subject boomed significantly during the 1970s and has maintained a reasonably strong following, up to the present day.

To understand the essence of the amorphous state demands some appreciation of the atomic structure of a typical material for which, because of its practical significance, we choose amorphous silicon (though similar models can, of course, be constructed for other materials—see, for example, fig. 1.3 in Street 1991). Figure 10.6 provides a two-dimensional illustration of the bonding of Si atoms in amorphous silicon, an example of a so-called 'continuous random network' which should be compared with Figure 1.3 in Chapter 1. Note that the fourfold bonding behaviour of silicon atoms which we discussed in Chapter 1 is maintained in the amorphous state but there is a degree of variability in bond length and bond angle which does not occur in the crystalline state. Two features of the structure are immediately obvious: first, there is a lack of long-range order—once we move a few atom distances away from any particular silicon atom, it is quite impossible to predict the positions of surrounding atoms—and, second, there exist 'dangling bonds' (i.e. unsatisfied bonds) which represent defect states in the network. Similar effects occur in the real three-dimensional structure, the lack of order and the presence of dangling bond states being typical of all amorphous semiconductors and implying important new electronic properties, not to be found in the ideal single crystal. Rather than specifying a crystal structure and lattice constant, as in crystalline materials, it is usual to describe amorphous materials in terms of a radial distribution function (RDF) which specifies the atomic density at a distance r from any particular atom. In amorphous silicon (usually written as a-Si), this peaks at distances $r_1 = 0.235$ nm and $r_2 = 0.35$ nm which are the same as for crystalline (tetrahedrally bonded) silicon but the peaks are no longer sharp, their width providing a measure of the spread in bond length and bond angle. Subsequent peaks become less and less distinct as the radial distance r increases further, confirming the lack of long-range order.

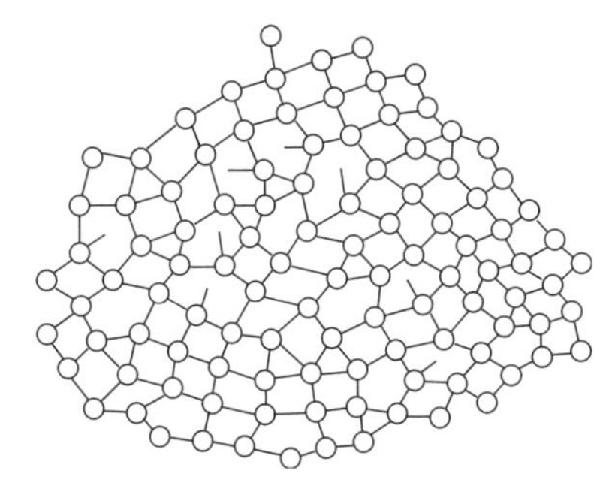

Figure 10.6. Artist's impression of a two-dimensional continuous random network of silicon atoms to illustrate the nature of the structure of amorphous silicon (which is, of course, three- dimensional). Note that both bond lengths and bond angles differ significantly from the regular crystalline structure of Figure 1.3. Note also that the structure contains a significant number of dangling bonds which represent defects in the material, introducing high densities of electron states within the band gap.

We can now examine typical electronic properties of amorphous materials in more detail. Considering first the most important parameter for any semiconductor, its band gap, it is usually found that an amorphous semiconductor is characterized by a gap which is surprisingly similar to that of its crystalline equivalent, though quite significantly dependent on the method used to deposit the film. In the case of a-Si, for example, the gap takes values between about 1.2 and 1.8 eV, compared with 1.12 eV for the crystalline case. Nor, because of disorder, is it possible to define a momentum quantum number k, as in a crystal, so the distinction between direct and indirect optical transitions is no longer meaningful and the shape of the absorption curve may therefore be very different in the amorphous case, as, indeed, it is for a-Si. However, for our purposes it is probably sufficient to appreciate that a fairly well-defined band gap does exist so we can talk about valence and conduction bands in much the same way as for a crystalline material.

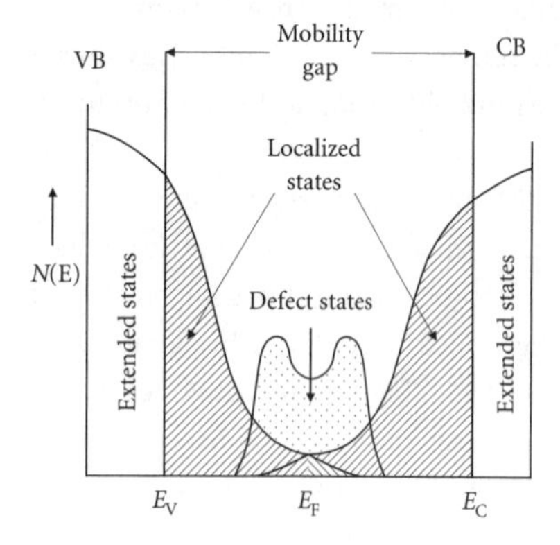

Figure 10.7. Schematic density of states (DOS) function for a typical amorphous semiconductor. Tail states extend from both valence and conduction bands into the energy gap and deep defect states occur near the centre of the gap. A demarcation can be made between extended and localized states which allows one to define a mobility gap (see text). The Fermi level is pinned near the centre of the gap, implying a very small degree of electron/hole occupancy of the extended states.

However, it would be a mistake to suppose that the energy gap is devoid of electron states, as would be true for an ideal crystal. We know that even crystals show evidence of deep states arising from defects and impurities—in the amorphous case this is also true but in a much exaggerated manner. In fact, the standard description of an amorphous semiconductor is in terms of a broad density of states function which is continuous throughout the gap. An example is shown in Figure 10.7, from which it is apparent that band edges are much less sharp than in the crystalline case, so-called band tail states extending quite a long way into the gap from both conduction and valence band edges. In addition, there is a high density of defect states deep in the gap which correspond to the dangling bonds apparent in Figure 10.6. These states tend to pin the Fermi level near the middle of the gap which implies very few free carriers in either of the bands. For example, if we were to use the familiar expression for free electron density $n = N_C \exp\{-(E_C - E_F)/kT\}$ with $(E_C - E_F) = 0.7$ eV we obtain $n \sim 10^{13}$ m^{-3} which is about seven orders less than appropriate to a very lightly doped pure crystalline semiconductor! Thus, conduction in band states is almost nonexistent but, before we assume that such samples inevitably show very high resistivities, we should be careful not to overlook the possibility of alternative conduction mechanisms involving the deep defect states. We shall come back to this in a moment.

Before we can say anything categorically about resistivity, we need to know something of the likely electron mobility and here again mobilities in amorphous materials tend to lag well behind the more familiar crystalline values. This should come as no surprise—the high carrier mobilities in crystalline semiconductors stem from the periodicity of the crystal lattice and it is only the presence of impurities, defects, and lattice vibrations (all features which partially destroy crystal perfection) which lead to finite mobilities. In the amorphous solid, there is no periodicity so we must surely expect low mobilities. However, this is less than half the story. If we look more carefully at Figure 10.7, it is clear that the electron (hole) states near the edge of the conduction (valence) band can be divided into two groups according to whether electrons (holes) are free or localized. Only electrons with energies greater than a certain value E_C are mobile (in the sense that free electrons in the conduction band of a crystalline semiconductor are mobile), those with energies less than E_C being localized. This sharp division between mobile and local states is known as an 'Anderson transition' after its discoverer P. W. Anderson of Bell Labs (in 1958). There is a corresponding valence band energy E_V which allows us to define a 'mobility gap' $(E_C - E_V)$ which plays a role similar to (though not identical to) the band gap in normal semiconductors. In particular, we should expect the optical and mobility gaps to differ slightly from one another, whereas in an ideal crystalline material, the electrical and optical gaps are identical.

We discuss amorphous semiconductor conductivity in Box 10.3, where three different regimes are distinguished, each having a characteristic temperature-dependence (see Figure 10.8). In this example, conduction is dominated at room temperature by electrons in the extended states above E_C with a thermal activation energy equal to $(E_C - E_F)$. Somewhere well below room temperature, the dominant process is one of 'thermally assisted hopping' in the localized tail states (with activation energy $(E_T - E_F)$, E_T being an effective tail state energy), while at lower temperatures still, where few phonons are available, another process, known as 'variable range hopping' takes over. In this latter case the temperature dependence is much weaker and is not strictly activated at all, being proportional to $\exp\{-(T_0/T)^{1/4}\}$. The hopping regimes involve transitions between localized states, which may sound somewhat anomalous but the nomenclature 'localized' applies strictly only at the absolute zero of temperature. Thus $\sigma_{tail} = 0$ at $T = 0$ but is finite at finite temperatures. Nevertheless, there is always an abrupt change of conductivity between states above and below E_C which is infinite at $T = 0$ but amounts typically to a factor of 10^2–10^3 at room temperature. As we show in Box 10.3, the conductivity in band states at room temperature may be as large as 600 $(\Omega\,m)^{-1}$ in amorphous silicon, if $(E_C - E_F) =$ 0.1 eV, but in most materials $(E_C - E_F)$ is considerably larger and the conductivity is therefore orders of magnitude less. In fact, it may often be the case that hopping conductivity in deep states actually exceeds band state conduction at room temperature—even though the corresponding mobility may be much smaller, the density of electrons (or holes) close to the Fermi level is very much greater and this latter effect often turns out to dominate. (Nevertheless, we should note that band state conduction will still become dominant at elevated temperatures because of the large activation energy involved. This simply implies a shift of the temperature axis in Figure 10.8.)

Finally, to return to our discussion of mobility, we note (see Box 10.3) that the mobility of electrons with energies just above E_C is expected to

Box 10.3. Conductivity in amorphous semiconductors

The theory of conductivity in amorphous semiconductors is more complicated than that for crystalline materials and is dominated by the presence of the mobility edge at E_C, see Figure 10.7 (we shall discuss conduction in the conduction band). Detailed calculations suggest that the conductivity in conduction band states above E_C varies with energy and, as energy approaches E_C from above it decreases to a limiting value called the 'minimum metallic conductivity' σ_{min} which depends on the spacing '*a*' between the atoms in the material according to:

$$\sigma_{min} \sim 10^4/a \quad (a \text{ in nanometre}). \tag{B10.20}$$

Thus, taking a to be equal to 0.3 nm, we obtain $\sigma_{min} = 3 \times 10^4 (\Omega\,\mathrm{m})^{-1}$. Note that, if we take the effective density of electron states to be $N \sim 10^{26}\,\mathrm{m}^{-3}$, this suggests an electron mobility for conduction above the band edge to be:

$$\mu_{band} = \sigma_{min}/eN, \tag{B10.21}$$

that is, $\mu_{band} \sim 2 \times 10^{-3}\,\mathrm{m^2\,V^{-1}\,s^{-1}}$. Another approach to deriving an order of magnitude for this is to suppose the mean free path for electron scattering to be approximated by the inter-atomic spacing '*a*'. This gives us:

$$\mu_{band} = ea/mv_{th} \tag{B10.22}$$

where v_{th} is the thermal velocity for electrons ($\sim 10^5\,\mathrm{m\,s^{-1}}$). Thus $\mu_{band} \sim 5 \times 10^{-4}\,\mathrm{m^2\,V^{-1}\,s^{-1}}$. Neither of these calculations should be taken too seriously but we may reasonably expect a value of band mobility in the region of $10^{-3}\,\mathrm{m^2\,V^{-1}\,s^{-1}}$ (i.e. some two orders smaller than a typical value for a crystalline material).

In practice, the electron density in these mobile states is extremely low because the Fermi level is pinned well below E_C and we must write the conductivity as:

$$\sigma_{ext} = \sigma_{band} \exp\{-(E_C - E_F)/kT\}. \tag{B10.23}$$

This is not the end of the story because, though conduction through localized states is not allowed at all at absolute zero temperature, it *is* allowed at finite temperatures by a process known as 'hopping conductivity'. Thus, electrons in band tail states below E_C may conduct by phonon-assisted hopping between adjacent sites. If the average energy of these band tail states is E_T, this hopping conductivity can be written as:

$$\sigma_{tail} = \sigma_T \exp\{-(E_T - E_F)/kT\}. \tag{B10.24}$$

The prefactor σ_T is smaller than σ_{band} but the exponential term is larger than that in equation (B10.23), particularly at low temperatures, so this process is often important at temperatures well below room temperature.

Finally, at very low temperatures, a modified form of hopping (known as 'variable range hopping') takes over, having a temperature dependence of the form:

$$\sigma_{var} = \sigma_V \exp\{-(T_0/T)^{1/4}\}, \tag{B10.25}$$

which implies a much slower temperature-dependence. The overall variation of conductivity with temperature is shown schematically in Figure 10.8.

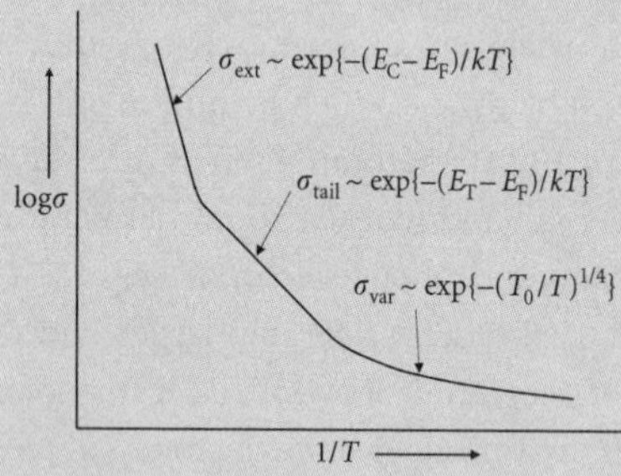

Figure 10.8. Typical temperature-dependence of conductivity σ for an amorphous semiconductor, showing three regimes: σ_{ext} represents conductivity in the extended electron states above E_C, σ_{tail} hopping conductivity in tail states (with phonon assistance), and σ_{var} variable range hopping at the Fermi level which is characterized by a rather small temperature-dependence. The three regimes can be recognized by their very different activation energies.

Box 10.3. Continued

Perhaps the most important result of this box is the indication that, even at room temperature, we must expect rather small values of conductivity. Even in a-Si:H, where the Fermi level can lie as close as about 0.1 eV below the band edge E_C, the maximum conductivity takes a value (from equation (B10.23) of about 600 $(\Omega\,m)^{-1}$, while in most other materials it is very much smaller than this. For comparison, we should remember that the conductivity in a typical crystalline semiconductor might be in the range $1–10^5\,(\Omega\,m)^{-1}$, depending on doping level. In other words, the very largest conductivity in amorphous materials lies somewhere near the middle of the range for doped crystalline materials but we should emphasize that this a-Si:H conductivity represents a quite unusually high value.

be of order $10^{-3}\,m^2V^{-1}s^{-1}$, a value some two orders of magnitude smaller than what we might expect for a crystalline semiconductor, reflecting the lack of periodicity referred to above, while hopping mobilities are found to be very much smaller still, being of order $10^{-7}–10^{-6}\,m^2V^{-}s^{-1}$.

Having thus outlined the nature of amorphous semiconductor electrical behaviour, we can now take up the story of commercial developments by concentrating specifically on a-Si. In fact, the development of a-Si represented a watershed in the sense that it was only when this material came of age that amorphous materials could seriously be considered for the *active* sector of electronic devices. As already remarked, in most amorphous semiconductors, the Fermi level was pinned near the centre of the mobility gap by deep states and all attempts to control conductivity by doping or field effects were fruitless. The early work on a-Si, deposited either by vacuum evaporation or by radio frequency sputtering from Si targets, simply confirmed such behaviour in this material, too. Then, in 1969, an essentially new semiconductor appeared on the scene, hydrogenated amorphous silicon (a-Si:H), grown by a process of plasma deposition from silane (SiH_4) gas. Interestingly enough, though its discovery certainly represented a major breakthrough, recognition came only slowly—indeed, the group at Standard Telephone Laboratories responsible for its discovery was almost immediately closed down by a sceptical management, the running then being taken up by Walter Spear's group at the University of Dundee in Scotland. And, in spite of some excellent work by Spear and his colleague Peter LeComber, almost a decade passed before a-Si:H began to achieve its practical potential. Given the worldwide interest in amorphous semiconductors and the large number of groups involved, one may be forgiven for wondering why it should have taken so long. Very simply, there was a tremendous amount to be done before the subtleties of its behaviour could be fully appreciated by the waiting commercial world. There was also a small matter of stability to be brought under control before the necessary financial investment for commercialization was (or could be expected to be) forthcoming. There was a purely technical problem, too. The Hall effect measurements which had made such an important contribution to understanding electrical conductivity in both single crystal and polycrystalline samples were rendered impotent in amorphous materials, where not even the sign of the Hall coefficient could be relied on to reflect the nature of the conduction process. It was necessary to work considerably harder to obtain unequivocal information on conductivity in a-Si:H than had ever been the case with conventional semiconductors.

The authors who first described the new deposition process were Chittick et al. of Standard Telephone Laboratories, Harlow in Essex. Being in the semiconductor industry, they were well aware that silane was used as a source of silicon for epitaxial deposition of single crystal films and was therefore available in high purity. The standard epitaxy process, however, involved substrate temperatures of order 1000°C which were incompatible with the use of glass substrates (necessary to the kind of large area circuitry envisaged for non-crystalline applications) so some alternative process was obviously needed. The answer was to excite the silane with a radio frequency discharge (or 'glow discharge') so as to decompose the molecules at temperatures below 600°C (indeed, even as low as room temperature). The apparatus consisted of a quartz tube, surrounded by an RF coil and containing a suitably heated substrate, over which flowed the silane gas. Films of amorphous silicon were grown at rates of a few microns per hour and their properties studied as a function of the growth conditions, particularly substrate temperature.

Three features of the resulting films were noteworthy: first, their conductivities were very much *less* than those of the more familiar evaporated or sputtered films, second, these conductivities varied considerably with deposition temperature and, third, it was found possible to modify the conductivity by doping with phosphorous (by adding small amounts of phosphine gas, PH_3, to the silane). This latter observation, though of considerable significance, seems to have been largely overlooked in the short term, perhaps because of a degree of puzzlement over the other two factors, which *were* taken seriously. In a word, while the conductivities of evaporated or sputtered films lay in the range 10^{-3}–10^{-1} $(\Omega\,m)^{-1}$, those of the new films varied between 10^{-8} and 10^{-3} $(\Omega\,m)^{-1}$ as the deposition temperature was increased from room temperature to 600°C. While the old films showed no sign of photoconductivity, the new ones were excellent photoconductors. While thermal annealing of old films had led to a reduction in dark conductivity, the new ones showed an increase—in fact, annealing a room temperature film at 400°C, for example, resulted in conductivity appropriate to a film grown at 400°C. Like many another innovative step, the initial response was one of total confusion—indeed, such were the differences between old and new samples that David Adler, in his book on amorphous semiconductors, published in 1971, made the comment: 'Comparing the experimental results on . . . glow discharge produced Si and evaporated Si is in most respects like comparing lettuce and cabbage.' (Certainly, they did respond very differently when cooked!) However, the Dundee group were convinced there was something important to be discovered in all this and proceeded determinedly to discover it. In 1973 Spear was able to present a paper at a conference in Garmisch–Partenkirchen which cleared away a lot of the fog. Hydrogenated amorphous silicon was beginning to look more than a little interesting.

The key to understanding lay with the density of states, and both direct and indirect evidence was accumulating to suggest that the essential feature of the new material was a strongly *reduced* density of defect states—that is, those states which lay close to the centre of the gap. This had two immediate effects—it reduced the overlap of the electronic wavefunctions associated with electrons or holes trapped in these deep states, which, in turn, decreased the probability for hopping conduction at the Fermi level. Glow discharge films deposited at relatively low temperatures thus became very much *less* conductive because this hopping process dominated conduction at room temperature (it could be recognized by its temperature-dependence). Second, the reduced density of states (DOS) allowed the Fermi level to move towards the conduction band and led to samples which were deposited at rather higher temperatures showing conduction in states above E_C (characterized by a relatively large thermal activation energy $(E_C - E_F)$). The higher the deposition temperature, the more the Fermi level moved and the greater the conductivity became. On the other hand, evaporated samples were characterised by a high DOS, conduction therefore being determined entirely by hopping (of holes) at the Fermi level. Because the DOS was so high, this process readily explained the high values of conductivity observed. It also explained why annealing caused a *reduction* in conductivity—annealing had the effect of reducing the DOS. However, a reduction of DOS in glow discharge samples allowed the Fermi level to move a little closer to E_C which *increased* the band conduction. Once it was established that these two conduction mechanisms played decisive roles, the lettuce/cabbage problem was effectively resolved—the result (if I may be forgiven for changing the metaphor from culinary to criminological) of some excellent detective work.

The next question needing an answer concerned the reason why there should be such a large difference in the density of defect states and again it took several years before definitive evidence could be assembled. As early as 1970, Marc Brodsky of IBM suggested that hydrogen might play an important role and the Dundee group pursued a similar line of thought. Silane contains hydrogen and it seemed quite likely that some of this would find its way into the film in the form of atomic hydrogen, capable of neutralizing Si dangling bonds. Brodsky also suggested that annealing could well have the effect of driving it out again. At the same time, there was evidence that oxygen influenced the conductivity of evaporated Si films, perhaps through a similar mechanism. Unfortunately, the first experiments designed to prove the idea were unsuccessful and led to a period of scepticism until Bill Paul's group at Harvard provided the first definitive evidence in favour of the hydrogen model. They had been studying sputtered a-Si films for several years, obtaining properties similar to those of evaporated films until, in 1974, they hit on the idea of incorporating hydrogen in their sputtering equipment. Immediately, they were able to reproduce the behaviour of the new glow discharge material in almost every detail, encouraging other groups around the world to take this up as a means of depositing what was rapidly becoming an exciting new material. More importantly, they showed, by measuring an infrared absorption spectrum characteristic of the Si—H bond, that hydrogen did, indeed, mop up dangling bonds and, from the absorption strength, they estimated that their a-Si:H contained approximately 10% hydrogen. Somewhat later, Fritzche in Chicago demonstrated similar results for glow discharge material and the immediate mystery was satisfactorily cleared up.

The way was now clear for exploitation. Crystalline semiconductors formed the basis of a multibillion dollar industry on the grounds that it was possible to control their electrical conductivities either by doping or by application of an electric field. The bipolar transistor depended on controlled *n*- and *p*-type doping, the MOS transistor on field control, while similar remarks could be made about light emitting diodes (LEDs), lasers, photodiodes, Gunn diodes, indeed, almost all the electronic devices so far invented. If amorphous materials were to contribute to the field of active devices, it would be necessary to achieve similar control over their conduction mechanisms, and a low density of gap states was an essential requirement. Spear and LeComber had already made use of the field effect in order to *measure* the density of states, a measurement in which the Fermi level was caused to move through these states by applying an electric field via a gate electrode, while the conductance of the film was monitored through what were essentially source and drain electrodes. In fact, it was one way in which conduction in band states above E_C had been clearly demonstrated. Looked at slightly differently, one could say that here was the first demonstration of a field effect transistor—a transistor which, as we shall see later, has established a hugely important commercial role. But, if the DOS was low enough to allow a strong field effect, it should also be low enough to allow conductivity control by doping. After all, the work of the STL group had already provided a strong hint—in their first paper they demonstrated a change in conductivity in one particular sample by a factor of nearly three orders of magnitude, as a result of phosphorus doping. What was now needed was a systematic investigation to establish the practical realization of a possible commercial process.

Spear and LeComber were not slow to provide it. In 1975 they published the first of several papers in which they mapped out the range of both *n*-type and *p*-type conductivities from 10^{-10} to 10^{0} $(\Omega\,m)^{-1}$, using PH_3 or B_2H_6 (di-borane) in the silane gas stream when depositing films at temperatures between 500°C and 600°C (it was already established that this temperature range resulted in minimum DOS). There was a precise correlation between the amount of dopant gas and the resulting conductivity, and measurement of the associated thermal activation energies showed that it was possible to move the Fermi level to within 0.2 eV of either band edge. No one could ask for a clearer demonstration of successful doping behaviour. It apparently provided the complete answer to the prophets of doom who had argued that doping could never be successful because pentavalent atoms such as phosphorus would be accommodated in the continuous random network in such a way that all five outer electrons were involved in bond formation, precluding the possibility of free electron donation.

The irony of the situation was, though, that the prophets of doom were very, very nearly right! It was a long time before the behaviour of dopants in a-Si:H could be properly understood but, following work by Bob Street and colleagues at Xerox Corporation, Palo Alto, during the 1980s, it became clear that most of the dopant atoms were, indeed, in threefold coordinated (i.e. non-doping) sites and only a small minority was in the fourfold coordination required for donor action. And, of these, something like 90% were effectively compensated by deep states, while only 10% of the remainder resulted in electrons in band states above E_C, the overall doping efficiency at room temperature being a mere 0.01%—that is, for every 10^4 phosphorus atoms in the solid, only one free electron was produced! This contrasts with the behaviour in crystalline silicon where the efficiency is close to 100% right up to the solubility limit, all the phosphorus atoms being forced into fourfold coordination by the exigency of the crystal lattice. Well, we knew that amorphous materials were quite unlike their crystalline counterparts, didn't we?

Where did all this leave the application interest? Actually, in not too depressed a condition. It was still possible to make *p-n* junctions which formed the basis of solar cells and it was clear that the field effect could be applied to making thin film transistors which eventually made possible the commercialization of LCTV displays. Series resistance in such devices was certainly a worry but, in practice, could be kept within acceptable levels. There was, however, one further cloud on the

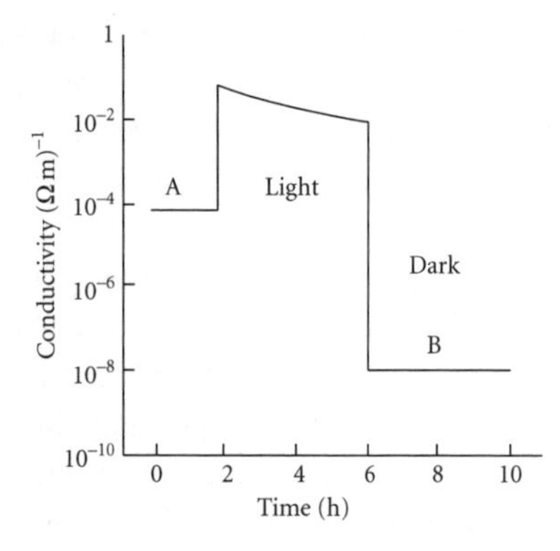

Figure 10.9. Plot of a-Si:H conductivity vs time to illustrate the Staebler–Wronski effect. A and B represent stable states in which conductivity remains constant against time. State B is reached following exposure to near-band-gap light of intensity 2000 W m^{-2} for 2 h. The return to state A is achieved by thermal annealing at 150°C. 'A' represents the thermodynamic equilibrium state of the material—the conductivity in state B depends on the total amount of illumination so state B is not an equilibrium state but can best be described as a quasi-equilibrium. (Following Staebler and Wronski 1977: fig. 1). Reprinted with permission from American Institute of Physics.

amorphous horizon, that of material instability. In 1977 Staebler and Wronski, working at the RCA Laboratories in Princeton, discovered the effect which bears their names. By studying both dark conductivity and photoconductivity over a period of many hours, they observed quite dramatic changes in the quasi-equilibrium state of an undoped sample of glow discharge amorphous silicon when it was irradiated with light of near-band-gap photon energy. The gist of their observations is contained in Figure 10.9, taken from their paper. The state labelled A in the diagram represents a standard condition reached by annealing the sample at 150°C, then cooling it to room temperature in the dark. The constant conductivity measured during the first 2 h indicates that this is a stable state. At this point, light is switched-on and the conductivity increases by three orders of magnitude but this new condition is obviously *not* stable, conductivity decaying by about one order during the 2 h period of light exposure. But what is even more startling is the change of dark conductivity after the light is switched-off. State B (which is also a stable state) is characterized by a conductivity almost four orders of magnitude smaller than that of state A! Finally, they found that, if the sample was annealed again at 150°C, it returned to state A and the cycle could be repeated. In a word, alternately thermal annealing and shining light onto the sample could effect dramatic changes in dark conductivity, shifting the sample between apparently stable states with hugely different electrical properties (it also changed the photoconductivity). Staebler and Wronski were working in a programme to develop amorphous silicon solar cells (devices which suffer long exposure to light in their daily function!), so they were inevitably somewhat concerned about these results!

What was going on? A clue was provided by measurement of the activation energy associated with dark conductivity—in state A, $E_A = 0.57$ eV, while in state B, $E_A = 0.87$ eV. Now, as it was already well established that this activation energy was given by $E_A = (E_C - E_F)$, it followed that exposure to light was changing the position of the Fermi level and, as we have also seen, E_F depended on the density of defect states. In other words, light exposure resulted in a significant change in DOS—it appeared that dangling bonds were being generated. Such a process must surely demand energy? Yes indeed—but energy was readily available from photon absorption, while thermal energy was available to return the sample to its original condition. This, in brief, was the origin of the Staebler–Wronski effect, and later work, summarized in Street (1991: ch. 6), showed that similar metastable changes could be effected by other mechanisms such as the injection of free charge carriers. This, in turn, was important in the development of TFTs. We shall deal with these practical concerns in the following sections but it is clear from this initial account that a-Si:H was not only a material of considerable interest for device applications but also a challenge to the material specialist with an interest in understanding the fine details of its behaviour. Suffice it to say that in the twenty-first century scientists are still arguing about a precise model for such metastability. Anyone wishing to pursue the topic in detail might like to consult one of the most recent papers (Powell et al. 2002) which attempts to construct a detailed atomic model to explain the considerable amount of data which has accrued over the years.

10.4 Solar cells

The solar cell, which provides direct conversion of sunlight (or other suitable radiation source) into electric power, already has a long history. In the 1930s selenium photocells demonstrated the effect but with only miniscule efficiency, less than 1% of the light energy appearing in electrical form. The birth of the solar cell really occurred, as an offshoot of the junction transistor programme, at Bell Labs in 1954, when Gerald Pearson and Calvin Fuller made a shallow diffused *p-n* junction in a thin slice of crystalline silicon and demonstrated conversion efficiencies of 6%. Bell made quite a song about its 'Solar Battery' in the popular press and emphasized how valuable it would be in providing electric power for telephone systems in remote localities. (And anyone acquainted with emergency procedures along the French autoroute system will recognize this as an application still alive and well today.) This, indeed, has proved one of its successes but there have been many others, including the provision of power for spacecraft, for pocket calculators and watches and, at the other end of the scale, an (experimental) source of considerably greater power as an adjunct to the national grid. Anyone concerned with environmental issues, however, must be disappointed by the thought that, even after nearly 50 years of development, its contribution to the overall world power consumption is still tiny. Nevertheless, as we shall see later, there seems every likelihood of a major improvement in this contribution within the next decade or two, a key fact being that: 'in only a few days, the earth receives more energy from the sun than from all the fuel burnt over the whole of human history. Three weeks of sunshine offsets all known fossil fuel reserves' (Green 2000). In spite of the western world's infatuation with oil, it must surely be possible to utilize such abundant energy riches in the service of mankind as a whole.

The reason why progress has been slow is, of course, one of cost. There was an immediate burst of interest following Bell's demonstration of its solar battery, when it was quickly appreciated just how widely clean, silent solar electricity could *in principle* be applied to mankind's benefit but, almost as immediately, came the realization that the price was too high. Solar cells, even when efficiencies rapidly improved to about 15%, were just too expensive for any but the most specialized applications. Fortunately, perhaps, one such application came early into prominence in the form of the space race. The first solar powered satellite, Vanguard 1 was launched in 1958 and it was followed by an ever-increasing assortment of progeny. Transatlantic satellite communication systems took off in 1963 with the launch of Telstar and Cold War competition led to the first men being landed on the moon in 1969, both aided and abetted by photovoltaic solar power. It was an almost ideal application for solar cells, price being very much a secondary consideration and sunlight being available in abundance. A healthy (if modest) industry for the supply of silicon cells for satellites grew rapidly and the technology was soon firmly established, reliability being the key issue, rather than price. Reliability, indeed, is one of the principal features associated with silicon technology and represents an important parameter in the complex reasoning which has insinuated itself into discussion of future terrestrial applications. The future of the satellite industry, however, appears secure—over 1000 solar powered satellites are likely to be launched during the first decade of the twenty-first century.

On the ground, enthusiasms have waxed and waned according to political far-sightedness (or lack of it!). The Middle East oil crisis of the early 1970s stimulated considerable thought about and no little research into possible alternative energy sources. The cost of making solar cells came into prominence and encouraged efforts to reduce it, first by improvements in silicon technology and, second, by the introduction of thin film technology. The major problem was one of investment. Photovoltaic energy generation on a scale appropriate to challenging the established fossil fuel industry demanded economies of scale which could only be realized by a huge increase in the size of the market, while such an increase could only be anticipated on the back of considerable government subsidies for both research and (more importantly) commercial take up. Without political eggs, commercial chickens had little chance of being hatched (except, perhaps, in the very long term). Political commitment on the necessary scale was always likely to waver. In the United States, the (1976–80) Carter administration was in favour, the Star-Wars-obsessed Reagan administration of 1980–8 appeared much less so, though the nuclear scares of Three Mile Island (1979) and Chernobyl (1986) tended to push the pendulum back again. The two Bush administrations appear to have stayed loyal to the oil lobby, while Clinton showed greater sympathy with environmental issues which clearly favour solar power. In Europe, France made its firm commitment to nuclear generation some time ago, while Germany, often in the vanguard where 'green' issues are concerned, is committed to a major programme of solar development. The United Kingdom and Denmark appear to favour wind and wave power over solar (though the solar resource is vastly greater than both). Japan, on the other hand, has chosen photovoltaics. (Such a brief, and over-simplified summary of the international scene makes clear that opinions are, to say the least, divided—a sure indication of the complexity of the issues involved.)

The net effect of these fluctuating emphases and de-emphases was a small, though steady terrestrial programme which was able to supply reliable products to those (numerous) areas of the world which lacked any form of electricity supply and which were remote enough to inhibit (on grounds of cost) any extension of a national grid. Domestic power for running a radio or TV set and a refrigerator, and lighting was provided by a household package, including, say, a 2 kW solar panel, together with storage battery and charge control electronics. Assuming a solar input of order 1 kW m^{-2}, the household panel would have an area of perhaps 10 m^2, small enough to be located on even a modest dwelling. Water pumping and other minor industrial processes could be satisfied by a small central facility (5–10 kW), run by the local community. It has been estimated that over 2 million people in the developing world now rely on solar PV modules for their electricity, though there are, perhaps, 2 *billion* still without an electricity supply of any kind.

In the developed world the trade-off is very different. Any kind of domestic supply must compete in price with that of the electricity grid which is firmly based on the use of fossil fuels. In Box 10.4 we illustrate the kind of calculation which determines the economic trade-off in using solar cells to generate some proportion of domestic power. The conclusion is that, at their current cost of about $3 per watt, solar cells are roughly a factor 2, too expensive for solar energy to break even with established fossil fuel generating plant. However, the price of solar cells has been decreasing consistently for the last 25 years, as shown in Figure 10.10 (taken from the book by Martin Green—Green 2000). Green has plotted the average selling price for solar cells against the accumulated total number of cells produced by the industry (measured in megawatts of generating capacity). Interestingly, this log–log plot

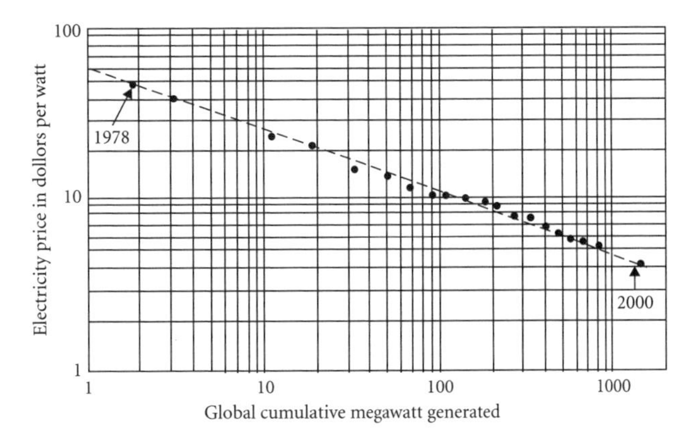

Figure 10.10. Log–log plot of solar cell selling price vs the accumulated solar cell generating capacity for the period 1978–2000. The price of solar cells has reduced by approximately 20% for every doubling of the accumulated generation capacity. Extrapolation suggests that the selling price will reach $1 per watt when the capacity reaches 10^5 MW. (From Green 2000: 11). Courtesy of Martin Green.

Box 10.4. The 'true' cost of solar electricity

The key to wider use of solar electric power is one of cost, so in this box we provide an outline of the kind of argument used to justify (or not!) the use of solar cells in a typical 'western' environment. On the face of it, the balance is a simple one—does solar electricity cost more or less per kilowatt hour than present day grid power? Complexity, however, creeps in when one has to consider the nature of any investment in solar panels. To a very large extent the cost of solar electricity is contained in a single large initial investment which can be expected to provide power continuously over a period of some 20 years, with only minimal running or replacement cost. Perhaps the issues are best brought out by considering a personal decision. I could, in principle, install a solar panel on my house roof in England to supply a fraction of the electricity I use annually. By adding a bit of electronics to convert the DC output from the cells into a 230 V AC supply and a powermeter, I can use the instantaneous power generated either to run my domestic appliances or to feed current back into the grid, thus avoiding the additional complexities associated with storage batteries. Provided the amount of power fed back to the grid constitutes only a small fraction of the total grid supply, this represents a very satisfactory solution to the storage problem. Why do I not go ahead with such a scheme right away in the common interests of saving myself money and reducing the environmental pollution associated with burning fossil fuels?

Suppose that I have £2000 to spend. What can I buy with that? At present prices for solar panels of approximately £2 per watt ($3 per watt), I can buy myself a module providing 1 kW peak power. To obtain this requires 5 kW peak insolation (assuming cells with 20% efficiency). In England the average insolation level is about 100 $\mathrm{W\,m^{-2}}$ (peak level 500 $\mathrm{W\,m^{-2}}$) so I would need an area of roof of about 10 $\mathrm{m^2}$ which presents no problem. How much energy will such a unit generate for me over its anticipated lifetime of 20 years? The average power delivered would be about 200 W, corresponding to $200 \times 24 = 4.8 \times 10^3$ W h or 4.8 kW h per day. Over 20 years this amounts to the grand sum of roughly 35,000 kW h so I appear to be generating this amount of energy for an outlay of just £2000—a unit cost of 5.7p, which happens to correspond almost exactly to the price I pay for my grid supply! Wonderful?

Well, no! What I have forgotten here is the possibility that, if I have £2000 to spare, I could simply invest it in the Stock Exchange and make a lot more money, without the hastle of having workmen crawl all over my house to install the solar equipment. Suppose I invest the money at an interest rate of 8% above inflation. My £2000, over 20 years yields interest of about £9000, which is to be compared with the money saved in solar electricity of just £2000! Not so wonderful, after all! However, this is not quite fair. If I think a little more deeply about it, I realize that I would save £100 in electricity costs each year, so I could invest £100 for 19 years, another £100 for 18 years, etc., yielding a net total of nearly £4600 at the end of 20 years—much better, but still only half what I could have made by investing my £2000.

The alternative way to approach the problem is to ask the question: how much must the cost of solar modules be reduced before the above sums break even? The answer is simple enough. If I could buy the same solar panels for £1000, I would still make the same saving on electricity, whereas investing the money would yield only half the original amount. In other words, if solar cells cost £1 per watt, rather than £2 per watt, I would be as well advised to install them as to invest my money in the Stock Exchange. And, at the same time I would be helping the environment.

Of course, there are numerous points of argument in the above calculation (not least the rate of interest!) and it overlooks the fact that installation of the units would add to their fundamental cost, as would the DC–AC converter. This implies that the module cost must come down somewhat further than estimated above—let us say to $1 per watt. Nevertheless, it illustrates the kind of calculation necessary to reach any decision about whether to proceed in a serious way with investment in solar electricity.

turns out to be approximately linear—the price decreasing by about 20% for every doubling of the capacity. (Note that the world's total energy consumption is running at about 10^7 MW—some 10^4 times present solar capacity so there is still plenty of scope for future expansion.) Figure 10.10 certainly demonstrates the importance of investment—the principal hope for solar electricity lies in encouraging the growth of generating capacity, which can only happen if there is an increase in demand, which, at present, depends on government subsidy. (Encouragingly enough, several governments do have quite ambitious plans to stimulate this necessary growth.) Figure 10.10 allows us to make an informed guess as to when the cost of solar electricity will reach the critical $1 per watt level. It corresponds to an accumulated capacity of nearly 10^5 MW (1% of world demand) and can be expected in roughly 15 years' time. Therefore, assuming we can legitimately extrapolate from past experience, there seems to be every hope that solar PV power will begin to play a major role in supplying the world's energy demands as from the year 2020. Will this be the last bastion to fall to the all-engulfing advance of semiconductors? Perhaps the drug industry will hold out!

Needless to say, this brief introduction to solar electricity does no more than scratch the surface (we have said nothing, for example, about the important subject of concentrator cells)—for more detail the reader should consult the rapidly growing literature—for example, Green (2000), Markvart (2000), Street (1999), Kazmerski (1998). For the present, we must now return to the subject of semiconductor devices.

The mode of operation of a solar cell has much in common with the *p-n* junction photodetector which we met in Chapters 8 and 9, though with very different design rules. The photodetector is a low power, low noise device, designed to convert ultrasmall, rapidly changing signals into recognizable electrical replicas, while the solar cell deals with watts, kilowatts, or even megawatts of very slowly varying radiation power which must be converted into electrical energy with maximum efficiency, but at minimum cost. Nevertheless, the function of their respective junctions is the same—to separate photogenerated holes and electrons so as to create an electric current in an external circuit. The principal difference is that the solar cell must be designed for maximum *power* output—that is, it is the product of output current and output voltage which matters—whereas the photodetector must be optimized in terms of signal/noise ratio.

In view of the fact (see our introductory paragraph) that the first serious solar cell was made from *crystalline* silicon, the reader may be forgiven for wondering why we should choose to introduce the topic in this chapter which concerns itself with *non*-crystalline materials and devices. The truth is that solar cells have been developed based on crystalline, polycrystalline, and amorphous materials so there is *some* logic in our choice of venue for their discussion. It certainly seems

(a)

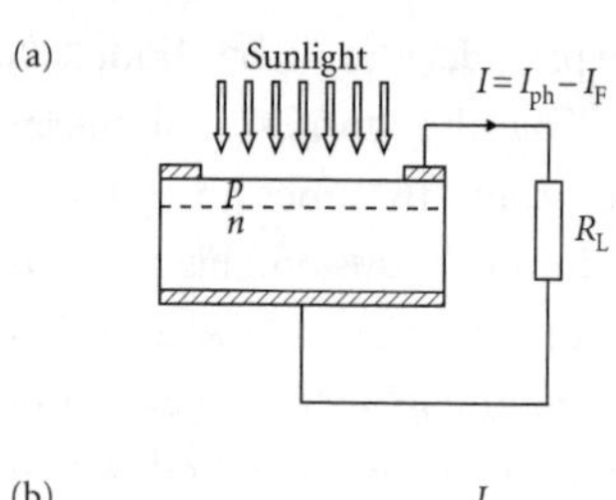

(b)

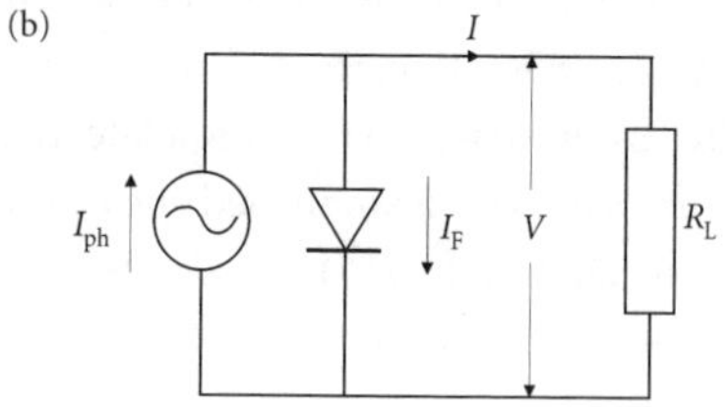

(c)

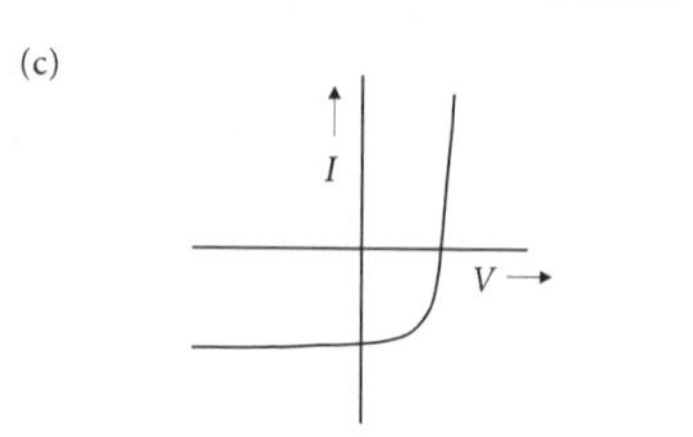

(d)

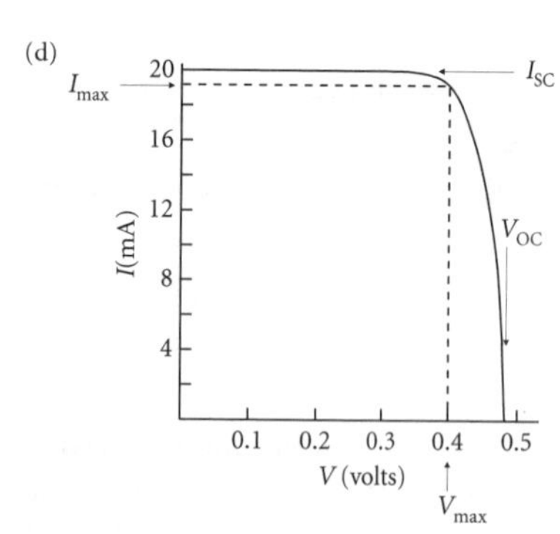

Figure 10.11. (a) Schematic diagram of a *p-n* junction solar cell driving a net current I through a resistive load R_L. (b) The simplified equivalent circuit of the cell. Photocurrent I_{ph} results from the separation of photogenerated holes and electrons by the junction field (note that in diode terminology I_{ph} is a reverse current). I_F is the forward current through the diode resulting from the voltage V which is dropped across the load resistor. (c) Current–voltage characteristics of the diode under illumination, together with in (d) an inverted version of the fourth quadrant. In this final diagram the current I is positive, consistent with the convention used in parts (a) and (b). I_{SC} is the cell short-circuit current, while V_{OC} is the open-circuit voltage. I_{max} and V_{max} correspond to the maximum output power available from the cell (see example in Figure 10.12).

more appropriate to collect all the information together in one place, rather than spreading it over several different (however appropriate) chapters. As we shall see, there is an optimum band gap for an efficient solar cell which lies in the region of 1.0–1.7 eV and this determines to a very large extent those materials which have been investigated. However, there are other factors, such as cost per unit area, availability of materials, energy expended in actually *making* the cell, cell stability, useful working life, etc., which are essential ingredients in the decision-making pie. One such is bound up with the question we have already raised concerning the suitability of polycrystal or amorphous materials for large area devices. If other factors could be equal, the basic cheapness of a technology involving deposition of thin films of semiconductors on glass or plastic substrates offers an economic appeal which can never be totally overlooked, particularly when one is concerned to collect sunlight over a large area for high power generation. It follows that a number of non-crystalline materials have been at the forefront of solar cell research, chief of which are amorphous silicon (a-Si:H), polycrystalline silicon, Cu_2S/CdS heterojunctions, CdTe/CdS heterojunctions, and $CuInSe_2$/CdS heterojunctions. On the other hand, it has to be said that single crystal cells of Si, GaAs, or InP have so far demonstrated the greater efficiencies and reliabilities. As in many aspects of commercial life, the ultimate trade-offs are often complex and controversial. While it is true that over 90% of the market has so far gone to silicon in its various forms (crystalline, polycrystalline, and amorphous), the door is still wide open to other material technologies if appropriate improvements can be obtained.

However, before delving any deeper into the nature of commercial decision-making, we should first acquaint ourselves with the working of a basic *p-n* junction solar cell in order to appreciate something of the technical possibilities and difficulties. Figure 10.11 provides a simplified representation of solar cell operation and Box 10.5 contains some simple mathematics to illustrate the derivation of power output and efficiency for such a cell. Solar radiation falling on the top surface of the cell is absorbed, generating holes and electrons in the vicinity of the junction which are swept across the depletion region, producing a photocurrent I_{ph}, proportional to the incident photon flux. The photovoltage developed across the load resistor sets up an opposing current I_F through the diode and it is the *net* current I (the difference between I_{ph} and I_F) which represents the electrical output. By suitable choice of load resistor it is possible to optimize the output power, as can be seen from Figure 10.12, optimum values of I and V being indicated on the current–voltage characteristic in Figure 10.11(d). For the example chosen in Box 10.5, the output power is just under 8 mW when the solar input is 500 W m^{-2}, representing a conversion efficiency of about 15%.

Box 10.5. Modus operandi of the solar cell

The operation of a *p-n* junction solar cell is illustrated in Figure 10.11. Part (a) of this diagram provides a schematic of the cell connected to a resistive load R_L, through which it drives a current I. The net current I is made up of two parts, a photocurrent I_{ph} which results from absorption of photons near the *p-n* junction, the photogenerated holes and electrons being separated by the junction field, and a diode forward current I_F which results from the fact that a voltage V appears across the load resistor. Thus, I_{ph} (which is a *reverse* current) is given by the expression:

$$I_{ph} = \eta N_{ph} eA, \tag{B10.26}$$

where η is the quantum efficiency, N_{ph} is the number of photons per unit area per second impinging on the cell, and A is the cell area. Note that I_{ph} is controlled by the intensity of solar radiation and does not depend on the details of the cell material, etc. The forward current I_F is given by the standard current voltage characteristic of the diode which, for an ideal diode is:

$$I_F = I_S[\exp(eV/kT) - 1]. \tag{B10.27}$$

I_S being the diode reverse saturation current which is typically of order 10^{-11}–10^{-9} A. As is clear from the equivalent circuit in Figure 10.11(b), the *net* current through the load is:

$$I = I_{ph} - I_F = I_{ph} - I_S[\exp(eV/kT) - 1]. \tag{B10.28}$$

The fact that V appears across the load resistor R_L and I flows through it, implies the relation:

$$I = V/R_L. \tag{B10.29}$$

It is straightforward, now, to combine equations (B10.28) and (B10.29) to obtain an equation for the voltage V across the load resistor:

$$\exp(eV/kT) + V/I_S R_L = I_{ph}/I_S + 1 \approx I_{ph}/I_S. \tag{B10.30}$$

We can solve this transcendental equation graphically to discover how V depends on R_L and, finally derive a plot of output power P_{out} vs R_L from the relation:

$$P_{out} = V^2/R_L. \tag{B10.31}$$

In Figure 10.12 we have plotted P_{out} vs R_L for a typical diode, using the following parameters: $I_S = 10^{-10}$ A, $A = 1\ \text{cm}^2 = 10^{-4}\ \text{m}^2$, $kT/e = 0.025$ V, $\eta = 0.5$, $I_{ph} = 20$ mA. This last figure is based on a solar input of 500 W m^{-2}, corresponding to $N_{ph} = 2.5 \times 10^{21}\ \text{m}^{-2}$ at a wavelength of 1 μm. The obvious conclusion to draw from Figure 10.12 is the need to match the load resistance correctly to obtain maximum output power. Taking $R_L(\text{max}) = 21\ \Omega$, we obtain $V_{max} = 0.40$ V, while the corresponding current $I_{max} = 19.2$ mA. These figures may be compared with those of the open circuit voltage $V_{OC} = 0.48$ V and short-circuit current $I_{SC} = I_{ph} = 20$ mA (obtained from equation (B10.28)).

If we write P_{max} in terms of V_{OC} and I_{SC} as:

$$P_{max} = V_{max} I_{max} = FF \times V_{OC} I_{SC}, \tag{B10.32}$$

Box 10.5. Continued

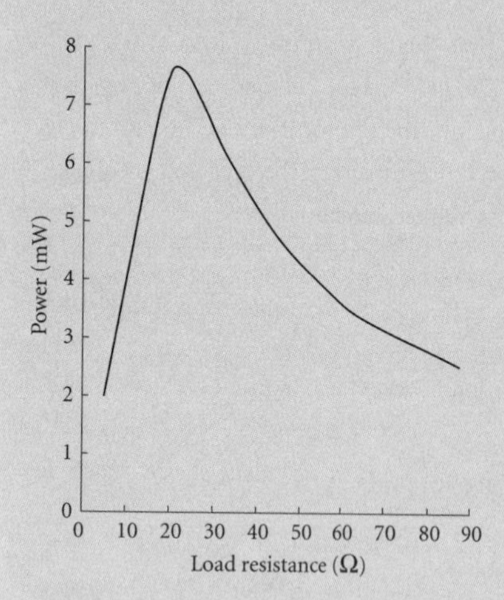

Figure 10.12. Plot of output power P from a p-n junction diode solar cell as a function of the load resistance R_L. The parameters used to obtain this graph are: $I_S = 10^{-10}$ A, $A = 1$ cm^2, $kT/e = 0.025$ V, $\eta = 0.5$, $I_{ph} = 20$ mA. The maximum output power is obtained for a load resistance of 21 Ω, for which $V_{max} = 0.40$ V and $I_{max} = 19.2$ mA (see Box 10.4).

it is easy to see that the 'fill-factor' $FF = (0.4 \times 19.2)/(0.48 \times 20) = 0.80$. This is illustrated graphically in Figure 10.11(d), where V_{max} and I_{max} define the 'maximum power rectangle'.

Finally, we can work out the conversion efficiency of our cell. Knowing that it delivers 7.7 mW of electrical power for an input radiation power of 50 mW (500 W m^{-2} over an area of 10^{-4} m^2), we obtain an efficiency of about 15%. (A perfect diode would yield an efficiency of close to 30%, which represents the ideal target figure against which all real cells are to be compared.)

It is convenient to consider one further feature of cell design here—that of series resistance. If the resistivity of the cell material is high, it has the effect of introducing an additional resistance R_S into the equivalent circuit, in series with the load resistor and this will dissipate power which detracts from that dissipated in the load. Clearly, the condition for this loss to be negligible is that R_S should be very much smaller than R_L. Achieving this requires fairly heavy doping of the semiconductor material and careful design of the top contacts. Figure 10.11(a) shows that current must flow sideways through the upper p-type layer to reach the contacts, implying the need for compromise between layer thickness, doping, and contact stripe separation. It will also be apparent that the high resistivity associated with amorphous and polycrystalline materials makes this aspect of design more difficult.

To obtain this figure of 15% we assumed a quantum efficiency of 0.5, which implies that an ideal cell would have an efficiency of 30%, and this serves as a target figure to which all real cells may aspire. Our assumption of $\eta = 0.5$ suggests that only 50% of the incoming photons actually result in electrons (or holes) which contribute to the current. In fact, this is an oversimplification. There are several factors which act to reduce the overall efficiency. First, some fraction of the incident photons will be screened from the semiconductor surface by the (essential) metallic top contact to the cell. Second, some further fraction will be reflected from the top surface of the cell. If the surface is left uncoated, this fraction will be typically about 30%, because of the high refractive indices of most semiconductors, but a good anti-reflection coating can reduce this to minor proportions. Third, the absorption coefficient may be low enough (as it is in indirect materials like silicon) that many of the photons are absorbed some distance below the p-n junction. This requires that the resulting minority carriers must diffuse back to the junction if they are to contribute to the photocurrent. Not all of them will make the journey successfully, before recombining. Fourth, minority carriers generated close to the top surface of the cell may be lost by recombination via surface states. Fifth, there is a subtle problem which arises when an absorbed photon has a photon energy significantly larger than the band

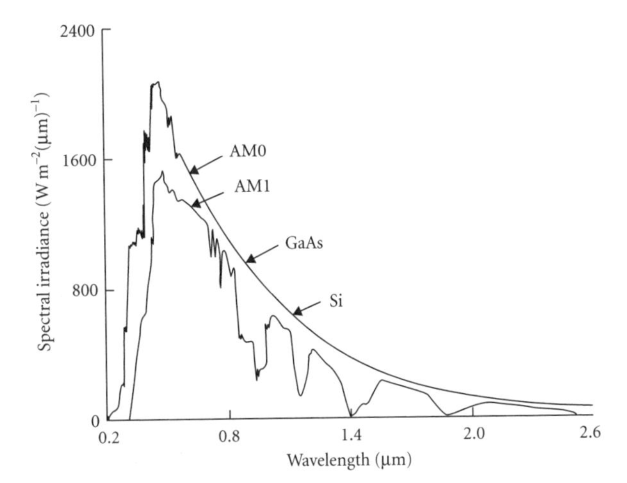

Figure 10.13. Spectral distribution of solar radiation for AM0 (outer space) and AM1 (at the earth's surface on a clear day) conditions. The difference between the two spectra represents the absorption and scattering due to the earth's atmosphere. The integrated radiance for AM0 is 1350 W m^{-2}, while for AM1 it is 925 W m^{-2}.

gap of the semiconductor. No matter what the photon energy, it produces only one electron–hole pair and this implies that *energy* is lost (actually as heat) when the hot carriers thermalize to the band edges. In fact, this loss process is not strictly a *quantum* efficiency loss—the quantum efficiency is 100%—but it clearly represents a contribution to the overall energy inefficiency. We have nevertheless (rather sloppily) lumped it in with the other quantum efficiency factors. Apparently, there is considerable scope for good cell design in order to minimize the effect of these various loss factors but, for the moment, we shall consider only one of them in detail, the choice of semiconductor band gap.

To appreciate the importance of band gap, it is necessary to consider the spectrum of radiation reaching the cell, which is illustrated in Figure 10.13 for two cases, that of outer space (relevant to satellite usage and referred to as 'air mass zero', AM0) and that of the earth's surface when the sun is overhead on a clear day ('air mass one', AM1). The difference is due to absorption and scattering by the earth's atmosphere, the sharp bites taken out of the AM1 spectrum representing the absorption bands of the various constituents of our air. The integrated energy over the whole AM1 spectrum is a little less than 1 kW m^{-2}, a figure which provides a useful approximation for general use—but one must be careful not to overlook the effect of cloud cover and latitude! A more practical figure is the average irradiance (averaged over a complete year) which varies, modestly enough, between a maximum of 300 W m^{-2} in one or two rather warm regions of the earth to approximately 100 W m^{-2} for the north of Scotland, Hudson Bay area of Canada, Central Russia, Northern Japan, Southern New Zealand, and Southern Patagonia, for example. However, it is the shape of the distribution which is important in determining the optimum semiconductor band gap. A narrow gap material would obviously absorb nearly all the solar radiation falling on it but would suffer badly from the hot electron

Table 10.1. Band gaps of solar cell materials

$CuInSe_2$	1.05 eV Direct
Si	1.12 eV Indirect
InP	1.34 eV Direct
GaAs	1.43 eV Direct
CdTe	1.49 eV Direct
Cu_2S	~1.2 eV Indirect
CdS	2.49 eV Direct
a-Si:H	~1.6 eV

loss mentioned above. On the other hand, a wide gap would absorb very little radiation at all. Assuming an AM1 radiation spectrum, the happy medium is represented by band gaps within the range 1.0–1.7 eV, with a peak efficiency of 30% for $E_g = 1.3$ eV. Some relevant band gaps are listed in Table 10.1.

It should be emphasized, perhaps, that the above argument is based on the use of a single *p-n* junction which implies a choice of one absorbing material. Significantly better efficiencies can be expected for the case of 'tandem cells' which employ two or more different materials with a suitable range of band gaps, the wide gap materials being presented first to the radiation and absorbing the high energy photons, followed by narrower gap materials which absorb progressively lower energy photons, thus largely avoiding the hot electron problem. While providing improved efficiencies (experimental values of over 30% have already been demonstrated), such cells, containing multiple junctions, are more complicated to make and consequently more expensive. Like so much else in this fascinating subject, it is not yet clear what their impact will be on the future of photovoltaic energy conversion.

So much for generalities—we must now return to the history of solar cell development. The first silicon cells were, as we have seen, developed from bulk single crystal material using a technology very similar to that developed for microelectronics. A single crystal boule was prepared in the standard way and 300 μm thick slices cut from it with a wire saw, which were then polished, etched, diffused (to form the junction), provided with an anti-reflection coating, and contacted top and bottom, the top contact typically consisting of 100 μm wide stripes spaced 3 mm apart to provide an acceptable compromise between shadowing and contact resistance. Not only did the early work make use of microelectronic techniques, it also depended on the self-same material—indeed, much of this consisted of scrap silicon from integrated circuit rejects or unwanted offcuts, etc which was reprocessed, and it has been a feature of solar cell development that the volume of material used has tended to track that of the integrated circuit industry over quite a number of years. (Though, if large-scale electricity generation does eventually become a reality, this linkage will surely have to be broken.) There were just two problems with the early cells—their prices were high and their efficiencies low. The latter difficulty soon yielded to the application of relatively straightforward design criteria—by 1960 the best cell efficiencies were close to 15%—but prices were to drop rather more slowly.

Two factors were particularly unsatisfactory from the economic viewpoint. First, the high purity, high quality single crystals grown by the Czochralski method were inevitably expensive because of the various purification and processing steps required. Second, the production of square slices (from cylindrical boules) by sawing was a tedious and wasteful business, even when multiwire saws were introduced during the 1960s. The first attempt to overcome these difficulties involved the replacement of Czochralski boules with so-called 'multi-crystalline silicon' which was produced already as rectangular blocks simply by melting the silicon in appropriately shaped graphite crucibles (usually with quartz liners to minimize contamination of the silicon). This large grain polycrystalline material was of marginally lower quality than that of a Czochralski crystal but resulted in cells whose efficiencies were reduced by no more than 1%. However, the sawing into 300 μm thick slices was still a major source of production cost and numerous attempts were made to eliminate it. Most successful of these was the 'edge-defined film-fed growth' (EFG) method (see Figure 10.14) which produced long thin ribbons of silicon, requiring only to be cut into appropriate lengths for cell processing. It was introduced by Bates and co-workers of Tyco Laboratories in 1972 and has been modified by others to produce multiple ribbons from a single apparatus, laser cutting having replaced sawing for their separation. In spite of the many alternative approaches taken by research during the ensuing 25 years (which we shall examine in a moment), the majority of solar modules produced commercially in 1998 were based on one or other of these bulk silicon technologies.

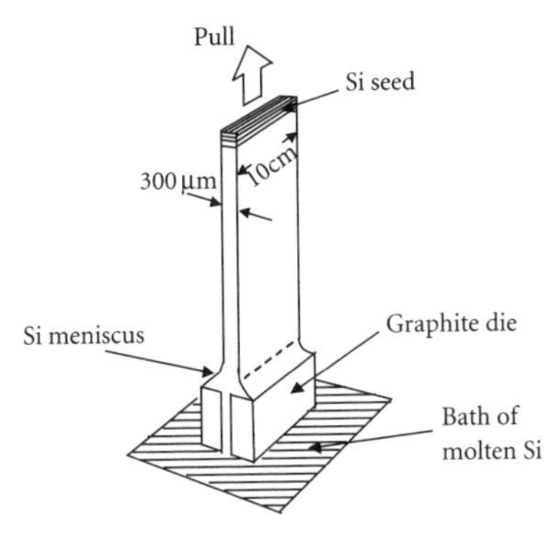

Figure 10.14. Schematic drawing of the EFG method for producing silicon ribbon. Molten silicon is drawn upwards through the narrow capillary slot in the graphite die to supply a small reservoir from which the ribbon is pulled, somewhat along the same lines as occurs in the Czochralski crystal pulling method. Ribbons of cross-section 10 cm × 300 μm may be pulled to a length of several metres.

This is not to say, however, that nothing changed. A great many detailed improvements were effected which allowed efficiencies to creep steadily upwards. In 1962 it had been discovered that improved radiation resistance could be obtained by using lithium doping of *p*-type starting material, causing a switch from the earlier *p*-on-*n* structures to *n*-on-*p*, involving a phosphorus diffusion (rather than boron) to form the junction. This became standard procedure from then onwards. (Techniques were later established for performing the diffusion in 'conveyor belt' type furnaces to speed production.) The next significant step came in 1969 when a 'back surface field' was introduced to reflect minority electrons back towards the junction. (This involved the inclusion of a heavily doped *p*-type region under the back contact.) Then, in the early 1970s an improved diffusion process was introduced to allow the use of a shallow *n*-type region (this was often referred to as the 'violet cell') and, in 1974 a major innovation took place (the 'black cell') in the form of a textured (or serrated) top surface to reduce reflection losses. It was made by using selective etching to form an array of tiny pyramids which resulted in incoming radiation being reflected sideways to strike neighbouring pyramids and thus enhance the probability of light being collected into the cell. It also had the effect of containing light within the cell, once captured, and, together with a reflecting back contact, enabled even a relatively thin cell to absorb all the radiation captured. This offered the possibility for making silicon cells with thicknesses of only a few tens of microns, rather than the standard 300 μm in common use. About this time too the vertical junction cell was demonstrated, giving yet more efficient collection of minority carriers. A modified version of this, the 'buried contact cell' was proposed by Green and Wenham at the University of New South Wales, in 1984—the top metallic contacts were buried within narrow grooves laser-cut into the top surface of the cell—once again leading to improved efficiency. Finally, in an attempt to achieve the best possible efficiencies from single silicon cells, this same group chose to work with floating zone single crystal silicon and obtained values as high as 24.7% in 4 cm^2 cells, beginning to approach the 30% which represents the theoretical limit for a single silicon junction. Clearly, a great deal had happened in the detailed design of silicon cells and a gradual increase in performance resulted from it. Large area cells based on single crystal material have achieved efficiencies of 16.7%, while the best results from multi-crystalline silicon are even slightly better, at 17.2%. Commercial modules usually show efficiencies a few per cent lower than small-cell values.

Another thing which happened during this period was the introduction and development of the single crystal GaAs cell. The band gap of GaAs is close to the optimum value of 1.3 eV and it also has an advantage over Si of being a direct gap material, needing a thickness of only a few microns to absorb virtually all the above-band-gap radiation falling on it. As we know, it also forms high quality heterojunctions with AlGaAs alloys which may be used as wide gap windows to allow incoming radiation to pass through to the GaAs below, thus avoiding any problems with surface recombination at the top surface of the cell. Epitaxial techniques were clearly appropriate here and liquid phase epitaxy was used to good effect by several laboratories in making efficient GaAs cells. Alferov at the Ioffe Institute in Russia was a leading figure in these developments and already in 1972 he had achieved efficiencies of nearly 15% from heterojunction cells which later (1986) were successfully used on the Russian Mir space station. In addition, GaAs has been used as the basis for very efficient tandem cells. A triple junction, lattice-matched structure of InGaP/GaAs/Ge recently achieved an efficiency of 31% which represents an important result for satellite modules where the additional cost of such elaborate structures is not of major significance. Another area in which such high efficiencies may be exploited is that of concentrator systems where, again, the cell cost is far less significant than in conventional systems.

The push for improved efficiency was obviously impressive but the push towards *cheaper* modules implied the use of smaller volumes of semiconductor and suggested a move from bulk to thin film materials. Indeed, this was well appreciated in the 1970s when several thin film materials were investigated for possible solar cell applications. Both polycrystalline Si and GaAs were studied, together with a number of thin film heterostructures such as CdS/Cu_2S, CdS/CdTe, CdS/InP, and CdS/$CuInSe_2$, but of most immediate commercial importance was the arrival on the solar cell scene of amorphous silicon. We saw in Section 10.3 that the discovery of a-Si:H led to the achievement of controlled doping with P and B which allowed the development of *p-n* and *p-i-n* junctions. It was also clear that the optical band gap lay in the region of 1.6–1.7 eV, well within the range appropriate to solar cell applications, and the absence of any *k*-selection rule implied that absorption above the band edge would be relatively strong. A thickness of only 1 μm was sufficient to absorb nearly all the radiation with photon energies greater than 1.8 eV. The additional advantage that it could readily be deposited on large area substrates such as glass or sheet steel made it an obvious choice for potentially cheap solar cells and this was the first serious application of a-Si:H to be explored. The pioneering work at the RCA, David Sarnoff Research Center was led by Dave Carlson who first observed the photovoltaic effect in glow-discharge a-Si:H in 1974. Carlson is also the author of a useful review article on amorphous silicon solar cells (Carlson 1984). (See also Street 1991: ch. 10.)

By the end of 1974, both *p-i-n* and Schottky barrier (SB) cells had been demonstrated, though with only minimal efficiencies due to poor contacts. This was soon rectified but it gradually became clear that the precise details of the deposition process could be critical in determining the suitability of the material for solar cells. The efficient separation of minority carriers in the *i*-region of a *p-i-n* device depended on the product of carrier mobility and lifetime ($\mu\tau$) and both parameters were affected by material quality, the density of deep states and that of band tail states both being important, as were those of a variety of impurity atoms. Tail states were also influential in limiting the doping efficiency of the end regions which was crucial to obtaining a satisfactory open circuit voltage. Attention came to be concentrated on deposition temperatures in the range 200–300°C, though it was found necessary to optimize conditions empirically for each individual growth kit. Deep states were also involved in the Staebler–Wronski effect which led to a gradual deterioration in cell performance under steady solar irradiation, the increased density of states causing a reduction in lifetime τ. In practical terms, cell efficiency decays typically from 10 to 7% over a period of 20 years.

All these studies took time, and cell performance improved only slowly. By 1980, efficiencies were in the range 6–7% both at RCA and elsewhere (including an innovative step by Madan et al. at Energy Conversion Devices, Troy, MI who introduced fluorine into the material mix by growing from SiF_4 and H_2). Further advance was based on the introduction of alloy materials such as a-Si/Ge:H or a-Si/C:H, the former having a band gap smaller than a-Si:H, the latter larger. In 1981, Tawada et al. at Osaka University made use of a-Si/C:H to obtain a cell efficiency of 7.5% while, in 1982, Nakamura et al. of Mitsubishi Electric achieved 8.5% with a stack of three junctions involving all three materials. The 10% 'barrier' was finally breached late in 1982 by Catalano et al. at RCA using a *p*-type a-Si/C:H window layer as top contact in a glass/SnO_2/$p^{\star}$-i-n/Ag structure, radiation being incident through the glass substrate. (A thin film of tin oxide served as a transparent conducting contact to the *p*-type a-Si/C:H layer.) Today, the efficiency of the best single junction cell is about 12% (compared with an estimated theoretical limit of 14%), while stacked cells have achieved close to 14%, against a theoretical limit of about 20%.

In spite of the US lead in research, amorphous silicon solar cells were first commercialized in Japan, in 1980, first Sanyo, then Fuji Electric employing them to power pocket calculators. It was something of an inspired step to recognize that a-Si:H was well suited to converting radiation from domestic lighting (as well as sunlight) into battery power. The application to watches followed a year or two later. At the time of writing, amorphous silicon cells have a healthy 13% of the world market of some $5 billion.

Several other thin film cells have been proposed at different times, the first of which, based on a CdS/Cu_2S heterojunction, being quite widely investigated in the 1960s. An early method of fabrication involved evaporating a thin film of CdS onto a suitable metal substrate, then dipping the structure into a hot solution of cuprous chloride to form the desired Cu_2S layer. Typical areas were about 50 cm^2. Later, a spray technology was also developed which allowed continuous, production line fabrication and led to predictions of electricity generating costs well below those associated with fossil fuels. However, there were numerous problems with instability and this particular structure seems to have disappeared from the present day scene. Nevertheless, similar polycrystalline thin film structures based on CdS windows have shown greater promise. Two which have survived over the years first came into prominence round about 1975—CdS/CdTe and CdS/CuInSe (CuInSe sometimes being modified by the addition of small amounts of Ga and S). More recently, attention has been focused on the use of thin

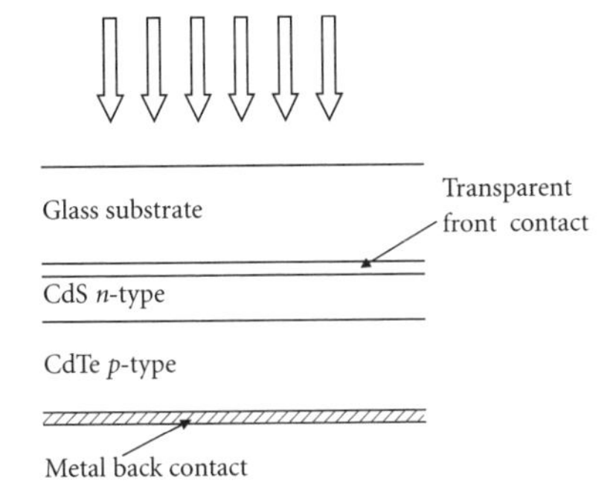

Figure 10.15. Example of the structure used in making a thin film CdS/CdTe solar panel. All the layers are deposited on a cheap glass substrate and include a conducting transparent layer such as SnO_2 or indium tin oxide (ITO) acting as the front contact. CdS can only be prepared in *n*-type form so the CdTe must be doped lightly *p*-type. Note that the various layer thicknesses are not to scale.

film polycrystalline silicon and it now seems likely that any one of these three approaches may challenge a-Si:H for the role of thin film champion of the solar world.

It had been known from the 1950s that CdTe possessed a direct band gap appropriate for solar cell use but for many years there were problems in preparing it with the necessary quality. Deep level defect states acted as efficient recombination centres and had to be eliminated by a lengthy war of attrition. It was then necessary to develop inexpensive methods of depositing thin films on glass substrates which maintained adequate material quality and to combine these with a CdS window and transparent electrical contact layer, leading to the structure shown in Figure 10.15. Because CdS can only be prepared in *n*-type, it was necessary to dope the CdTe lightly *p*-type. Note, too, that the CdS, in addition to its function in allowing long wavelengths to reach the CdTe, serves to absorb shorter wavelengths and add its own contribution to the photocurrent. Recent research has led to cell efficiencies as high as 16% which promise well for possible application to large area PV power generators. Several commercial firms are now poised to exploit it.

$CuInSe_2$, like CdTe, can be doped both *n*- and *p*-type and has a direct band gap suitable for absorbing solar radiation. In fact, it has probably the highest absorption coefficient of any of the materials considered for this application. The structure used is identical to that shown in Figure 10.15, apart from the obvious substitution of $CuInSe_2$ for CdTe. The commonly used back contact is molybdenum which makes an adequate ohmic contact to heavily doped *p*-type $CuInSe_2$ but implies a doping gradient within the material because efficient carrier collection by the *p-n* junction requires light doping (in order for the depletion region to be as wide as possible). It has also been found necessary to passivate grain boundaries within the $CuInSe_2$ by heating the cell in air (oxygen appears to be important here in much the same way as it was for improving the performance of polycrystalline PbS infrared detectors). Commercial production is now based on an inverse fabrication process where Cu and In are sputtered onto the molybdenum back contact, then selenized by passing H_2Se gas over the structure at 400°C. CdS and a contact layer of ZnO are then deposited. In this case, glass is used simply as an encapsulation. A practical detail is the use of Ga to aid adhesion of the $CuInSe_2$ to the molybdenum and, at the same time, to increase its band gap slightly. As for the CdS/CdTe cells, efficiencies in research structures have reached 16%. Large scale trials have also demonstrated promising stability over several years and commercial production is now at the point of taking off.

More recently, the development of thin film polycrystalline silicon cells is also showing considerable promise, following the realization of light trapping techniques which cause light to be reflected back and forth within the cell, giving effective light paths of order 20 times the

Table 10.2. Efficiencies of thin film solar cells

Material	Commercial products (%)	Best large area (%)	Best R&D (%)	Theoretical limit (%)
a-Si:H	5–8	10	13	20
Poly-Si	11	12	16	25
$CuInSe_2$	8	14	16	21
CdTe	7	11	16	28

actual cell thickness. This allows silicon layers of only a few tens of microns to absorb most of the above-band-gap radiation while reducing the amount of semiconductor material required. It also considerably reduces the required minority carrier diffusion lengths needed for efficient carrier collection, thus easing the problem of obtaining adequate material quality. Suitable thin films have been deposited by liquid 'epitaxy' from solution of Si in molten tin and progress appears to have been made towards developing an appropriate large-scale process. Research results for these silicon cells are comparable to those obtained on CdS cells, setting up an intriguing four horse race to dominate the large scale future of PV generation. One can only wait with interest to see which will run out winner. Table 10.2, taken from Markvart 2000, sets out the overall position at the turn of the millenium. Whereas, at present, crystalline silicon takes 39% and multicrystalline silicon 44% of the market, perhaps this situation will soon be changed out of all recognition by one or more of these thin film modules. Perhaps we shall at last see solar electricity becoming a major contributor to the world's energy needs but whether this will still be based on silicon or on one of the alternatives remains an intriguing question.

10.5 Liquid crystal displays

There can be little doubt that the development of large area display panels based on liquid crystals represents one of the most significant innovations to impact upon twentieth-century lifestyles. From its inception in the 1920s, TV display was based on some form of cathode ray tube (CRT), a particularly significant development being that of the shadow mask tube at RCA in the 1950s which brought colour into millions of domestic lives. This and other modifications of the basic CRT concept totally dominated TV display until eventually challenged by liquid crystals during the 1980s, a clear proof of the excellence of CRT performance. It was not that the CRT was beyond criticism, far from it—the tube suffered from requiring dangerously high voltages to drive it (which, incidentally, served as an embarrassing source of

X-rays), from providing a screen which was never quite flat, and from the inevitable large bulge behind the screen associated with the electron optics. For many years research laboratories round the world sought to develop an effective replacement in the form of a thin, flat, low voltage display panel, having adequate brightness which could also be conveniently addressed with the necessary picture information. It also had to be cheap! Flat tubes using novel scanning techniques were investigated, gas discharge panels were developed and various solid-state electroluminescent displays were tried but none of them was able to mount a serious commercial challenge to the ubiquitous CRT. Not until the arrival of the liquid crystal pixel addressed by way of a TFT was the dominance of the CRT so much as put under pressure. Nowadays we are perfectly well accustomed to seeing LCDs in aircraft seats, automobiles, computer screens, mobile phones and, of course, domestic TV sets and the price premium is very well affordable. We can even look forward to a time when it will be the CRT which is the minority interest. How did it happen?

The first serious use of liquid crystals for display came in the 1960s in the form of digital readout for pocket calculators, wristwatches, and measuring instruments. It followed many years of study. The basic concept of liquid crystal materials actually dates to the end of the nineteenth century when it was realized that a state of matter somewhere between solid and liquid could exist, characterized by some degree of ordering of the constituent molecules. (In a solid, molecules are spatially fully ordered, whereas in a liquid their alignments are totally random—in a liquid crystal there may be some preferential direction about which molecules show a degree of alignment.) It gradually came to be understood that several different types of liquid crystals existed, the three main ones being smectic (soap-like), nematic (thread-like), and cholesteric (like compounds of cholesterol) and, within each category there might be several variations, some of which showed ferroelectric behaviour, constituting a considerable complexity which is still in process of being unravelled. A convenient summary can be found in the book by Smith (1995: ch. 5), where many useful references also appear. For our purposes it will suffice to describe the properties of just one type of liquid crystal, the so-called 'twisted nematic' form which has been widely used in display applications.

The molecules of a nematic liquid crystal are long chain organic molecules which have a strong tendency to align themselves parallel to one another in the bulk of the sample where the alignment axis can be controlled by the application of an electric field. (In some cases alignment is parallel, in others at right angles to the field.) However, at the interface with a solid surface the molecules can be forced to align themselves along a preferred direction determined by the surface structure of the solid. This may be controlled by making fine scratches (a process known as

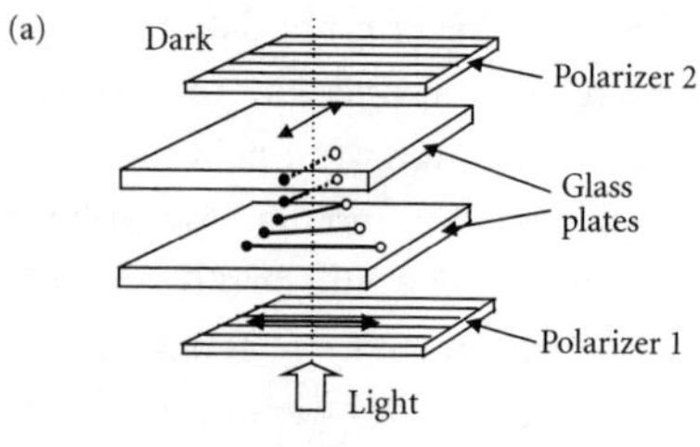

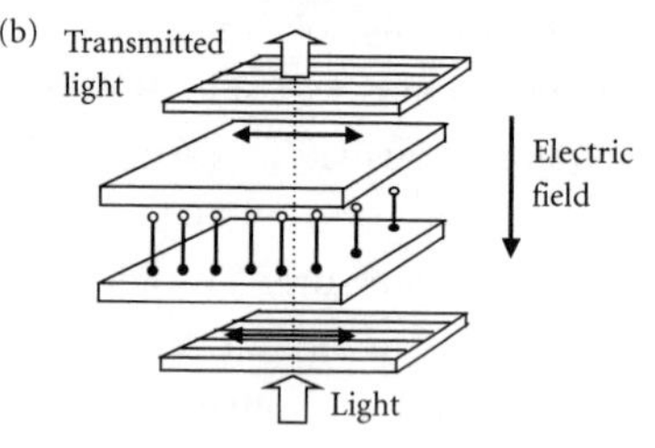

Figure 10.16. Illustration of the light switching behaviour of a twisted nematic liquid crystal, confined between a pair of glass plates. In (a) the liquid crystal molecules are aligned in orthogonal directions at the top and bottom plates by means of fine scratches on their inner surfaces. The plane of polarization of the light beam is rotated through 90° by the liquid crystal so no light is transmitted through the upper polarizer. In (b) an applied electric field is used to align the molecules normal to the plates so that the polarization plane of the light no longer rotates and most of the light is therefore transmitted through the upper polarizer. In order to provide the electric field it is necessary to deposit transparent electrodes of ITO (not shown in the figure) on the glass plates.

'rubbing') or by evaporating a surface coating at an angle to the surface. Generally speaking the lowest energy state is achieved when the molecules are parallel to the scratch direction. Suppose, now, that the liquid crystal material is confined between a pair of glass plates separated by a distance of about 10 μm and that the inner surfaces of the plates have been rubbed in mutually perpendicular directions. The molecules at the top and bottom surfaces will be aligned at right angles to one another, while those in the bulk will try to align themselves parallel to their nearest neighbours, the result being a smooth rotation of the alignment direction about an axis normal to the plates (see Figure 10.16(a)). Light incident on the bottom surface—which has passed through a sheet of polarizing material, such that its polarization direction coincides with the alignment direction at the bottom plate—will pass into the liquid crystal, where its polarization locks onto the liquid crystal molecular orientation and is rotated through 90° in passing through the film. If a second polarizer is placed immediately above the upper plate in the same orientation as the lower polarizer, no light will be transmitted. However, if an electric field is applied between the plates, as shown in Figure 10.16(b), the liquid crystal molecules in the bulk of the film will align themselves along the field direction and there will no longer be any rotation of the light polarization direction. Light will therefore be transmitted. The overall structure acts as an electrically switchable light modulator, or SML (spatial light modulator). The one additional requirement, not shown in Figure 10.16, is for a pair of transparent electrodes, in the form of ITO films, deposited on the glass plates.

Several thousand organic compounds were discovered which showed liquid crystal behaviour but rather few were found suitable for practical displays. Not only should the material have a high ohmic resistance, in order to minimize current leakage through the display element, but it had also to work over a suitable temperature range near room temperature and be stable under irradiation with visible light and under electric stress. It was the stimulus provided by simple display needs in the 1960s which led to the development of suitable compounds, work at RCA, at Hull University, and at RSRE Malvern in England being particularly noteworthy. Initially, the target was to make seven bar reflective numerical displays in which the total number of elements amounted to a hundred or less. Displaying any number required each bar to be either black (non-reflecting) or white so the necessary electro-optic characteristic was the sharp switching form shown in Figure 10.17(a). For all applied voltages less than V_1, the element was 'off', while for all voltages greater than V_2, it was 'on'. How, though, was the switching voltage to be applied to each element? For small displays there was no difficulty in providing each with its own wire (actually an evaporated metal film), connected directly to the top electrode which was shaped in the form of the desired bar, appropriate signal voltages greater than

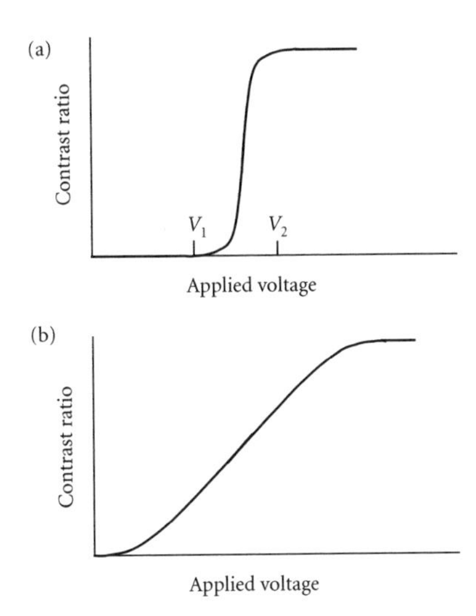

Figure 10.17. Switching characteristics for twisted nematic liquid crystal cells. In (a) is shown a sharp characteristic suitable for displays where the only requirement is for the element to be either fully 'on' or fully 'off'. In (b) we illustrate the type of switching behaviour appropriate to imaging applications where it is necessary for the pixel brightness to be controllable by the applied signal voltage.

V_2 being used to turn each one on, when required. In the case of the twisted nematic type of display element, this demanded voltages of about 5 V, consistent with the use of integrated circuit drivers, thus providing a very simple and convenient display technology.

So successful were these numerical displays that thoughts soon turned to enlarging them to make more sophisticated and more flexible versions—indeed, as we know, many researchers were already seeking the ultimate sophistication of an *imaging* panel which, of course, represented an altogether greater challenge. The number of display elements (pixels) in a typical black and white TV picture amounted to about half a million—one and a half million for colour! Once the pixel count exceeded a few hundred, it became unrealistic to think in terms of individual address wires (a simple sum, assuming a metal linewidth of only 10 μm reveals that most of the display area would be blacked out by metal!) some radical new approach being necessary. This took the form of so-called crossbar addressing in which each pixel lies at the crossover point of a matrix of orthogonal wires. By applying a voltage of $+V$ to the nth horizontal wire and $-V$ to the mth vertical wire, the voltage across the pixel at the point (n, m) amounts to $2V$, while that at all other pixels is no greater than V. Provided $V < V_1$ (in Figure 10.17(a)) and $2V > V_2$, such a scheme can be employed successfully to switch pixels on or off over the whole display area (an even better scheme is to use voltages of $+2V$ for the rows and either $-V$ or $+V$ for the columns) but there is an additional requirement that information must be scanned in, using a procedure known as multiplexing. The first row is switched on for a time t_L (the line time) during which the signal information for all the pixels in that row is sent simultaneously down the appropriate columns, then the first row is switched-off, the second row is switched-on and the appropriate signal information supplied in the same manner as for row one. This procedure is repeated for all the rows in the display, the frame time t_F being equal to Nt_L, where N is the number of rows in the matrix (lines in the display) and where, for a flicker-free display, t_F must be less than the eye integration time of 1/25 s. Notice that each pixel is on for a fraction of time $t_L/t_F = 1/N$ which has implications for the effective brightness of the display.

Detailed consideration of the RMS voltage which existed on any particular pixel under such an addressing scheme led to the conclusion that, as the number of display lines increases, the sharpness of the switching characteristic in Figure 10.17(a) must increase in sympathy. However, real liquid crystal materials are limited in this respect and, in practice, this limited the maximum number of lines to something less than a hundred, depending on the details of the particular display format. What was more, there was a further requirement on any image panel that it must cope with grey scale—one was not only concerned with whether a pixel was on or off but what was its instantaneous

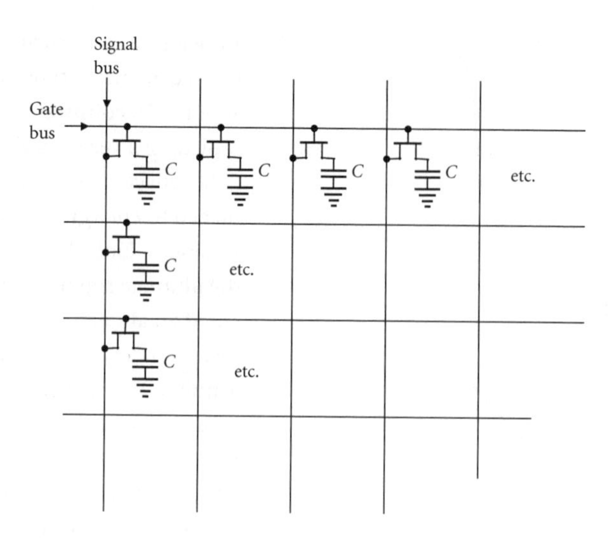

Figure 10.18. Matrix addressed liquid crystal display, using a TFT at each pixel point. The Gate bus is used to switch on a line of TFTs for a time t_L, the line time. During this period the signal buses are used to address each element in the line, setting the desired brightness by charging the liquid crystal capacitor C to its appropriate voltage. Gate lines are addressed in turn, the complete scan taking place in a frame time $t_F = Nt_L$, where N is the number of lines.

brightness—and this required a switching characteristic which looked like that shown in Figure 10.17(b). It was necessary that the contrast could be controlled by the signal voltage in a more or less linear fashion and this was clearly not compatible with the need for a sharp switching curve. It therefore led to an impasse which demanded yet another totally new approach and it was at this point that semiconductors came into the story.

The answer to this new dilemma was to use a switching device at each pixel in the form of a TFT, leading to the modified multiplexing scheme shown in Figure 10.18. The horizontal lines were now used to control the voltage on the gate of the transistor, thus switching it on or off, while the vertical columns supplied brightness information to the pixel by way of the source–drain, resistor R_{SD}. Note that the pixel in Figure 10.18 is represented as a capacitor C which can readily be calculated from the pixel area, the separation of the transparent electrodes and the dielectric constant of the liquid crystal material. In operation, when the transistor is on, the signal voltage charges the capacitor C through the 'on' resistance of the channel R_{on}, the brightness of the pixel being set by the magnitude of the signal voltage. It is clearly necessary that the charging process be completed within the line time t_L (which requires R_{on} to be small) and it is desirable for the charge to remain on the capacitor for the rest of the frame time t_F (which requires that R_{off} be large), in order to maximize the effective display brightness. As we show in Box 10.6, this implies a switching ratio R_{off}/R_{on} greater than a thousand (let us say of order 10^4 to be safe) with $R_{on} \sim 10^7\ \Omega$ and $R_{off} \sim 10^{11}\ \Omega$, and provides a basic specification for the TFTs. Note that the leakage resistance of the liquid crystal cell, itself, must also be high, at least comparable with R_{off}.

Finally, we should also note that, in the interest of brightness, LCTV displays were designed to operate in transmission mode, rather than

Box 10.6. Switching ratio for TFTs in matrix addressed displays

The use of a TFT to control the charging of each liquid crystal element in an LCTV display implies a minimum switching ratio between the on and off values of the transistor channel resistance. The pixel capacitor C must be fully charged within the line time t_L, while the capacitor must hold its charge for the whole frame time t_F. These two conditions can be expressed mathematically as follows:

$$R_{on}C < t_L \tag{B10.33}$$

and

$$R_{off}C > t_F = Nt_L, \tag{B10.34}$$

where N is the number of lines in the display. It follows directly that the switching ratio is given by:

$$R_{off}/R_{on} > N. \tag{B10.35}$$

For a typical TV display having 500 lines, this implies, in practice, that the ratio should be at least 5×10^3.

To discover what the absolute values of R_{on} and R_{off} should be, we need to calculate the value of C and this requires a value for the pixel area A. Again, in a typical TV display of area 0.1 m^2, having 1 million pixels, the area of each pixel is roughly 10^{-7} m^2. We can then calculate C according to:

$$C = A\varepsilon\varepsilon_0/d, \tag{B10.36}$$

where the relative dielectric constant of the liquid crystal material is about 10 and the thickness of the cell is about 10 μm. The resulting value of C is about 1 pF.

It then follows that R_{on} should be less than about 10^7 Ω and R_{off} greater than about 10^{11} Ω. Note, too, that the leakage resistance of the capacitor C must be at least as high as R_{off}, a condition which is better expressed in terms of the resistivity of the liquid crystal material. We require that the RC time constant of the cell be long compared with t_F. Thus:

$$\varepsilon\varepsilon_0\rho > t_F, \tag{B10.37}$$

which gives $\rho > 10^8$ Ω m. Not only does the liquid crystal have to perform as a light switch, it must be a good insulator too—and remain so over a long period of time even when repeatedly electrically stressed.

This account of transistor performance in terms of its source–drain resistance is convenient but, in the case of the on state, may need modification if the transistor runs into saturation, when it becomes essentially a constant current source. The condition that the capacitor C can be fully charged within the line time can then be written as:

$$Q = CV = I_{sat}t_L, \tag{B10.38}$$

from which we obtain a minimum value for $I_{sat} = CV/t_L \sim 3 \times 10^{-7}$ A. In other words, the saturated transistor should be capable of supplying a current of about a microampere.

the reflection mode used for simple numerical readout, and this implied the need for a suitable back-light. This took the form of a suitably shaped fluorescent tube, colour being introduced by the use of three different colour filters at appropriate picture elements. (Though white LEDs may offer a preferred source in future.)

Once this basic format for a display panel was accepted, it was then necessary to choose a suitable TFT which could be made in large numbers, with reproducible characteristics, over an area of perhaps 0.1 m^2. It was tantamount to making an integrated circuit some 300 times larger than anything available in crystalline silicon! (Though, to be fair, the individual devices were also rather larger.) It demanded an evaporator capable of handling a large substrate (typically glass or quartz) and producing uniform thin films over this area, together with the ability to define the various elements of the transistor using photolithography or some other appropriate technology over a very much larger area than had previously been contemplated. It was also essential that the transistor characteristics should remain stable under prolonged use. In short, there were several serious challenges to be met.

Following its invention by Paul Weimer of RCA in 1962, there had been considerable interest in the TFT. Weimer, himself demonstrated the device's potential using glass substrates, evaporated CdS as the semiconductor and evaporated SiO as the gate insulator. He used gate lengths in the range 5–50 μm, measured transconductances as high as 25 $mA\,V^{-1}$ and observed oscillation frequencies as high as 17 MHz. Though there were hints of trapping effects which might influence device stability and marked sensitivity to visible light, there was already talk of its rivalling the Si MOSFET (metal oxide semiconductor field effect transistor) as an integrated circuit element. A whole galaxy of industrial research laboratories took up the challenge to develop practical devices and circuits, with performance at microwave frequencies being one long-term objective. A wide range of semiconductor materials was studied, including polysilicon, CdS, CdSe, GaAs, InAs, InSb, PbS, PbTe, and Te, though efforts gradually became focused on Si, CdS, and CdSe which offered channel electron mobilities as high as 0.005 $m^2\,V^{-1}\,s^{-1}$ (for comparison, when a-Si:H eventually matured, it was characterized by mobility values of order 10^{-4} $m^2\,V^{-1}\,s^{-1}$ or less). By 1970, when thoughts were turning to large area display panels, there was considerable experience in the fabrication and study of such devices and it was no surprise when, in 1973, Brody and colleagues at Westinghouse published details of a matrix-addressed liquid crystal monochrome display using CdSe TFTs at each pixel point. Their panel was 6 in.^{2} and contained 14,000 pixels, the resolution of 20 lines per inch being some four times less than standard TV practice. It was, nevertheless, perfectly adequate to point the way ahead—the large area capability of TFT technology having been clearly demonstrated.

The proposal to use a-Si:H TFTs in place of CdSe devices was made by the Dundee group in 1976 and work at RSRE Malvern demonstrated the viability of the idea, though it was not until the early 1980s that practical TV-type display screens were realized, in Europe, America, and, particularly, in Japan. (Detailed discussion of the relevant issues will be found in Ast 1984, LeComber and Spear 1984, and van Berkel 1992.) The race was clearly on to develop commercially viable flat panel full colour display screens which eventually materialized towards the end of the 1980s in the form of miniature screens for portable TV sets and for personal use in aircraft. I remember well from my own involvement with the Philips programme how completely out of range even a small, 22 in. TV screen appeared at that time—performing the necessary deposition and lithography steps on even 6 in. panels represented a major challenge in itself. Quite recently, I understand that Philips has announced a 50 in. direct view LCTV screen—the achievement of 15 years of development having exceeded most people's wildest dreams. But this is to run very considerably ahead of our discussion—why, for instance did a-Si:H trigger such a staggering worldwide activity?

The original argument put forward from Dundee (see LeComber and Spear 1984) was that an elemental material like silicon could be expected to provide greater uniformity, reproducibility, and stability than compounds such as CdSe, all features vital to any serious commercial activity. In practice, as we already know, a-Si:H does suffer from instability in regard to the Staebler–Wronski effect. Nevertheless, the other two advantages have certainly been achieved, whereas stoichiometry is always a problem with II-VI compounds and very difficult to control over large area deposition. However, there is a further subtle advantage for a-Si:H (which was also referred to in the Dundee proposal) in relation to the required values of channel resistance and switching ratio. The electron mobility in a-Si:H is considerably smaller than that found for polycrystalline materials such as CdSe (or, for that matter, poly-Si). A typical value for a-Si:H might be $3 \times 10^{-5}\,\mathrm{m^2\,V^{-1}\,s^{-1}}$, compared with $3 \times 10^{-3}\,\mathrm{m^2\,V^{-1}\,s^{-1}}$ for CdSe, which (for transistors with similar dimensions) implies two orders of magnitude difference in channel resistance. Though this might at first thought sound like a serious disadvantage for a-Si:H, it turns out to be an *advantage* when applied to screens with small pixel size (which implies a small pixel capacitance), TFT dimensions being restricted both by the pixel and by the challenge of performing photolithography over relatively large areas. Full-colour TV display in a 6-in. panel implies pixels of about 100 μm linear dimension, allowing channel *widths* up to 100 μm, while channel lengths (and widths) are restricted to being no *smaller* than about 5 μm. The small capacitance is easily charged, so the value of the 'on' resistance may be relatively large and correctly matched by an a-Si:H TFT with channel width of about 20 μm. On the other hand, the smallest

CdSe TFT (5 × 5 μm) yields an 'on' resistance some 25 times smaller than is required, which is no disadvantage, in itself, but demands a very large switching ratio to achieve the required 'off' resistance, whereas a-Si:H achieves this with relatively little difficulty. In other words, the a-Si:H TFT is much better matched to the demands of display addressing than is its CdSe counterpart. (It should be said, however, that this argument can be countered by the addition of a storage capacitor in parallel with the liquid crystal element, though it does add to the complexity of processing.) On such subtleties can whole industries wax or wane!

Having chosen to use a-Si:H TFTs, it was necessary to design a suitable device structure which could easily be made with a modest number of process steps. First, came the choice of substrate and this, it was widely agreed, had to be glass which satisfied the two essentials of transparency and cheapness. This, in turn, required that all processing should involve temperatures well below the glass softening temperature, a criterion readily met in the case of the a-Si:H TFT where deposition temperatures of about 300°C were typical. Figure 10.19 illustrates a suitable structure for the TFT, together with its placement in relation to the liquid crystal element. Processing began with deposition of a chromium gate electrode which was then patterned using photolithography, followed by deposition of a silicon nitride gate insulator, undoped a-Si:H active layer, and n^+ a-Si:H contact layer (all in the same run). Following appropriate patterning of the device, metal source and drain contacts were deposited and patterned to complete the structure. Note how the drain metalization projects through the Si-N layer to make connection with the ITO contact film used to apply the electric field to the liquid crystal pixel. Needless to say, it was essential to ensure uniform deposition of gate insulator and active layers over the whole display area, uniform not only in thickness but also in electrical characteristics. Various minor modifications of this basic structure have been used by different manufacturers to optimize performance or convenience in fabrication but Figure 10.19 provides all the information necessary for understanding the modus operandi.

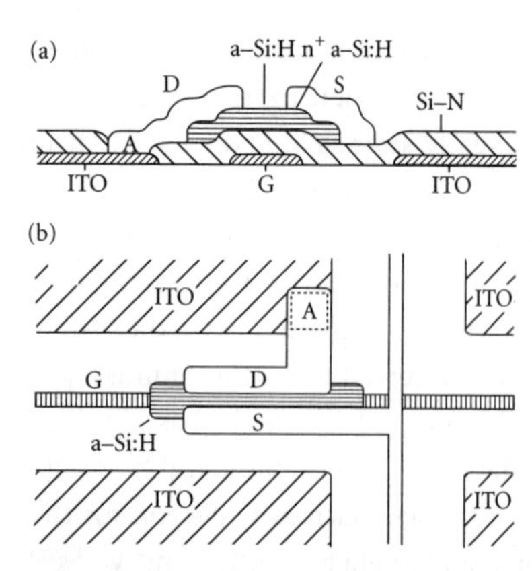

Figure 10.19. Design of a typical a-Si:H TFT for multiplexing a LCTV display panel. In (a) is shown a cross-section through the transistor, demonstrating the method by which the drain of the TFT is connected (at point A) to the ITO film used to apply an electric field to the pixel. In (b) is shown the layout of the transistor in relation to the pixel and the bus bars used to address the pixel. Typical TFT dimensions might be: gate length 5 μm, gate width 20 μm. It is clear that such a TFT takes up only a small fraction of the pixel area. (From LeComber and Spear 1984: fig. 3). Reprinted with permission from Elsevier.

In operation, a positive voltage in the range of 5–10 V applied to the gate electrode induces a corresponding density of electrons in the channel, thus controlling the channel conductance. Typical switching ratios of up to 10^6 between the off state and the fully on state have been measured on practical devices, more than adequate for the application. As we suggested earlier, the presence of deep electron states within the a-Si band gap hinders the movement of the Fermi level from midgap to conduction band edge and it is only because of their relatively low density in a-Si:H that such attractive device properties have been obtained. Looking at these results from an alternative viewpoint suggests that it should be possible to derive data on the distribution of the deep states from the measured TFT transfer characteristic and several groups have done this. Indeed, it represents another example of the field effect measurements originally used to confirm the low density of states in a-Si:H, and has led to rather better understanding of these measurements. The interesting fact to emerge (see van Berkel 1992) is that it is often impossible to derive a unique DOS function from the measured transfer characteristic, which probably explains the wide variation in values obtained during much of the early work. There is also evidence that the derived DOS depends quite strongly on the choice of insulator, SiN_x or SiO_y, suggesting a significant contribution from the insulator–semiconductor interface region. Indeed, the quality of this interface is crucial to device performance (as was the case in the early work on the silicon MOSFET).

In the context of system performance, photoconductivity and stability represent two very important device properties and have been much studied. By the very nature of LCTV display, using a backlight, it is impossible to avoid the influence of light on TFT characteristics—photoconductivity leading to a reduced switching ratio—so it is necessary to minimize the effect by thinning the a-Si:H active layer. This works well in practice, though it puts greater pressure on process control. Stability is rather more complex and it is difficult to give any brief summary which has much validity. Two main effects can be singled out, electron trapping in the insulator and deep state creation in the semiconductor, both of which cause shifts in the so-called 'flat band voltage' (the gate voltage at which the semiconductor bands remain energetically flat all the way to the interface). The second of these effects has much in common with the Staebler–Wronski effect, except that the movement of the Fermi level in this case results from application of a gate voltage, whereas in the Staebler–Wronski effect it results from absorption of light. The reader will doubtless appreciate that the obvious success of the a-Si:H TFT in LCTV display suggests that these instability mechanisms have not proved of more than minor significance and there is one salient reason for this. In operating conditions the transistor is switched-on for a line time, whereas it is off for a frame time, some thousand times longer, so any unwanted changes which do occur have ample opportunity for reversal before the next on period.

So much for the display, itself—there is, however, another important aspect of its operation which we cannot afford to overlook. The method of matrix addressing which we described earlier requires driver circuits to send the appropriate signals to each pixel point. The sequence of gate voltages which switch on each line of transistors in turn is relatively simple and operates at a frequency of about 10 kHz which can be handled by a-Si:H transistors. This allows the line switching function to be performed by fully integrated circuitry. On the other hand, the brightness instructions supplied to the various source electrodes, containing the necessary video information are not only complex in format but (perhaps more importantly) involve frequencies in the order of 10 MHz. This is far too fast for a-Si:H TFTs to deal with, so early display panels used a hybrid approach in which the driver circuits were realized in conventional silicon integrated circuits mounted along the edge of the panel. This certainly worked but was unsatisfactory in adding to fabrication complexity and, right from the start, it was clear that a fully integrated solution was very much to be preferred. (Brotherton 1995 points out that full integration would reduce the number of external connections from 1300 to about 20!) The difficulty lay with the low electron mobility in a-Si:H which was fundamental and irrevocable—any solution had to be based on a semiconductor with a much greater mobility and the obvious one to try was polysilicon. As we have seen already, *p*-Si may have an electron mobility two orders of magnitude greater than that of a-Si:H and this would be just adequate for the driver function, provided suitable transistors could be made from it. The question was: how to deposit *p*-Si in the appropriate place along the edge of the display?

Polysilicon was not at all a new material in 1985. As we saw in our earlier discussion of polycrystalline materials in Section 10.2, its study during the 1970s had considerably advanced the understanding of their electrical properties while the use of *p*-Si gates in crystalline MOSFETs was a well-established technology. Consequently, the technology for depositing thin films by the thermal decomposition of silane (SiH_4) was widely available. However, this process involved substrate temperatures of order 650°C, somewhat too high for use with glass substrates, and it tended to produce fine grain material with low electron mobility ($\mu_e \sim 5 \times 10^{-4}\,m^2\,V^{-1}\,s^{-1}$) so was not very promising for this particular application. Several attempts to reduce the deposition temperature resulted in little improvement in measured mobilities and it became apparent in the early 1980s that some radical alternative would be necessary. This took the form of recrystallization of a-Si films deposited by standard procedures and involved two very different methods, solid phase crystallization and laser crystallization. The former involved a rather lengthy thermal treatment of a-Si films at temperatures of about 600°C and, though it certainly succeeded in increasing the electron mobility to values as high as $5 \times 10^{-3}\,m^2\,V^{-1}\,s^{-1}$, it was far from attractive to the display technologist. Laser crystallization of a-Si using an excimer (UV) laser eventually proved far more satisfactory and was intensively studied during the 1990s.

Excimer lasers are high power, short pulse devices which are well suited to crystallizing thin films, the optical absorption length in a-Si being only a few nanometers at UV wavelengths. Short, high power pulses result in surface melting of the film with relatively little heat diffusion into the substrate. Using a capping film of SiO_2 (which is transparent to the laser radiation) allows satisfactory crystallization of the a-Si while keeping the glass temperature below 400°C. The structure of the resulting film is characterized by a surface region having large grain size while leaving a smaller grain sub-film underneath and this structure was found to yield high mobilities, with μ_e being as high as $2.5 \times 10^{-2}\ m^2 V^{-1} s^{-1}$ in some cases. However, to achieve this, required careful control of the laser power and spot scanning geometry, too high a power resulting in a degraded material with finer grain and low mobility. Optimum material, combined with a deposited SiO_2 gate insulator, showed promising TFT behaviour, except that there was evidence of problems with free carrier trapping at grain boundaries, in dangling bond deep states, distributed throughout the band gap. In a sense, this was a reflection of the behaviour of a-Si itself and the treatment turned out to be similar—hydrogen passivation of the dangling bonds in a hydrogen plasma, though it proved to be an unexpectedly complex procedure. Finally, towards the end of the 1990s, it was, at last, possible to integrate the video driving circuitry into the display panel, a triumph which took nearly 15 years of intensive study.

This was not the end of the story, however, for, all along, those concerned had been intent on developing panels based entirely on polysilicon transistors, not only as drivers but also at the pixel. There was one final hurdle to surmount in order to do this. As we have seen, it is important for the off resistance of the pixel transistor to be as high as possible and this was compromised by a peculiar source–drain leakage current which depended strongly on the electric field near the drain. As the millennium approached, this, too, was overcome by careful design of transistor geometry and it was then possible to make complete display panels based on laser-crystallized polysilicon transistors, raising the intriguing question as to the long-term future of the a-Si:H technology which had borne the brunt of the initial challenge and which had supported the development of a whole industry. What, indeed, will become of it? Commerce has little sentiment and we may yet see it becoming redundant. At the time of writing, it is much too soon to comment—after all, the CRT itself still has a major role in TV display, in spite of the success of LCDTV—and, as I have said so many times already, this book is concerned with history, rather than crystal gazing! In any case, nothing which may or may not happen in the future can detract from the merit of the pioneering work which led to the first commercially successful flat panel displays. This undoubtedly represented a genuine revolution in display technology and will continue to have an impact on consumer satisfaction well into the twenty-first century.

As a kind of post-script, I should probably add to this discussion the possibility of the liquid crystal cell being replaced by organic light emitting diodes (OLEDs) which have now reached adequate brightness levels and operating lifetimes but this would take us into even greater realms of speculation. Organic semiconductors probably have a bright future in many optoelectronic applications but that constitutes another story altogether and one which I leave to be told by those involved in their development during the past decade.

10.6 Porous silicon

Silicon has so obviously dominated the field of solid state electronics that it seems appropriate to bring our story to a conclusion by describing yet one further facet of its many-sided personality. We have noted on several occasions that silicon's indirect band gap has prevented it from making an impact in the field of light emission but even this apparent truism turns out to need qualification. Not only can silicon, in one of its several guises, emit radiation with surprisingly high efficiency but, even more surprisingly, this radiation may lie in the visible part of the spectrum—that is, with photon energy considerably larger than the semiconductor band gap. While the reader may be forgiven for feeling sceptical, it is a fact that visible silicon LEDs do actually exist. How could such an apparent contradiction come about?

First, it is important to recognize that we are concerned here with silicon in a rather special form, that known as 'porous silicon'—crystalline silicon which has been treated in an electrochemical bath to modify its mechanical structure by creating large numbers of tiny voids. These may occur in a variety of shapes and sizes and the degree of porosity (the fraction of the resulting material which consists of voids) may be varied over a wide range. Perhaps more importantly, the remaining silicon takes the form of extremely fine filaments in the shape of a skeleton which represents a so-called 'nanostructure' and it is the properties of this nanostructure which control the observed light emission. Many theoretical models of the emission process have been put forward but one essential feature appears to be that of quantum confinement. The skeleton consists essentially of fine silicon wires with dimensions significantly smaller than the exciton Bohr radius and the confinement energy associated with this degree of localization is sufficient to account for the large shift of energy from the bulk silicon band gap of 1.12 eV to that appropriate to visible radiation, 1.5–2.5 eV. (For much more detail and an outline of other possible applications of porous silicon see the emis Data Review edited by Leigh Canham—Canham 1997.)

Porous silicon was discovered, rather by accident (how often have we heard this phrase?) at Bell Labs in 1956 when Arthur and Ingeborg Uhlir were studying the possible use of electrolysis for obtaining smooth, passivated silicon surfaces (electropolishing). In the first instance, it was seen as nothing more than an interesting nuisance—it was clearly not what they were looking for—though later (1981) it was used at NTT in Japan for certain aspects of silicon device processing. Nevertheless the general level of interest was low—according to Cullis et al. (1997), less than 200 published papers were concerned with porous silicon (*p*-Si) during the 35 years up to 1990, which contrasts starkly with the figure of over 1500 published between 1990 and 1996, the reason for this dramatic surge being the discovery of relatively efficient, room temperature visible photoluminescence by Leigh Canham of RSRE in 1990.

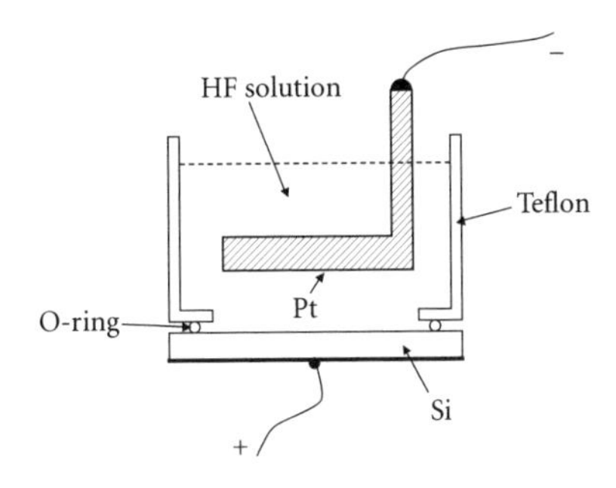

Figure 10.20. Sketch of an electrolytic cell suitable for forming porous silicon. A silicon wafer is sealed with an O-ring to the base of the cell so as to be immersed in the HF electrolyte while current is passed through the cell from a platinum cathode.

The formation of *p*-Si is associated with electrochemical dissolution of silicon in the presence of HF-based solutions (usually aqueous but occasionally ethanoic) and is readily achieved using a cell of the type illustrated in Figure 10.20. A suitable silicon wafer is sealed onto the bottom of the cell so as to make uniform contact with the solution and current passed through the solution from a platinum cathode, *p*-Si being obtained under low voltage conditions where the current density is below about 2 mA cm^{-2}. Typical depths of *p*-Si lie within the range 0.2–2.0 μm. The processes involved are complex and the final result depends on solution strength, current, time, and substrate doping but there is usually a porosity gradient from the surface downwards, with so-called 'microporous' material near the surface (dimension <2 nm) while below this there occurs 'mesoporous' (2–50 nm) or 'macroporous' (>50 nm) material. The degree of porosity increases with time and can be further controlled by etching in HF solution for several hours post electrolysis. The drying process is also important as it can introduce considerable strain which degrades the structure of the ultrafine skeleton and with it the luminescence. Yet again, storage may result in further degradation, presumably as a result of oxidation and other contamination.

Visible luminescence is associated with high porosity samples and one of the key experiments performed by Canham demonstrated that the emission wavelength could be smoothly decreased by continuous etching, to increase the porosity further. In this manner he was able to shift the wavelength from 950 to 750 nm over a period of 6 h, a result clearly consistent with a model of quantum confinement (the finer the skeleton structure, the shorter the wavelength). He also showed that at low temperatures the luminescence was characterized by phonon sidebands which corresponded to phonon energies consistent with those of crystalline silicon. This was crucial in supporting the quantum confinement model against several alternatives which depended on the presence of amorphous silicon, oxides, or other more complex molecules (which would yield different phonon modes).

Later work (1990–3) has shown that there are at least four types of luminescence in *p*-Si, an infrared band (1100–1500 nm), a 'slow' blue–red band (400–800 nm), a 'fast' blue–green band (~470 nm), and a UV band (~350 nm). Only the slow band (S band) is thought to result from quantum confinement and it is only this band which has excited commercial interest on account of its high efficiency—values of internal quantum efficiency as high as 85% have been measured at 50 K, while external efficiency at room temperature has been recorded at 3%. Measured recombination lifetimes are about 10 μs for the S band which, again, are consistent with the quantum wire model—the lifetime being characteristic of single crystal silicon. The high efficiencies are a consequence of the high quality of the original silicon sample, containing relatively few non-radiative recombination centres and the effect of exciton localization within the wire. Random variation in wire thickness implies that excitons collect in wider regions of the wire (where they have minimum confinement energy) and this localization minimizes the probability of their meeting up with a recombination centre. The fact that linewidths are rather large (though reducing as the temperature is lowered) and that broadening is inhomogeneous is also consistent with such a model.

Interest in *p*-Si luminescence stems, of course, from the possibility of combining radiative effects with conventional silicon electronic circuitry, integration of this kind having recently been demonstrated. However, this depends crucially on the development of efficient and reliable *p*-Si LEDs. For application to display, it has been estimated that external efficiencies greater than 1% are needed, while for the important application to optical coupling between electronic circuits, even higher performance is required ($\eta_{ext} > 10\%$). How do current LEDs measure up? It is never quite so easy to realize such high efficiency in an LED as obtained in photoluminescence. The first LEDs were reported in 1991–2, soon after Canham's observation of efficient photo luminescence. Encouraging results were obtained using electrolyte contacts to the *p*-Si layer, external efficiencies approaching 1% being reported, though it was realized that this could represent no more than an existence theorem—practical LEDs must employ solid contacts. A typical structure for such a device is shown in Figure 10.21, consisting of a thin *p*-Si layer supported on and contacted through the silicon substrate. The top contact took the form of a semi-transparent evaporated gold film or, better still, an ITO film, acting as a Schottky barrier contact. Later approaches made use of a *p*-*n* junction in the original substrate which was subsequently porosified, and a number of other contact materials have also been tried, including the use of Bragg mirrors to form an optical microcavity. However, the best efficiencies realized to date are about 0.2% and there are serious problems with stability. Many devices degrade severely within hours unless they are kept under vacuum—oxidation apparently plays a major role—and there is obviously a great deal of work yet to be done before viable commercial applications can be envisaged. Another serious problem is

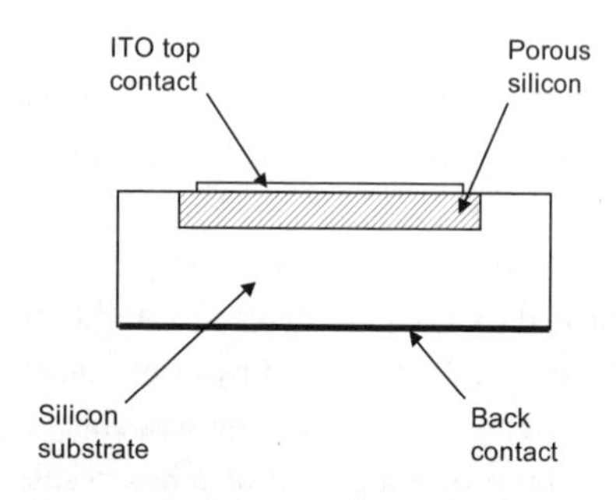

Figure 10.21. Structure of a simple Schottky barrier porous silicon electroluminescent diode. The ITO contact provides a transparent Schottky contact to the porous silicon, while the silicon wafer acts as the back contact.

the current inability to modulate diodes at the gigahertz frequencies demanded by optical interconnects. It is still possible that success will be achieved at a future date, of course, but some significant breakthrough is clearly needed. In spite of a decade of exciting progress, the holy grail of silicon light emitting devices remains a tantalizing, but frustrating gleam in the technologist's eye.

In a sense, this may seem an unfortunate conclusion on which to conclude but it is, perhaps, salutary to recognize that the quite remarkable success achieved by semiconductor electronics and optoelectronics during the past 50 years has involved more than a few failures, many frustrations and numerous lengthy struggles against apparently insuperable odds. It would be much less of a success story had it not been so. It would also be a rash individual who would write off silicon in any application, such has been its capacity to fight back—but, as you now know very well, this account deals only with history. The future will surely take good care of itself.

Bibliography

Adler, D. (1971) *Amorphous Semiconductors*, Butterworth, London.

Ast, D. G. (1984) *Semiconductors and Semimetals*, Vol. 21, p. 115. (eds. R. K. Willardson, and A. C. Beer), Academic Press, New York.

van Berkel, C. (1992) *Amorphous and Microcrystalline Semiconductor Devices*, Vol. 2, p. 397, (ed. J. Kanicki) Artech House, Norwood, MA.

Brotherton, S. D. (1995) *Semiconductor Sci. and Technol.*, 10, 721.

Bube, R. H. (1960) *Photoconductivity of Solids*, Wiley, New York.

Canham, L. T. (ed.) (1997) *Properties of Porous Silicon*, emis Data Review No. 18, INSPEC, The Institute of Electrical Engineers, London.

Carlson, D. E. (1984) *Semiconductors and Semimetals*, Vol. 21(d), p. 7 (eds. R. K. Willardson and A. C. Beer) Academic Press, New York.

Cullis, A. G., Canham, L. T., and Calcott, P. D. J. (1997) *J. Appl. Phys.*, 82, 909.

Green, M. A. (2000) *Power to the People: Sunlight to Electricity Using Solar Cells*, University of New South Wales Press, Sydney.

Kazmerski, L. L. (1998) *Photovoltaics: A Review of Cell and Module Technologies*, Vol. 1, p. 71 Renewable and Sustainable Energy Reviews, Pergamon Press, NY.

LeComber, P. G. and Spear, W. E. (1984) *Semiconductors and Semimetals*, Vol. 21, p. 89 (eds. R. K. Willardson and A. C. Beer) Academic Press, New York.

Markvart, T. (2000) (ed.) *Solar Electricity*, 2nd edn, John Wiley, Chichester.

Mott, N. F. and Davis, E. A. (1971) *Electronic Processes in Non-Crystalline Materials*, 2nd edn, Clarendon Press, Oxford.

Orton, J. W. and Powell, M. J. (1980) *Rep. Prog. Phys.*, 43, 1263.

Orton, J. W., Goldsmith, B. J., Chapman, J. A., and Powell, M. J. (1982) *J. Appl. Phys.*, 53, 1602.

Pankove, J. I. (1984) *Semiconductors and Semimetals*, Vol. 21 (Parts a,b,c and d) (eds. R. K. Willardson and A. C. Beer) Academic Press, New York.

Powell, M. J., Deane, S. C., and Wehrspohn, R. B. (2002) *Phys. Rev.*, B66, 155212.

Smith, S. D. (1995) *Optoelectronic Devices*, Prentice Hall, London.

Staebler, D. L. and Wronski, C. R. (1977) *Appl. Phys. Lett.*, 31, 292.
Street, R. A. (1991) *Hydrogenated Amorphous Silicon*, Cambridge University Press.
Street, W. (1999) IEEE Spectrum (January) Institute of Electrical and Electronic Engineers, New York, pp. 62–67.
Sze, S. M. (1969) *Physics of Semiconductor Devices*, Ch. 8. Wiley, New York.
Sze, S. M. (1985) *Semiconductor Devices: Physics and Technology*, Ch. 7. Wiley, New York.